Endocrinology Research and Clinical Developments

Natriuretic Peptides

Physiology, Molecular Biology, and Clinical Implications

ENDOCRINOLOGY RESEARCH AND CLINICAL DEVELOPMENTS

Additional books in this series can be found on Nova's website under the Series tab.

Additional e-books in this series can be found on Nova's website under the e-book tab.

ENDOCRINOLOGY RESEARCH AND CLINICAL DEVELOPMENTS

NATRIURETIC PEPTIDES

PHYSIOLOGY, MOLECULAR BIOLOGY, AND CLINICAL IMPLICATIONS

KAILASH N. PANDEY
EDITOR

nova publishers
New York

Library of Congress Cataloging-in-Publication Data

ISBN: 978-1-63321-080-6

Library of Congress Control Number: 2014939980

Published by Nova Science Publishers, Inc. † New York

CONTENTS

PREFACE

Since the pioneer discovery of de Bold and his colleagues in 1981, the advances in the field of natriuretic peptides and their receptor have been experienced in a continuous growth fashion. Now, the natriuretic peptide field has proved itself as a mature discipline in the biomedical arena. In the past thirty-four years, countless studies have provided an energetic impetus to explore this emerging area with respect to the cellular, molecular, and clinical aspects of natriuretic peptides and their receptors. Elucidation of the underlying mechanisms by which natriuretic peptides and their receptors elicited biological, physiological, and pathophysiological responses have provided hallmark significance and novelty to delineate both cGMP-dependent and independent signaling cascades. After ligand binding to the receptor, numerous sequential steps take place, which have been regarded important events for controlling overall hormone responsiveness, receptor activity, and transmembrane signaling to achieve ultimate hormone-receptor function(s). This book discusses the current knowledge and recent developments to navigate cellular, molecular, and clinical aspects of natriuretic peptides and their receptors in health and disease.

State-of-the-art technology in cell and molecular biology have been applied and also being continued to delineate the studies of natriuretic peptides and their receptors in terms of expression, function, and regulation both *in vitro* and *in vivo* systems. Concurrently at the same time, the field has experienced and undergone a resurgence, endured in large part by the advent of molecular biological approaches to generate the transgenic and gene-targeted (gene-knockout and gene-duplication) mutant mouse models. The genetically engineered transgenic and gene-targeted mouse models have clearly demonstrated that the absence of the functional natriuretic peptides and/or their cognate receptors could lead to not only elevated high blood pressures but cause severe hypertensive heart disease in the experimental animals similar to those seen in untreated human hypertensive patients. The genetically engineered mouse models have been useful to reveal not only the actual functions of natriuretic peptides and their receptors genes but more critically their modifying interactions with other genes. The current and past studies have led to the increasing importance of natriuretic peptides and their receptors in health and disease in bedside diagnosis, treatments, and therapeutics advancements. The book presents a comprehensive look and consists of assessments of the past and current status and future prospects of the ongoing activities in fourteen chapters of the natriuretic peptides and their receptor systems. After the discovery of three different but structurally-related natriuretic peptides and three specific cognate receptors, in particular, their combined therapeutic as well as diagnostic potentials in hypertension and cardiovascular diseases have initiated the need of an overall comprehensive compilation of the current status of this field.

In spite of tremendous progress and growth, still there have been reluctance to the advancement and full consideration of the natriuretic peptides and their receptors as the therapeutic treatment options. It is with this goal and spirit that the contributions of renowned and pioneering scientists and investigators have been compiled in this book to highlight more focused and detailed information regarding the growth of the field of natriuretic peptides and their specific receptors. The book comprises the contributions from eminent investigators and scientists who are nationally and internationally recognized in their area of research expertise, and presents a wide range of reviews spanning the research works done on the family members of natriuretic peptides and their receptors. Each chapter provides a comprehensive summary of the accomplished experimental findings and future perspectives and direction of research to be performed in this vital and important cardiovascular field.

The chapters in this book address key aspects of the natriuretic peptides and their receptors that high light the key areas in this field. Together, these chapters represent an overview of the status of the natriuretic peptides and their receptors from the past to the current accomplishments and exciting future investigations. I gratefully acknowledge and thank all the contributors for their genuine and painstaking efforts and quality of works they have presented. I am also indebted to the contributors for their support, encouragement and cooperation throughout the development and progress of this book. The interaction and correspondence with many of these colleagues made this undertaking a remarkable memorable assignment, which at time was an ambitious task yet enjoyable learning experience and educational accomplishments.

I sincerely thank and offer my deep appreciation to numerous investigators and researchers who have carefully conducted and published a large volumes of works on the natriuretic peptides and their receptor system, but due to space limitation their contributions was not listed in this book. I sincerely thank all the present and pervious members of my laboratory for their numerous and valuable contributions. Finally, I would like to recognize my wife Kamala Pandey for her efforts in helping me in the preparation of this book.

Kailash N. Pandey, Ph.D.
Department of Physiology, SL-39
Tulane University Health Sciences Center,
School of Medicine,
1430 Tulane Ave, New Orleans, LA 70112
Tel: (504) 988-1628
Fax: (504) 988-2675
Email: kpandey@tulane.edu

In: Natriuretic Peptides
Editor: Kailash N. Pandey

ISBN: 978-1-63321-080-6

Chapter 1

NATRIURETIC PEPTIDES

Tsuneo Ogawa* and Adolfo J. de Bold
Cardiovascular Endocrinology Laboratory, University of Ottawa Heart Institute, Ottawa, Ontario, Canada

ABSTRACT

The cardiac natriuretic peptides (cNPs) ANF and BNP mediate the endocrine function of the heart. From the hemodynamic point of view the endocrine heart serves to regulate cardiac preload and afterload through to its influence in the regulation of extracellular fluid. However, many other targets have been characterized that are not primarily related to hemodynamics. In addition to their various roles in the physiological state, cNPs serve as useful biomarkers of cardiac disease and as therapeutic agents. cNPs are coordinately regulated in cardiovascular diseases such as congestive heart failure but are differently regulated in disorders having a significant inflammatory component such as in acute cardiac allograft rejection, sepsis and autoimmune myocarditis. BNP plasma levels are more influenced by inflammation compared to ANF. Because inflammation and hemodynamic changes coexist in many cardiovascular disorders the measurement of both ANF and BNP in individual patients may be advisable.

1. HISTORICAL CONTEXT

Approximately 50 years ago electron microscopic studies demonstrated storage granules in atrial cardiomyocytes. The granules, named atrial specific granules, were of unknown content but were morphologically similar to those found in polypeptide-hormone producing cells [72,80,81]. About 20 years later, de Bold et al., [34] showed that injection of atrial extracts to bioassay rats produced a strong natriuresis and diuresis as well as hypotension accompanied by an increase in hematocrit. Subsequently, in 1983, a polypeptide factor named

* Corresponding Author: Adolfo J. de Bold, PhD; Cardiovascular Endocrinology Laboratory; University of Ottawa Heart Institute; 40 Ruskin Street, Ottawa, Ontario, Canada, K1Y 4W7; Tel: +1 613 761 4265; FAX: +1 613 761 1597; adebold@ottawaheart.ca.

atrial natriuretic factor (ANF) or atrial natriuretic peptide (ANP) with those properties was isolated, purified and sequenced by us from rat atrial extracts [36,45,78]. The following year ANF was isolated from human atria [76]. The association of ANF with the atrial specific granules was demonstrated by both tissue fractionation studies and by immunocytochemistry [33,39]. The presence of natriuretic activity in the atria of mammals and in both the atria and ventricles of non-mammalian vertebrates was subsequently demonstrated corresponding with the morphological distribution of atrial specific-like granules in various species [38]. Later other natriuretic peptides were isolated thus demonstrating the existence of a family of natriuretic peptides.

2. The Natriuretic Peptide (NP) Family

The NPs share in common a 17 amino acid ring structure formed by an intra-molecular disulfide linkage (Table 1). This ring structure is essential for biological activity [67]. The natriuretic peptide brain natriuretic peptide (BNP), isolated from porcine brain in 1988 [168] is a 32-amino acid peptide in humans. It has a similar ring structure to that of ANF. Although originally isolated from porcine brain, BNP is most abundantly expressed in the atria of the heart and considered to be, together with ANF, a cardiac hormone or cardiac natriuretic peptides (cNPs). Both these hormones are functionally considered to be the main endocrine products of the heart. ANF and BNP are co-stored in atrial specific granules [39]. A third "natriuretic" peptide; C-type natriuretic peptide (CNP), is a 22 amino acid peptide isolated from porcine brain in 1990 [86] that lacks the carboxyl terminal of ANF and BNP and hence it lacks natriuretic properties. CNP is produced in a constitutive manner in the brain and pituitary gland [86], and in the endothelium, smooth muscle cells of the intima, media, and vasa vasorum of arteries and in macrophages in normal and atheromatous coronary artery in humans [21] as well as in other tissues [88]. Urodilatin, an N-terminal 4-amino acid extended form of ANF, has been identified in urine and considered to be a renal ANF [154]. In addition to the natriuretic peptides mentioned above, dendroaspis natriuretic peptide (DNP) [156] and vasonatrin peptide (VNP) [99] have been identified in the green mamba snake and in the eel, respectively.

Table 1. Human natriuretic peptide sequences

ANF	SLRRSS	CFGGRMDRIGAQSGLGC	NSFRY
BNP	SPKMVQGSG	CFGRKMDRISSSSGLGC	KVLRRH
CNP	GLSKG	CFGLKLDRIGSMSGLGC	
Urodilatin	TAPRSLRRSS	CFGGRMDRIGAQSGLGC	NSFRY

Amino acid sequences are shown in a single character code. The sequences surrounded by a rectangle represent the disulfide ring structure.

3. Function of cNPs

From the hemodynamic point of view the endocrine heart serves to regulate preload and afterload due to its influence in the regulation of extracellular fluid volume through actions on its principal targets including the kidney, the renin-angiotensin-aldosterone system and arterial and sympathetic tones. However many other targets have been characterized (for a review see [109,110]. The biological actions of ANF and BNP are mediated through the natriuretic peptide receptor type A (NPR-A). The importance of the ANF/BNP/NPR-A system is shown by the fact that genetic disruption of the ANF, BNP or NPR-A genes results in the impairment of cardiorenal homeostasis. The NPR-A knockout model shows that absence of NPR-A leads to hypertension, cardiac hypertrophy, interstitial fibrosis and sudden death [140]. ANF knockout mice show salt-dependent hypertension but do not develop cardiac hypertrophy at baseline. Cardiac interstitial fibrosis can be seen with pressure overload or volume overload [47,74,119]. Mice lacking BNP have multifocal fibrotic lesions in ventricles but do not develop hypertension or cardiac hypertrophy [174].

Interaction of natriuretic peptides ligands with their receptors leads to an increase in intracellular cyclic guanosine monophosphate (cGMP) and hence to interactions with cGMP-dependent protein kinases, cGMP-regulated cyclic nucleotide phosphordiesterases and cGMP-gated ion channels [96]. The principal biological actions of ANF and BNP are shown in Table 2.

Table 2. The principal biological actions of ANF and BNP

Kidney	Glomerular filtration ↑
	Collecting duct sodium reabsorption ↓
Nervous system	Sympathetic nerve activity ↓
	Salt and water appetite ↓
	Vasopressin ↓
Heart	Myocyte hypertrophy ↓
	Cardiac fibrosis ↓
Vascular	Vascular smooth muscle proliferation ↓
	Vascular smooth muscle tone ↓
	Vascular regeneration ↑
RAAS	RAAS ↓
Adipose Tissue	Lipolysis ↑
Immune system	Proinflammatory cytokines↓

RAAS: Renin angiotensin aldosterone system.

In the kidneys ANF and BNP increase glomerular filtration rate and filtration fraction dilating afferent arterioles and constricting efferent arterioles leading to an increase in glomerular capillary hydraulic pressure. This results in an enhancement of the driving force for ultrafiltration. ANF and BNP also inhibit Na^+ reabsorption at the level of collecting ducts [114,181]. These renal actions of ANF and BNP underlie the strong diuresis and natriuresis that may be caused by these peptides.

ANF and BNP have multiple actions on nerve activity. ANF reduces cardiac and pulmonary chemo- and baroreceptor activity, which leads to the suppression of sympathetic outflow to the heart. A decrease in sympathetic activity together with the increase in vagal afferent activity leads to a reduction in heart rate and cardiac output [35]. ANF and BNP also decrease salt and water appetite, thus leading to a reduction of extracellular fluid volume [176]. ANF inhibits the neurosecretion of vasopressin from the posterior pituitary thus decreasing water reabsorption at the collecting duct of the kidney [152].

ANF and BNP possess growth-suppressing and anti-proliferative properties in the heart. Growth-promoting effects generated by norepinephrine are inhibited by ANF treatment on both cardiomyocytes and fibroblasts [20]. BNP inhibits cardiac intersitium remodeling through its antifibrotic effects [174]. BNP also inhibits transforming growth factor-induced fibrosis in association with cGMP-dependent protein kinase activity and the MEK/ERK pathway [77].

cNPs have multiple effects on the vascular system. ANP antagonizes the growth of vascular smooth muscle cells mediated by guanylyl cyclase linked receptors [71] and is a direct smooth muscle relaxant [70]. cNPs also promote endothelial regeneration and could improve endothelial cells function after injury in atherosclerosis [90].

cNPs antagonize RAAS by inhibiting aldosterone synthesis and secretion, and by suppressing renin release [11,18,48].

cNPs exert potent lipolytic effects in isolated human fat cells as well as in preadipocytes through cGMP-mediated phosphorylation [50,93,157,158]. Cardiac cachexia often seen in the heart failure, during which cNPs are increased, may be explained by the lipolytic effects of cNPs. ANP inhibits the proliferation of human visceral preadipocytes [153].

4. Genes Encoding cNPs

Two distinct genes, *Nppa* and *Nppb*, encode ANF and BNP, respectively. In humans, both genes are located on the distal short arm of chromosome 1 (1p36.2). *Nppb* is located ~8 kb from *Nppa* [173,175]. In the mouse and rat, these genes reside on chromosome 4 and 5, respectively.

The structure of cNP-encoding genes are similar. They have three exons separated by two introns (Figure 1, 2). The signal sequences are located in exon 1 and the majority of the protein-coding sequences, which also include the bioactive portion of the molecule, reside in exon 2. Exon 3 encodes the terminal tyrosine and 3'-untranslated region [44,138,169]. ANF has a high degree of homology across mammals [172]. A single nucleotide codon substitution results in a change of an isoleucine residue in a position 110 in rats, mice and rabbits into a methionine residue for humans, dogs and cows [35]. In contrast to ANF, BNP is comparatively poorly conserved although mature BNP from sheep, cow, pig and dog share sequence homologies [1].

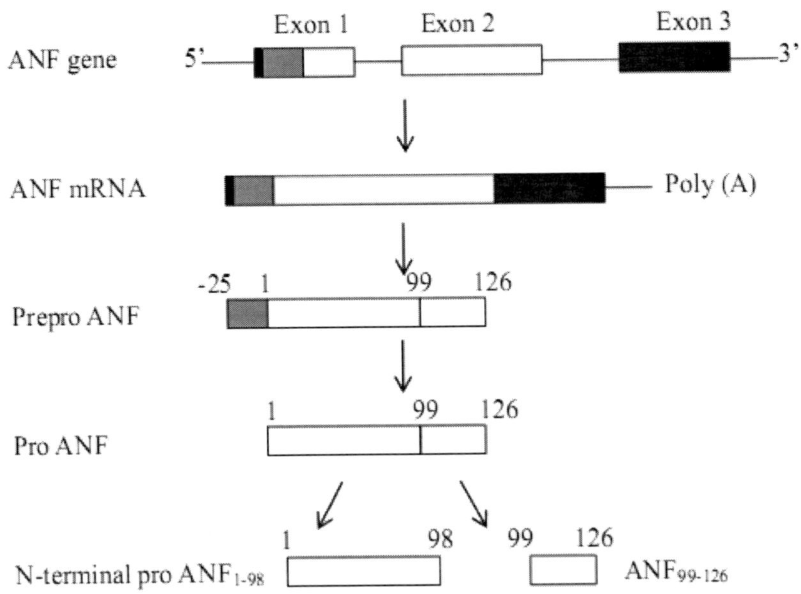

Figure 1. Schematic representation of ANF gene and biosynthesis pathway.

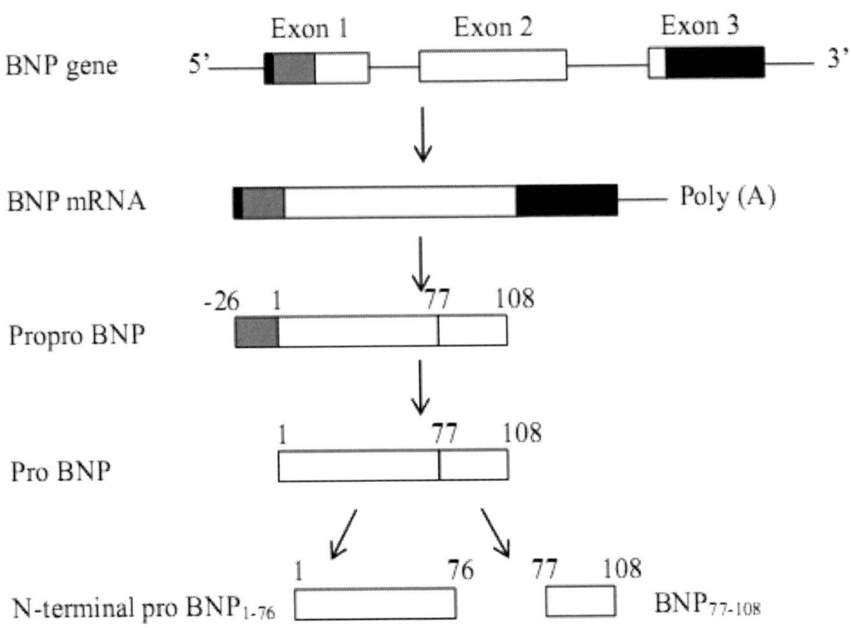

Figure 2. Schematic representation of BNP gene and biosynthesis pathway.

5. cNPs Gene Promoters

The 5' flanking regions of the cNP genes contain *cis*-acting elements to which various transcription factors are known to bind and thus regulate the level of expression.

Some elements are common to ANF and BNP but other are specific for their respective genes.

5.a. cis-Acting Sequences Found in Both cNP Promoters

There are two GATA-4 binding sites in the ANF gene promoter and three GATA-4 binding site in BNP promoter [24]. Patients with heterozygous GATA-4 mutations shows various congenital heart diseases such as atrial septal defect (ASD), ventricular septal defect (VSD), atrioventricular septal defect, double outlet right ventricle (DORV), dextrocardia and pulmonary stenosis [28] suggesting that GATA-4 is essential for normal heart development. GATA-4 is also involved in pathological conditions because the elevation of ANF and BNP genes in response to AII, ET-1, isoproterenol and stretch are mediated by GATA-4 [55,64].

AP-1/CRE-like binding sites are seen in both ANF and BNP genes. AP-1 transcription factor and the products of two proto-oncogenes, c-*fos* and c-*jun* bind to this site [147]. C-*fos* and c-*jun* have been shown to regulate ANF gene expression associated with cardiac hypertrophy [89]. The AP-1/CRE-like binding site plays important roles in the basal regulation of BNP [56,94].

5b. cis-Acting Sequences Found Exclusively in the ANF Gene Promoter

The 5' flanking region of ANF gene contains two CArG boxes that bind serum response factor, play important roles in cardiovascular development and diseases [115]. Nkx2.5 is an important transcription factor for cardiac development because mutation of Nkx2.5 accounts for 4% of congenital heart diseases such as ASD, VSD, DORV, Tetralogy of Fallot, aortic stenosis and Ebstein anomaly [28]. Two Nkx2.5 response elements, which are the binding site of Nkx2.5, can be seen in ANF promoter [163]. Tbx5, another transcription factor, binds to the T-box binding elements found in ANF promoter [15,68]. Tbx5 and Nkx2.5 synergistically activate ANF gene transcription. Mutation of Tbx5 results in the Holt-Oram syndrome, characterized by various cardiac defect and conduction abnormalities [8].

5.c. cis-Acting Sequences Found Exclusively in the BNP Gene Promoter

The proximal human BNP promoter contains two M-CAT elements and members of the transcriptional enhancer factor-1 (TEF-1) family binds to them [191]. Mutation of this sequence significantly depresses basal and phenilephrine (PE)-induced gene expression of rat BNP [179]. Three shear stress-responsiveness elements (SSRE) are found in BNP promoter. Cyclic mechanical stretch increased in NF-κB binding to the SSRE, and mutation of the SSRE or co-transfection with constitutive suppressor of NF-κB activity caused a maximum of 40% decrease in strain-activated BNP [95]. In association with GATA-4, the nuclear factor of activated T cells (NF-AT) family transcription factors mediate PE-induced activation of the human BNP promoter [117].

6. BIOSYNTHESIS OF CNPS

Human ANF and BNP are synthesized as 151- and 134-amino acid preprohormones, respectively. The removal of signal peptides results in proANF (ANF_{1-126}) and proBNP (BNP_{1-108}) [35]. ANF is stored in the specific granules of atrial cardiomyocytes as proANF and is further processed during secretion at the monobasic cleavage site into the circulating biological active ANF_{99-126} and the N-terminal-proANF_{1-98} (NT-proANF). proBNP, is cleaved in the trans-Golgi network generating the C-terminal circulating form BNP_{77-108}, which is stored in the atrial granules together with proANF [39].

The final processing step involved in the maturation of ANF appears to be undertaken by serine proteases. Corin, a type II transmembrane serine protease, that cleaves ANF_{1-126} into ANF_{1-98} during secretion [189]. However, a pool of ANF_{99-126} with high turnover exists within the atrial cardiomyocytes that may contribute significantly to the circulating pool (de Bold et al., unpublished information) without the intervention of a membrane protease. proBNP is also processed by furin, a subtilisin/Kex2p-like endoprotease [126].

7. SECRETION OF CNPS

Many of the agonists of cNPs secretion such as endothelin-1 (ET-1), angiotensin II (AII) and phenylephrine (PE) signal through G protein-coupled signaling and stimulate the RAS/c-Raf-1 signaling pathway and the phosphorinositide pathway leading to the recruitment and activation of effector molecules such as those found in the mitogen-activated protein kinase (MAPK) cascades [37,92]. Ligands such as ET-1, PE and AII typically signal through receptors coupled to Gq proteins, although other subclasses may also be used [42,66]. We have shown that Gi/o inhibition by pertussis toxin (PTX) abolishes stretch-secretion coupling without affecting baseline secretion through a mechanism that is independent of Gq signaling suggesting that PTX sensitive Gi/o protein is involved in the stretch induced secretion in atria [9]. By immunocytochemistry, ANF and Gao are seen co-localized in atrial granules suggesting a role for this G protein in the secretion of ANF. The decline in secretion of ANF and BNP following stretch is not due to NP pool depletion but due to the decreased sensitivity to the stimulus [135]. ET-1 and PE increase ANF and BNP in a more gradual manner than stretch and the decline of the secretion is also more gradual compared to mechanical stretch [14,16]. The protein synthesis inhibitor cycloheximide, partially but not completely abolishes basal NPs secretion *in vitro* [135] but monensin, an ionophore that impairs protein sorting and transport in the trans-Golgi network and that inhibits vesicle formation without directly affecting protein synthesis, completely inhibits stimulated secretion without affecting baseline release.. Double-label pulse-chase analysis investigations demonstrate that cardiomyocytes preferentially release newly synthesized hormone in a manner that does not entirely rely one protein synthesis [106]. These findings suggest that a constitutive-like release, which is based on exocytosis of vesicles budding from immature granules, is the main release mechanism of operating in atrial cardiocytes. This constitutive-like release is independent of the regulated and constitutive pathways [6,7].

The abnormally high levels of cNPs observed in several clinical entities is often ascribed to myocardial stretch but this view ignores the stimulatory contribution of neuroendocrine activation. Experimentally we have shown that both mechanical and neuroendocrine stimuli contribute to high cNP circulating levels in an aortic banding model [134].

8. Origin of Circulating cNPs

The greater part of ANF and BNP gene expression in mammals is found in the atria of heart. Other sites, such as the cardiac ventricles, aortic arch as well as extra cardiac sites including hypothalamus, pituitary gland and lung also produce cNPs at much lower levels than atria [51,134,149].

It is not unusual to find BNP defined as a ventricular hormone while ANF is define as an atrial hormone. This is not correct. In certain pathophysiological conditions ANF and BNP plasma levels increase significantly and while ANF remains higher than BNP, the latter increases only *relatively* more than ANF [107,108,190].

The biochemical evidence supports an overwhelming atrial source of cNPs. In the rat atria for example, the concentrations of ANF and BNP are about 30,000 and 300 pmol/g respectively. In the ventricle the cNP concentrations for ANF and BNP is about 5 and 0.5 pmol/g respectively. There are three clear conclusions that may be drawn from these data. Firstly, not only BNP but also ANF is present in the ventricles. Secondly, in both atria and ventricles, ANF is more abundant than BNP. Thirdly, even taking into account the differences in mass between the atrial and ventricular chambers (approximately 1:10) it is not possible to conclude that BNP might be a ventricular hormone. Similar conclusions can be reached when evaluating transcript abundance for atrial and ventricular ANF and BNP [10,190].

Cardiac catheterization data in humans with left ventricular hypertrophy suggest that the abnormal circulating levels of cNPs are significantly derived from atrial sources [121]. Importantly, in humans, replacement of the failing ventricle [107] as it is done in orthotopic heart transplantation, does not result in normalization of either BNP or ANF plasma levels even after normalization of intracardiac pressures and the RAAS occurs. From the pathophysiological point of view, these findings show that while there may be a good correlation between parameters such as ventricular hypertrophy or failure or ischemic events and NP secretion, the influence of these parameters on atrial function and hence, on atrial NP production should not be ignored.

9. NPs Receptors

The biological effects of ANF and BNP are predominantly mediated through the guanylyl cyclase-coupled receptor NPR-A, which is widely distributed throughout the body, including the kidneys, vascular smooth muscle, adrenal glands, brain and heart [100,103,155]. NPR-B, a second guanylyl cyclase-coupled receptor, is associated with CNP signaling [87]. Three molecular domains are recognized in the NPR-A and NPR-B receptors: an extracellular ligand-binding domain that binds to NPs, a short transmembrane domain and an intracellular domain that acts as a docking site for other proteins [91]. Binding of NPs to their receptors results in an increase in intracellular cGMP. Systemic administration of ANF induces elevated urinary and circulating levels of cGMP [54].

Intracellular cGMP can interact with various enzymes and ion channels including cGMP-dependent protein kinases, cGMP-gated ion channels and cGMP-regulated cyclic nucleotide phosphordiesterases [96]. The changes in activity of these enzymes or channels result in vasorelaxant effect on vascular muscle cells, reduction in cytosolic calcium concentration, muscle relaxation, inhibition of Na^+ reabsorption in the inner medullary collecting ducts and various other effects including those related to cardiac function, adrenal steroidogenesis and phototransduction [97,101,165].

A third natriuretic peptide receptor: NPR-C, is regarded as a clearance receptor although G-protein signaling has been associated with it [3,122]. The 37-amino acid cytoplasmic segment of NPR-C is characteristic of several receptors involved in peptide and macromolecule clearance. NPR-C is found most abundantly in the glomerular and vascular structures of the kidney, and also in the adrenals, lungs, brain, heart and the vascular wall [22,49]. NPR-C recognizes an eight-amino acid linear fragment of the ANF molecule. This structure requirement is much less stringent than the 17-amino acid disulfide-bridge ring necessary for ANF binding to the NPR-A receptor. NPR-C internalization occurs at a rate of 5%-7%/min with a constitutive process [29]. Since receptor internalization occurs much faster than ANF dissociation from the NPR-C receptor, efficient ligand transport and degradation can be achieved. Following ANF NPR-C internalization, ANF is hydrolyzed by lysosomes and NPR-C is recycled to the cell surface, making the clearance receptor available for removing natriuretic peptides from circulation.

The binding potency of the NPRs is ranked as follows: NPR-A = ANF ≥ BNP > CNP; NPR-B = CNP ≥ ANF > BNP; NPR-C = ANF ≥ CNP > BNP [167].

10. Metabolic Clearance of NPs

In addition to NPR-C, cNPs are cleared form the circulation via proteolytic cleavage by neutral endopeptidase (NEP) 24.11 [57]. NEP 24.11 is expressed mainly in the proximal renal tubules of the kidney as well as throughout the vascular endothelium, lungs and hearts. Under pathophysiological conditions, when NPR-C is thought to be saturated, NEP 24.11 contributes significantly to the removal of NPs [62]. The half-life of ANF is between 3 and 5 min, depending on the species while the half-life of BNP is as long as 20 min [167]. This difference can be explained by the reduced affinity of BNP to NPR-C as well as slower rate of hydrolysis by NEP 24.11 compared to ANF [69,170].

11. cNPs as Biomarkers

Because plasma ANF and BNP are increased in various pathological conditions such as heart failure, myocardial infarction, hypertension, left ventricular hypertrophy and pulmonary hypertension [17,31,46,123,124,129,144,151], plasma levels of cNPs are useful for the diagnosis and prognosis of these diseases. In addition to $ANF_{99\text{-}126}$ and $BNP_{77\text{-}108}$, the biologically inactive N-terminal fragments, NT-proANP and NT-proBNP also circulate in plasma. Since the N-terminal prohormones are more stable and the half-lives of the N-terminal prohormones in the plasma are longer than those of the C-terminal prohormones

[143,178] the usefulness of NT-proANP and NT-proBNP as biomarkers has been evaluated. As biomarkers of cardiovascular diseases, BNP predicts disease state and prognosis somewhat better than ANF and NT-proANP [60,141,188]. NT-proBNP and BNP provide interchangeable information although NT-proBNP seems to be more advantageous for diagnosing mild heart failure or asymptomatic left ventricular dysfunction [166]. At present BNP and NT-proBNP are established diagnostic biomarkers for heart failure [105,184], ventricular remodeling after acute myocardial infarction [123,128], left ventricular hypertrophy [182,188] and pulmonary hypertension [125].

An immunoassay that detects the midregional fragment of proANF (MRproANP) has been developed [118]. In a study with 797 chronic congenital heart failure patients, MR-proANP outperformed BNP and NT-proBNP in the prediction of death [116]. In another study of 251 patients with dyspnea, BNP and MR-proANP provides similar diagnostic information and were clinically useful as an aid in the diagnosis of acutely destabilized heart failure [41].

Because the cNPs are powerful diagnostic biomarkers in patients with heart failure, the biomarker-guided care with either BNP or NT-proBNP has been investigated. However, the effectiveness of the biomarker-guided care is not consistent probably because target BNP levels vary widely. A study using as a target a BNP level of 100 pg/ml [75] demonstrated useful performance while a study of target BNP level of 450 pg/ml [160] did not yield satisfactory results. Therefore achievement of a proper target concentration (~125 pg/ml for BNP or ~1000 pg/ml for NT-proBNP) is necessary to perform a successful biomarker-guided care [73].

12. cNPs as Therapeutic Agents

Because cNPs possess diuretic, natriuretic and hypotensive activity, ANF and BNP have been applied in the treatment of various cardiovascular conditions. Since ANP and BNP bind to the same receptor, NPR-A, it would be expected that the therapeutic effects of these peptides would be similar, but some differences between ANF and BNP are apparent in clinical settings.

Infusion of carperitide, a recombinant form of human ANF to 3777 acute heart failure patients at a median dose of 0.085 μg/kg/min for a median duration of 76 hours resulted in an improvement in 82% of patients [171]. Another study, in which carperitide was infused to 1832 acute heart failure patients at an initial dose of 0.025 – 0.05μg/kg/min demonstrated that 83.2% of the patients achieved recovery of the acute phase of the disease [130]. The 277 acute myocardial infarction patients who were administered carperitide for 72 hour showed significantly smaller infarct size, higher ejection fraction and lower cardiac death compared to 292 control groups [82]. A renal-protective effect of ANF was demonstrated in patients who underwent cardiac catheterization: the prevalence of contrast-induced nephropathy, defined as a 25% increase or at least 0.5 mg/dl increase in serum creatinine from baseline within 48 hours, was significantly lower in the carperitide group compared to control group [120]. These studies demonstrated that clinical use of ANF to cardiovascular disease is effective and safe.

The effectiveness and safety of nesiritide, a recombinant form of human BNP, has been questioned. Analysis of 5 randomized studies that included 1269 patients demonstrated that administration of nesiritide at a dose of less than 0.03 μg/kg/min to acutely decompensated heart failure patients significantly increased the risk of worsening renal function compared to a control group [150]. The 3496 acute heart failure patients who were administered nesiritide at a dose of 0.01 μg/kg/min demonstrated no significant change regarding the death rate, re-hospitalization rate and renal function compared to the 3511 placebo group. The nesiritide group was associated with a non-significant effect on dyspnea and an increase in hypotension. On the basis of these results, the authors did not recommend the routine use of nesiritide to acute heart failure patients [131]. Another study however found that subcutaneous administration of BNP at a dose of 10 μg/kg twice a day for 8 weeks to NYHA II to III heart failure patients improved Minnesota Living with Heart Failure score and LV filling pressure and decreased left atrial volume index with preserved glomerular filtration rate [25].

13. cNPs and Inflammation

Although the cNPs share most biological properties some differences between ANF and BNP have been reported in certain settings. Gene expression of BNP was induced more rapidly than that of ANF in the left ventricles following coronary artery ligation in rats [58]. Mice with genetically reduced production of ANF show salt-sensitive hypertension [74] but mice with targeted disruption of the BNP gene develop multifocal fibrotic lesions in the ventricles without systemic hypertension or ventricular hypertrophy [174]. Although the stimuli for production and secretion of ANF and BNP have been considered nearly identical, a discoordinated change of ANF and BNP was found in cardiac allograft rejection episodes as described below.

13.a. Plasma BNP in Acute Cardiac Allograft Rejection

While investigating changes in cNPs plasma levels in human cardiac transplant patients we found that during an acute rejection episode, BNP but not ANF plasma levels increased significantly over their levels obseved prior to the rejection episode [108]. Successful treatment of the rejection episode resulted in a decrease of BNP plasma levels to values observed prior to acute rejection.

In a study of 69 consecutive allograft cardiac recipients, we found that plasma BNP and NT-proBNP levels in patients with endomyocardial biopsies graded 3 according to the International Society of Heart and Lung Transplantation (ISHLT) were significantly higher than those in ISHLT grade 0 to 2. Plasma ANF did not change with increasing ISHLT grade [136]. Others also have reported an association between plasma BNP and cardiac cellular rejection [2,5,59]. Similar findings were seen in pediatric cardiac allograft recipients [98,148]. Finally, plasma BNP or NT-proBNP levels in post cardiac transplantation predicted cardiac mortality strongly and independently [52,112].

Other studies have not found a significant relationship between plasma BNP and endomyocardial biopsy rejection grade. No significant differences were seen in the biopsy score between high BNP (BNP > 150 pg/ml) and low BNP groups (BNP < 150 pg/ml) [142]. In another study, the rejection history based on ISHLT grade classification was not independently correlated with plasma BNP levels [84]. Plasma BNP levels were found not to differ according to the presence or degree of histological rejection [132]. No significant differences in NT-proBNP levels between patients with or without rejection were found in another study [32]. Plasma BNP levels were significantly higher in patients with ISHLT grade 2 or higher compared to non-rejecting recipients within three month after transplant. However, after the third month, BNP levels were similar in patients with and without rejection [65]. Some of these investigations found an association with pulmonary artery pressure or pulmonary capillary wedge pressure [65,84,132,142]. It is possible that an increase in pulmonary pressure led to a variable degree of tricuspid valve insufficiency resulting in right ventricular failure hence increasing BNP gene expression and plasma BNP levels. Thus, these changes may disturb the association between plasma BNP and rejection. Other factors such as time post-transplant or renal function may also affect plasma BNP levels possibly disturbing the association between BNP and rejection [84].

In an investigation on the relationship between BNP levels and hemodynamic change in rejecting and non-rejecting patients, we reported that cardiac filling pressures were correlated with plasma BNP or NT-proBNP only in ISHLT grade 0 and 1 cases, and that these correlation did not exist for ISHLT grade greater than 2 [136].

This finding suggests that in this category, BNP plasma levels are determined by factors other than or in addition to the hemodynamic status of the patient. In multivariable linear regression studies examining the outcome of increasing seriousness of rejection BNP predicted a new episode of rejection independent of hemodynamic measurements [186]. These results suggest that the elevation of BNP during an allograft rejection episode is not the result of hemodynamic change.

It is important to note that in our investigations [108] the increase in BNP plasma levels observed during acute cardiac allograft rejection was evident when comparing the individual patient values with their plasma BNP baseline value, i.e., prior to rejection. That is to say that a threshold absolute value for BNP plasma levels indicative of acute rejection could not be established across the patient population. Another study reported that individual changes in plasma BNP levels over time were more helpful than absolute BNP levels for detecting ISHLT grade 2R or greater rejection [83]. This further suggests that for BNP levels to be useful as a prognostic and diagnostic aid in cardiac allograft rejection, the history of BNP plasma values must be evaluated during follow up visits.

In most studies reported to date BNP plasma levels remained elevated for a relatively long time after cardiac allograft transplantation even in the absence of acute rejection episodes. A study found that in non-rejecting transplanted patients plasma BNP declined slowly following the transplant surgery without reaching normal values during one year of follow up observation [79,136]. In another study it was shown that NT-proBNP levels in transplanted patients without complication continued to decrease for the 6 months after transplant [32]. The elevation in plasma BNP levels is not due to the surgical procedure. Plasma BNP levels after coronary artery bypass grafting, for example, return to near normal values within one week after surgery, while plasma BNP levels after heart transplant remain elevated for one week and did not return to normal values [53]. Cardiac allograft vasculopathy (CAV) could be related to these findings. As opposed to atherosclerosis, CAV is a diffuse and pan-arterial process, characterized by intimal thickening and luminal stenosis of coronary artery, which account for major morbidity and mortality [30]. Elevated plasma BNP levels were associated with CAV diagnosed by coronary angiography or by biopsy [84,111,186].

13.b. BNP and Cytokines

Because successful treatment of a rejection episode using the anti-T cell monoclonal antibody OKT3 decreased the elevated plasma BNP level to pre-rejection values [108] we hypothesized that cytokines activated during the inflammatory process may be involved in the observed increase in BNP blood levels during acute rejection. It was found that in neonatal rat ventricular cardiomyocytes (NRVC) cultures, tumor necrosis factor alpha (TNFα) and interleukin one beta (IL-1β) but not interferon gamma or interleukin 6, induced a significant increase in BNP mRNA, and secretion, whereas, ANF mRNA levels and secretion did not change. Inhibition of p38 MAP kinase signaling using SB203580 abolished IL-1β- and TNF-α-stimulated increase in BNP mRNA, promoter activity and secretion. The conditioned medium obtained from unidirectional mixed lymphocyte reaction (MLR) culture, which is *in vitro* model of transplantation immunity, also increased BNP but not ANF secretion; an effect completely abolished by SB203580 pre-treatment [102].

In the aforementioned study on transplant patients conducted in our laboratory, plasma IL-1β or TNF-α levels did not increase during acute rejection episodes and further, IL-1β and TNF-α blood levels did not correlate with BNP levels [136] nor does the conditioned medium from MLR cultures have detectable amounts of IL-1β or TNF-α suggesting that substances other than pro-inflammatory cytokines are responsible for the selective upregulation of BNP in inflammation. Nevertheless, studies of biopsy samples using RT-PCR or immunohistochemistry have demonstrated that IL-1β or TNF-α gene expression was increased during cardiac rejection suggesting that plasma levels of these cytokines do not reflect accurately the *in situ* level at the graft [145,159]. To investigate whether other cytokines or factors in addition to TNFα and IL-1β could be related to the upregulation of BNP during cardiac rejection, plasma samples taken from rejecting and non-rejecting patients were analyzed by cytokine array [113]. Regulated on Activation, Normal T Expressed and Secreted (RANTES), Neutrophil Activating Protein-2 (NAP-2), and Insulin Growth Factor Binding Protein-1 (IGFBP-1) had significant correlations with BNP plasma levels during acute allograft rejection as diagnosed by endomyocardial biopsy. IGFBP-1 and RANTES induced BNP but not ANF secretion in NRVC cultures. In a separate study using 316 biopsy samples, the expression of RANTES measured by RT-PCR was found to increase significantly in ISHLT grade 2 and grade 3 rejection and returned to normal after treatment with pulse steroid therapy [43].

Another study, which measured the expression of 5000 genes in biopsy samples taken from 28 clinically quiescent patients found that elevated plasma BNP was associated with genes related to cardiac structure remodeling, vascular injury, inflammation and alloimmune processes [112].

13.c. BNP and the Immune System

Because OKT3 decreased plasma BNP in transplant patients we investigated the possibility that BNP may in turn, have immunomodulatory properties on MLR cultures. Gene expression of the GC-A receptor in lymphocytes taken from MLR was significantly increased compared to that in the lymphocyte taken from individual Brown Norway or Lewis rats, which were not activated by MLR [40]. Addition of either ANF or BNP to the MLR cultures decreased cell proliferation suggesting that both ANF and BNP have immunomodulatory functions. Several observations from other laboratories are compatible with this notion. BNP is synthesized and secreted from macrophages and T-cells infiltrating the failing myocardium [13] and was found to have immunomodulatory activity in human monocyte cell line as measured by the production of reactive oxygen and nitrogen species, leukotriene B_4, prostaglandin E_2, IL-10 and cell motility [26]. In addition, BNP directly induced apoptosis of $CD8^{+}T$ cells taken from cardiac transplant patients via a caspase 3-associated mechanisms [161]. These studies suggest that even though both ANF and BNP can modulate proliferation of cells of the immune system, it is BNP that may carry out this function by virtue of the selective upregulation of this peptide by various neurohumoral species, including various cytokines and related substances involved in the immune response.

Studies investigating circulating BNP levels in rheumatoid arthritis (RA) patients have shown that NT-proBNP, C-reactive protein and TNFα levels in RA patients without clinical heart failure were higher than control groups suggesting that the increase in NT-proBNP in RA patients is related to inflammation [164]. Armstrong et al. reported that RA patients without a history of cardiac disease had higher plasma BNP levels than a control group [4]. Taken together, the above studies link the endocrine function of the heart to the immune system.

13.d. BNP Gene Regulation at the Transcriptional Level in Inflammation

In NRVC cultures, the upregulation of BNP gene expression caused by TNF-α and IL-1β can be prevented by actinomycin D, which suggests that the effect of cytokines on BNP gene expression is performed at transcriptional level [102]. In accordance with these studies, NRVC transfected with -2.2-kb rat promoter coupled to a luciferase reporter gene, TNF-α and IL-1β stimulated the activation of -2.2-kb rat promoter, and SB203580, a p38 MAP kinase inhibitor completely suppressed this effect [63]. In this study, mutation of the M-CAT element at -97-bp reduced IL-1β stimulation of the human BNP promoter activity by 60% suggesting that there are unknown elements which stimulate IL-1β- based activities. Mutation of other elements such as GATA at -85-bp and M-CAT at -124-bp did not affect IL-1β stimulation of BNP promoter activity. Another pro-inflammatory cytokine, IL-6, also stimulated BNP synthesis and secretion in neonatal cardiomyocyte culture, and this elevation was inhibited by actinomycin D, but not cycloheximide, suggesting that this regulation is due to an increase in transcription [187].

13.e. BNP in Autoimmune Myocarditis

Immunization of rats with myosin heavy chain produces an acute cardiac autoimmune response resembling giant cell myocarditis. In this myocarditis model, macrophages and $CD4^+$ T cells are key components for the trigger of the disease [85]. In this model, TNFα and IL-1β mRNA are induced in the ventricles 19 days after the immunization, heart weight/body weight ratio increases and mono nuclear cells infiltration and necrosis are seen microscopically [139]. At this time we found that plasma BNP levels increase significantly in the absence of changes in ANF plasma levels in a manner reminiscent of what we previously observed in acute allograft rejection in humans suggesting that an increase in plasma BNP accompanied by unchanging circulating levels of ANF is indicative of inflammation *in vivo* [137].

13.f. BNP in Sepsis and Following Lipolysaccharide (LPS) Treatment

Several reports have indicated that plasma ANF or BNP levels are increased in sepsis and in septic shock patients [27,127,162,185]. The increase in plasma cNPs are partly due to cardiac dysfunction because plasma ANF or BNP were inversely correlated with cardiac function [23,61,146]. However, in other reports where cardiac function was preserved, plasma BNP levels were elevated [19,104,162]. Administration of LPS, the main bacterial component responsible for the immune response to Gram negative bacterial infection, to healthy men increased plasma NT-proBNP without changing heart rate and blood pressure [183]. Also in the sepsis patient without systolic myocardial dysfunction, plasma BNP was correlated with CRP and IL-1[162]. These findings are a further indication that elevated plasma BNP occurs in association with inflammation with or without cardiac dysfunction. As a predictor of cardiac death in septic patients, plasma BNP was a better diagnostic parameter for the prediction of death than plasma ANF [192]. The changes in cNPs blood levels have been useful as predictors of outcomes. Plasma NT-proBNP, for example, was a predictor of nonsurvivors in severe sepsis patients [12].

To study the mechanism whereby plasma BNP changes in septic patients, we administered LPS to rats, and plasma and cardiac gene expression of NPs were investigated. Administration of LPS to rats increased both plasma ANF and BNP levels although the increase in plasma BNP was always larger than the increase in ANF [133]. BNP gene expression was increased without a corresponding change in ANF expression in both atria and ventricles. In this study, the correlation among mRNA data in 16 rats which included controls, and rats treated with low dose LPS, high dose LPS and LPS plus the MAPK inhibitor SB203580 was calculated. ANF mRNA was not correlated with BNP mRNA. The results also revealed that gene expression of MCP-1, MMP-8, TIMP-1 CINC-1, TNFα, ICAM-1 and MIP-3 correlated with gene expression of BNP although none of these mRNA levels correlated with ANF mRNA. These results confirm that BNP, not ANF is related to inflammation process. Transcriptional activation of the BNP gene by LPS has been found to be mediated through GATA elements [180].

14. The Effect of Inflammation on the Use of cNPs As Biomarkers

While ANF and BNP are differently regulated in disorders involving inflammation such as in acute cardiac allograft rejection, sepsis and experimental autoimmune myocarditis these cNPs are coordinately regulated in other cardiovascular diseases such as congestive heart failure, ischemic heart disease and hypertensive heart diseases. In congestive heart failure plasma TNFα, IL-1β and IL-6 levels increase with increasing New York Heart Association functional class [177] indicating that the inflammation process is involved in most cardiovascular disorders that display hemodynamic and hormonal changes. The fact that BNP is more influenced by inflammation compared to ANF may be one reason why BNP appears as a marginally better biomarker for diagnostic and prognostic purposes in various cardiovascular disorders. When using cNPs as biomarkers for cardiovascular diseases, this concept should be taken into account. Because inflammation and hemodynamic changes coexist in most cardiovascular disorders the measurement of both ANF and BNP in individual patients may be advisable. Plasma ANF increases relatively larger than those of BNP may be indicative of hemodynamic deterioration while the opposite observation may be indicative of worsening inflammation. Because these two pathologies are totally different and so are the indicated therapeutic interventions, measurement of ANF and BNP simultaneously may help to evaluate patient status more accurately.

References

[1] Aitken GD, Raizis AM, Yandle TG, George PM, Espiner EA and Cameron VA. The characterization of ovine genes for atrial, brain, and C-type natriuretic peptides. *Domest Anim. Endocrinol.* 16: 115-121, 1999.

[2] Almenar L, Hervas I, Martinez-Dolz L, Rueda J, Arnau MA, Osa A, Mateo A, Dicenta F and Palencia M. The value of brain natriuretic peptide for the diagnosis of heart transplant rejection. *Transplant Proc.* 34: 174-175, 2002.

[3] Anand-Srivastava MB and Trachte GJ. Atrial natriuretic factor receptors and signal transduction mechanisms. *Pharmacological Reviews* 45: 455-497, 1993.

[4] Armstrong DJ, Gardiner PV and O'Kane MJ. Rheumatoid arthritis patients with active disease and no history of cardiac pathology have higher brain natriuretic peptide (BNP) levels than patients with inactive disease or healthy control subjects. *Ulster Med. J.* 79: 82-84, 2010.

[5] Arnau-Vives MA, Almenar L, Hervas I, Osa A, Martinez-Dolz L, Rueda J, Zorio E, Martinez-Ortiz DU, Perez JL, Mateo A and Palencia M. Predictive value of brain natriuretic peptide in the diagnosis of heart transplant rejection. *J. Heart Lung Transplant* 23: 850-856, 2004.

[6] Arvan P and Halban PA. Sorting ourselves out: seeking consensus on trafficking in the beta-cell. *Traffic* 5: 53-61, 2004.

[7] Arvan P, Kuliawat R, Prabakaran D, Zavacki AM, Elahi D, Wang S and Pilkey D. Protein discharge from immature secretory granules displays both regulated and constitutive characteristics. *J. Biol. Chem.* 266: 14171-14174, 1991.

[8] Basson CT, Bachinsky DR, Lin RC, Levi T, Elkins JA, Soults J, Grayzel D, Kroumpouzou E, Traill TA, Leblanc-Straceski J, Renault B, Kucherlapati R, Seidman JG and Seidman CE. Mutations in human TBX5 [corrected] cause limb and cardiac malformation in Holt-Oram syndrome. *Nat. Genet.* 15: 30-35, 1997.

[9] Bensimon M, Chang A, Kuroski-de Bold ML, Ponce A, Carreras D and de Bold AJ. Participation of G Proteins in Natriuretic Peptide Hormone Secretion from Heart Atria. *Endocrinology* 145: 5313-5321, 2004.

[10] Bianciotti LG and de Bold AJ. Modulation of cardiac natriuretic peptide gene expression following endothelin type A receptor blockade in renovascular hypertension. *Cardiovasc. Res.* 49: 808-816, 2001.

[11] Brenner BM, Ballermann BJ, Gunning ME and Zeidel ML. Diverse biological actions of atrial natriuretic peptide. [Review]. *Physiol. Rev.* 70: 665-699, 1990.

[12] Brueckmann M, Huhle G, Lang S, Haase KK, Bertsch T, Weiss C, Kaden JJ, Putensen C, Borggrefe M and Hoffmann U. Prognostic value of plasma N-terminal pro-brain natriuretic peptide in patients with severe sepsis. *Circulation* 112: 527-534, 2005.

[13] Bruggink AH, De Jonge N, van Oosterhout MF, Van Wichen DF, de Koning E, Lahpor JR, Kemperman H, Gmelig-Meyling FH and de Weger RA. Brain natriuretic peptide is produced both by cardiomyocytes and cells infiltrating the heart in patients with severe heart failure supported by a left ventricular assist device. *J. Heart Lung Transplant* 25: 174-180, 2006.

[14] Bruneau BG and de Bold AJ. Selective changes in natriuretic peptide and early response gene expression in isolated rat atria following stimulation by stretch or endothelin-1. *Cardiovasc. Res.* 28: 1519-1525, 1994.

[15] Bruneau BG, Nemer G, Schmitt JP, Charron F, Robitaille L, Caron S, Conner DA, Gessler M, Nemer M, Seidman CE and Seidman JG. A murine model of Holt-Oram syndrome defines roles of the T-box transcription factor Tbx5 in cardiogenesis and disease. *Cell* 106: 709-721, 2001.

[16] Bruneau BG, Piazza LA and de Bold AJ. BNP gene expression is specifically modulated by stretch and ET- 1 in a new model of isolated rat atria. *Am. J. Physiol.* 273: H2678-86, 1997.

[17] Buckley MG, Markandu ND, Miller MA, Sagnella GA and MacGregor GA. Plasma concentrations and comparisons of brain and atrial natriuretic peptide in normal subjects and in patients with essential hypertension. *Journal of Human Hypertension* 7: 245-250, 1993.

[18] Burger AJ. A review of the renal and neurohormonal effects of B-type natriuretic peptide. *Congest Heart Fail* 11: 30-38, 2005.

[19] Burjonroppa SC, Tong AT, Xiao LC, Johnson MM, Yusuf SW and Lenihan DJ. Cancer patients with markedly elevated B-type natriuretic peptide may not have volume overload. *Am. J. Clin. Oncol.* 30: 287-293, 2007.

[20] Calderone A, Thaik CM, Takahashi N, Chang DLF and Colucci WS. Nitric oxide, atrial natriuretic peptide, and cyclic GMP inhibit the growth-promoting effects of norepinephrine in cardiac myocytes and fibroblasts. *J. Clin. Invest.* 101: 812-818, 1998.

[21] Casco VH, Veinot JP, Kuroski de Bold ML, Masters RG, Stevenson MM and de Bold AJ. Natriuretic peptide system gene expression in human coronary arteries. *J. Histochem. Cytochem.* 50: 799-809, 2002.

[22] Cerra MC. Cardiac distribution of the binding sites for natriuretic peptides in vertebrates. [Review]. *Cardioscience* 5: 215-224, 1994.

[23] Charpentier J, Luyt CE, Fulla Y, Vinsonneau C, Cariou A, Grabar S, Dhainaut JF, Mira JP and Chiche JD. Brain natriuretic peptide: A marker of myocardial dysfunction and prognosis during severe sepsis. *Crit. Care Med.* 32: 660-665, 2004.

[24] Charron F, Paradis P, Bronchain O, Nemer G and Nemer M. Cooperative interaction between GATA-4 and GATA-6 regulates myocardial gene expression. *Mol. Cell Biol.* 19: 4355-4365, 1999.

[25] Chen HH, Glockner JF, Schirger JA, Cataliotti A, Redfield MM and Burnett JC, Jr. Novel protein therapeutics for systolic heart failure: chronic subcutaneous B-type natriuretic peptide. *J. Am. Coll. Cardiol.* 60: 2305-2312, 2012.

[26] Chiurchiu V, Izzi V, D'Aquilio F, Carotenuto F, Di NP and Baldini PM. Brain Natriuretic Peptide (BNP) regulates the production of inflammatory mediators in human THP-1 macrophages. *Regul. Pept.* 148: 26-32, 2008.

[27] Chua G and Kang-Hoe L. Marked elevations in N-terminal brain natriuretic peptide levels in septic shock. *Crit. Care* 8: R248-R250, 2004.

[28] Clark KL, Yutzey KE and Benson DW. Transcription factors and congenital heart defects. *Annu. Rev. Physiol.* 68: 97-121, 2006.

[29] Cohen D, Koh GY, Nikonova LN, Porter JG and Maack T. Molecular determinants of the clearance function of type C receptors of natriuretic peptides. *J. Biol. Chem.* 271: 9863-9869, 1996.

[30] Colvin-Adams M and Agnihotri A. Cardiac allograft vasculopathy: current knowledge and future direction. *Clin Transplant* 25: 175-184, 2011.

[31] Cowie MR, Struthers AD, Wood DA, Coats AJ, Thompson SG, Poole-Wilson PA and Sutton GC. Value of natriuretic peptides in assessment of patients with possible new heart failure in primary care. *Lancet* 350: 1349-1353, 1997.

[32] Cuppoletti A, Roig E, Perez-Villa F, Marin JL, Orus J, Vallejos I, Rivera A and Botta C. Value of NT-proBNP determinations in the follow-up of heart transplantation. *Transplant Proc.* 37: 4033-4035, 2005.

[33] de Bold AJ. Tissue fractionation studies on the relationship between an atrial natriuretic factor and specific atrial granules. *Can. J. Physiol. Pharmacol.* 60: 324-330, 1982.

[34] de Bold AJ, Borenstein HB, Veress AT and Sonnenberg H. A rapid and potent natriuretic response to intravenous injection of atrial myocardial extracts in rats. *Life Sci.* 28: 89-94, 1981.

[35] de Bold AJ and Bruneau BG. Natriuretic Peptides. In: *Handbook of Physiology, Section 7: The Endocrine System, Volume III: Endocrine Regulation of Water and Electrolyte Balance*, edited by Fray JCS and Goodman MH. American Physiological Society by Oxford University Press, 2000, 2000.

[36] de Bold AJ and Flynn TG. Cardionatrin I - a novel heart peptide with potent diuretic and natriuretic properties. *Life Sci.* 33: 297-302, 1983.

[37] de Bold AJ, Ma KK, Zhang Y, de Bold ML, Bensimon M and Khoshbaten A. The physiological and pathophysiological modulation of the endocrine function of the heart. *Can. J. Physiol. Pharmacol.* 79: 705-714, 2001.

[38] de Bold AJ and Salerno TA. Natriuretic activity of extracts obtained from hearts of different species and from various rat tissues. *Can. J. Physiol. Pharmacol.* 61: 127-130, 1983.

[39] de Bold ML and de Bold AJ. Immunocytochemical demonstration of cardionatrins in human atrial myocardium. In: *Immunocytochemistry in tumor diagnosis.*, edited by Russo J. Boston: Martinus Nijhoss Publishing, 1985.

[40] de Bold ML, Etchepare A, Martinuk A and de Bold AJ. Cardiac hormones ANF and BNP modulate proliferation in the unidirectional mixed lymphocyte reaction. *J. Heart Lung Transplant* 29: 323-326, 2010.

[41] Dieplinger B, Gegenhuber A, Kaar G, Poelz W, Haltmayer M and Mueller T. Prognostic value of established and novel biomarkers in patients with shortness of breath attending an emergency department. *Clin. Biochem.* 43: 714-719, 2010.

[42] Eguchi S, Hirata Y, Imai T and Marumo F. Endothelin receptor subtypes are coupled to adenylate cyclase via different guanyl nucleotide-binding proteins in vasculature. *Endocrinology* 132: 524-529, 1993.

[43] Fahmy NM, Yamani MH, Starling RC, Ratliff NB, Young JB, McCarthy PM, Feng J, Novick AC and Fairchild RL. Chemokine and receptor-gene expression during early and late acute rejection episodes in human cardiac allografts. *Transplantation* 75: 2044-2047, 2003.

[44] Flynn TG, Davies PL, Kennedy BP, de Bold ML and de Bold AJ. Alignment of rat cardionatrin sequences with the preprocardionatrin sequence from complementary DNA. *Science* 228: 323-325, 1985.

[45] Flynn TG, de Bold ML and de Bold AJ. The amino acid sequence of an atrial peptide with potent diuretic and natriuretic properties. *Biochem. Biophys. Res. Commun.* 117: 859-865, 1983.

[46] Foy SG, Crozier IG, Richards AM, Nicholls MG, Turner JG, Frampton CM and Ikram H. Neurohormonal changes after acute myocardial infarction. Relationships with haemodynamic indices and effects of ACE inhibition. *European Heart Journal* 16: 770-778, 1995.

[47] Franco V, Chen YF, Oparil S, Feng JA, Wang D, Hage F and Perry G. Atrial natriuretic peptide dose-dependently inhibits pressure overload-induced cardiac remodeling. *Hypertension* 44: 746-750, 2004.

[48] Franco-Saenz R, Atarashi K, Takagi M and Mulrow PJ. Effect of atrial natriuretic factor on renin and aldosterone. [Review]. *J. Cardiovasc. Pharmacol.* 13 Suppl 6: S31-S35, 1989.

[49] Fuller F, Porter JG, Arfsten AE, Miller J, Schilling JW, Scarborough RM, Lewicki JA and Schenk DB. Atrial Natriuretic Peptide Clearance Receptor. *J. Biol. Chem.* 263: 9395-9401, 1988.

[50] Galitzky J, Sengenes C, Thalamas C, Marques MA, Senard JM, Lafontan M and Berlan M. The lipid-mobilizing effect of atrial natriuretic peptide is unrelated to sympathetic nervous system activation or obesity in young men. *J. Lipid Res.* 42: 536-544, 2001.

[51] Gardner DG, Vlasuk GP, Baxter JD, Fiddes JC and Lewicki JA. Identification of atrial natriuretic factor gene transcripts in the central nervous system of the rat. *Proc. Natl. Acad. Sci. USA* 84: 2175-2179, 1987.

[52] Gardner RS, Chong KS, Murday AJ, Morton JJ and McDonagh TA. N-terminal brain natriuretic peptide is predictive of death after cardiac transplantation. *Heart* 92: 121-123, 2006.

[53] Geny B, Follenius M, Epailly E, Charpentier A, Brandenberger G, Eisenmann B, Haberey P and Piquard F. Transient reduction without normalization of brain natriuretic

peptide early after heart transplantation. *J. Thorac. Cardiovasc. Surg.* 115(2): 473-475, 1998.

[54] Gerzer R, Witzgall H, Tremblay J, Gutkowska J and Hamet P. Rapid increase in plasma and urinary cyclic GMP after bolus injection of atrial natriuretic factor in man. *J. Clin. Endocrinol. Metab.* 61: 1217-1219, 1985.

[55] Grepin C, Dagnino L, Robitaille L, Haberstroh L, Antakly T and Nemer M. A hormone-encoding gene identifies a pathway for cardiac but not skeletal muscle gene transcription. *Mol. Cell Biol.* 14: 3115-3129, 1994.

[56] Grepin C, Robitaille L, Antakly T and Nemer M. Inhibition of transcription factor GATA-4 expression blocks in vitro cardiac muscle differentiation. *Mol. Cell Biol.* 15: 4095-4102, 1995.

[57] Hall C. Essential biochemistry and physiology of (NT-pro)BNP. *Eur. J. Heart Fail* 6: 257-260, 2004.

[58] Hama N, Itoh H, Shirakami G, Nakagawa O, Suga S, Ogawa Y, Masuda I, Nakanishi K, Yoshimasa T, Hashimoto Y and et al., Rapid ventricular induction of brain natriuretic peptide gene expression in experimental acute myocardial infarction. *Circulation* 92: 1558-1564, 1995.

[59] Hammerer-Lercher A, Mair J, Antretter H, Ruttmann E, Poelzl G, Laufer G, Puschendorf B and Hangler H. B-type natriuretic peptide as a marker of allograft rejection after heart transplantation. *J. Heart Lung Transplant* 24: 1444, 2005.

[60] Hammerer-Lercher A, Neubauer E, Muller S, Pachinger O, Puschendorf B and Mair J. Head-to-head comparison of N-terminal pro-brain natriuretic peptide, brain natriuretic peptide and N-terminal pro-atrial natriuretic peptide in diagnosing left ventricular dysfunction. *Clin. Chim. Acta* 310: 193-197, 2001.

[61] Hartemink KJ, Twisk JW and Groeneveld AB. High circulating N-terminal pro-B-type natriuretic peptide is associated with greater systolic cardiac dysfunction and nonresponsiveness to fluids in septic vs nonseptic critically ill patients. *J. Crit. Care* 26: 108.e1-108.e8, 2011.

[62] Hashimoto Y, Nakao K, Hama N, Imura H, Mori S, Yamaguchi M, Yasuhara M and Hori R. Cleance Mechanisms of Atrial and Brain Natriuretic Peptides in Rats. *Pharmaceutical Research* 11: 60-64, 1994.

[63] He Q and LaPointe MC. Interleukin-1beta regulation of the human brain natriuretic peptide promoter involves Ras-, Rac-, and p38 kinase-dependent pathways in cardiac myocytes. *Hypertension* 33: 283-289, 1999.

[64] He Q, Mendez M and LaPointe MC. Regulation of the human brain natriuretic peptide gene by GATA-4. *Am. J. Physiol. Endocrinol. Metab.* 283: E50-E57, 2002.

[65] Hervas I, Arnau MA, Almenar L, Perez-Pastor JL, Chirivella M, Osca J, Bello P, Osa A, Marti JF, Vera F and Mateo A. Ventricular natriuretic peptide (BNP) in heart transplantation: BNP correlation with endomyocardial biopsy, laboratory and hemodynamic measures. *Lab. Invest.* 84: 138-145, 2004.

[66] Hilal-Dandan R, Ramirez MT, Villegas S, Gonzalez A, Endo-Mochizuki Y, Brown JH and Brunton LL. Endothelin ET_A receptor regulates signaling and ANF gene expression via multiple G protein-linked pathways. *Am. J. Physiol.* 272: H130-H137, 1997.

[67] Hirata Y, Takata S, Kawahara Y, Takai Y, Chino N, Kimura T and Sakakibara S. Molecular mechanism of action by atrial natriuretic peptide in rat vascular smooth muscle cells. *Jap Circ. J.* 52: 1430-1435, 1988.

[68] Hiroi Y, Kudoh S, Monzen K, Ikeda Y, Yazaki Y, Nagai R and Komuro I. Tbx5 associates with Nkx2-5 and synergistically promotes cardiomyocyte differentiation. *Nat. Genet.* 28: 276-280, 2001.

[69] Holmes SJ, Espiner EA, Richards AM, Yandle TG and Frampton C. Renal, endocrine, and hemodynamic effects of human brain natriuretic peptide in normal man. *J. Clin. Endocrinol. Metab.* 76: 91-96, 1993.

[70] Hughes A, Thom S, Goldberg P, Martin G and Sever P. Direct effect of alpha-human atrial natriuretic peptide on human vasculature *in vivo* and in vitro. *Clin Sci* 74: 207-211, 1988.

[71] Itoh H, Pratt RE, Ohno M and Dzau VJ. Atrial natriuretic polypeptide as a novel antigrowth factor of endothelial cells. *Hypertension* 19: 758-761, 1992.

[72] Jamieson JD and Palade GE. Specific granules in atrial muscle cells. *J. Cell Biol.* 23: 151-172, 1964.

[73] Januzzi JL, Jr. The role of natriuretic peptide testing in guiding chronic heart failure management: review of available data and recommendations for use. *Arch. Cardiovasc. Dis.* 105: 40-50, 2012.

[74] John SW, Krege JH, Oliver PM, Hagaman JR, Hodgin JB, Pang SC, Flynn TG and Smithies O. Genetic decreases in atrial natriuretic peptide and salt-sensitive hypertension. *Science* 267: 679-681, 1995.

[75] Jourdain P, Jondeau G, Funck F, Gueffet P, Le Helloco A, Donal E, Aupetit JF, Aumont MC, Galinier M, Eicher JC, Cohen-Solal A and Juilliere Y. Plasma brain natriuretic peptide-guided therapy to improve outcome in heart failure: the STARS-BNP Multicenter Study. *J. Am. Coll. Cardiol.* 49: 1733-1739, 2007.

[76] Kangawa K and Matsuo H. Purification and complete amino acid sequence of alpha-human atrial natriuretic polypeptide (alpha-hANP). *Biochem. Biophys. Res. Commun.* 118: 131-139, 1984.

[77] Kapoun AM, Liang F, O'Young G, Damm DL, Quon D, White RT, Munson K, Lam A, Schreiner GF and Protter AA. B-type natriuretic peptide exerts broad functional opposition to transforming growth factor-beta in primary human cardiac fibroblasts: fibrosis, myofibroblast conversion, proliferation, and inflammation. *Circ. Res.* 94: 453-461, 2004.

[78] Kennedy BP, Marsden JJ, Flynn TG, de Bold AJ and Davies PL. Isolation and nucleotide sequence of a cloned cardionatrin cDNA. *Biochem. Biophys. Res. Commun.* 122: 1076-1082, 1984.

[79] Kirchhoff WC, Gradaus R, Stypmann J, Deng MC, Tian TD, Scheld HH, Breithardt G and Brisse B. Vasoactive peptides during long-term follow-up of patients after cardiac transplantation. *J. Heart Lung Transplant* 23: 284-288, 2004.

[80] Kisch B. Electron microscopy of the atrium of the heart I. Guinea pig. *Exp. Med. Surg.* 14: 99-112, 1956.

[81] Kisch B. A significant electron microscopic difference between the atria and the ventricles of the mammalian heart. *Exp. Med. Surg.* 21: 193-221, 1963.

[82] Kitakaze M, Asakura M, Kim J, Shintani Y, Asanuma H, Hamasaki T, Seguchi O, Myoishi M, Minamino T, Ohara T, Nagai Y, Nanto S, Watanabe K, Fukuzawa S, Hirayama A, Nakamura N, Kimura K, Fujii K, Ishihara M, Saito Y, Tomoike H and Kitamura S. Human atrial natriuretic peptide and nicorandil as adjuncts to reperfusion

treatment for acute myocardial infarction (J-WIND): two randomised trials. *Lancet* 370: 1483-1493, 2007.

[83] Kittleson MM, Skojec DV, Wittstein IS, Champion HC, Judge DP, Barouch LA, Halushka M, Hare JM, Kasper EK and Russell SD. The change in B-type natriuretic peptide levels over time predicts significant rejection in cardiac transplant recipients. *J. Heart Lung Transplant* 28: 704-709, 2009.

[84] Klingenberg R, Koch A, Gleissner C, Schnabel PA, Haass M, Remppis A, Katus HA and Dengler TJ. Determinants of B-type natriuretic peptide plasma levels in the chronic phase after heart transplantation. *Transpl. Int.* 18: 169-176, 2005.

[85] Kodama M, Zhang S, Hanawa H and Shibata A. Immunohistochemical characterization of infiltrating mononuclear cells in the rat heart with experimental autoimmune giant cell myocarditis. *Clin. Exp. Immunol.* 90: 330-335, 1992.

[86] Kojima M, Minamino N, Kangawa K and Matsuo H. Cloning and sequence analysis of a cDNA encoding a precursor for rat C-type natriuretic peptide (CNP). *Febs. Lett.* 276: 209-213, 1990.

[87] Koller KJ, Lowe DG, Bennett GL, Minamino N, Kangawa K, Matsuo H and Goeddel DV. Selective activation of the B natriuretic peptide receptor by C- type natriuretic peptide (CNP). *Science* 252: 120-123, 1991.

[88] Komatsu Y, Nakao K, Suga S, Ogawa Y, Mukoyama M, Arai H, Shirakami G, Hosoda K, Nakagawa O, Hama N and et al., C-type natriuretic peptide (CNP) in rats and humans. *Endocrinology* 129: 1104-1106, 1991.

[89] Kovacic-Milivojevic B, Wong VS and Gardner DG. Selective regulation of the atrial natriuretic peptide gene by individual components of the activator protein-1 complex. *Endocrinology* 137: 1108-1117, 1996.

[90] Kuhn M. Endothelial actions of atrial and B-type natriuretic peptides. *Br. J. Pharmacol.* 166: 522-531, 2012.

[91] Kumar R, Grammatikakis N and Chinkers M. Regulation of the atrial natriuretic peptide receptor by heat shock protein 90 complexes. *J. Biol. Chem.* 276: 11371-11375, 2001.

[92] Kyosseva SV. Mitogen-activated protein kinase signaling. *Int. Rev. Neurobiol.* 59: 201-220, 2004.

[93] Lafontan M, Sengenes C, Galitzky J, Berlan M, De G, I, Crampes F, Stich V, Langin D, Barbe P and Riviere D. Recent developments on lipolysis regulation in humans and discovery of a new lipolytic pathway. *Int. J. Obes. Relat. Metab. Disord* 24 Suppl 4: S47-S52, 2000.

[94] LaPointe MC. Molecular regulation of the brain natriuretic peptide gene. *Peptides* 26: 944-956, 2005.

[95] Liang F, Kovacic-Milivojevic B, Chen S, Cui J, Roediger F, Intengan H and Gardner DG. Signaling mechanisms underlying strain-dependent brain natriuretic peptide gene transcription. *Can. J. Physiol. Pharmacol.* 79: 640-645, 2001.

[96] Lincoln TM and Cornwell TL. Intracellular cyclic GMP receptor proteins. *FASEB J.* 7: 328-338, 1993.

[97] Lincoln TM, Dey N and Sellak H. Invited review: cGMP-dependent protein kinase signaling mechanisms in smooth muscle: from the regulation of tone to gene expression. *J. Appl. Physiol.* 91: 1421-1430, 2001.

[98] Lindblade CL, Chun DS, Darragh RK, Caldwell RL, Murphy DJ and Schamberger MS. Value of plasma B-type natriuretic peptide as a marker for rejection in pediatric heart transplant recipients. *Am. J. Cardiol.* 95: 909-911, 2005.

[99] Loretz CA, Pollina C, Kaiya H, Sakaguchi H and Takei Y. Local synthesis of natriuretic peptides in the eel intestine. *Biochem. Biophys. Res. Commun.* 238: 817-822, 1997.

[100] Lowe DG, Chang MS, Hellmiss R, Chen E, Singh S, Garbers DL and Goeddel DV. Human atrial natriuretic peptide receptor defines a new paradigm for second messenger signal transduction. *EMBO J.* 8: 1377-1384, 1989.

[101] Lucas KA, Pitari GM, Kazerounian S, Ruiz-Stewart I, Park J, Schulz S, Chepenik KP and Waldman SA. Guanylyl cyclases and signaling by cyclic GMP. *Pharmacol. Rev.* 52: 375-414, 2000.

[102] Ma KK, Ogawa T and de Bold AJ. Selective upregulation of cardiac brain natriuretic peptide at the transcriptional and translational levels by pro-inflammatory cytokines and by conditioned medium derived from mixed lymphocyte reactions via p38 MAP kinase. *J. Mol. Cell Cardiol.* 36: 505-513, 2004.

[103] Maack T, Okolicany J, Koh GY and Price DA. Functional properties of atrial natriuretic factor receptors. *Semin. Nephrol.* 13: 50-60, 1993.

[104] Maeder M, Ammann P, Kiowski W and Rickli H. B-type natriuretic peptide in patients with sepsis and preserved left ventricular ejection fraction. *Eur. J. Heart Fail* 7: 1164-1167, 2005.

[105] Maisel AS, Krishnaswamy P, Nowak RM, McCord J, Hollander JE, Duc P, Omland T, Storrow AB, Abraham WT, Wu AH, Clopton P, Steg PG, Westheim A, Knudsen CW, Perez A, Kazanegra R, Herrmann HC and McCullough PA. Rapid measurement of B-type natriuretic peptide in the emergency diagnosis of heart failure. *N. Engl. J. Med.* 347: 161-167, 2002.

[106] Mangat H and de Bold AJ. Stretch-induced atrial natriuretic factor release utilizes a rapidly depleting pool of newly synthesized hormone. *Endocrinology* 133: 1398-1403, 1993.

[107] Masters RG, Davies RA, Keon WJ, Walley VM, Koshal A and de Bold AJ. Neuroendocrine response to cardiac transplantation. *Can. J. Cardiol.* 9: 609-617, 1993.

[108] Masters RG, Davies RA, Veinot JP, Hendry PJ, Smith SJ and de Bold AJ. Discoordinate Modulation of Natriuretic Peptides During Acute Cardiac Allograft Rejection in Humans. *Circulation* 100: 287-291, 1999.

[109] McGrath MF and de Bold AJ. Determinants of natriuretic peptide gene expression. *Peptides* 26: 933-943, 2005.

[110] McGrath MF, de Bold ML and de Bold AJ. The endocrine function of the heart. *Trends Endocrinol. Metab.* 16: 469-477, 2005.

[111] Mehra MR, Uber PA, Potluri S, Ventura HO, Scott RL and Park MH. Usefulness of an elevated B-type natriuretic peptide to predict allograft failure, cardiac allograft vasculopathy, and survival after heart transplantation. *Am. J. Cardiol.* 94: 454-458, 2004.

[112] Mehra MR, Uber PA, Walther D, Vesely M, Wohlgemuth JG, Prentice J, Tayama D and Billingham M. Gene expression profiles and B-type natriuretic peptide elevation in heart transplantation: more than a hemodynamic marker. *Circulation* 114: I21-I26, 2006.

[113] Meirovich YF, Veinot JP, de Bold ML, Haddad H, Davies RA, Masters RG, Hendry PJ and de Bold AJ. Relationship between natriuretic peptides and inflammation: proteomic evidence obtained during acute cellular cardiac allograft rejection in humans. *J. Heart Lung Transplant* 27: 31-37, 2008.

[114] Melo LG, Steinhelper ME, Pang SC, Tse Y and Ackermann U. ANP in regulation of arterial pressure and fluid-electrolyte balance: lessons from genetic mouse models. *Physiol. Genomics* 3: 45-58, 2000.

[115] Miano JM. Role of serum response factor in the pathogenesis of disease. *Lab Invest* 90: 1274-1284, 2010.

[116] Moertl D, Berger R, Struck J, Gleiss A, Hammer A, Morgenthaler NG, Bergmann A, Huelsmann M and Pacher R. Comparison of midregional pro-atrial and B-type natriuretic peptides in chronic heart failure: influencing factors, detection of left ventricular systolic dysfunction, and prediction of death. *J. Am. Coll. Cardiol.* 53: 1783-1790, 2009.

[117] Molkentin JD, Lu JR, Antos CL, Markham B, Richardson J, Robbins J, Grant SR and Olson EN. A calcineurin-dependent transcriptional pathway for cardiac hypertrophy. *Cell* 93: 215-228, 1998.

[118] Morgenthaler NG, Struck J, Thomas B and Bergmann A. Immunoluminometric assay for the midregion of pro-atrial natriuretic peptide in human plasma. *Clin. Chem.* 50: 234-236, 2004.

[119] Mori T, Chen YF, Feng JA, Hayashi T, Oparil S and Perry GJ. Volume overload results in exaggerated cardiac hypertrophy in the atrial natriuretic peptide knockout mouse. *Cardiovasc. Res.* 61: 771-779, 2004.

[120] Morikawa S, Sone T, Tsuboi H, Mukawa H, Morishima I, Uesugi M, Morita Y, Numaguchi Y, Okumura K and Murohara T. Renal protective effects and the prevention of contrast-induced nephropathy by atrial natriuretic peptide. *J. Am. Coll. Cardiol.* 53: 1040-1046, 2009.

[121] Murakami Y, Shimada T, Inoue S, Shimizu H, Ohta Y, Katoh H, Nakamura K and Ishibashi Y. New insights into the mechanism of the elevation of plasma brain natriuretic polypeptide levels in patients with left ventricular hypertrophy. *Can. J. Cardiol.* 18: 1294-1300, 2002.

[122] Murthy KS and Makhlouf GM. Identification of the G protein-activating domain of the natriuretic peptide clearance receptor (NPR-C). *J. Biol. Chem.* 274: 17587-17592, 1999.

[123] Nagaya N, Nishikimi T, Goto Y, Miyao Y, Kobayashi Y, Morii I, Daikoku S, Matsumoto T, Miyazaki S, Matsuoka H, Takishita S, Kangawa K, Matsuo H and Nonogi H. Plasma brain natriuretic peptide is a biochemical marker for the prediction of progressive ventricular remodeling after acute myocardial infarction. *Am. Heart J.* 135: 21-28, 1998.

[124] Nagaya N, Nishikimi T, Okano Y, Uematsu M, Satoh T, Kyotani S, Kuribayashi S, Hamada S, Kakishita M, Nakanishi N, Takamiya M, Kunieda T, Matsuo H and Kangawa K. Plasma brain natriuretic peptide levels increase in proportion to the extent of right ventricular dysfunction in pulmonary hypertension. *J. Am. Coll. Cardiol.* 31: 202-208, 1998.

[125] Nagaya N, Nishikimi T, Uematsu M, Satoh T, Kyotani S, Sakamaki F, Kakishita M, Fukushima K, Okano Y, Nakanishi N, Miyatake K and Kangawa K. Plasma brain

natriuretic peptide as a prognostic indicator in patients with primary pulmonary hypertension. *Circulation* 102: 865-870, 2000.

[126] Nakayama K. Furin: a mammalian subtilisin/Kex2p-like endoprotease involved in processing of a wide variety of precursor proteins. *Biochem. J.* 327 (Pt 3): 625-635, 1997.

[127] Nikolaou NI, Goritsas C, Dede M, Paissios NP, Papavasileiou M, Rombola AT and Ferti A. Brain natriuretic peptide increases in septic patients without severe sepsis or shock. *Eur. J. Intern. Med.* 18: 535-541, 2007.

[128] Nishikimi T, Maeda N and Matsuoka H. The role of natriuretic peptides in cardioprotection. *Cardiovasc. Res.* 69: 318-328, 2006.

[129] Nishikimi T, Yoshihara F, Morimoto A, Ishikawa K, Ishimitsu T, Saito Y, Kangawa K, Matsuo H, Omae T and Matsuoka H. Relationship between left ventricular geometry and natriuretic peptide levels in essential hypertension. *Hypertension* 28: 22-30, 1996.

[130] Nomura F, Kurobe N, Mori Y, Hikita A, Kawai M, Suwa M and Okutani Y. Multicenter prospective investigation on efficacy and safety of carperitide as a first-line drug for acute heart failure syndrome with preserved blood pressure: COMPASS: Carperitide Effects Observed Through Monitoring Dyspnea in Acute Decompensated Heart Failure Study. *Circ. J.* 72: 1777-1786, 2008.

[131] O'Connor CM, Starling RC, Hernandez AF, Armstrong PW, Dickstein K, Hasselblad V, Heizer GM, Komajda M, Massie BM, McMurray JJ, Nieminen MS, Reist CJ, Rouleau JL, Swedberg K, Adams KF, Jr., Anker SD, Atar D, Battler A, Botero R, Bohidar NR, Butler J, Clausell N, Corbalan R, Costanzo MR, Dahlstrom U, Deckelbaum LI, Diaz R, Dunlap ME, Ezekowitz JA, Feldman D, Felker GM, Fonarow GC, Gennevois D, Gottlieb SS, Hill JA, Hollander JE, Howlett JG, Hudson MP, Kociol RD, Krum H, Laucevicius A, Levy WC, Mendez GF, Metra M, Mittal S, Oh BH, Pereira NL, Ponikowski P, Tang WH, Tanomsup S, Teerlink JR, Triposkiadis F, Troughton RW, Voors AA, Whellan DJ, Zannad F and Califf RM. Effect of nesiritide in patients with acute decompensated heart failure. *N. Engl. J. Med.* 365: 32-43, 2011.

[132] O'Neill JO, McRae AT, III, Troughton RW, Ng K, Taylor DO, Yamani MH, Young JB and Starling RC. Brain natriuretic peptide levels do not correlate with acute cellular rejection in De Novo orthotopic heart transplant recipients. *J. Heart Lung Transplant* 24: 416-420, 2005.

[133] Ogawa T and de Bold AJ. Uncoordinated regulation of atrial natriuretic factor and brain natriuretic peptide in lipopolysaccharide-treated rats. *Biomarkers* 17: 140-149, 2012.

[134] Ogawa T, Linz W, Stevenson M, Bruneau BG, Kuroski de Bold ML, Chen JH, Eid H, Scholkens BA and de Bold AJ. Evidence for load-dependant and load-independant determinants of cardiac natriuretic peptide production. *Circulation* 93: 2059-2067, 1996.

[135] Ogawa T, Vatta M, Bruneau BG and de Bold AJ. Characterization of natriuretic peptide production by adult heart atria. *Am. J. Physiol. Heart Circ. Physiol.* 276: H1977-H1986, 1999.

[136] Ogawa T, Veinot JP, Davies RA, Haddad H, Smith SJ, Masters RG, Hendry PJ, Starling R, Kuroski de Bold ML, Ponce A, Ma KK, Williams K and de Bold AJ. Neuroendocrine Profiling of Humans Receiving Cardiac Allografts. *J. Heart Lung Transplant* 24: 1046-1054, 2005.

[137] Ogawa T, Veinot JP, Kuroski de Bold ML, Georgalis T and de Bold AJ. Angiotensin II receptor antagonism reverts the selective cardiac BNP upregulation and secretion observed in myocarditis. *Am. J. Physiol. Heart Circ. Physiol.* 294: H2596-H2603, 2008.
[138] Oikawa S, Imai M, Ueno A, Tanaka S, Noguchi T, Nakazato H, Kangawa K, Fukuda A and Matsuo H. Cloning and sequence analysis of cDNA encoding a precursor for human atrial natriuretic polypeptide. *Nature* 309: 724-726, 1984.
[139] Okura Y, Yamamoto T, Goto S, Inomata T, Hirono S, Hanawa H, Feng L, Wilson CB, Kihara I, Izumi T, Shibata A, Aizawa Y, Seki S and Abo T. Characterization of cytokine and iNOS mRNA expression in situ during the course of experimental autoimmune myocarditis in rats. *J. Mol. Cell Cardiol.* 29: 491-502, 1997.
[140] Oliver PM, Fox JE, Kim R, Rockman HA, Kim HS, Reddick RL, Pandey KN, Milgram SL, Smithies O and Maeda N. Hypertension, cardiac hypertrophy, and sudden death in mice lacking natriuretic peptide receptor A. *Proc. Natl. Acad. Sci. USA* 94(26): 14730-14735, 1997.
[141] Omland T, Aakvaag A, Bonarjee VV, Caidahl K, Lie RT, Nilsen DW, Sundsfjord JA and Dickstein K. Plasma brain natriuretic peptide as an indicator of left ventricular systolic function and long-term survival after acute myocardial infarction. Comparison with plasma atrial natriuretic peptide and N-terminal proatrial natriuretic peptide. *Circulation* 93: 1963-1969, 1996.
[142] Park MH, Scott RL, Uber PA, Harris BC, Chambers R and Mehra MR. Usefulness of B-type natriuretic peptide levels in predicting hemodynamic perturbations after heart transplantation despite preserved left ventricular systolic function. *Am. J. Cardiol.* 90: 1326-1329, 2002.
[143] Pemberton CJ, Johnson ML, Yandle TG and Espiner EA. Deconvolution analysis of cardiac natriuretic peptides during acute volume overload. *Hypertension* 36: 355-359, 2000.
[144] Ramos H and de Bold AJ. Gene Expression, Processing and Secretion of Natriuretic Peptides: Physiologic and Diagnostic Implications. *Heart Failure Clinics* 2: 255-268, 2006.
[145] Ramsperger-Gleixner M, Spriewald BM, Tandler R, Kondruweit M, Amann K, Weyand M and Ensminger SM. Increased transcript levels of TNF-alpha, TGF-beta, and granzyme B in endomyocardial biopsies correlate with allograft rejection. *Exp. Clin. Transplant* 9: 387-392, 2011.
[146] Roch A. What does high NT-proBNP mean in septic shock patients? A part of the puzzle. *Crit. Care* 11: 122, 2007.
[147] Rosenzweig A and Seidman CE. Atrial natriuretic factor and related peptide hormones. *Annu. Rev. Biochem.* 60: 229-255, 1991.
[148] Rossano JW, Denfield SW, Kim JJ, Price JF, Jefferies JL, Decker JA, Smith EO, Clunie SK, Towbin JA and Dreyer WJ. B-type natriuretic peptide is a sensitive screening test for acute rejection in pediatric heart transplant patients. *J. Heart Lung Transplant* 27: 649-654, 2008.
[149] Ruskoaho H. Cardiac hormones as diagnostic tools in heart failure. *Endocr. Rev.* 24: 341-356, 2003.
[150] Sackner-Bernstein JD, Skopicki HA and Aaronson KD. Risk of worsening renal function with nesiritide in patients with acutely decompensated heart failure. *Circulation* 111: 1487-1491, 2005.

[151] Sagnella GA, Markandu ND, Shore AC and MacGregor GA. Raised circulating levels of atrial natriuretic peptides in essential hypertension. *Lancet* 1(8474): 179-181, 1986.
[152] Samson WK. Atrial natriuretic factor inhibits dehydration and hemorrhage-induced vasopressin release. *Neuroendocrinology* 40: 277-279, 1985.
[153] Sarzani R, Marcucci P, Salvi F, Bordicchia M, Espinosa E, Mucci L, Lorenzetti B, Minardi D, Muzzonigro G, Dessi-Fulgheri P and Rappelli A. Angiotensin II stimulates and atrial natriuretic peptide inhibits human visceral adipocyte growth. *Int. J. Obes. (Lond)* 32: 259-267, 2008.
[154] Saxenhofer H, Raselli A, Weidmann P, Forssmann WG, Bub A, Ferrari P and Shaw SG. Urodilatin, a natriuretic factor from kidneys, can modify renal and cardiovascular function in men. *Am. J. Physiol.* 259: F832-F838, 1990.
[155] Schulz S, Singh S, Bellet RA, Singh G, Tubb DJ, Chin H and Garbers DL. The primary structure of a plasma membrane guanylate cyclase demonstrates diversity within this new receptor family. *Cell* 58: 1155-1162, 1989.
[156] Schweitz H, Vigne P, Moinier D, Frelin C and Lazdunski M. A new member of the natriuretic peptide family is present in the venom of the green mamba (Dendroaspis angusticeps). *J. Biol. Chem.* 267: 13928-13932, 1992.
[157] Sengenes C, Berlan M, De G, I, Lafontan M and Galitzky J. Natriuretic peptides: a new lipolytic pathway in human adipocytes. *FASEB J.* 14: 1345-1351, 2000.
[158] Sengenes C, Bouloumie A, Hauner H, Berlan M, Busse R, Lafontan M and Galitzky J. Involvement of a cGMP-dependent pathway in the natriuretic peptide-mediated hormone sensitive lipase phosphorylation in human adipocytes. *J. Biol. Chem.* 278: 48617-48626, 2003.
[159] Seto T, Kamijo S, Wada Y, Yamaura K, Takahashi K, Komatsu K, Otsu Y, Terasaki T, Fukui D, Amano J, Taniguchi S, Sagara J and Ito K. Upregulation of the apoptosis-related inflammasome in cardiac allograft rejection. *J. Heart Lung Transplant* 29: 352-359, 2010.
[160] Shah MR, Califf RM, Nohria A, Bhapkar M, Bowers M, Mancini DM, Fiuzat M, Stevenson LW and O'Connor CM. The STARBRITE trial: a randomized, pilot study of B-type natriuretic peptide-guided therapy in patients with advanced heart failure. *J. Card Fail* 17: 613-621, 2011.
[161] Shaw SM, Critchley WR, Puchalka CM, Williams SG, Yonan N and Fildes JE. Brain natriuretic peptide induces CD8+ T cell death via a caspase 3 associated pathway--implications following heart transplantation. *Transpl. Immunol.* 26: 119-122, 2012.
[162] Shor R, Rozenman Y, Bolshinsky A, Harpaz D, Tilis Y, Matas Z, Fux A, Boaz M and Halabe A. BNP in septic patients without systolic myocardial dysfunction. *Eur. J. Intern Med.* 17: 536-540, 2006.
[163] Small EM and Krieg PA. Molecular regulation of cardiac chamber-specific gene expression. *Trends Cardiovasc. Med.* 14: 13-18, 2004.
[164] Solus J, Chung CP, Oeser A, Avalos I, Gebretsadik T, Shintani A, Raggi P, Sokka T, Pincus T and Stein CM. Amino-terminal fragment of the prohormone brain-type natriuretic peptide in rheumatoid arthritis. *Arthritis Rheum* 58: 2662-2669, 2008.
[165] Stanton BA. Molecular mechanisms of ANP inhibition of renal sodium transport. *Can. J. Physiol. Pharmacol.* 69: 1546-1552, 1991.
[166] Steiner J and Guglin M. BNP or NTproBNP? A clinician's perspective. *Int. J. Cardiol.* 129: 5-14, 2008.

[167] Stoupakis G and Klapholz M. Natriuretic peptides: biochemistry, physiology, and therapeutic role in heart failure. *Heart Dis.* 5: 215-223, 2003.

[168] Sudoh T, Kangawa K, Minamino N and Matsuo H. A new natriuretic peptide in porcine brain. *Nature* 332: 78-81, 1988.

[169] Sudoh T, Maekawa K, Kojima M, Minamino N, Kangawa K and Matsuo H. Cloning and sequence analysis of cDNA encoding a precursor for human brain natriuretic peptide. *Biochem. Biophys. Res. Commun.* 159: 1427-1434, 1989.

[170] Suga S, Nakao K, Hosoda K, Mukoyama M, Ogawa Y, Shirakami G, Arai H, Saito Y, Kambayashi Y, Inouye K and Imura H. Receptor selectivity of natriuretic peptide family, atrial natriuretic peptide, brain natriuretic peptide, and C-type natriuretic peptide. *Endocrinology* 130: 229-239, 1992.

[171] Suwa M, Seino Y, Nomachi Y, Matsuki S and Funahashi K. Multicenter prospective investigation on efficacy and safety of carperitide for acute heart failure in the 'real world' of therapy. *Circ. J.* 69: 283-290, 2005.

[172] Takei Y. Structural and functional evolution of the natriuretic peptide system in vertebrates. *Int. Rev. Cytol.* 194: 1-66, 2000.

[173] Takei Y. Does the natriuretic peptide system exist throughout the animal and plant kingdom? *Comp. Biochem. Physiol. B. Biochem. Mol. Biol.* 129: 559-573, 2001.

[174] Tamura N, Ogawa Y, Chusho H, Nakamura K, Nakao K, Suda M, Kasahara M, Hashimoto R, Katsuura G, Mukoyama M, Itoh H, Saito Y, Tanaka I, Otani H and Katsuki M. Cardiac fibrosis in mice lacking brain natriuretic peptide. *Proc. Natl. Acad. Sci. U S A* 97: 4239-4244, 2000.

[175] Tamura N, Ogawa Y, Yasoda A, Itoh H, Saito Y and Nakao K. Two cardiac natriuretic peptide genes (atrial natriuretic peptide and brain natriuretic peptide) are organized in tandem in the mouse and human genomes. *J. Mol. Cell Cardiol.* 28: 1811-1815, 1996.

[176] Tarjan E, Denton DA and Weisinger RS. Atrial natriuretic peptide inhibits water and sodium intake in rabbits. *Regul. Pept.* 23: 63-75, 1988.

[177] Testa M, Yeh M, Lee P, Fanelli R, Loperfido F, Berman JW and LeJemtel TH. Circulating levels of cytokines and their endogenous modulators in patients with mild to severe congestive heart failure due to coronary artery disease or hypertension. *J. Am. Coll. Cardiol.* 28: 964-971, 1996.

[178] Thibault G, Murthy KK, Gutkowska J, Seidah NG, Lazure C, Chretien M and Cantin M. NH2-terminal fragment of rat pro-atrial natriuretic factor in the circulation: identification, radioimmunoassay and half-life. *Peptides* 9: 47-53, 1988.

[179] Thuerauf DJ and Glembotski CC. Differential effects of protein kinase C, Ras, and Raf-1 kinase on the induction of the cardiac B-type natriuretic peptide gene through a critical promoter-proximal M-CAT element. *J. Biol. Chem.* 272: 7464-7472, 1997.

[180] Tomaru KK, Arai M, Yokoyama T, Aihara Y, Sekiguchi KK, Tanaka T, Nagai R and Kurabayashi M. Transcriptional activation of the BNP gene by lipopolysaccharide is mediated through GATA elements in neonatal rat cardiac myocytes. *J. Mol. Cell Cardiol.* 34: 649-659, 2002.

[181] Vanderheyden M, Bartunek J and Goethals M. Brain and other natriuretic peptides: molecular aspects. *Eur. J.. Heart Fail* 6: 261-268, 2004.

[182] Vasan RS, Benjamin EJ, Larson MG, Leip EP, Wang TJ, Wilson PW and Levy D. Plasma natriuretic peptides for community screening for left ventricular hypertrophy and systolic dysfunction: the Framingham heart study. *JAMA* 288: 1252-1259, 2002.

[183] Vila G, Resl M, Stelzeneder D, Struck J, Maier C, Riedl M, Hulsmann M, Pacher R, Luger A and Clodi M. Plasma NT-proBNP increases in response to LPS administration in healthy men. *J. Appl. Physiol.* 105: 1741-1745, 2008.

[184] Wang TJ, Larson MG, Levy D, Benjamin EJ, Leip EP, Omland T, Wolf PA and Vasan RS. Plasma natriuretic peptide levels and the risk of cardiovascular events and death. *N. Engl. J. Med.* 350: 655-663, 2004.

[185] Witthaut R, Busch C, Fraunberger P, Walli A, Seidel D, Pilz G, Stuttmann R, Speichermann N, Verner L and Werdan K. Plasma atrial natriuretic peptide and brain natriuretic peptide are increased in septic shock: impact of interleukin-6 and sepsis-associated left ventricular dysfunction. *Intensive Care Med.* 29: 1696-1702, 2003.

[186] Wu AH, Johnson ML, Aaronson KD, Gordon D, Dyke DB and Koelling TM. Brain natriuretic peptide predicts serious cardiac allograft rejection independent of hemodynamic measurements. *J. Heart Lung Transplant* 24: 52-57, 2005.

[187] Xia WJ, Huang YY, Chen YL, Chen SL and He JG. Acute myocardial ischemia directly modulates the expression of brain natriuretic peptide at the transcriptional and translational levels via inflammatory cytokines. *Eur. J. Pharmacol.* 670: 7-12, 2011.

[188] Yamamoto K, Burnett JC, Jr., Jougasaki M, Nishimura RA, Bailey KR, Saito Y, Nakao K and Redfield MM. Superiority of brain natriuretic peptide as a hormonal marker of ventricular systolic and diastolic dysfunction and ventricular hypertrophy. *Hypertension* 28: 988-994, 1996.

[189] Yan W, Wu F, Morser J and Wu Q. Corin, a transmembrane cardiac serine protease, acts as a pro-atrial natriuretic peptide-converting enzyme. *Proc. Natl. Acad. Sci. U S A* 97: 8525-8529, 2000.

[190] Yokota N, Bruneau BG, Fernandez BE, Kuroski de Bold ML, Piazza LA, Eid H and de Bold AJ. Dissociation of cardiac hypertrophy, myosin heavy chain isoform expression, and natriuretic peptide production in DOCA-salt rats. *Am. J. Hypertens* 8: 301-310, 1995.

[191] Yoshida T. MCAT elements and the TEF-1 family of transcription factors in muscle development and disease. *Arterioscler Thromb. Vasc. Biol.* 28: 8-17, 2008.

[192] Yucel T, Memis D, Karamanlioglu B, Sut N and Yuksel M. The prognostic value of atrial and brain natriuretic peptides, troponin I and C-reactive protein in patients with sepsis. *Exp. Clin. Cardiol.* 13: 183-188, 2008.

In: Natriuretic Peptides
Editor: Kailash N. Pandey

ISBN: 978-1-63321-080-6

Chapter 2

CELLULAR, MOLECULAR AND CLINICAL ASPECTS OF NATRIURETIC PEPTIDES AND THEIR RECEPTORS IN HEALTH AND DISEASE

Kailash N. Pandey*

Department of Physiology, Tulane University Health Sciences Center, School of Medicine, New Orleans, LA, US

ABSTRACT

Atrial natriuretic factor (ANF), also known as atrial natriuretic peptide (ANP), is an endogenous and potent hypotensive hormone that elicits natriuretic, diuretic, vasorelaxant, anti-hypertrophic, anti-proliferative, and anti-inflammatory effects that are important in the control of blood pressure and cardiovascular homeostasis. Studies on ANP, brain natriuretic peptide (BNP), and C-type natriuretic peptide (CNP), as well as their cognate receptors, have greatly increased our knowledge of the control of hypertension and cardiovascular disorders. The principal locus involved in the regulatory action of ANP and BNP is the guanylyl cyclase/natriuretic peptide receptor-A (GC-A/NPRA), which produces intracellular second messenger cGMP. Cellular, biochemical, molecular, and clinical studies have helped delineate the functions of natriuretic peptide (NP) hormones, their receptors, and signaling mechanisms. Gene-targeted (gene-knockout and gene-duplication) and transgenic mouse models have advanced our understanding of the importance of NPs and their receptors at the physiological and pathophysiological levels in normal and disease states. Importantly, ANP and BNP are used as clinical markers of cardiac events and heart failure; however, their therapeutic potential for the diagnosis and treatment of hypertension, heart failure, and stroke have just begun to be realized. We are now at the initial stage of the exponential phase of molecular therapeutics and pharmacogenomic advancement of the functional aspects of natriuretic peptides and their receptor systems for the treatment, prevention, and control

* Address for Correspondence: Dr. Kailash N. Pandey, Department of Physiology, SL-39 Tulane University Health Sciences Center, School of Medicine, 1430 Tulane Ave, New Orleans, LA 70112, Tel: (504) 988-1628, Fax: (504) 988-2675, Email: kpandey@tulane.edu.

of hypertension, heart failure, and other cardiovascular and neurological disorders. Research needs to be extended in this important cardiovascular field.

Keywords: Natriuretic peptides, guanylyl cyclase/natriuretic peptide receptors, gene-targeting, high blood pressure, cardiac hypertrophy, heart failure

1. INTRODUCTION

The pioneer discovery by de Bold and colleagues, that atrial extracts contain natriuretic and diuretic activity, led them to identify and characterize atrial natriuretic factor/peptide (ANF/ANP) in the heart [1,2]. The natriuretic peptides (NPs) are critical in the control of renal, cardiovascular, endocrine, and skeletal homeostasis [3-11]. ANP, the first member of the natriuretic peptide hormone family, exhibits natriuretic, diuretic, and vasorelaxant properties, along with anti-proliferative, anti-inflammatory, and anti-hypertrophic effects, all of which are aimed to directed to the reduction of body fluid and the maintenance of blood pressure and cardiovascular homeostasis [5,8]. Subsequently, two other members of this family, B-type, or brain natriuretic peptide (BNP), and C-type, natriuretic peptide (CNP), were identified, but each is derived from a separate gene [12-14]. All three natriuretic peptide hormones have highly homologous structure, bind to specific cell surface receptors, and elicit discrete biological functions [4,15].

This family of endogenous peptide hormones, including ANP, BNP, CNP, and urodilatin, are considered to have an integral role in hypertension and cardiovascular regulation [4,6,8,9,11,16]. Interestingly, it has been suggested that the natriuretic peptides not only regulate blood pressure, but also maintain antagonistic action to renin and angiotensin II (Ang II), exert an anti-mitogenic effect, inhibit myocardial hypertrophy and fibrosis, and act in endothelial cell function, cartilage growth, immunity, and mitochondrial biogenesis [6-8,17-20]. ANP and BNP are also increasingly used to screen and diagnose cardiac etiologies of shortness of breath and congestive heart failure in patients with emergency conditions [21].

Three subtypes of natriuretic peptide receptors have been identified: NP receptor-A (NPRA), NP receptor-B (NPRB), and NP receptor-C (NPRC). Both NPRA and NPRB contain intrinsic guanylyl cyclase catalytic domain, and are also referred to, respectively, as receptor guanylyl cyclase-A (GC-A) and receptor guanylyl cyclase-B (GC-B) [6]. Interestingly, both ANP and BNP activate NPRA, which produces second-messenger cGMP in response to hormone binding. CNP activates NPRB, which also produces cGMP. However, but all three natriuretic peptides indiscriminately bind to NPRC, which lacks a guanylyl cyclase domain and does not produce cGMP [22-24]. The discovery of structurally related natriuretic peptides indicated that the physiological control of blood pressure and cardiovascular homeostasis is complex and multifactorial. A combination of cellular, biochemical, molecular, and pharmacological aspects of NPs and their prototype receptors have demonstrated the hallmark functions of their physiological and pathophysiological effects, including renal, cardiovascular, neuronal, skeletal, and immunological influences on health and disease [6,8,9,20,25,26].

Both NPRA and NPRB have also been designated as guanylyl cyclase/natriuretic peptide receptor-A (GC-A/NPRA) and guanylyl cyclase/natriuretic peptide receptor-B (GC-B/NPRB). GC-A/NPRA, is a principal locus involved in the regulatory action of ANP and

BNP [6,8,27,28]. Gaining insight into the intricacies of ANP/NPRA/cGMP signaling is of pivotal importance for understanding both receptor biology and the disease state arising from abnormal hormone-receptor interplay. It has been suggested that binding of ANP and BNP to the extracellular domain of NPRA causes a conformational change, thereby transmitting a signal to the intracellular carboxyl-terminal guanylyl cyclase catalytic domain, which then generates the second-messenger cGMP in target cells and tissues [15,23,29]. Recent studies have focused on biochemical and molecular elucidation of the nature and mode of functioning of GC-A/NPRA [8,9,26,30]. Cultured cells *in vitro* and gene-targeted (gene-knockout and gene-duplication) mouse models *in vivo* have been used to gain understanding of the NPs and their cognate receptors to delineate the normal and abnormal control of cellular, physiological, and pathophysiological processes. Although there has been widespread appreciation of the functional roles of natriuretic peptides and their receptors in renal, cardiovascular, endocrine, neuronal, and skeletal homeostasis, in-depth research is still needed to achieve full delineation of their potential molecular targets in diseases states. It is expected that future studies on natriuretic peptides and their receptors will yield new therapeutic targets and novel loci for the control and treatment of hypertension, cardiovascular events, and neurological disorders.

2. Structural Properties of Natriuretic Peptides

ANP, the initial member of the natriuretic peptide hormone family, is largely synthesized and secreted from the atrium, which led to classification of the heart as an endocrine organ. The primary structure deduced from cDNA sequences indicates that ANP is synthesized as the 152-amino acid preproANP molecule, which contains sequences of active peptide in its carboxyl-terminal region, and the major form of biologically active ANP is a 28-amino acid circulatory hormone [31]. Initially, different lengths of sequences of ANP were synthesized for studies of the structure-activity relationship. It was found that a 17-amino-acid disulfide-ring conformation of the ANP molecule is essential for its biological activities [4,11]. Disruption of the ring structure of ANP abolishes its functional activity [32]. The carboxyl-terminal sequences extending from the disulfide-ring structure to Asn-Phe-Arg-Tyr are also required for the full biological and physiological activities of ANP. Similarly, BNP is synthesized as a 134-amino acid preprohormone, which yields a 108-amino acid prohormone molecule. Processing of the proBNP molecule yields a 75-residue amino-terminal-BNP (NT-proBNP) and a 32-residue biologically active circulating BNP hormone [33,34]. CNP, which is thought to be synthesized as a 103-amino-acid prohormone, is cleaved to a 53-residue peptide by the protease furin, and subsequently it is processed to yield a 22-amino-acid biologically active molecule [35].

The amino acid sequence of ANP is almost identical across mammalian species; the exception occurs at position 10. Isoleucine is at this position in rats, mice, and rabbits but in humans, dogs, and bovines a methionine occurs at this position. Both BNP and CNP were initially isolated from porcine brain homogenates on the basis of their potent vasorelaxant properties [33,36]. Soon it was established that BNP is predominantly synthesized and secreted from the heart ventricles [37]. CNP, which was initially localized in the central nervous system and endothelial cells, is considered to be a noncirculatory hormone [38]. As discussed earlier, all three NPs (ANP, BNP, and CNP) are synthesized from the large precursor molecules. The mature bioactive peptides contain a 17-amino-acid loop bridged by an intramolecular disulfide bond (Figure 1). Interestingly, 11 of these amino acids in the ring structure are identical in ANP, BNP, and CNP; however, the amino- and carboxyl-terminus sequences vary in length and composition. Among different species, ANP and BNP exhibit the most variability in primary structure, while CNP is highly conserved across species. In addition, a 32-amino acid peptide, urodilatin (URO) was discovered, which is present in the urine [39,40]. URO is not detected in the circulation and appears to be a unique intrarenal natriuretic peptide having with largely unexplored biological and physiological functions [41]. D-type natriuretic peptide (DNP), an additional member in the NP hormone family, was initially isolated as a 38-amino acid biologically active peptide molecule from the venom of the green mamba, *Dendroaspis angusticeps* [42,43]. Biochemical, immunohistochemical, and molecular studies have suggested that three specific natriuretic peptides (ANP, BNP, CNP) and their three distinct receptor subtypes (NPRA, NPRB, NPRC) have widespread tissue and cell distributions, indicating the pleotropic actions at both systemic and local levels (Table 1). However, the mechanisms of action of NPs in relation to their structure are not well understood and remain to be investigated.

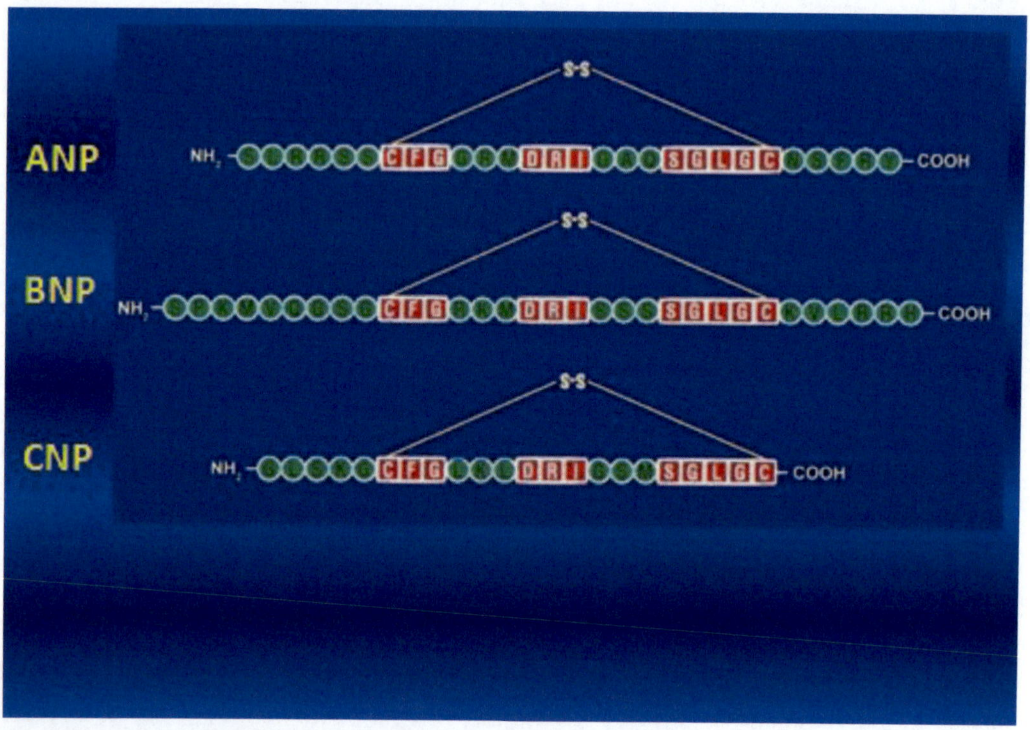

Figure 1. Amino acid sequence of the natriuretic peptide hormones. Comparison of amino-acid sequence of human atrial natriuretic peptide (ANP), brain natriuretic peptide (BNP), and C-type natriuretic peptide (CNP) with conserved amino acid residues, which are represented by shaded boxes. The lines joining the two cysteine residues in ANP, BNP, and CNP indicate a 17-amino acid disulfide bridge, which makes the ring structure and is essential for the physiological activity of three natriuretic peptides. S-S, disulfide bridge.

Table 1. The tissue and cell distribution of natriuretic peptides (ANP, BNP, and CNP) and their receptors (NPRA, NPRB, and NPRC) and gene-knockout phenotypes

Ligand/ Receptor Gene	Tissue-specific distribution	Cell-specific distribution	Gene-knockout phenotype in mice
NPRA (Npr1)	Kidney, adrenal glands, brain, heart, liver, lung, olfactory, ovary, pituitary gland, placenta, testis, thymus, vascular beds, liver, ileum	Renal epithelial and mesangial cells, vascular smooth muscle cells, endothelial cells, Leydig cells, granulosa cells, fibroblasts, Neuroblastoma, LLCPk-1, MDCK cells	High blood pressure, hypertension, cardiac hypertrophy and fibrosis, inflammation, volume overload, reduced testosterone
NPRB (Npr2)	Adrenal glands, brain, cartilage, fibroblast, heart, lung, ovary,pituitary gland, placenta, testis, thymus, vascular beds	Vascular smooth muscle cells, fibroblasts, chondrocytes	Dwarfism, decreased adiposity, female sterility, seizures, vascular complication
NPRC (Npr3)	Kidney, heart, brain liver, vascular bed, intestine	Vascular smooth muscle cells, endothelial cells, mesangial cells, fibroblasts	Bone deformation, skeletal over-growth, long bone overgrowth
ANP (Nppa)	Atrium, ventricle, brain, kidney, testis, ovary	Cardiac myocytes, Leydig cells, granulosa-lutean cells, neuronal cells	High blood pressure, hypertension, cardiac hypertrophy
BNP (Nppb)	Atrium, ventricle, brain	Cardiacmyocytes, ventricular cells, neuronal cells	Ventricular fibrosis, skeletal abnormalities, vascular complications
CNP (Nppc)	Vascular endothelium, aorta, brain, heart, testis	Endothelial cells, neuronal cells	Inhibition of long bone growth, dwarfism, abnormal chondrocyte differentiation

The abbreviation used are; NPRA, natriuretic peptide receptor-A; Npr1, coding for guanylyl cyclase/natriuretic peptide receptor-A; NPRB, natriuretic peptide receptor-B; Npr2, coding for guanylyl cyclase/natriuretic peptide receptor-B; NPRC, natriuretic peptide receptor-C; Npr3, coding for natriuretic peptide clearance receptor; ANP, atrial natriuretic peptide; Nppa, coding for pro-atrial natriuretic peptide; BNP, B-type natriuretic peptide; Nppb, coding for pro-brain natriuretic peptide; CNP, C-type natriuretic peptide; Nppc, coding for pro-C-type natriuretic peptide.

ANP is usually characterized as having a fast rate of clearance from rate in the circulation; its half-life ranges from 0.5 to 3.5 min in experimental animals. [44-46] However, the half-life of ANP in human subjects ranges between approximately 2 and 2.5 min [44]. The clearance of BNP in humans occurs with both short and long half-life, from approximately 3 to 4 min to 20 to 23 min. [47,48]. On the other hand, CNP has been reported to have a half-life of 2 to 3 min in humans and approximately 1.5 to 2.0 min in experimental animals [49,50]. Interestingly, the design and synthesis of chimeric natriuretic peptides has led to the production of biologically active peptide molecules representing single-chemical entities with the combined structural and functional properties of different natriuretic peptides [51]. The chimeric natriuretic peptides induce the actions of at least two different natriuretic peptide molecules and often reduce undesirable and adverse biological effects fast-acting native peptides. One such chimeric molecule is CD-NP, produced by a fusion of 22-amino acid CNP with the 15-amino-acid C-terminus of DNP [51]. Thus, the chimeric hormone CD-NP exhibits vasorelaxant properties, reduces cardiac volume overload, and exerts renoprotective effects. At the same time, the 15-amino-acid C-terminus of DNP is highly resistant to neutral endopeptidase (NEP), making CD-NP a more stable and desirable molecule than are naturally occurring natriuretic peptides [51-53].

3. Natriuretic Peptides Synthesis and Secretion

The three natriuretic peptides, ANP, BNP, and CNP, have highly homologous structure, but distinct sites of synthesis and secretion. Both ANP and BNP are predominantly synthesized in the heart. ANP concentrations range from 50- to 100-fold higher than BNP. It is generally thought that following processing of the preprohormone to prohormone molecule, cleavage and secretion of mature and biologically active 28-amino acid ANP molecule occurs predominantly in response to atrial stretch and distension. The cardiac atrium is the primary site of synthesis and secretion for both ANP and BNP but the ventricle also produces both hormones, although at respective level 100- to 1000-fold less than does the atrium.

The differences in natriuretic peptide concentrations also correlate with mRNA levels [54]. Interestingly, the expression of ANP and BNP increases dramatically in both the atrium and ventricle in cardiac hypertrophy and congestive heart failure [48]. Nevertheless, the ventricle becomes the primary site of synthesis and release of BNP in cardiac disease states. In patients with severe congestive heart failure, concentrations of both ANP and BNP increase higher than the control values. However, the increase in BNP concentration is 10- to 100-fold greater than the corresponding increase in the ANP level [48]. It believed that ANP and BNP elicit distinct physiological and pathophysiological effects in patients with cardiac events. In essence, ANP and BNP show similar hemodynamic responses, however, because of its longer half-life, BNP, as compared with ANP, has extended action and enhanced rather than blunts natriuretic and diuretic responses compared with ANP [55,56].

The atrium expresses 50- to 100- fold or even higher ANP mRNA levels than do extracardiac tissues [57]. Although a high levels of ventricular ANP is present in the developing embryo and fetus, both mRNA and peptide levels of ANP decline rapidly during the prenatal period [58]. However, ANP gene expression in the ventricle is reinduced postnatally in response to phenylephrine administration, after-load stress, and myocardium infarction [59]. Previous studies have suggested that the mRNA levels of BNP are markedly lower than ANP levels in the heart. However, the BNP concentrations are higher in the ventricle than in either neonatal or adult rat hearts. Nevertheless, in adult hearts the reduction in ventricular expression of BNP is far less than the reduction of ANP [60]. Although the circulating BNP levels are far lower than that ANP levels in normal subjects, the increase in

BNP concentrations in plasma can surpass the level of ANP in patients with congestive heart failure [48,61,62]. Because of the longer half-life and stability of NT-proBNP, its measurement of BNP and ANP. Moreover, the NT-proBNP is considered to be a predictive of death after cardiac transplantation [63]. On the other hand CNP does not seem to behave as a cardiac hormone and its levels are extremely low in the circulation are extremely low [64]. CNP is largely present in the central nervous system-and in vascular endothelial cells [65-69].

Interestingly, DNP is present in the venom of the green mamba. On the other hand, URO is identical to the C-terminal sequence of pro-ANP, appears to be present only in the urine [39,40]. It was initially purified from human urine and is presumed to be synthesized only in the kidney [70]. It is thought that URO is not present in the circulation, but is a unique intrarenal natriuretic peptide hormone with unexplored physiological and pathophysiological significance [41,70]. Immunohistological staining indicates that URO is largely present in cortical tubules around the collecting ducts of the kidneys [71-73].

4. Topology and Structure of Natriuretic Peptide Receptors

During initial studies to identify ANP receptors, the specific binding of radiolabeled-ANP (^{125}I-ANP) to plasma membranes, intact cells of various types, and cultured cells was done, Only one high-affinity receptor type, with a Kd value of 10-100 pM, was reported [4,74-76]. Cross-linking and photoaffinity labeling studies showed the existence of ANP receptors with a wide range of molecular weights (60-180 kDa). Based on the biological activities of different ANP analogs, ANP receptors were classified as biologically active or silent (clearance) receptors [24,74,75,77,78]. Subsequently, photoaffinity and cross-linking identified different subtypes of ANP receptors, as well as different guanylyl cyclase activities, which appeared to be specific to different cell types [74-76]. Using photoaffinity labeling and the guanylyl cyclase activity assay, three distinct types of natriuretic peptide receptors were classified in different cell types [75]. Biochemical, immunohistochemical, and molecular biological data indicated that natriuretic peptides and their receptors are quite widespread in their tissue distributions and the extent of their varying levels of guanylyl cyclase activity [4,16,74-76,79,80].

Initially, GC-A/NPRA was cloned and sequenced from rat brain [81], human placenta [82], and mouse testis [29]. Three different subtypes of NP receptors, NPRA, NPRB, and NPRC, constitute the natriuretic peptide receptor family; however, these receptors are variable in their ligand specificity, guanylyl cyclase activity, and signal transduction efficacy. Both GC-A/NPRA and GC-B/NPRB are 135-kDa transmembrane proteins; ligand binding to these receptors generates second-messenger cGMP [15,23,29,81,83]. In contrast, NPRC lacks the intracellular GC catalytic domain and, by default, has been termed a natriuretic peptide clearance receptor [24,78]. Both ANP and BNP selectively stimulate NPRA, whereas CNP primarily binds and activates NPRB. However, all three NPs show affinity to NPRC [69,83,84]. It has been suggested that ANP binding to its receptor *in vivo* requires chloride, which probably exerts a chloride-dependent feedback-control mechanism on receptor function [85,86].

The general structural topology of GC-A/ NPRA and GC-B/NPRB is consistent with that in the GC-receptor family, containing at least four distinct regions: an extracellular ligand-binding domain, a single transmembrane spanning region, an intracellular protein-kinase-like homology domain (protein-KHD), and the intracellular carboxyl-terminal GC catalytic domain (Figure 2). Overall, the NPRB domain is structurally similar to that of NPRA but has binding selectivity and specificity for CNP [13,81]. Indeed, NPRA is the biologically dominant form of the natriuretic peptide receptors; it is found in several peripheral and visceral organs and mediates most of the known physiological actions of both ANP and BNP hormones. Using a homology-based cDNA library screening system, additional members of the GC-receptor family have been identified, but their specific ligands and/or activators have not been found [3,7,23,87].

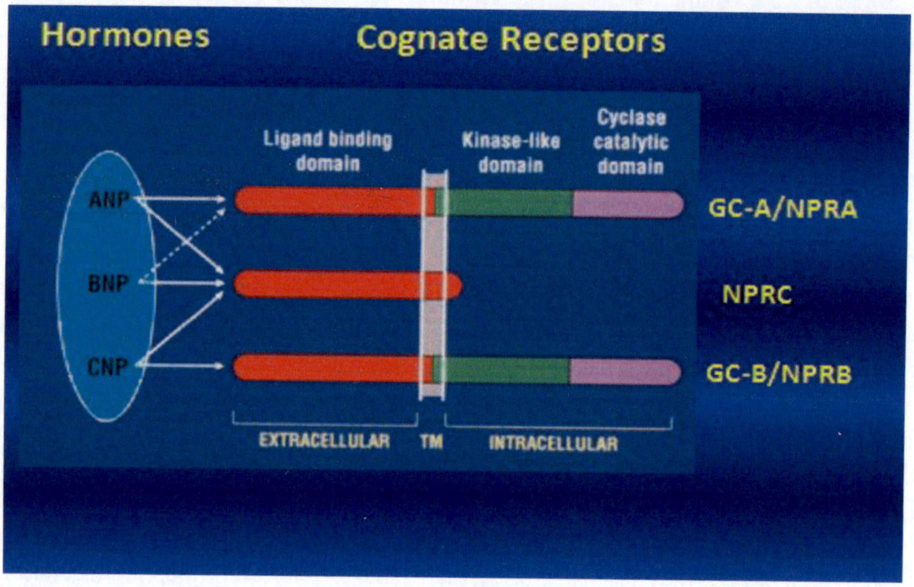

Figure 2. Hormone binding and structure of transmembrane natriuretic peptide receptor sub-types. The diagram depicts that a specific natriuretic peptide receptor activates a specific receptor sub-type on the plasma membrane. The solid lines indicate the receptors with their preferred ligand. The extracellular ligand-binding domains, transmembrane domain, and intracellular protein kinase-like homology domain (protein-KHD) and guanylyl cyclase (GC) catalytic domain of GC-A/NPRA and GC-B/NPRB are shown. Both guanylyl cyclase receptors, GC-A/NPRA and GC-B/NPRB have been indicated to generate intracellular second messenger cGMP. The NPRC contains extracellular ligand-binding domain, a transmembrane region, and short intracellular cytoplasmic tail, which does not contain GC catalytic domain. GC-A/NPRA, guanylyl cyclase/natriuretic peptide receptor-A; GC-B/NPRB, guanylyl cyclase/natriuretic peptide receptor-B, NPRC, natriuretic peptide clearance receptor.

The intracellular region of NPRA is divided into two domains: the protein-KHD is the 280-amino acid region immediately following the transmembrane domain; distal to this is the GC catalytic domain of the receptor. More than 80% of the conserved amino acid residues that have been found in all protein kinases are considered to be present in NPRA [88]. It has been suggested that the GC catalytic domain of NPRA consists of a 250-amino acid region at the carboxyl-terminal end of the receptor molecule. Initial studies showed that deletion of the carboxyl-terminal region of NPRA resulted in a protein that bound to ANP but did not contain GC activity [6,15,89,90].

Molecular structure modeling studies based on the crystal structure of the adenylyl cyclase II C_2 (AC II C_2) homodimer [91] predicted that the active sites of guanylyl and adenylyl cyclases are closely related [92,93]. The GC catalytic active site of murine GC-A/NPRA includes a 31-amino-acid sequence (974-1,004 residues) at the carboxyl-terminal end of the receptor molecule. The crystal structure of the soluble guanylyl cyclase (CYG12) catalytic domain from unicellular green algae (*Chlamydomonas renharditi)* has been reported to contain a 188-amino acid catalytic core region [94]. Similarly, a 202-amino-acid catalytic core of membrane-bound guanylyl cyclase (Cya2) has been shown to be present in prokaryotic cyanobacterium, (*Synechocysts*) [95]. Early findings demonstrated, for the first time, the structure of the catalytic domain of any guanylyl cyclase after the adenylyl cyclase structure, which was reported almost ten years before the crystal structures of soluble and membrane-bound guanylyl cyclases (GYG12 and Cya2) were evident. It is predicted that guanylyl cyclases and all known predicted adenylyl cyclases belong to the class-III nucleotide cyclase family of receptors and share a high-sequence similarity. Based on the comparisons of amino acid sequence, the CYG12 seems to be homologous with mammalian soluble cyclases. However, Cya^2 is homologous with the members of the particulate guanylyl cyclase receptors.

To delineate the mechanisms of GC-coupled transmembrane receptors (NPRA and NPRB), the production of the soluble form of receptors became essential. The first soluble form of the truncated ligand-binding domain of NPRA was expressed in *Escherichia coli* [89]. Later, the extracellular hormone-binding domain of NPRA was expressed in soluble form in COS-7 cells [96]. Solubilization of the extracellular domain of NPRA provided a starting point to produce the crystal structure of the ligand-binding domain of this receptor protein [97]. NPRA was crystalized as a dimer of two receptor molecules, which tended to dimerize spontaneously at high concentrations [96]. The NPRA contains a specific structural component near the transmembrane spanning domain, which was also designated as a juxtamembrane signaling motif of this receptor protein [98]. The juxtamembrane region seems to be highly conserved among the family of GC-receptors. Interestingly, the hormone-induced structural changes in NPRA differ from those that occurs n NPRC [99]. In NPRC, ligand binding causes each receptor molecule of the dimer to bend between the two subdomains of the receptor, which creates a clamping motion to capture the ligand molecule. It is possible that the conformational flexibility at the intramolecular hinge region in the NPRC is a critical factor for the broad specificity of this receptor molecule, which bind all three NPs, including ANP, BNP, and CNP [99].

Although GC-A/NPRA contains a single cyclase-catalytic active site per polypeptide molecule, based on modeling data two polypeptide chains are required to activate the functional receptor molecule [97]. The transmembrane GC receptors are predicted to function as homodimers [86,100]. It has been suggested that the dimerization region of GC-A/NPRA is located between the protein-KHD and GC catalytic domains. It has been predicted that this region form an amphipathic alpha helix structure [86]. It appears that NPRB is mainly localized in the brain and vascular tissues, and is thought to mediate the actions of CNP in vascular beds and the central nervous system [13]. The third member of the natriuretic peptide receptor family, NPRC, consists of a large extracellular domain of 496 amino acids, a single transmembrane domain, and a very short 37-amino-acid cytoplasmic tail that bears no homology to any other known receptor protein domain [24].

5. Intracellular Trafficking and Downregulation of NP Receptors

Studies have shown that the specific binding of ^{125}I-ANP to plasma membranes of various types of tissues and intact cultured cells exhibited a high-affinity receptor species with a Kd value of 10-100 pM (Table 2). After ligand binding, NPRA is ligand-dependently internalized and redistributed into subcellular compartments [74,75,101-106]. The endocytosis and intracellular sequestration of NPRA involves a series of sequential sorting steps through which ligand-receptor complexes are eventually endocytosed, sequestered, and degraded in lysosomes. However, a population of receptor recycle back to the plasma membrane and intact ligand is released to the cell exterior [90,102,106-108]. The recycling of internalized receptor to the plasma membrane and the release of intact ANP to the cell exterior occur simultaneously leading to degradation of the majority of ligand-receptor complexes [90,102,105,108]. Degradation of the majority of ligand-receptor complexes of both NPRA and NPRC occurs in the lysosomes (Figure 3). Those findings provided direct evidence that treatment of cells with unlabeled ANP accelerates the internalization of surface receptors [74,102,105]. It has been suggested that the GDAY (Gly^{920} –Asp^{921}-Ala^{922}-Tyr^{923}) sequence motif in the carboxyl terminal domain of GC-A/NPRA serves as an internalization signal for endocytosis and intracellular trafficking of this receptor protein [105]. The amino acid residues Gly^{920} and Tyr^{923} seem to be the important elements in the GDAY internalization signal motif of NPRA. However, it is thought that the residue Asp^{921} provides an acidic environment for efficient signaling of the GDAY motif during the internalization and trafficking of NPRA. Mutation of Asp^{921} to alanine did not have a major effect on receptor internalization, but did significantly attenuate the recycling of internalized receptors back to the cell surface [105]. The mutations of Gly^{920} and Tyr^{923} to alanine inhibited the internalization of NPRA, although these residues did not have a discernible effect on the recycling process of the internalized receptors. The tyrosine-based sequence modulates the early internalization process of NPRA, whereas Asp^{921} in the GDAY sequence motif mediates recycling or later sorting of the receptor. Thus, two overlapping motifs within the GDAY sequence in the carboxyl-terminal domain of NPRA exert different but specific effects on endocytosis and trafficking of this receptor protein.

It has also been demonstrated that guanylyl cyclase-B/natriuretic peptide receptor-B (GC-B/NPRB) is also internalized and recycled back to the cell surface [109]. The trafficking of NPRB occurs ligand-dependently in response to stimulation with CNP. The sequestration of NPRB has been suggested to involve a clathrin-dependent mechanism. Similarly, the internalization of NPRA also involves micro-RNA and clathrin-dependent pathways [110].

Internalization and sequestration of hormone receptors have been implicated as having important roles in the desensitization and down-regulation of various receptor proteins [105,106,111].

Our studies have demonstrated that ANP-dependent downregulation of NP receptors occurs in rat thoracic aortic smooth muscle (RTASM) cells expressing NPRC and human embryonic kidney-293 (HEK-913) cells expressing a high density of recombinant NPRA [107,108,112]. The exposure of RTASM cells to 10 nM ANP caused a substantial decrease in ^{125}I-ANP binding, leading to a loss of 50%-55% of total receptor number without changing the affinity of the binding sites. RTASM cells, which were pretreated with various concentration of ANP, markedly decreased the surface binding of ^{125}I-ANP to the receptor in a dose-dependent manner. However, micromolar concentrations of ANP produced almost complete loss of cell-surface binding of ligand to the receptor protein [112]. Pretreatment of recombinant HEK-293 cells expressing high-density NPRA with unlabeled ANP caused a substantial time- and dose-dependent decrease in ^{125}I-ANP binding capacity of NPRA [108]. The treatment of cells with ANP (10 nM) reduced receptor binding by 55%-60%, while a 1μM ANP concentration produced complete loss of cell-surface binding capacity of NPRA. However, it should be noted that RTASM cells mainly contain NPRC and only a minor amount of NPRA [75,112].

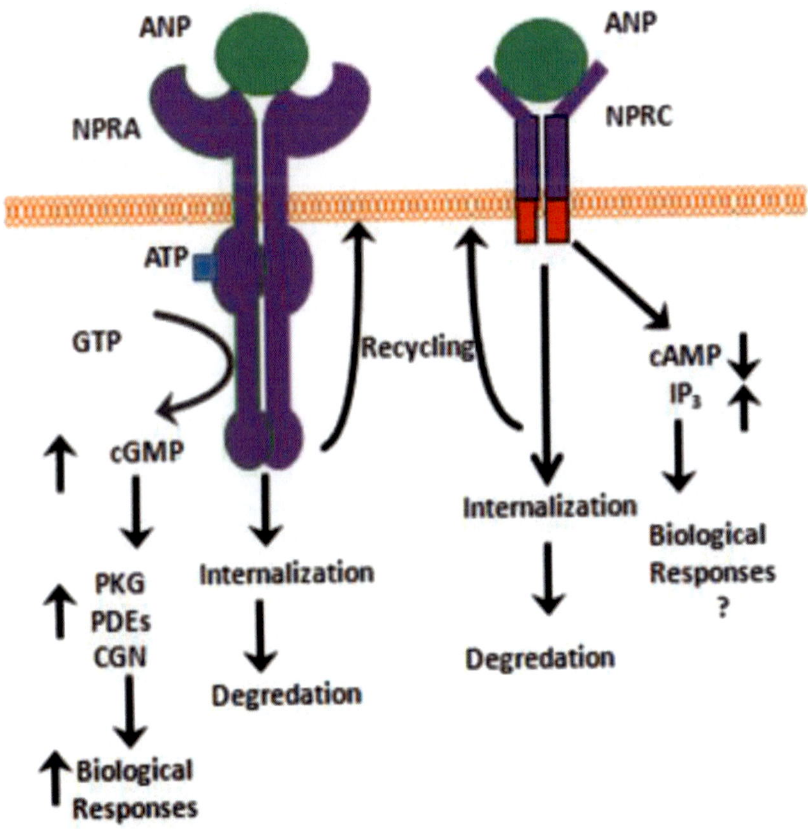

Figure 3. Diagrammatic representation of ligand-dependent internalization activation and post-binding events of NPRA and NPRC. ANP binding at the amino-terminus of NPRA leads to increased production of intracellular cGMP, which activates cGMP-dependent protein kinases (PKGs), cGMP-dependent phosphodiesterases (PDEs), and cGMP-mediated ion channels to activate physiological functions. Ultimately, ligand-receptor complexes of NPRA are internalized into the intracellular compartments and a large population of ligand-receptor complexes are degraded into the lysosomes. A small population of receptor is recycled back to the plasma membrane. The binding of ANP to NPRC lowers intracellular accumulation of cAMP; however, increases the formation of inositoltrisphosphate (IP3). The ligand-bound NPRC complexes are internalized and degraded in the lysosomes, however, a small population ligand-receptor complexes are also recycled back to the plasma membranes.

Table 2. ANP-dependent binding parameters of GC-A/NPRA and intracellular accumulation of cGMP in different cell types

Cell Type	ANP-dependent Intracellular cGMP (fold stimulation)	Ligand binding parameters of NPRA	
		Kd value (Molar)	B_{max} (receptor site/cell)
Granulosa cells	30	10 pM – 100 pM	0.5 X 10^5
Glomerulosa cells	50	100 pM – 1 pM	2 X 10^5
HEK 293 cells	2,000	100 pM – 1 nM	1.2 X 10^6
MA-10 cells	1,500	100 pM – 1 nM	1 X 10^6
MDCK cells	50	10 pM – 100 pM	0.5 X 10^5
Primary Leydig cells	60	10 pM – 100 pM	0. 5 X 10^5
N4TG1 cells	30	1 pM – 100 pM	0.5 X 10^5
RTASM cells	10	1 pM – 100 pM	0.2 X 10^5

HEK-293 cells, human embryonic Kidney-293 cells; Kd, dissociation constant; Bmax, receptor density; MA-10 cells, Leydig tumor cells; MDCK cells, Maiden -Darby kidney epithelial cells; N4TG1 cells, neuroblastoma cells; RTASM, rat thoracic aortic smooth muscle cells.

The expression of mRNA levels of NPRA is nevertheless regulated by various intrinsic and extrinsic factors, including glucocorticoids [113], transforming growth factor-β [114], human chorionic gonadotropin hormone [115], angiotensin II (Ang II) [116,117], and retinoic acid [118,119]. Endogenous transcription factors such as Ets-1 and p300 exert remarkable stimulating effects on *Npr1* (coding for GC-A/NPRA) gene transcription and expression [119]. At the protein level, Ang II inhibits the GC activity of NPRA [116,120,121]. Similarly, at the receptor level, NPRA is downregulated by ANP, as well as 8-bromo-cGMP [90,105,106,108,122]. The downregulation of NPRC has also been shown in cultured vascular smooth muscle cells, which predominantly contain this receptor protein [75,112,123-128]. Various investigators have reported the results of detailed kinetic studies using vascular smooth muscle cells to explore the metabolic processing of ANP involving NPRC, which does not contain GC catalytic activity [112,123,124,129-133]. However, the precise function of NPRC and its transmembrane signaling mechanisms still remains to be established.

6. Intracellular Signaling Mechanisms of NP Receptors

Studies have now established that ANP markedly and dose-dependently increases second-messenger cGMP in target cells and tissues [75,107,108,134,135]. The production of cGMP is mediated by ligand binding to the amino-terminal region of the extracellular domain of GC-coupled receptors. It is likely that these receptors allosterically regulate increases in the specific activity of the cytoplasmic GC catalytic domains of both NPRA and NPRB [3,6,7,90,136,137]. The nonhydrolyzable analogs of ATP mimicked the ANP effect, suggesting that ATP can allosterically regulate the GC catalytic activity of NPRA [87,138-140]. Earlier findings suggested that under natural conditions the protein-KHD negatively regulates the catalytic moiety of NPRA. Initially, this model was considered to be the standard mark in explaining the signal transduction mechanism of GC-coupled natriuretic peptide receptors [141]. However, that model has not been supported by the studies of other investigators, suggesting that deletion of the protein-KHD does not increase the basal GC activity of NPRA [87,136]. Nevertheless, ATP was predicted to be obligatory for the transduction activities of both GC-coupled NPRA and NPRB receptor proteins [87,136,142].

Previously, it was suggested that NPRA exists in a phosphorylated state and that the addition of ANP caused a decrease in phosphate contents as well as a reduction in the ANP-dependent GC catalytic activity of this receptor protein [137,143]. The apparent mechanism of desensitization of NPRA is in contrast with that in many other cell-surface receptors, which appear to be desensitized by phosphorylation [144-151]. Initial studies also suggested that ANP seems to stimulate phosphorylation of its GC-coupled receptor protein [152-157]. Recently, it has been indicated that cGMP-dependent protein kinase (PKG), a serine/threonine protein kinase, is capable of phosphorylating NPRA *in vitro* [158]. It also has been suggested that PKG is recruited to the plasma membrane after ANP treatment and increases the GC catalytic activity of NPRA.

It also has been suggested that glycosylation is essential for the ligand-binding and functional activity of NPRA [159-161]. The glycosylation sites in GC-coupled receptors appear to be important for proper folding and stability of the receptor proteins [161-163]. However, the exact role of glycosylation sites in ligand-binding properties of the receptor has not yet been defined. It should be noted that there is no appreciable conservation of the precise position of glycosylation sites within the members of the GC receptor family. The binding of ANP and BNP to GC-A/NPRA leads to increased production of intracellular second-messenger cGMP, which stimulates three known cGMP effector molecules: cGMP-dependent protein kinases (PKGs), cGMP-dependent phosphodiesterases (PDEs), and cGMP-dependent ion channels [7,164-169]. The activation of these effector molecules leads to a wide range of biological and physiological functions of ANP/NPRA/cGMP signaling mechanisms, including excretion of salt and water, vasorelaxation, anti-inflammation, anti-fibrosis, and anti-hypertrophic mechanisms (Figure 4). Overall, the signaling mechanisms of NPs and their receptors lead to combined physiological functions, which together provide renoprotection, cardioprotection, and vascular protection in the context of pathophysiological conditions.

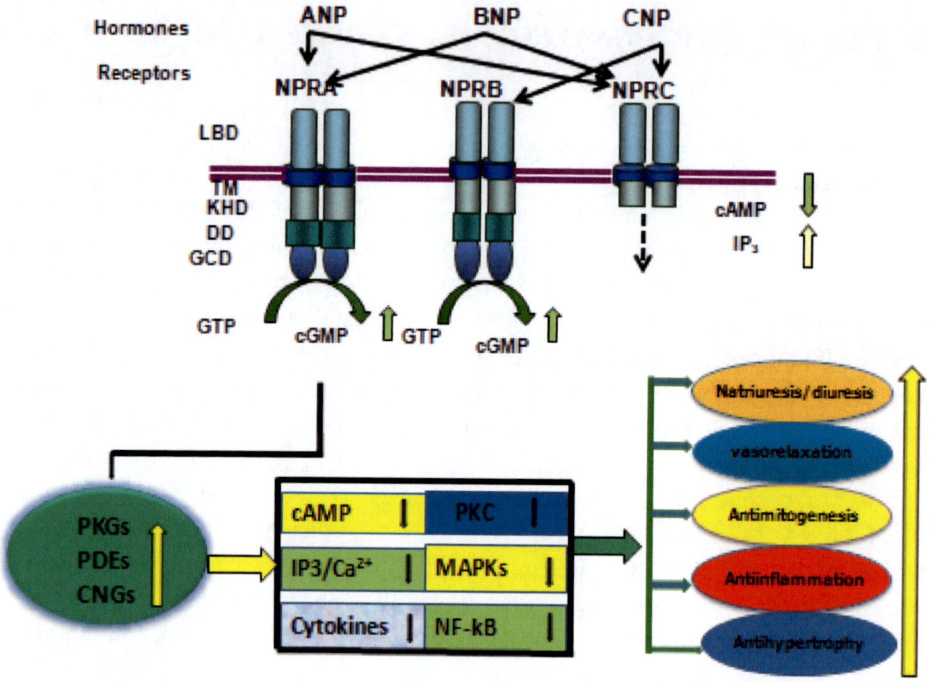

Figure 4. Diagram represents the ligand specificity and physiological function(s) of GC-A/NPRA. The ligand-binding to NPRA generates second messenger cGMP from the hydrolysis of GTP. An increased level of intracellular cGMP is produced, which activates three known cGMP effecter molecules namely; cGMP-dependent protein kinases (PKGs), cGMP-dependent phosphodiesterases (PDEs), and cGMP-dependent ion-gated channels (CNGs). The ANP/NPRA/cGMP signaling may antagonize a number of pathways including; intracellular formation of cAMP, Ca2+, IP3; cytokine expression; and the activation of protein kinase C (PKC) and mitogen-activated protein kinases (MAPKs). The resulting signaling cascade can mimic the physiological responses of ANP/NPRA. LBD, ligand binding domain; TM transmembrane region; protein-KHD, protein kinase-like homology domain; and GCD, guanylyl cyclase catalytic domain; DD, dimerization domain of NPRA and NPRB. The ligand binding region, transmembrane domain, and small intracellular tail region of NPRC are indicated.

7. Functional Role of NP Receptors in Kidneys and Vasculature

ANP actions facilitate the excretion of salt and water, as well as increases in glomerular filtration rate (GFR) and renal plasma flow (RPF) in the kidneys [7,170-177]. Early studies showed that renal sites of ANP actions include inner medullary collecting duct cells, glomeruli, tubular epithelial cells, and mesangial cells [90,178-183]. The intracellular action of ANP in renal cells stimulates GC activity and reduces adenylyl cyclase and phospholipase C activity, sodium influx, and calcium concentrations [4,184]. Increased production of the second messenger cGMP in response to ANP affecting renal functions correlates well with the effects of dibutyryl-cGMP, which prevents mesangial cell contraction in response to the actions of vasoconstrictors such as endothelin, vasopressin, Ang II, and agonists of adrenergic system [181]. It is important to note that ANP/cGMP antagonizes most of the actions of vasoconstrictors such as Ang II (Table 3).

Table 3. Antagonistic actions of ANP on angiotensin II-dependent physiological functions

Physiological Function	Ang II	ANP
Aldosterone release	Stimulation	Inhibition
Blood vessels tone	Contraction	Relaxation
CNS-mediated hypertension	Stimulation	Inhibition
Estradiole	Unknown	Stimulation
Neurotransmission	Stimulation	Inhibition
Renin Release	Inhibition	Inhibition
Testosterone	Unknown	Stimulation
Vasopressin Release	Stimulation	Inhibition
Water drinking	Stimulation	Inhibition

Stimulatory and inhibitory actions of atrial natriuretic peptide (ANP) or angiotensin II (Ang II) - dependent cellular and physiological responses in target tissues and cells.

The most compelling evidence of a role for cGMP in mediating the renal effects of ANP was obtained with the specific selective NPRA antagonists A71915 and HS-121-1, which abolished the renal effect of infused ANP and decreased urinary cGMP [185,186]. Earlier studies established that the effect of ANP in the kidneys is largely mediated by increased production of second-messenger cGMP through the activation of NPRA.

A combination of hemodynamic effects and tubular transport processes may be responsible for ANP/NPRA-induced natriuresis and diuresis. Evidence suggest that ANP/NPRA/cGMP signaling exerts both direct and indirect effects on tubules, including the proximal tubules, as well as cortical and innermedullary collecting ducts [4,176,187,188]. The ANP/ NPRA system induces diuresis by increasing GFR and RPF, which are hemodynamic parameters [4,175,189-191]. However, it has been suggested that ANP-induced natriuresis is also caused by direct inhibition of tubular transport processes, independent of alterations in GFR [187]. Interestingly, it is thought that ANP-induced natriuresis and diuresis favor a combination of hemodynamic and tubular effects [192]. The function of ANP in mediating renal and vascular effects was also demonstrated with selective NPRA antagonists to eliminate the effect of ANP [193,194]. The intrarenal actions of ANP/NPRA signaling at nephron sites exerts a combined effect, leading to increased excretion of sodium and water, enhanced renal hemodynamic function, and reduced arterial pressure (Table 4).

Table 4. Intrarenal action of atrial natriuretic peptide at different sites of the nephron in the kidney

Nephron Sites	Biological Parameters	Signaling action
Cortical collecting duct	VAP-induced hydro-osmosis	Decrease
Glomerulus	Glomerular filtration rate	Increase
Inner medullary collecting duct	Na^+-K^+-$2Cl^-$ co-transporter	Decrease
Juxtaglomerulus	Renin secretion	Decrease
Proximal tubule	Na^+ and water transport	Decrease
Proximal tubule	Na^+ absorption	Decrease
Renal artery	Renal blood flow	Increase
Renal medulla	Solute concentration	Decrease
Vasorecta	Blood flow, hydraulic pressure	Increase

The atrial natriuretic peptide (ANP) action in the kidneys include; 1) Decrease in the vasopressin (VAP)-induced hydro-osmosis in collecting duct, 2) Increase in glomerular filtration rate (GFR) and renal plasma flow (RBF), 3) Decrease in sodium reabsorption in the proximal tubules and collecting ducts, 4) decrease solute concentration in medulla, 5) Decrease Na+-K+-2cl- co-transporter in the inner medullary collecting duct, 6) Decrease renin secretion in juxtaglomerular cells, and 7) Increase in blood flow and hydraulic pressure.

Recent studies have suggested that in *Xenopus* 2F3 cells epithelial sodium channels (ENaCs) are regulated by ANP/NPRA-dependent second messenger cGMP [195]. Immunofluorescence confocal microscopy has indicated that NPRA is expressed at the apical membrane of 2F3 cells. Studies have demonstrated that ANP or 8-(4-chlorophenylthio)-cGMP, but not CNP or C-ANP_{4-23}, decrease the open probability of EnaC. Moreover, there are indications that NPRA, but not NPRB or NPRC, is involved in regulating ENaC activity in *Xenopus* 2F3 cells. It has been suggested that decreased ENaC activity is dependent on cGMP. Similarly, increasing concentrations of intracellular cGMP inhibited ENaC activity in primary human cystic fibrosis bronchial epithelial cells [196].

ANP markedly lowers renin secretion from the kidneys; it also reduces plasma renin concentrations [172,177,197,198]. Strong experimental data have established that ANP/NPRA signaling is important in regulating renal function by its vasodilatory and natriuretic responses and its tissue-specific ability to counteract the renin-angiotensin-aldosterone system (RAAS) [6,8,172,198]. Attempts have been made to define physiological responses in the kidney by infusing exogenous ANP [199]. Cardiac appendectomy has been used to prevent ANP release; however, the problem in this setting is that the missing normal cardiac functions resulted in a lack of physiological reflexes that are normally elicited by atrial components [200]. Additional approaches have been used to include monoclonal antibodies against circulating ANP; however, problems occurred in relation to the nonspecific effect of the antibody or antigen complexes. One classical approach has been used to specifically inhibit the intracellular signaling pathway of NPRA to block cGMP production. Two NPRA antagonists, A71915 and HS-140-1, have been shown to diminish the effect of ANP by antagonizing the effect of NPRA, but these compounds do not completely inhibit the activity of NPRA [201,202]. Activation of natriuretic peptides (ANP, BNP) enhances the pressure-natriuresis relationship and reduces atrial pressure. It has also been suggested that chloride-mediated feedback control of NPRA occurs in the kidney and has a role in ANP-mediated natriuresis [85]. Initial studies showed that ANP suppresses renin activity and other RAAS components, and decreases blood pressure [177,197]. More recent studies have shown that the hallmark function of the ANP/NPRA system is to lower blood pressure and antagonize RAAS [4,16].

Gene-targeting of *Npr1* (coding for GC-A/NPRA) in mice with gene disruption or gene duplication has provided strong evidence of the physiological roles of the ANP/NPRA/cGMP system in kidney function [175,176,198,203-213]. Gene-targeting strategies in mice have provided novel approaches to the study of the physiological responses corresponding to gene-dosage *in vivo* [8,9,214,215]. Studies with *Npr1* gene-disrupted mice have demonstrated that the absence of NPRA at birth allows greater renin and Ang II levels and increased renin mRNA expression than occur in with the wild-type mice [198]. However, in adult *Npr1* homozygous null mutant mice both circulating and kidney renin and Ang II levels were dramatically decreased as compared with levels in wild-type control mice. This decrease in renin activity in adult null mutant mice is a consequence of progressive elevation in arterial pressure leading to inhibition of renin synthesis and release from kidney juxtaglomerular cells.

It has been suggested that increased levels of ANP released into the plasma in response to blood volume expansion are mainly responsible for the natriuretic and diuretic responses [199,216]. Evidence suggests that both ANP and acute blood volume expansion act on the kidney through a similar saturable mechanism [205,217]. Our studies have examined the detailed quantitative contributions and possible mechanisms mediating the responses of blood volume expansion. For this purpose we have used the varying numbers of *Npr1* gene copies by determining the renal plasma flow RPF, GFR, urine flow, and sodium excretion patterns following blood volume expansion in *Npr1* homozygous null mutant ($Npr1^{-/-}$; 0-copy), wild-type ($Npr1^{+/+}$; 2-copy), and gene-duplicated ($Npr1^{++/++}$; 4-copy) mice in a *Npr1* gene-dose-dependent manner [217]. Using whole blood, hemodilution did not occur and plasma protein levels were not reduced. Our studies have also demonstrated that disruption of the *Npr1* gene drastically affects kidney function and leads to renal abnormalities in *Npr1* gene-disrupted null mutant mice as compared with wild-type mice [209,210,217].

Although blood volume expansion stimulated the release of ANP in all three *Npr1* genotypes of mice, significant functional responses (RPF, GFR, and sodium excretion) occurred only in wild-type (2-copy) and gene-duplicated (4-copy) mice, not in homozygous null mutant (0-copy) mice. These findings demonstrated that the ANP/NPRA axis is primarily responsible for mediating the renal hemodynamic and sodium excretory responses to intravascular blood volume expansion. ANP responses to volume expansion led to significantly less excretion of sodium and water in null mutant 0-copy mice and significantly greater excretory responses, along with reduced tubular reabsorption, in gene-duplicated (4-copy) mice as compared with wild-type (2-copy) mice. Similarly, during the blood volume expansion, urinary cGMP concentration was significantly lower in *Npr1* gene-disrupted (0-copy) null mutant mice and greater responses occurred in 4-copy gene-duplicated mice. Those findings established that NPRA is a hallmark receptor that has a critical role in mediating the natriuresis, diuresis, and renal hemodynamic responses to acute blood volume expansion.

The ANP/NPRA system inhibits aldosterone synthesis and release from adrenal glomerulosa cells [4,218]. suggesting that this ANP action could be physiologically important and probably accounts for natriuretic and diuretic responsiveness. The biochemical and cellular effects of ANP within the adrenal glomerulosa cells include activation of GC activity and potassium channel conductance; whereas T-type calcium channel conductance and adenylyl cyclase activity are suppressed [184]. The NPRA antagonist HS-142-1 eliminated ANP effects to suppress the aldosterone synthesis and increase cGMP production in bovine adrenal glomerulosa cells, providing strong evidence of the involvement of NPRA in inhibiting ANP-dependent aldosterone synthesis and secretion [219]. However, the correlative evidence that cGMP suppresses aldosterone secretion was challenged by the observation that aldosterone release was maintained in the presence of cGMP analogs such as dibutyryl-cGMP or 8-bromo-cGMP [4,220]. Further studies suggested that ANP acts via NPRA in the adrenal gland to increase cGMP concentrations, resulting in activation of cAMP-dependent phosphodiesterases, which decrease both cAMP and aldosterone levels in adrenal glands [6, 221].

Cultured vascular smooth muscle cells treated with ANP have always shown an increase in the intracellular accumulation of cGMP [75,76]. The correlative evidence between ANP-induced cGMP production and vasodilation has suggested the role of cGMP as the second messenger of dilator responses to ANP [4,9,105,184,194,222,223]. ANP, as well as cGMP analogs, have been found to reduce agonist-induced increases in cytosolic Ca^{2+} and inositoltrisphosphate (IP3) concentrations in target cells [224-227]. It has been reported that cGMP activates sarcolemmal Ca^{2+}-ATPase; this mechanism may be important in ANP-induced decreases in cytosolic Ca^{2+} in vascular smooth muscle cells [228, 229]. Nevertheless, it is anticipated that the ultimate effect of ANP in vascular smooth muscle cells is caused by the production of cGMP and activation of cGMP-dependent protein kinases [194,223,225]. Indeed, more studies are needed to delineate the biochemical and molecular basis of the actions of natriuretic peptides and receptors in the vascular bed, especially in vascular smooth muscle and endothelial cells.

8. Anti-Mitogenic and Anti-Proliferative Effects of NP Receptors

Natriuretic peptides act as the suppressor of cell proliferation and growth in a variety of cell types. including the kidneys [73,183,230-232], heart [233], neurons [234,235], and thymus [236], as well as the vasculature [69,223,230,235,237-241], and fibroblasts [242] . Because cGMP analogs can mimic the anti-proliferative action of natriuretic peptides, it is considered that they exert antimitogenic effects largely through second- messenger cGMP [183,223,225,232,241]. ANP has been shown to inhibit collagen synthesis in cardiac fibroblasts [243] and to inhibit hypertrophy in primary cultures of cardiac myocytes [244-

246]. Similarly, PKG has been shown to suppress extracellular matrix production in vascular smooth muscle cells [247]. The expression of ANP and BNP genes are greatly augmented in hypertrophied hearts, which supports the notion that autocrine and/or paracrine effects of ANP/BNP signal are important for acting against pathological cardiac hypertrophy and fibrosis in disease states [245,248-250]. Both GC-linked NP receptors (NPRA and NPRB), as well as GC-unlinked NP receptor (NPRC), have been suggested to play a role in ANP-dependent anti-mitogenic responses [183,232,241,251]. It has been demonstrated that ANP inhibits mitogen-activated protein kinases (MAPKs) stimulated by Ang II, platelet-derived growth factor (PDGF), epidermal growth factor (EGF), and vascular endothelial growth factor (VEGF) in different tissues and cell types [183,223,232,251]. Nevertheless, the involvement of specific ANP receptor subtypes in the inhibitory effects of ANP on agonist-stimulated MAPK activity favors the role of NPRA. Clearly, more experiments are needed to delineate the underlying mechanisms of the anti-proliferative effect of ANP in target cells. In vascular smooth muscle and mesangial cells, ANP has been shown to inhibit different subtypes of MAPKs activities, including extracellular-regulated kinases (Erk1/2), p38 mitogen-activated protein kinase ($p38^{MAPK}$), and c-jun N-terminal kinase (JNK), as well as downstream signaling molecules c-fos and c-jun after stimulation with mitogenic agents such as PDGF and VEGF [183,223].

The ANP/NPRA signaling cascade exerts a cGMP-dependent inhibitory effect involving NPRA on MAPK activity in kidney mesangial cells [73,90,104,183]. However, in astroglial cells, ANP has been shown to inhibit Erk1/2 activity through NPRC [252]. Other studies, in contrast, have indicated that des- (Cys^{105}-Cys^{121}) -ANP, a ligand selective to NPRC, does not inhibit basal or serum-stimulated MAPKs and DNA synthesis in fibroblasts. However, CNP, which acts through NPRB, potently and cGMP-dependently inhibited MAPK activity in fibroblasts [242]. It has also been indicated that cGMP-dependent signaling mechanisms of NPRA probably are activated at the level of gene transcription [233], although the exact mechanism of this activation remains to be elucidated.

In addition to its antimitogenic effect, ANP induces apoptosis in cultured vascular smooth muscle cells and neonatal rat cardiac myocytes [253,254]. The ANP-induced apoptotic effect was mimicked by 8-bromo-cGMP, a membrane-permeable analog of cGMP, and by nitroprusside, an activator of soluble guanylyl cyclase. The effect of ANP has been shown to be potentiated by zaprinast, a cGMP-specific phosphodiesterase inhibitor. Norepinephrine, a myocyte growth effector, inhibited ANP-induced apoptosis via activation of β-adrenergic receptor and intracellular elevation of cAMP [254]. The existence of a complementary ANP-mediated mechanism to inhibit cell growth and proliferation is not surprising because the inhibition of cell proliferation is often accompanied by an increased probability of apoptosis, whereas growth-promoting agents also tend to promote cell proliferation. For instance, Ang II inhibits apoptosis whereas, in contrast, ANP and nitric oxide, both potently inhibit cell growth and proliferation and induce apoptosis [254,255]. It has been suggested that the anti-apoptotic molecule Bcl-2 homologue Mcl-1 serves as an important target in ANP-induced apoptosis of cardiac myocytes. Interestingly, Mcl-1 is expressed at high levels in cardiac tissues [254,256]. The Bcl-2 homologue Mcl-1, which was initially identified as a marker was upregulated during the differentiation of monocytoid cell line ML-1 cells [257,258]. ANP is considered to be a direct relaxant of smooth muscle cells and a potent regulator of cell growth and proliferation in the vasculature. The anti-growth paradigm could potentially work through negative regulation of the activities of MAPKs

[90,183,223]. ANP has been reported to antagonize growth-promoting effects in target cells, but there is controversy about the mechanism of the anti-growth paradigm of ANP and the involvement of specific ANP receptor subtypes (NPRA and/or NPRC) in target cells [90,241,252,259,260].

9. Role of NP Receptors in Hypertension and Cardiac Dysfunction

9.1. Blood Pressure Homeostasis

ANP/NPRA signaling lowers blood volume and blood pressure by releasing salt and water through the kidneys, as well as vasorelaxation of vascular smooth muscle cells. Various genetically engineered mouse models have been produced to delineate the functions of ANP, BNP, and NPRA in physiological and pathophysiological conditions of high blood pressure and hypertension. The genetic defects that reduce the signaling of ANP/NPRA can be considered candidate contributors to essential hypertension and cardiovascular homeostasis [7]. ANP-deficient mice demonstrated that a defect in proANP synthesis can increase blood pressure and cause hypertension [203]. Thus, a genetic mouse model with disruption of the *Nppa* gene (coding for proANP) has provided strong support for the central role of ANP in hypertension.

In *Nppa* homozygous null mutant (*Nppa*$^{-/-}$) mice, blood pressure was elevated by the administration of 8-10 mmHg when they were fed standard- or intermediate-salt diets. Heterozygous (*Nppa*$^{+/-}$) mice on a standard-salt diet showed a normal amount of ANP and normal blood pressure. However, *Nppa*$^{+/-}$ mice on a high-salt diet were hypertensive; their blood pressure was elevated by 27 mmHg. These findings indicated that mutation in the *Nppa* allele can lead to salt-sensitive hypertension even when the plasma ANP level is not significantly decreased. Transgenic mice overexpressing ANP developed sustained hypotension; their mean arterial pressure (MAP) was lower by 25-30 mmHg than that of their nontransgenic siblings [261,262]. Somatic delivery of the ANP gene in spontaneously hypertensive rats (SHR) induced a sustained reduction in systemic blood pressure.[263] Overexpression of ANP in hypertensive mice lowered systolic blood pressure, suggesting the use of ANP gene therapy for the treatment of human hypertension [264].

In null mutant *Npr1*$^{-/-}$ (0-copy) mice, disruption of the *Npr1* gene increases blood pressure by 35-40 mmHg as compared with that in wild-type *Npr1*$^{+/+}$ (2-copy) mice [198,209-211,249]. In contrast, increased expression of *Npr1* in gene-duplicated *Npr1*$^{++/+}$ and *Npr1*$^{++/++}$ (3- and 4-copy) mice significantly reduced blood pressure, corresponding to the increasing number of *Npr1* gene copies [175,208,212,213,265]. We have examined the possible mechanisms mediating the responses of varying numbers of *Npr1* gene copies by determining urine flow, sodium excretion, RPF, GFR, and blood pressure after blood volume expansion in *Npr1* 0-copy, 2-copy, and 4-copy mice in an *Npr1* gene-dose-dependent manner [175]. During the volume expansion with whole blood infusion, the MAP increased in *Npr1* gene-disrupted (0-copy), wild-type (2-copy), and gene-duplicated (4-copy) mice. However, MAPs were always significantly higher in *Npr1* gene-disrupted 0-copy mice and significantly lower in gene-duplicated 4-copy mice than in wild-type 2-copy mice. The GFR was 25% to

35% lower in 0-copy mice and 35% to 45% higher in 4-copy mice than in control wild-type mice [175]. Similarly, RPF was 25% to 30% lower in 0-copy null mutant mice and 45% to70% higher in gene-duplicated 4-copy mice than in 2-copy wild-type control mice. Homozygous null mutant (0-copy) mice retained significantly higher levels of sodium and water; however, gene-duplicated (4-copy) mice showed drastically reduced levels of sodium and water as compared with wild-type (2-copy) mice. These findings demonstrated that the ANP/NPRA signaling axis is primarily responsible for mediating the renal hemodynamic and sodium excretory responses to intravascular blood volume expansion.

The absence of NPRA in mice also leads to altered renin, Ang II, and aldosterone levels [175,188,198,208,211,249,265,266]. Our studies have shown that disruption of the *Npr1* gene in mice provokes kidney fibrosis, remodeling, and significant expression of proinflammatory and profibrotic cytokines and genes of the nuclear factor kappa B (NF-kB) pathway [209,210]. Significant increases in plasma creatinine concentrations and urinary albumin excretion, together with a reduced creatinine clearance rate, suggested the onset of renal insufficiency in *Npr1* null mutant mice. Interestingly, treatment of *Npr1* 0-copy mice with NF-kB inhibitors reversed renal hypertrophic and fibrotic responses in *Npr1* null mutant mice [209].

Gene-targeting studies have demonstrated that ANP/NPRA signaling inhibits aldosterone synthesis and release from adrenal glomerulosa cells [198,208], which may account for its ANP-dependent natriuretic and diuretic effects in the kidneys. Furthermore, studies with *Npr1* gene-disrupted mice suggest that the absence of NPRA at birth allows greater renin, Ang II, and aldosterone levels and increased renin mRNA expression than occurs in wild-type mice [198,208,265].

However, at 3-16 weeks of age, circulating renin and Ang II levels were dramatically decreased in *Npr1* null mutant mice as compared with wild-type mice. The decrease in renin activity in adult *Npr1* null mutant mice is likely due to progressive elevation of arterial pressure leading to inhibition of renin synthesis and release from the kidney juxtaglomerular cells [175]. On the other hand, adrenal renin content and renin mRNA levels, as well as Ang II and aldosterone concentrations, were elevated in adult homozygous null mutant mice as compared with wild-type mice [198,208]. Indeed, ANP/NPRA signaling appears to oppose almost all actions of Ang II in both physiological and phathophysioogical disease states [6,106].

Earlier, it was reported that the blood pressure of homozygous *Npr1* null mutant mice remained elevated and unchanged in response to either minimal- or high-salt diets [204]. In contrast, we and others have found that disruption of the *Npr1* gene results in chronic elevation of blood pressure in mice fed a high-salt diet [208,212,265]. However, adrenal Ang II and aldosterone levels were decreased in *Npr1* gene-duplicated mice fed a high salt-diet, whereas a low-salt diet stimulated adrenal Ang II and aldosterone levels in both gene-disrupted and gene-duplicated mice [208,265]. The high-salt diet suppressed adrenal Ang II and aldosterone levels in *Npr1* gene-disrupted mice and wild-type mice, but not in *Npr1* gene-duplicated (3-copy and 4-copy) mice. Our findings indicate that NPRA signaling is protective against the effects of a high-salt in *Npr1* gene-duplicated mice compared with *Npr1* gene-disrupted mice [208].

9.2. Cardiac Remodeling and Homeostasis

Accumulating evidence suggests that ANP and BNP exert their cardioprotective functions not only as circulating peptide hormones, but also as the local autocrine and paracrine factors. Plasma levels of both ANP and BNP are markedly elevated under the pathophysiological conditions of cardiac dysfunction, including diastolic dysfunction, congestive heart failure, pulmonary embolism, fibrosis, and cardiac hypertrophy [169,249,265,267-271]. Expression of both ANP and BNP is significantly increased in proportion to the severity of cardiac remolding and dysfunction in experimental animal models and humans [7,169,270,272,273]. Concentrations of ANP and BNP are markedly increased in the cardiac tissues and plasma of patients with congestive heart failure (CHF); both of these peptide hormones appear to reduce the preload and afterload of the heart [169,267,272]. In humans, high plasma levels of ANP and BNP tend to predict heart failure [272,274]. In patients with severe CHF, concentrations of both ANP and BNP increase, but the BNP concentration rises 10- to 50-fold higher than does ANP.[48] ANP and BNP genes are overexpressed in hypertrophied hearts, suggesting that the autocrine and/or paracrine effects of natriuretic peptides predominate and might endogenously protect against maladaptive pathological cardiac hypertrophy and dysfunction [249,267,270,271,275,276].

There are indications that ventricular expression of ANP and BNP is more closely associated with local cardiac hypertrophy and fibrosis than are plasma ANP levels and systemic blood pressures [249,271]. Since the half-life of BNP is longer than that of ANP, diagnostic evaluations of natriuretic peptides have favored BNP, which is considered to be an important indicator of CHF in patents with chest pain [169]. However, NT-proBNP seems to be a stronger indicator of the risk of cardiovascular events [277-280]. Expression levels of *Nppa* and *Nppb* genes (coding, respectively) for proANP and proBNP) are greatly stimulated in hypertrophied and failing hearts, suggesting that the autocrine and/or paracrine effects of NPs predominate and serve as an endogenous protective mechanisms against maladaptive cardiac events and dysfunction [249,275,281]. Functional alterations of the *Nppa* promoter are linked with cardiac hypertrophy in the progeny of rats generated by crossing the Wistar Kyoto (WKY) and Wistar Kyoto-derived hypertensive (WKYH) rats [282].

Npr1 gene-disrupted mice lacking NPRA develop cardiac hypertrophy and fibrosis independent of blood pressure [249,250,271]. Genetic disruption of *Npr1* in mice increases the cardiac mass and greatly enhances the incidence of cardiac hypertrophy and fibrosis [211,249,265,270,271,283-285]. Our studies have demonstrated that *Npr1* gene disruption in mice not only increases the expression of hypertrophic marker genes, proinflammatory cytokines, and matrix metalloproteinases (MMP-2, MMP-9), but also enhances the expression and activation of transcription factors, NF-kB, and activating protein-1 (AP-1), which seem to be associated with cardiac hypertrophy, fibrosis, and extracellular matrix remodeling [7,8,249,271].

The *Npr1* gene-disruption activates NF-kB and AP-1, leading to cardiac and renal disease conditions (Figure 5). *In-vitro* studies using cardiac fibroblasts have shown that the ANP/NPRA signaling system antagonizes Ang II-induced collagen systhesis by suppressing the activities of MMP-2 and MMP-9, as well as nuclear translocation of NF-kB [286]. Expression of sarcolemmal or endoplasmic reticulum Ca^{2+}-ATPase-2a (SERCA-2a) is progressively decreased in the hypertrophied hearts of *Npr1* null mutant mice [249,250].

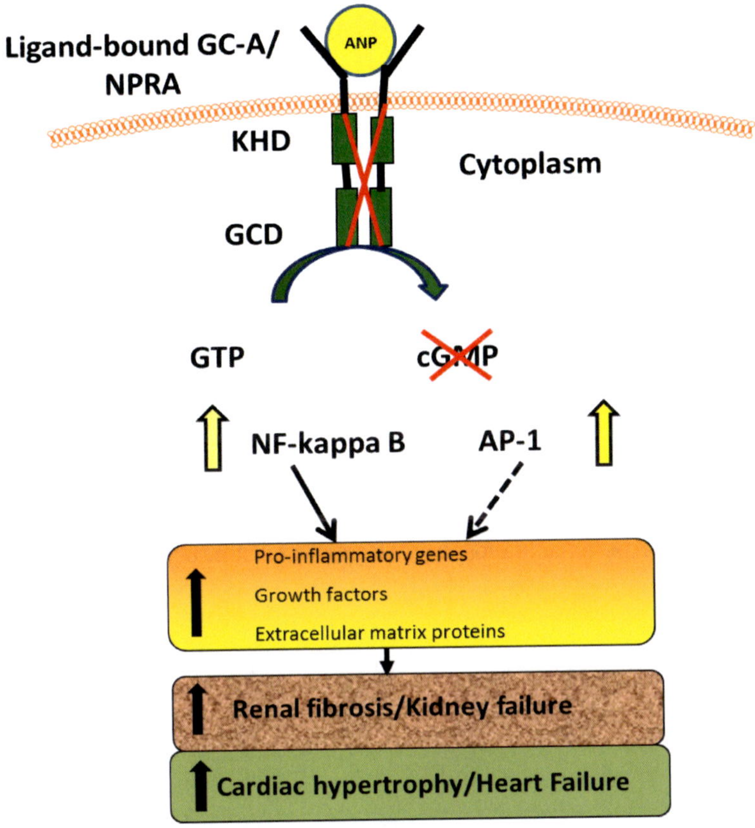

Figure 5. Schematic representation of the impact of NPRA deficiency in renal and cardiac remodeling and disease conditions. The genetic disruption of Npr1 gene (coding of NPRA) leads the activation of transcription factors, nuclear factor kappa B (NF-kB), and activated protein-1 (AP-1). The activation of NF-kB and AP-1 cascades stimulates the expression of proinflammatory genes, growth factors, and extracellular matrix proteins, which lead to renal fibrosis with kidney disease and cardiac hypertrophy with heart failure conditions. GC-A/NPRA, guanylyl cyclase/natriuretic peptide receptor-A; ANP, atrial natriuretic peptide; KHD, kinase-like homology domain; GCD, guanylyl cyclase domain; NF-kB, nuclear factor kappa B; AP-1, activating protein-1.

Expression of both angiotensin-converting enzyme (ACE) and Ang II receptor type A (AT1a) were found to be greatly increased in *Npr1* null mutant mice as compared with wild-type control mice [271]. Our studies and those of others have indicated that NPRA signaling antagonizes Ang II and AT1a receptor-mediated cardiac remodeling and provides an endogenous protective mechanism in failing hypertrophied hearts [265,271,287,288]. Smooth muscle and endothelial cell-specific arteries from *Npr1* gene-disrupted mice exhibited significant arterial hypertension [289]. Selective inactivation of the *Npr1* gene by homologous lox/Cre-mediated recombination exhibited mild cardiac hypertrophy, but ANP levels were significantly increased [290]. Circulating levels of ANP and BNP are also dramatically elevated in the initial stages of myocardial infarction [291-293]. In experimental myocardial infarction in *Npr1* gene-disrupted mice, both ANP and BNP levels were markedly elevated, leading to a higher incidence of heart failure and significantly greater mortality than occurred in wild-type mice [249,271,283]. The activation of ANP/cGMP signaling has been shown to have enhanced recovery of vascular blood flow [294].

10. Therapeutic Implications of Natriuretic Peptides

Intraperitoneal injection of recombinant ANP (carperitide) accelerated the recovery of blood flow with increasing capillary density in ischemic conditions [295,296]. Intraperitoneal injection of carperitide the recovery of accelerated blood flow induced by streptozotocin, doing so at increasing capillary density in hyperglycemic mice [296]. Similarly, recombinant BNP (Nesirtide) elicited natriuretic and vasodilatory action in heart failure patients [297]. Using cultured neonatal myocytes and fibroblasts, exogenous injection of ANP and its antagonist showed that ANP has anti-hypertrophic and anti-fibrotic functions [298]. Laser Doppler perfusion imaging experiments showed that the recovery of blood flow in ischemic limbs was significantly inhibited in *Npr1* gene-disrupted mice as compared with wild-type mice [299]. Furthermore, vascular regeneration in response to critical hind limb ischemia was severely impaired in *Npr1* gene-disrupted mice [300]. The physiological and pathophysiological functions of ANP, BNP, and NPRA signaling are associated with protective mechanisms in various organ systems, including the kidneys, heart, lungs, central nervous system, gonads, adrenal glands, and vasculature (Table 5).

11. Genetics of *Nppa*, *Nppb*, and *Npr1*

Recent genetic and clinical studies have indicated that *Nppa*, *Nppb*, and *Npr1* gene polymorphisms are associated with hypertension and cardiovascular disorders in humans [276,301-303]. It has been suggested that patients with monogenic forms of hypertension have rare genetic mutations [304,305]. Several pathways, including natriuretic peptides, RAAS, and the adrenergic system are considered to regulate blood pressure and cardiovascular events. Interestingly, the genetic determinants in these pathways contributing to interindividual differences in blood pressure regulation have been linked with only the natriuretic peptides and their receptors [301]. A single nucleotide polymorphism (SNP) at the *Nppa-Nppb* locus was associated with increased concentrations of plasma ANP and BNP and lower systolic and diastolic blood pressures. An association between *Nppa* promoter polymorphism (-C66UG) and left ventricular hypertrophy (LVH) exists in Italian hypertensive patients, indicating that individuals carrying a copy of the *Nppa* variant allele exhibit marked decreases in proANP levels associated with LVH [302]. Studies on the association between a microsatellite marker in *Npr1* promoter and LVH have indicated that ANP/NPRA signaling contributes to ventricular remodeling in essential hypertension in humans [302]. The relationship between high blood pressure and cardiovascular risk is continuous; thus, in the absence of ANP/NPRA signaling, even small increases in blood pressure confer excessive and detrimental effects.

Epidemiological studies have demonstrated that substantial heritability of blood pressure and cardiovascular risks indicates a role for genetic factors [306]. Mutation of a single allele in the human *Npr1* gene was found to decrease the expression of NPRA by nearly 70% [307]. Studies have suggested that a single allele mutation is associated with increased susceptibility to hypertension and left ventricular cardiac hypertrophy. In humans, a "4-minus" haplotype in the 3'-untranslated region of the *Npr1* gene promoter is associated with an increased level of

NT-proBNP [303]. The "4-minus" haplotype constitutes 4C repeats at nucleotide position 14,319 and a 4-bp deletion of AGAA at nucleotide position 14,649 of the *Npr1* gene.

Table 5. Physiological and pathophysiological effects associated with the activation of ANP/NPRA signaling in different target organs

Organs	Functional Parameters	Effect of ANP/NPRA
Adrenal Glands	Aldosterone	Decrease
Kidney	Natriuresis	Increase
	Diuresis	Increase
	Hypertrophy	Decrease
	Remodeling	Decrease
	Renin release	Decrease
Heart	Cardiomyocyte size	Decrease
	LVDS	Decrease
	LVDD	Decrease
	Fractional shortening	Increase
	Cardiac hypertrophy	Decrease
	Fibroblast growth	Decrease
Lung	Smooth muscle relaxation	Increase
Vasculature		
	Smooth muscle relaxation	Increase
	Endothelial permeability	Increase
	Intravascular volume	Decrease
Testis	Testosterone	Increase
Ovary	Estradiol	Increase
CNS	Thrust	Decrease
	Sympathetic activity	Decrease

ANP, natriuretic peptide; NPRA, natriuretic peptide receptor-A; LVDS, Lift ventricular dimension systolic; LVDD, left ventricular dimension diastolic; and CNS, central nervous system. Up arrow indicates increase and down arrow indicates decrease of functional parameters.

It has been speculated that the causal mechanism for high blood pressure may be *Npr1* mRNA instability, leading to decreased translational products of the receptor protein [308]. Consequently, this could elicit a feedback mechanism whereby diminished function of NPRA/cGMP signaling caused by the defect in the *Npr1* gene could provoke compensatory enhanced expression and release of ANP and/or BNP.

Taken together, a positive association exists among *Nppa*, *Nppb*, and *Npr1* gene polymorphisms, essential hypertension, and left ventricular mass index. It is predicted that human subjects carrying certain variants of *Nppa, Nppb,* and/or *Npr1* gene are particularly susceptible to hypertension, cardiac hypertrophy, and congestive heart failure. Furthermore, it has been reported that allelic variations in *Nppa*, *Nppb*, and the *Npr1* gene in humans are associated with a family history of hypertension and cardiovascular disorders, including myocardial infarction, left ventricular cardiac mass index, and septal wall thickening. More studies are needed to characterize the most functionally significant markers of *Nppa*, *Nppb,* and *Npr1* variants in a larger human population, since common variants associated with blood pressure regulation and cardiac events have not yet been fully established.

12. Current and Future Perspectives

In the past thirty-three years, a large body of literature has provided a new perspective on delineating cellular, biochemical, molecular, and genetic information relating to the regulation and function of natriuretic peptides and their receptors in hypertension and cardiovascular homeostasis. Currently, ANP and BNP are considered to be the markers of congestive heart failure. However, we still need to determine their therapeutic potential for the treatment of cardiovascular diseases such as hypertension, renal insufficiency, cardiac hypertrophy, congestive heart failure, and stroke. Recent genetics studies have used molecular approaches to delineate the genomic functions affected by decreasing and/or increasing numbers of *Nppa*, *Nppb,* and *Npr1* gene copies by gene-targeting techniques such as gene disruption (knockout) and/or gene duplication (gene dosage) in mice. Comparative analyses of the biochemical and physiological phenotypes of *Npr1* gene-disrupted and gene-duplicated mutant mice have high potential to answer fundamental questions about the importance of ANP/NPRA/cGMP signaling in disease states by genetically altering *Npr1* gene copy numbers and product levels in intact animals *in vivo* with identical genetic background. An intriguing finding was that a common genetic variant at the *Nppa* and *Nppb* locus is associated with circulating ANP and BNP concentrations, contributing to individual variation in blood pressure regulation.

Future studies will lead to a better understanding of the genetic basis of *Npr1* gene function in regulating high blood pressure, strokes, and cardiovascular events. It also remains to be seen whether continuing research on ANP, BNP, and GC-A/NPRA can facilitate the rapid clinical use of these agents to treat patients with high blood pressure and cardiovascular disorders. The results of future investigations of natriuretic peptides and their receptors should be of great value in resolving the genetic complexities related to hypertension and congestive heart failure. Overall, studies should be aimed at providing a unique perspective on delineating the genetic and molecular basis of *Npr1* gene expression, regulation, and function in both normal and disease states in experimental animals as well as humans. The resulting knowledge should yield new therapeutic targets for treating hypertension and preventing hypertension-related cardiovascular diseases, stroke, and other pathological conditions. Above all, future clinical studies are needed to characterize functionally significant markers of *Nppa, Nppb*, and *Npr1* gene variants in a large human populations. In the clinical environment, human recombinant ANP and BNP might be used for heart failure therapy, although careful future investigations are required. Progress in this field of research will greatly strengthen and advance our knowledge of genetic and molecular approaches to diverse physiological functions and pathophysiological events. During the next decade, researchers should expect to define the potential clinical implications of ongoing investigations on the pharmacogenomics of NPs and their receptors.

ACKNOWLEDGMENTS

I sincerely thank all the present and previous members of my laboratory for their numerous valuable contributions and Mrs. Kamala Pandey for her assistance in the preparation of this manuscript. The research work in the author's laboratory was supported by the National Institutes of Health grants (HL57531 and HL62147).

REFERENCES

[1] de Bold AJ, Borenstein HB, Veress AT, Sonnenberg H. A rapid and potent natriuretic response to intravenous injection of atrial myocardial extract in rats. *Life Sci.* 1981;28:89-94.
[2] de Bold AJ, Salerno TA. Natriuretic activity of extracts obtained from hearts of different species and from various rat tissues. *Canadian journal of physiology and pharmacology* 1983;61:127-30.
[3] Drewett JG, Garbers DL. The family of guanylyl cyclase receptors and their ligands. *Endocr. Rev.* 1994;15:135-62.
[4] Brenner BM, Ballermann BJ, Gunning ME, Zeidel ML. Diverse biological actions of atrial natriuretic peptide. *Physiol. Rev.* 1990;70:665-99.
[5] McGrath MF, de Bold ML, de Bold AJ. The endocrine function of the heart. *Trends Endocrinol. Metab.* 2005;16:469-77.
[6] Pandey KN. Biology of natriuretic peptides and their receptors. *Peptides* 2005;26:901-32.
[7] Pandey KN. Emerging Roles of Natriuretic Peptides and their Receptors in Pathophysiology of Hypertension and Cardiovascular Regulation. *J. Am. Soc. Hypertens.* 2008;2:210-26.
[8] Pandey KN. The functional genomics of guanylyl cyclase/natriuretic peptide receptor-A: perspectives and paradigms. *The FEBS journal* 2011;278:1792-807.
[9] Kishimoto I, Tokudome T, Nakao K, Kangawa K. Natriuretic peptide system: an overview of studies using genetically engineered animal models. *The FEBS journal* 2011;278:1830-41.
[10] de Bold AJ, Ma KK, Zhang Y, de Bold ML, Bensimon M, Khoshbaten A. The physiological and pathophysiological modulation of the endocrine function of the heart. *Canadian journal of physiology and pharmacology* 2001;79:705-14.
[11] de Bold AJ. Atrial natriuretic factor: a hormone produced by the heart. *Science* 1985;230:767-70.
[12] LaPointe MC. Molecular regulation of the brain natriuretic peptide gene. *Peptides* 2005;26:944-56.

[13] Schulz S. C-type natriuretic peptide and guanylyl cyclase B receptor. *Peptides* 2005;26:1024-34.
[14] Rosenzweig A, Seidman CE. Atrial natriuretic factor and related peptide hormones. *Annu. Rev. Biochem.* 1991;60:229-55.
[15] Koller KJ, de Sauvage FJ, Lowe DG, Goeddel DV. Conservation of the kinaselike regulatory domain is essential for activation of the natriuretic peptide receptor guanylyl cyclases. *Mol. Cell Biol.* 1992;12:2581-90.
[16] Levin ER, Gardner DG, Samson WK. Natriuretic peptides. *N. Engl. J. Med.* 1998;339:321-8.
[17] Garbers DL, Chrisman TD, Wiegn P, et al. Membrane guanylyl cyclase receptors: an update. Trends in endocrinology and metabolism: *TEM* 2006;17:251-8.
[18] Gardner DG. Natriuretic peptides: markers or modulators of cardiac hypertrophy? Trends in endocrinology and metabolism: *TEM* 2003;14:411-6.
[19] Richards AM. Natriuretic peptides: update on Peptide release, bioactivity, and clinical use. *Hypertension* 2007;50:25-30.
[20] Vollmer AM. The role of atrial natriuretic peptide in the immune system. *Peptides* 2005;26:1087-94.
[21] Vasan RS, Benjamin EJ, Larson MG, et al. Plasma natriuretic peptides for community screening for left ventricular hypertrophy and systolic dysfunction: the Framingham heart study. *JAMA* 2002;288:1252-9.
[22] Koller KJ, Goddel DV. Molecular biology of the natriuretic peptides and their receptors. *Circulation* 1992;86:1081-8.
[23] Garbers DL. Guanylyl cyclase receptors and their endocrine, paracrine, and autocrine ligands. Cell 1992;71:1-4.
[24] Fuller F, Porter JG, Arfsten AE, et al. Atrial natriuretic peptide clearance receptor. Complete sequence and functional expression of cDNA clones. *The Journal of biological chemistry* 1988;263:9395-401.
[25] Kuhn M. Cardiac and intestinal natriuretic peptides: insights from genetically modified mice. *Peptides* 2005;26:1078-85.
[26] Misono KS, Philo JS, Arakawa T, et al. Structure, signaling mechanism and regulation of the natriuretic peptide receptor guanylate cyclase. *The FEBS journal* 2011;278:1818-29.
[27] Lucas KA, Pitari GM, Kazerounian S, et al. Guanylyl cyclases and signaling by cyclic GMP. *Pharmacol. Rev.* 2000;52:375-414.
[28] Tremblay J, Desjardins R, Hum D, Gutkowska J, Hamet P. Biochemistry and physiology of the natriuretic peptide receptor guanylyl cyclases. *Mol. Cell Biochem.* 2002;230:31-47.
[29] Pandey KN, Singh S. Molecular cloning and expression of murine guanylate cyclase/atrial natriuretic factor receptor cDNA. *The Journal of biological chemistry* 1990;265:12342-8.
[30] Potter LR. Natriuretic peptide metabolism, clearance and degradation. *The FEBS journal* 2011;278:1808-17.
[31] Maki M, Takayanagi R, Misono K, Pandey KN, Tibbetts C, Inagami T. Structure of rat atrial natriuretic factor precursor deduced from cDNA sequence. *Nature* 1984;309:722-4.

[32] Misono KS, Fukumi H, Grammer RT, Inagami T. Rat atrial natriuretic factor: complete amino acid sequence and disulfide linkage essential for biological activity. *Biochemical and biophysical research communications* 1984;119:524-9.

[33] Sudoh T, Minamino N, Kangawa K, Matsuo H. Brain natriuretic peptide-32: N-terminal six amino acid extended form of brain natriuretic peptide identified in porcine brain. *Biochemical and biophysical research communications* 1988;155:726-32.

[34] Seilhamer JJ, Arfsten A, Miller JA, et al. Human and canine gene homologs of porcine brain natriuretic peptide. *Biochemical and biophysical research communications* 1989;165:650-8.

[35] Wu C, Wu F, Pan J, Morser J, Wu Q. Furin-mediated processing of Pro-C-type natriuretic peptide. *The Journal of biological chemistry 2003*;278:25847-52.

[36] Sudoh T, Minamino N, Kangawa K, Matsuo H. C-type natriuretic peptide (CNP): a new member of natriuretic peptide family identified in porcine brain. *Biochemical and biophysical research communications* 1990;168:863-70.

[37] Philips RA, Ardeljan M, Shimabukuro S, et al. Normalisation of left ventricular mass and associated changes in neurohormones and atrial natriuretic peptide after 1 year of sustained nifedipine therapy for severe hypertension. *J. Am. card* 1991 17 1595-602.

[38] Suga S, Nakao K, Hosoda K, et al. Phenotype-related alteration in expression of natriuretic peptide receptors in aortic smooth muscle cells .*Circ. Res.* 1992 71 34-9.

[39] Feller SM, Mägert HJ, Schulz-Knappe P, Forssmann WG. Urodilatin (hANF 95-126) - Characteristics of a new atrial natriuretic factor peptide. In: Struthers AD, ed. Atrial Natriuretic Factor. Oxford: UK Blackwell; 1990 209-26.

[40] Schulz-Knappe P, Forssmann K, Herbst F, Hock D, Pipkorn R, Forssmann WG. Isolation and structural analysis of "urodilatin", a new peptide of the cardiodilatin-(ANP)-family, extracted from human urine. Klin Wochenschr 1988;66:752-9.

[41] Saxenhofer H, Raselli A, Weidmann P, et al. Urodilatin, a natriuretic factor from kidneys, can modify renal and cardiovascular function in men. *The American journal of physiology* 1990;259:F832-8.

[42] Schweitz H, Vigne P, Moinier D, Frelin CH, Lazdunski M. A new member of the natriuretic peptide family is present in the venom of the green mamba (Dendroaspis angusticeps). *J. Biol. Chem.* 1992;267:13928-32.

[43] Lisy O, Jougasaki M, Heublein DM, et al. Renal actions of synthetic dendroaspis natriuretic peptide. *Kidney Int.* 1999;56:502-8.

[44] Nakao K, Sugawara A, Morii N, et al. The pharmacokinetics of alpha-human atrial natriuretic polypeptide in healthy subjects. *Eur. J. Clin. Pharmacol.* 1986;31:101-3.

[45] Ruskoaho H. Atrial natriuretic peptide: synthesis, release, and metabolism. *Pharmacol.Rev.* 1992;44:479-602.

[46] Yandle TG, Richards AM, Nicholls MG, Cuneo R, Espiner EA, Livesey JH. Metabolic clearance rate and plasma half life of alpha-human atrial natriuretic peptide in man. *Life sciences* 1986;38:1827-33.

[47] Holmes SJ, Espiner EA, Richards AM, Yandle TG, Frampton C. Renal, endocrine, and hemodynamic effects of human brain natriuretic peptide in normal man. *J. Clin. Endocrinol. Metab.* 1993;76:91-6.

[48] Mukoyama M, Nakao K, Hosoda K, et al. Brain natriuretic peptide as a novel cardiac hormone in humans. Evidence for an exquisite dual natriuretic peptide system, atrial natriuretic peptide and brain natriuretic peptide. *J. Clin. Invest.* 1991;87:1402-12.

[49] Charles CJ, Espiner EA, Richards AM, Nicholls MG, Yandle TG. Biological actions and pharmacokinetics of C-type natriuretic peptide in conscious sheep. *Am. J. Physiol.* 1995;268:R201-7.

[50] Hunt PJ, Richards AM, Espiner EA, Nicholls MG, Yandle TG. Bioactivity and metabolism of C-type natriuretic peptide in normal man. *J. Clin. Endocrinol. Metab.* 1994;78:1428-35.

[51] Lisy O, Huntley BK, McCormick DJ, Kurlansky PA, Burnett JC, Jr. Design, synthesis, and actions of a novel chimeric natriuretic peptide: CD-NP. *J. Am. Coll. Cardiol.* 2008;52:60-8.

[52] Lee CY, Lieu H, Burnett JC, Jr. Designer natriuretic peptides. *J. Investig. Med.* 2009;57:18-21.

[53] Dickey DM, Potter LR. Dendroaspis natriuretic peptide and the designer natriuretic peptide, CD-NP, are resistant to proteolytic inactivation. *Journal of molecular and cellular cardiology* 2011;51:67-71.

[54] Kojima M, Minamino N, Kangawa K, Matsuo H. Cloning and sequence analysis of cDNA encoding a precursor for rat brain natriuretic peptide. *Biochemical and biophysical research communications* 1989;159:1420-6.

[55] Yoshimura M, Yasue H, Morita E, et al. Hemodynamic, renal, and hormonal responses to brain natriuretic peptide infusion in patients with congestive heart failure. *Circulation* 1991;84:1581-8.

[56] Omland T, Aakvaag A, Bonarjee VV, et al. Plasma brain natriuretic peptide as an indicator of left ventricular systolic function and long-term survival after acute myocardial infarction. Comparison with plasma atrial natriuretic peptide and N-terminal proatrial natriuretic peptide. *Circulation* 1996;93:1963-9.

[57] Gardner DG, Deschepper CF, Ganong WF, et al. Extra-atrial expression of the gene for atrial natriuretic factor. *Proceedings of the National Academy of Sciences of the United States of America* 1986;83:6697-701.

[58] Cameron VA, Aitken GD, Ellmers LJ, Kennedy MA, Espiner EA. The sites of gene expression of atrial, brain, and C-type natriuretic peptides in mouse fetal development: temporal changes in embryos and placenta. *Endocrinology* 1996;137:817-24.

[59] Larsen TH, Saetersdal T. Regional appearance of atrial natriuretic peptide in the ventricles of infarcted rat hearts. *Virchows Arch. B Cell Pathol. Incl. Mol. Pathol.* 1993;64:309-14.

[60] Glembotski CC. Cellular and Molecular Biology of B-type natriuretic peptide In: W.K.Samson, Levin ER, eds. In: Contemp Endocrinol Natriuretic peptides in Health and Disease. Totowa, NJ: Humana Press; 1997:95-106.

[61] Hanford DS, Thuerauf DJ, Murray SF, Glembotski CC. Brain natriuretic peptide is induced by alpha 1-adrenergic agonists as a primary response gene in cultured rat cardiac myocytes. *The Journal of biological chemistry* 1994;269:26227-33.

[62] Hanford DS, Glembotski CC. Stabilization of the B-type natriuretic peptide mRNA in cardiac myocytes by alpha-adrenergic receptor activation: potential roles for protein kinase C and mitogen-activated protein kinase. *Mol. Endocrinol.* 1996;10:1719-27.

[63] Gardner RS, Chong KS, Murday AJ, Morton JJ, McDonagh TA. N-terminal brain natriuretic peptide is predictive of death after cardiac transplantation. *Heart* 2006;92:121-3.

[64] Igaki T, Itoh H, Suga S, et al. C-type natriuretic peptide in chronic renal failure and its action in humans. *Kidney Int. Suppl.* 1996;55:S144-7.
[65] Suga S, Itoh H, Komatsu Y, et al. Cytokine-induced C-type natriuretic peptide (CNP) secretion from vascular endothelial cells--evidence for CNP as a novel autocrine/paracrine regulator from endothelial cells. *Endocrinology* 1993;133:3038-41.
[66] Tamura N, Ogawa Y, Yasoda A, Itoh H, Saito Y, Nakao K. Two cardiac natriuretic peptide genes (atrial natriuretic peptide and brain natriuretic peptide) are organized in tandem in the mouse and human genomes. *J. Mol. Cell Cardiol.* 1996;28:1811-5.
[67] Chen HH, Burnett JC, Jr. C-type natriuretic peptide: the endothelial component of the natriuretic peptide system. *J. Cardiovasc. Pharmacol.* 1998;32 Suppl. 3:S22-8.
[68] Ogawa Y, Nakao K, Nakagawa O, et al. Human C-type natriuretic peptide. Characterization of the gene and peptide. *Hypertension* 1992;19:809-13.
[69] Suga S, Nakao K, Itoh H, et al. Endothelial production of C-type natriuretic peptide and its marked augmentation by transforming growth factor-beta. Possible existence of "vascular natriuretic peptide system". *J. Clin. Invest.* 1992;90:1145-9.
[70] Goetz KL. Renal natriuretic peptide (urodilatin?) and atriopeptin: evolving concepts. *The American journal of physiology* 1991;261:F921-32.
[71] Forssmann WG, Meyer M, Schulz-Knappe P. Urodilatin: from cardiac hormones to clinical trials. *Exp. Nephrol.* 1994;2:318-23.
[72] Meyer M, Richter R, Brunkhorst R, et al. Urodilatin is involved in sodium homeostasis and exerts sodium-state-dependent natriuretic and diuretic effects. *The American journal of physiology* 1996;271:F489-97.
[73] Sugimoto T, Kikkawa R, Haneda M, Shigeta Y. Atrial natriuretic peptide inhibits endothelin-1-induced activation of mitogen-activated protein kinase in cultured rat mesangial cells. *Biochemical and biophysical research communications* 1993;195:72-8.
[74] Pandey KN, Inagami T, Misono KS. Atrial natriuretic factor receptor on cultured Leydig tumor cells: ligand binding and photoaffinity labeling. *Biochemistry* 1986;25:8467-72.
[75] Pandey KN, Pavlou SN, Inagami T. Identification and characterization of three distinct atrial natriuretic factor receptors. Evidence for tissue-specific heterogeneity of receptor subtypes in vascular smooth muscle, kidney tubular epithelium, and Leydig tumor cells by ligand binding, photoaffinity labeling, and tryptic proteolysis. *The Journal of biological chemistry* 1988;263:13406-13.
[76] Leitman DC, Andresen JW, Catalano RM, Waldman SA, Tuan JJ, Murad F. Atrial natriuretic peptide binding, cross-linking, and stimulation of cyclic GMP accumulation and particulate guanylate cyclase activity in cultured cells. *The Journal of biological chemistry* 1988;263:3720-8.
[77] Kuno T, Andresen JW, Kamisaki Y, et al. Co-purification of an atrial natriuretic factor receptor and particulate guanylate cyclase from rat lung. T*he Journal of biological chemistry* 1986;261:5817-23.
[78] Maack T, Suzuki M, Almeida FA, et al. Physiological role of silent receptors of atrial natriuretic factor. *Science* 1987;238:675-8.
[79] Pandey KN. Vascular action natriuretic peptide receptor. Contemporary Endocrinology In: Sowers J, ed. Endocrinology of the Vasculature Totawa, NJ: *Humana Press* 1996:255-67.

[80] Pandey KN. Physiology of the natriuretic peptides Gonadal function. . . In: W. K. Samson aERL, ed. Contemporary Endocrinology: Natriuretic Peptides in Health and Disease. Totawa NJ: *Humana Press*, ; 1997:171-91.

[81] Chinkers M, Garbers DL, Chang MS, et al. A membrane form of guanylate cyclase is an atrial natriuretic peptide receptor. *Nature* 1989;338:78-83.

[82] Chang MS, Lowe DG, Lewis M, Hellmiss R, Chen E, Goeddel DV. Differential activation by atrial and brain natriuretic peptides of two different receptor guanylate cyclases. *Nature* 1989;341:68-72.

[83] Khurana ML, Pandey KN. Receptor-mediated stimulatory effect of atrial natriuretic factor, brain natriuretic peptide, and C-type natriuretic peptide on testosterone production in purified mouse Leydig cells: activation of cholesterol side-chain cleavage enzyme. *Endocrinology* 1993;133:2141-9.

[84] Koller KJ, Lowe DG, Bennett GL, et al. Selective activation of the B natriuretic peptide receptor by C-type natriuretic peptide (CNP). Science 1991;252:120-3.

[85] Misono KS. Atrial natriuretic factor binding to its receptor is dependent on chloride concentration: A possible feedback-control mechanism in renal salt regulation. *Circ. Res.* 2000;86:1135-9.

[86] Misono KS, Ogawa H, Qiu Y, Ogata CM. Structural studies of the natriuretic peptide receptor: a novel hormone-induced rotation mechanism for transmembrane signal transduction. *Peptides* 2005;26:957-68.

[87] Sharma RK. Membrane guanylate cyclase is a beautiful signal transduction machine: overview. *Mol. Cell Biochem.* 2010;334:3-36.

[88] Hunter T. Protein kinases and phosphatases: the yin and yang of protein phosphorylation and signaling. *Cell* 1995;80:225-36.

[89] Pandey KN, Kanungo J. Expression of extracellular ligand-binding domain of murine guanylate cyclase/atrial natriuretic factor receptor cDNA in Escherichia coli. Biochemical and biophysical research communications 1993;190:724-31.

[90] Pandey KN, Kumar R, Li M, Nguyen H. Functional domains and expression of truncated atrial natriuretic peptide receptor-A: the carboxyl-terminal regions direct the receptor internalization and sequestration in COS-7 cells. *Mol. Pharmacol.* 2000;57:259-67.

[91] Zhang G, Liu Y, Ruoho AE, Hurley JH. Structure of the adenylyl cyclase catalytic core. *Nature* 1997;386:247-53.

[92] Sunahara RK, Beuve A, Tesmer JJ, Sprang SR, Garbers DL, Gilman AG. Exchange of substrate and inhibitor specificities between adenylyl and guanylyl cyclases. *J. Biol. Chem.* 1998;273:16332-8.

[93] Tucker CL, Hurley JH, Miller TR, Hurley JB. Two amino acid substitutions convert a guanylyl cyclase, RetGC-1, into an adenylyl cyclase. *Proc. Natl. Acad. Sci. U S A* 1998;95:5993-7.

[94] Winger JA, Derbyshire ER, Lamers MH, Marletta MA, Kuriyan J. The crystal structure of the catalytic domain of a eukaryotic guanylate cyclase. BMC Struct Biol 2008;8:42.

[95] Rauch A, Leipelt M, Russwurm M, Steegborn C. Crystal structure of the guanylyl cyclase Cya2. *Proc. Natl. Acad. Sci. U S A* 2008;105:15720-5.

[96] Misono KS, Sivasubramanian N, Berkner K, Zhang X. Expression and purification of the extracellular ligand-binding domain of the atrial natriuretic peptide (ANP) receptor: monovalent binding with ANP induces 2:2 complexes. *Biochemistry* 1999;38:516-23.

[97] van den Akker F, Zhang X, Miyagi M, Huo X, Misono KS, Yee VC. Structure of the dimerized hormone-binding domain of a guanylyl-cyclase-coupled receptor. *Nature* 2000;406:101-4.

[98] Huo X, Abe T, Misono KS. Ligand binding-dependent limited proteolysis of the atrial natriuretic peptide receptor: juxtamembrane hinge structure essential for transmembrane signal transduction. *Biochemistry* 1999;38:16941-51.

[99] He X, Chow D, Martick MM, Garcia KC. Allosteric activation of a spring-loaded natriuretic peptide receptor dimer by hormone. *Science* 2001;293:1657-62.

[100] De Lean A, McNicoll N, Labrecque J. Natriuretic peptide receptor A activation stabilizes a membrane-distal dimer interface. *J. Biol. Chem.* 2003;278:11159-66.

[101] Rathinavelu A, Isom GE. Differential internalization and processing of atrial-natriuretic-factor B and C receptor in PC12 cells. *Biochem. J.* 1991;276 (Pt 2):493-7.

[102] Pandey KN. Stoichiometric analysis of internalization, recycling, and redistribution of photoaffinity-labeled guanylate cyclase/atrial natriuretic factor receptors in cultured murine Leydig tumor cells. *The Journal of biological chemistry* 1993;268:4382-90.

[103] Pandey KN. Dynamics of internalization and sequestration of guanylyl cyclase/atrial natriuretic peptide receptor-A. *Canadian journal of physiology and pharmacology* 2001;79:631-9.

[104] Pandey KN. Intracellular trafficking and metabolic turnover of ligand-bound guanylyl cyclase/atrial natriuretic peptide receptor-A into subcellular compartments. *Mol. Cell Biochem.* 2002;230:61-72.

[105] Pandey KN. Internalization and trafficking of guanylyl cyclase/natriuretic peptide receptor-A. Peptides 2005;26:985-1000.

[106] Pandey KN. Ligand-mediated endocytosis and intracellular sequestration of guanylyl cyclase/natriuretic peptide receptors: role of GDAY motif. *Mol. Cell Biochem.* 2010;334:81-98.

[107] Pandey KN, Nguyen HT, Garg R, Khurana ML, Fink J. Internalization and trafficking of guanylyl (guanylate) cyclase/natriuretic peptide receptor A is regulated by an acidic tyrosine-based cytoplasmic motif GDAY. *Biochem. J.* 2005;388:103-13.

[108] Pandey KN, Nguyen HT, Sharma GD, Shi SJ, Kriegel AM. Ligand-regulated internalization, trafficking, and down-regulation of guanylyl cyclase/atrial natriuretic peptide receptor-A in human embryonic kidney 293 cells. *The Journal of biological chemistry* 2002;277:4618-27.

[109] Brackmann M, Schuchmann S, Anand R, Braunewell KH. Neuronal Ca2+ sensor protein VILIP-1 affects cGMP signalling of guanylyl cyclase B by regulating clathrin-dependent receptor recycling in hippocampal neurons. *J. Cell Sci.* 2005;118:2495-505.

[110] Somanna NK, Pandey AC, Arise KK, Nguyen V, Pandey KN. Functional silencing of guanylyl cyclase/natriuretic peptide receptor-A by microRNA interference: analysis of receptor endocytosis. *International journal of biochemistry and molecular biology* 2013;4:41-53.

[111] Barak LS, Tiberi M, Freedman NJ, Kwatra MM, Lefkowitz RJ, Caron MG. A highly conserved tyrosine residue in G protein-coupled receptors is required for agonist-mediated beta 2-adrenergic receptor sequestration. *J. Biol. Chem.* 1994;269:2790-5.

[112] Pandey KN. Kinetic analysis of internalization, recycling and redistribution of atrial natriuretic factor-receptor complex in cultured vascular smooth-muscle cells. Ligand-dependent receptor down-regulation. *Biochem. J.* 1992;288:55-61.

[113] Nuglozeh E, Mbikay M, Stewart DJ, Legault L. Rat natriuretic peptide receptor genes are regulated by glucocorticoids in vitro. *Life sciences* 1997;61:2143-55.

[114] Fujio N, Gossard F, Bayard F, Tremblay J. Regulation of natriuretic peptide receptor A and B expression by transforming growth factor-beta 1 in cultured aortic smooth muscle cells. *Hypertension* 1994;23:908-13.

[115] Gutkowska J, Jankowski M, Sairam MR, et al. Hormonal regulation of natriuretic peptide system during induced ovarian follicular development in the rat. *Biol. Reprod.* 1999;61:162-70.

[116] Arise KK, Pandey KN. Inhibition and down-regulation of gene transcription and guanylyl cyclase activity of NPRA by angiotensin II involving protein kinase C. *Biochemical and biophysical research communications* 2006;349:131-5.

[117] Garg R, Pandey KN. Angiotensin II-mediated negative regulation of Npr1 promoter activity and gene transcription. *Hypertension* 2003;41:730-6.

[118] Kumar P, Garg R, Bolden G, Pandey KN. Interactive roles of Ets-1, Sp1, and acetylated histones in the retinoic acid-dependent activation of guanylyl cyclase/atrial natriuretic peptide receptor-A gene transcription. *The Journal of biological chemistry* 2010;285:37521-30.

[119] Kumar P, Pandey KN. Cooperative activation of *Npr1* gene transcription and expression by interaction of Ets-1 and p300. *Hypertension* 2009;54:172-8.

[120] Haneda M, Kikkawa R, Maeda S, et al. Dual mechanism of angiotensin II inhibits ANP-induced mesangial cGMP accumulation. *Kidney Int.* 1991;40:188-94.

[121] Bottari SP, King IN, Reichlin S, Dahlstroem I, Lydon N, de Gasparo M. The angiotensin AT2 receptor stimulates protein tyrosine phosphatase activity and mediates inhibition of particulate guanylate cyclase. *Biochemical and biophysical research* communications 1992;183:206-11.

[122] Liang F, Schaufele F, Gardner DG. Functional interaction of NF-Y and Sp1 is required for type a natriuretic peptide receptor gene transcription. *J. Biol. Chem.* 2001;276:1516-22.

[123] Cahill PA, Redmond EM, Keenan AK. Vascular atrial natriuretic factor receptor subtypes are not independently regulated by atrial peptides. *J. Biol. Chem.* 1990;265:21896-906.

[124] Anand-Srivastava MB. Down-regulation of atrial natriuretic peptide ANP-C receptor is associated with alteration in G-protein expression in A10 smooth muscle cells. Biochemistry 2000;39:6503-13.

[125] Neuser D, Bellermann P. Receptor binding, cGMP stimulation and receptor desensitization by atrial natriuretic peptides in cultured A10 vascular smooth muscle cells. *FEBS Lett.* 1986;209: 347-51.

[126] Hirata Y, Hirose S, Takada S, Takagi Y, Matsubara H. Down-regulation of atrial natriuretic peptide receptor and cyclic GMP response in cultured rat vascular smooth muscle cells. *Eur .J. Pharmacol.* 1987;135:439-42.

[127] Roubert P, Lonchampt MO, Chabrier PE, Plas P, Goulin J, Braquet P. Down-regulation of atrial natriuretic factor receptors and correlation with cGMP stimulation in rat cultured vascular smooth muscle cells. *Biochem. Biophys. Res. Commun.* 1987;148:61-7.

[128] Hughes RJ, Struthers RS, Fong AM, Insel PA. Regulation of the atrial natriuretic peptide receptor on a smooth muscle cell. *Am. J. Physiol.* 1987; 253 C809-C16.

[129] Hirata Y, Takata S, Tomita M, Takaichi S. Binding, internalization, and degradation of atrial natriuretic peptide in cultured vascular smooth muscle cells of rat. *Biochem. Biophys. Res. Commun.* 1985;132:976-84.

[130] Napier M, Arcuri K, Vandlen R. Binding and internalization of atrial natriuretic factor by high-affinity receptors in A10 smooth muscle cells. *Arch. Biochem. Biophys.* 1986;248:516-22.

[131] Murthy KK, Thibault G, Cantin M. Binding and intracellular degradation of atrial natriuretic factor by cultured vascular smooth muscle cells. Mol Cell Endocrinol 1989;67:195-206.

[132] Nussenzveig DR, Lewicki JA, Maack T. Cellular mechanisms of the clearance function of type-C receptors of atrial natriuretic factor. *J. Biol. Chem.* 1990;265:20952-8.

[133] Cohen D, Koh GY, Nikonova LN, Porter JG, Maack T. Molecular determinants of the clearance function of type-C receptor of natriuretic peptides. *J. Biol. Chem.* 1996;271:9863-9.

[134] Pandey KN, Kovacs WJ, Inagami T. The inhibition of progesterone secretion and the regulation of cyclic nucleotides by atrial natriuretic factor in gonadotropin responsive murine Leydig tumor cells. *Biochemical and biophysical research communications* 1985;133:800-6.

[135] Waldman SA, Rapoport RM, Murad F. Atrial natriuretic factor selectively activates particulate guanylate cyclase and elevates cyclic GMP in rat tissues. *The Journal of biological chemistry* 1984;259:14332-4.

[136] Burczynska B, Duda T, Sharma RK. ATP signaling site in the ARM domain of atrial natriuretic factor receptor guanylate cyclase. *Mol. Cell Biochem.* 2007;301:93-107.

[137] Potter LR, Garbers DL. Protein kinase C-dependent desensitization of the atrial natriuretic peptide receptor is mediated by dephosphorylation. *J. Biol. Chem.* 1994;269:14636-42.

[138] Kurose H, Inagami T, Ui M. Participation of adenosine 5'-triphosphate in the activation of membrane-bound guanylate cyclase by the atrial natriuretic factor. *FEBS letters* 1987;219:375-9.

[139] Sharma RK. Evolution of the membrane guanylate cyclase transduction system. *Molecular and cellular biochemistry* 2002;230:3-30.

[140] Chinkers M, Garbers DL. The protein kinase domain of the ANP receptor is required for signaling. *Science* 1989;245:1392-4.

[141] Foster DC, Garbers DL. Dual role for adenine nucleotides in the regulation of the atrial natriuretic peptide receptor, guanylyl cyclase-A. *J. Biol. Chem.* 1998;273:16311-8.

[142] Duda T, Goraczniak RM, Sharma RK. Core sequence of ATP regulatory module in receptor guanylate cyclases. *FEBS letters* 1993;315:143-8.

[143] Potter LR, Garbers DL. Dephosphorylation of the guanylyl cyclase-A receptor causes desensitization. *The Journal of biological chemistry* 1992;267:14531-4.

[144] Sibley DR, Benovic JL, Caron MG, Lefkowitz RJ. Regulation of transmembrane signaling by receptor phosphorylation. *Cell* 1987 48:913-33.

[145] Hugnir RI, Greengard P. Regulation of neurotransmitter receptor desensitization by protein phosphorylation. *Neuron* 1990;5:555-67.

[146] Goodman OBJ, Krupnick JG, Santini F, et al. β-arrestin acts as a clathrin adaptor in endocytosis of the β2-adrenergic receptor. *Nature* 1996 383:447-50.

[147] Zhang G, Liu Y, Arnold ER, Hurley JH. Structure of the adenylyl cyclase catalytic core. *Nature* 1997;386:247-53.

[148] Lefkowitz RJ, Pitcher J, Krueger K, Daaka Y. Mechanisms of β-adrenergic receptor desensitization and resensitization. *Adv. Pharmacol.* 1998;42:416-20.

[149] Sorkin A, von Zastrow M. Signal transduction and endocytosis close encounters of many kinds. *Nature Rev. Mol. Cell Biol.* 2002;3:600-14.

[150] Kurose H, Lefkowitz RJ. Differential desensitization and phosphorylation of three cloned and transfected alpha 2-adrenergic receptor subtypes. *J. Biol. Chem.* 1994;269:10093-9.

[151] Langlet C, Langer I, Vertongen P, Gaspard N, Vanderwinden JM, Robberecht P. Contribution of the carboxyl terminus of the VPAC1 receptor to agonist-induced receptor phosphorylation, internalization, and recycling. *J. Biol. Chem.* 2005;280:28034-43.

[152] Pandey KN. Stimulation of protein phosphorylation by atrial natriuretic factor in plasma membranes of bovine adrenal cortical cells. Biochemical and biophysical research communications 1989;163:988-94.

[153] Ballerman BJ, Marala RB, Sharma RK. Characterization and regulation by protein kinase C of renal glomerular atrial natriuretic peptide receptor-coupled guanylate cyclase. *Biochem. Biophys. Res. Commun.* 1988;157:755-61.

[154] Duda T, Sharma RK. Regulation of guanylate cyclase activity by atrial natriuretic factor and protein kinase C. *Molecular and cellular biochemistry* 1990;93:179-84.

[155] Larose L, Rondeau JJ, Ong H, De Lean A. Phosphorylation of atrial natriuretic factor R1 receptor by serine/threonine protein kinases: evidences for receptor regulation. *Molecular and cellular biochemistry* 1992;115:203-11.

[156] Potthast R, Potter LR. Phosphorylation-dependent regulation of the guanylyl cyclase-linked natriuretic peptide receptors. *Peptides* 2005;26:1001-8.

[157] Potter LR, Hunter T. Identification and characterization of the phosphorylation sites of the guanylyl cyclase-linked natriuretic peptide receptors. *A and B. Methods* 1999;19:506-20.

[158] Airhart N, Yang YF, Roberts CT, Jr., Silberbach M. Atrial natriuretic peptide induces natriuretic peptide receptor-cGMP-dependent protein kinase interaction. *J. Biol. Chem.* 2003;278:38693-8.

[159] Fenrick R, Bouchard N, McNicoll N, De Lean A. Glycosylation of asparagine 24 of the natriuretic peptide receptor-B is crucial for the formation of a competent ligand binding domain. *Mol. Cell Biochem.* 1997;173:25-32.

[160] Fenrick R, McNicoll N, De Lean A. Glycosylation is critical for natriuretic peptide receptor-B function. *Mol. Cell Biochem.* 1996;165:103-9.

[161] Lowe DG, Fendly BM. Human natriuretic peptide receptor-A guanylyl cyclase. Hormone cross-linking and antibody reactivity distinguish receptor glycoforms. *J. Biol. Chem.* 1992;267:21691-7.

[162] Heim JM, Singh S, Gerzer R. Effect of glycosylation on cloned ANF-sensitive guanylyl cyclase. *Life sciences* 1996;59:PL61-8.

[163] Koller KJ, Lipari MT, Goeddel DV. Proper glycosylation and phosphorylation of the type A natriuretic peptide receptor are required for hormone-stimulated guanylyl cyclase activity. *The Journal of biological chemistry* 1993;268:5997-6003.

[164] Pfeifer A, Klatt P, Massberg S, et al. Defective smooth muscle regulation in cGMP kinase I-deficient mice. *Embo J.* 1998;17:3045-51.

[165] Kaupp UB, Seifert R. Cyclic nucleotide-gated ion channels. *Physiol. Rev.* 2002;82:769-824.

[166] Maurice DH, Palmer D, Tilley DG, et al. Cyclic nucleotide phosphodiesterase activity, expression, and targeting in cells of the cardiovascular system. Mol Pharmacol 2003;64:533-46.

[167] Rybalkin SD, Yan C, Bornfeldt KE, Beavo JA. Cyclic GMP phosphodiesterases and regulation of smooth muscle function. *Circ. Res.* 2003;93:280-91.

[168] Schlossmann J, Feil R, Hofmann F. Insights into cGMP signalling derived from cGMP kinase knockout mice. *Front. Biosci.* 2005;10:1279-89.

[169] Reinhart K, Meisner M, Brunkhorst FM. Markers for sepsis diagnosis: what is useful? *Crit. Care Clin.* 2006;22:503-19.

[170] Camarago MJ, Kleinert HD, Atlas SA, Sealey JE, Laragh JH, Maack T. Ca2+ -dependent hemodynamic and natriuretic effects of atrial extract in isolated rat kidney. *Am. J. Physiol.* 1984;246:F447-F56.

[171] Freeman RH, Davis JO, Vari RC. Renal response to atrial natriuretic factor in conscious dogs with caval constriction *Am. J. Physiol.* 1985;248:R495-R500.

[172] Meyer M, Forsmann WG. Renal actions of atrial natriuretic peptide. In: Samson WK, Levin ER, eds. Contemporary Endocrinology: Natriuretic Peptides in Health and Disease. Totawa, NJ: *Humana Press*; 1997:147-70.

[173] Villrreal D, Freeman RH. Natriuretic peptides and salt sensitivity. . In: Samson WK, Levin ER, eds. Contemporary Endocrinology: Natriuretic peptides in health and disease Totowa, NJ: *Humana Press Inc*; 1997:239-58.

[174] Melo LG, Steinhelper ME, Pang SC, Tse Y, Ackermann U. ANP in regulation of arterial pressure and fluid-electrolyte balance: lessons from genetic mouse models. *Physiol. Genomics* 2000;3: 45-58.

[175] Shi SJ, Vellaichamy E, Chin SY, Smithies O, Navar LG, Pandey KN. Natriuretic peptide receptor A mediates renal sodium excretory responses to blood volume expansion. *American journal of physiology Renal physiology* 2003;285:F694-702.

[176] Zhao D, Pandey KN, Navar LG. ANP-mediated inhibition of distal nephron fractional sodium reabsorption in wild-type and mice overexpressing natriuretic peptide receptor. *Am. J. Physiol. Renal. Physiol.* 2010;298:F103-8.

[177] Burnett JC, Jr., Granger JP, Opgenorth TJ. Effects of synthetic atrial natriuretic factor on renal function and renin release. *Am. J. Physiol.* 1984;247:F863-6.

[178] Nonguchi H, Knepper MA, Mangiello VC. Effects of atrial natriuretic factor on cyclic guanosine monophosphate and cyclic adenosine monophosphate accumulation in microdissected nephron segments from rats. *J. Clin. Invest.* 1987;79:500-7.

[179] Kremer S, Troyer D, Kreisberg J, Skorecki K. Interaction of atrial natriuretic peptide-stimulated guanylyl cyclase and vasopressin-stimulated calcium signaling pathways in the glomerular mesangial cells. Arch. *Biochem. Biophys.* 1988;260:763-7.

[180] Light DB, Schwiebert EM, Karlson KH, Stanton BA. Atrial natriuretic peptide inhibits a cation channel in renal inner medullary collecting duct cells. *Science* 1989;243:383-5.

[181] Appel RG. Mechanism of atrial natriuretic factor-induced inhibition of rat mesangial cell mitogenesis. . *Am. J. Physiol.* 1990 259 E312-E8.

[182] Cermak R, Kleta R, Forssmann WG, Schlatter E. Natriuretic peptides increase a K+ conductance in rat mesangial cells. *Pflugers Arch.* 1996;431:571-7.

[183] Tripathi S, Pandey KN. Guanylyl cyclase/natriuretic peptide receptor-A signaling antagonizes the vascular endothelial growth factor-stimulated MAPKs and downstream effectors AP-1 and CREB in mouse mesangial cells. *Molecular and cellular* biochemistry 2012;368:47-59.

[184] Anand-Srivastava MB, Trachte GJ. Atrial natriuretic factor receptors and signal transduction mechanisms. *Pharmacol. Rev.* 1993;45:455-97.

[185] von Geldern TW, Budzik GP, Dillon TP, et al. Atrial natriuretic peptide antagonists Biological evaluation and structural correlations. *Mol. Pharmacol.* 1990;38:771-8.

[186] Sano T, Morishita Y, Matsuda Y, Yamada K. Pharmacological profile of HS-142-1, a novel nonpeptide atrial natriuretic peptide antagonist of microbial origin I. Selective inhibition of the actions of natriuretic peptides in anesthetized rats *J. Pharmacol. Exp. Thr.* 1992;260:825-31.

[187] Sonnenberg H, Honrath U, Chong CK, Wilson DR. Atrial natriuretic factor inhibits sodium transport in medullary collecting duct. *Am. J. Physiol.* 1986;250:F963-6.

[188] Prieto MC, Das S, Somanna NK, Harrison-Bernard LM, Navar LG, Pandey KN. Disruption of Npr1 gene differentially regulates the juxtaglomerular and distal tubular renin levels in null mutant mice. *International journal of physiology, pathophysiology and pharmacology* 2012;4:128-39.

[189] Bianchi C, Gutkowska J, Thibault G, Garcia R, Genest J, Cantin M. Distinct localization of atrial natriuretic factor and angiotensin II binding sites in the glomerulus. *Am. J. Physiol.* 1986;251:F594-602.

[190] Cogan MG, Huang CL, Liu FY, Madden D, Wong KR. Effect of atrial natriuretic factor on acid-base homeostasis. *J. Hypertens. Suppl.* 1986;4:S31-4.

[191] Fried TA, McCoy RN, Osgood RW, Stein JH. Effect of atriopeptin II on determinants of glomerular filtration rate in the in vitro perfused dog glomerulus. *Am. J. Physiol.* 1986;250:F1119-22.

[192] Meyer M, Wiebe K, Wahlers T, et al. Urodilatin (INN:ularitide) as a new drug for the therapy of acute renal failure following cardiac surgery. *Clinical and experimental pharmacology & physiology* 1997;24:374-6.

[193] von Geldern TW, Budzik GP, Dillon TP, et al. Atrial natriuretic peptide antagonists: biological evaluation and structural correlations. *Mol. Pharmacol.* 1990;38:771-8.

[194] Kumar R, Cartledge WA, Lincoln TM, Pandey KN. Expression of guanylyl cyclase-A/atrial natriuretic peptide receptor blocks the activation of protein kinase C in vascular smooth muscle cells. Role of cGMP and cGMP-dependent protein kinase. *Hypertension* 1997;29:414-21.

[195] Guo LJ, Alli AA, Eaton DC, Bao HF. ENaC is regulated by natriuretic peptide receptor-dependent cGMP signaling. *Am. J. Physiol. Renal. Physiol.* 2013;304:F930-7.

[196] Poschet JF, Timmins GS, Taylor-Cousar JL, et al. Pharmacological modulation of cGMP levels by phosphodiesterase 5 inhibitors as a therapeutic strategy for treatment of respiratory pathology in cystic fibrosis. *American journal of physiology Lung cellular and molecular physiology* 2007;293:L712-9.

[197] Obana K, Naruse N, Naruse K, et al. Synthetic rat atrial natriuretic factor inhibits in vitro and *in vivo* rennin secretion in rats. *Endocrinol.* 1985;117:1282-4.

[198] Shi SJ, Nguyen HT, Sharma GD, Navar LG, Pandey KN. Genetic disruption of atrial natriuretic peptide receptor-A alters renin and angiotensin II levels. *Am. J. Physiol.* 2001;281:F665-73.

[199] Paul RV, Ferguson T, Navar LG. ANF secretion and renal responses to volume expansion with equilibrated blood. *Am. J. Physiol.* 1988 255: F936-F43.

[200] Schwab TR, Edwards BS, Heublein DM, Burnett JCJ. Role of atrial natriuretic peptide in volume-expansion natriuresis. *Am. J. Physiol.* 1986;251:R310-R3.

[201] Ohyama Y, Miyamoto R, Morishita Y, et al. Stable expression of natriuretic peptide receptors effects of HS-142-1, a non-peptide *ANP antagonist Biochem. Biophys. Res. Commun.* 1992;189 336-42.

[202] Delporte C, Poloczek P, Tastenoy M, Winard J, Christopher J. Atrial natriuretic peptide binds to ANP-R 1 receptors in neuroblastoma cells or is degraded extracellularly at the Ser-Phe bond. *Eur. J. Pharmacol.* 1992 227 247-56.

[203] John SW, Krege JH, Oliver PM, et al. Genetic decreases in atrial natriuretic peptide and salt-sensitive hypertension. *Science* 1995;267:679-81.

[204] Lopez MJ, Wong SK, Kishimoto I, et al. Salt-resistant hypertension in mice lacking the guanylyl cyclase-A receptor for atrial natriuretic peptide. *Nature* 1995;378:65-8.

[205] Kishimoto I, Dubois SK, Garbers DL. The heart communicates with the kidney exclusively through the guanylyl cyclase-A receptor: acute handling of sodium and water in response to volume expansion. *Proceedings of the National Academy of Sciences of the United States of America* 1996;93:6215-9.

[206] Matsukawa N, Grzesik WJ, Takahashi N, et al. The natriuretic peptide clearance receptor locally modulates the physiological effects of the natriuretic peptide system. *Pro. Natl. Acad. Sci. USA* 1999;96:7403-8.

[207] Holtwick R, Baba HA, Ehler E, et al. Left but not right cardiac hypertrophy in atrial natriuretic peptide receptor-deficient mice is prevented by angiotensin type 1 receptor antagonist losartan. *J. Cardiovasc. Pharmacol.* 2002;40:725-34.

[208] Zhao D, Vellaichamy E, Somanna NK, Pandey KN. Guanylyl cyclase/natriuretic peptide receptor-A gene disruption causes increased adrenal angiotensin II and aldosterone levels. *Am. J. Physiol. Renal. Physiol.* 2007;293:F121-7.

[209] Das S, Periyasamy R, Pandey KN. Activation of IKK/NF-kappaB provokes renal inflammatory responses in guanylyl cyclase/natriuretic peptide receptor-A gene-knockout mice. *Physiological genomics* 2012;44:430-42.

[210] Das S, Au E, Krazit ST, Pandey KN. Targeted disruption of guanylyl cyclase-A/natriuretic peptide receptor-A gene provokes renal fibrosis and remodeling in null mutant mice: role of proinflammatory cytokines. *Endocrinology* 2010;151:5841-50.

[211] Oliver PM, Fox JE, Kim R, et al. Hypertension, cardiac hypertrophy, and sudden death in mice lacking natriuretic peptide receptor A. *Proceedings of the National Academy of Sciences of the United States of America* 1997;94:14730-5.

[212] Oliver PM, John SW, Purdy KE, et al. Natriuretic peptide receptor 1 expression influences blood pressures of mice in a dose-dependent manner. *Proc. Natl. Acad. Sci. U S A* 1998;95:2547-51.

[213] Pandey KN, Oliver PM, Maeda N, Smithies O. Hypertension associated with decreased testosterone levels in natriuretic peptide receptor-A gene-knockout and gene-duplicated mutant mouse models. *Endocrinology* 1999;140:5112-9.

[214] Kim HS, Lu G, John SWM, Maeda N, Smithies O. Molecular phenotyping for analyzing subtle genetic effects in mice application to an angiotensinogen gene titration. *Proc. Natl. Acad. Sci. USA* 2002;99:4602-7.

[215] Takahashi N, Smithies O. Gene targeting approaches to analyzing hypertension. *J. Ame. Soc. Nephrol.* 1999;10:1598-605.

[216] Antunes-Rodrigues J, Machado BH, Andrade HA, et al. Carotid aortic and renal baroreceptors mediate the atrial natriuretic peptide release induced by blood volume expansion. *Proc. Natl. Acad. Sci. USA* 1992;89:6829-31.

[217] Shi SJ, Vellaichamy E, Chin SY, Smithies O, Navar LG, Pandey KN. Natriuretic peptide receptor A mediates renal sodium excretory responses to blood volume expansion. *Am. J. Physiol. Renal. Physiol.* 2003;285:F694-F702.

[218] Atarashi K, Mulrow PJ, Franco-Saenz R, Snajdar R, Rapp J. Inhibition of aldosterone production by an atrial extract. *Science* 1984;224:992-4.

[219] Oda S, T. S, Morishita Y, Matsuda Y. Pharmacological profile of HS-142-1, a novel nonpeptide atrial natriuretic peptide (ANP) antagonist of microbial origin II. Restoration by HS-142-1 of ANP-induced inhibition of aldosterone production in adrenal glomerulosa cells. *J. Pharmacol. Exp. Thr.* 1992;263:241-5.

[220] Barrett PO, Isales CM. The role of cyclic nucleotides in atrial natriuretic peptide-mediated inhibition of aldosterone secretion. *Endocrinology* 1988;122:2790-5

[221] MacFarland RT, Zelus BD, Beavo JA. High concentrations of a cGMP-stimulated phosphodiesterase mediate ANP-induced decreases in cAMP and steroidogenesis in adrenal glomerulosa cells. *J. Biol. Chem.* 1991;266:136-42.

[222] Cao L, Wu J, Gardner DG. Atrial natriuretic peptide suppresses the transcription of its guanylyl cyclase-linked receptor. *The Journal of biological chemistry* 1995;270:24891-7.

[223] Sharma GD, Nguyen HT, Antonov AS, Gerrity RG, von Geldern T, Pandey KN. Expression of atrial natriuretic peptide receptor-A antagonizes the mitogen-activated protein kinases (Erk2 and P38MAPK) in cultured human vascular smooth muscle cells. *Mol. Cell Biochem.* 2002;233:165-73.

[224] Hassid A. Atriopeptin II decreases cytosolic free Ca^{2+} in cultured vascular smooth muscle cells. *Am. J. Physiol.* 1986;251:C681-C6.

[225] Lincoln TM, Komalavilas P, Cornwell TL. Pleiotropic regulation of vascular smooth muscle tone by cyclic GMP-dependent protein kinase. *Hypertension* 1994;243:383-5.

[226] Khurana ML, Pandey KN. Atrial natriuretic peptide inhibits the phosphoinositide hydrolysis in murine Leydig tumor cells. *Mol. Cell Biochem.* 1996;158:97-105.

[227] Lincoln TM, Dey N, Sellak H. Invited review: cGMP-dependent protein kinase signaling mechanisms in smooth muscle: from the regulation of tone to gene expression. *J. Appl. Physiol.* 2001;91:1421-30.

[228] Rashatwar SS, Cornwell TL, Lincoln TM. Effect of 8-bromo cGMP on Ca^{2+}-ATPase by cGMP dependent protein kinase. *Proc. Natl. Acad. Sci. USA* 1987 84:5685-9.

[229] Cornwell TL, Lincoln TM. Regulation of intracellular Ca^{2+} levels in cultured vascular smooth muscle cells. Reduction of Ca^{2+} by atriopeptin and 8-bromo-cyclic GMP is mediated by cyclic GMP-dependent protein kinase. *J. Biol. Chem.* 1989 264:1146-55.

[230] Appel RG. Growth-regulatory properties of atrial natriuretic factor. *Am. J. Physiol.* 1992;262:F911-F8.

[231] Isono M, Haneda M, Maeda S, Omatsu-Kanbe M, Kikkawa R. Atrial natriuretic peptide inhibits endothelin-1-induced activation of JNK in glomerular mesangial cells. *Kidney International* 1998;53:1133-42.
[232] Pandey KN, Nguyen HT, Li M, Boyle JW. Natriuretic peptide receptor-A negatively regulates mitogen-activated protein kinase and proliferation of mesangial cells: role of cGMP-dependent protein kinase. *Biochemical and biophysical research* communications 2000;271:374-9.
[233] Cao L, Gardner DG. Natriuretic peptides inhibit DNA synthesis in cardiac fibroblast. *Hypertension* 1995;25:227-34.
[234] Hu RM, Levin ER. Astrocyte growth is regulated by neuropeptides through Tis 8 and basic fibroblast growth factor. *J. Clin. Invest.* 1994;93:1820-7.
[235] Biesiada E, Razandi M, Levin ER. Egr-1 activates basic fibroblast growth factor transcription. . J Biol Chem 1996;271 18576-81.
[236] Vollmer AM, Schmidt KN, Schulz R. Natriuretic peptide receptors on rat thymocytes Inhibition of proliferation by atrial natriuretic peptide. *Endocrinology* 1996;137:1706-13.
[237] Itoh H, Pratt RE, Dzau V. Atrial natriuretic polypeptide inhibits hypertrophy of vascular smooth muscle cells. *J. Clin. Invest.* 1990;86:1690-7.
[238] Morishita R, Gibbons GH, Pratt RE, et al. Autocrine and paracrine effects of atrial natriuretic peptide gene transfer on vascular smooth muscle and endothelial cellular growth. *J. Clin. Invest.* 1994;94:824-9.
[239] Suhasini M, Li H, Lohmann SM, Boss GR, Pilz RB. Cyclic-GMP-dependent protein kinase inhibits the Ras/mitogen-activated protein kinase pathway. *Mole Cell. Biol.* 1998;18 6983-94.
[240] Itoh H, Pratt RE, Ohno M, Dzau VJ. Atrial natriuretic polypeptide as a novel antigrowth factor of endothelial cells. *Hypertension* 1992;19:758-61.
[241] Hutchinson HG, Trindade PT, Cunanan DB, Wu CF, Pratt RE. Mechanisms of natriuretic-peptide-induced growth inhibition of vascular smooth muscle cells. *Cardiovasc. Res.* 1997;35:158-67.
[242] Chrisman TD, Garbers DL. Reciprocal antagonism coordinates C-type natriuretic peptide and mitogen-signaling pathways in fibroblasts. *J. Biol. Chem.* 1999 274 4293-9.
[243] Redondo J, Bishop JE, Wilkins MR. Effect of atrial natriuretic peptide and cyclic GMP phosphodiesterase inhibition on collagen synthesis by adult cardiac fibroblasts. *Brit. J. Pharm.* 1998;124 1455-62.
[244] Calderone A, Thaik CM, Takahashi N, Chang DLF, Colucci WS. Nitric oxide, atrial natriuretic peptide, and cyclic GMP inhibit the growth-promoting effects of norepinephrine in cardiac myocytes and fibroblasts. *J. Clin. Invest.* 1998;101: 812-8.
[245] Silberbach M, Gorenc T, Hershberger RE, Stork PJS, Steyger PS, Roberts CTJ. Extracellular signal-regulated protein kinase activation is required for the anti-hypertrophic effect of atrial natriuretic factor in neonatal rat ventricular myocytes. *J. Biol. Chem.* 1999;274: 24858-64.
[246] Horio T, Nishikimi T, Yoshihara F, Matsuo H, Takishita S, Kangawa K. Inhibitory regulation of hypertrophy by endogenous atrial natriuretic peptide in cultured cardiac myocytes. *Hypertension* 2000;35:19-24.

[247] Dey NB, Boerth NJ, Murphy-Ullrich JE, Chang PL, Prince CW, Lincoln TM. Cyclic GMP-dependent protein kinase inhibits osteopontin and thrombospondin production in rat aortic smooth muscle cells. *Circ. Res.* 1998 82 139-46.

[248] Masciotra S, Picard S, Deschepper CF. Cosegregation analysis in genetic crosses suggests a protective role for atrial natriuretic factor against ventricular hypertrophy. *Circ. Res.* 1999;84:1453-8.

[249] Vellaichamy E, Khurana ML, Fink J, Pandey KN. Involvement of the NF-kappa B/matrix metalloproteinase pathway in cardiac fibrosis of mice lacking guanylyl cyclase/natriuretic peptide receptor A. *The Journal of biological chemistry* 2005;280:19230-42.

[250] Pandey KN, Vellaichamy E. Regulation of cardiac angiotensin-converting enzyme and angiotensin AT1 receptor gene expression in Npr1 gene-disrupted mice. *Clinical and experimental pharmacology & physiology* 2010;37:e70-7.

[251] Prins BA, Weber MJ, Hu R-M, Pedram A, Daniels M, Levin ER. Atrial natriuretic peptide inhibits mitogen-activated protein kinase through the clearance receptor Potential role in the inhibition of astrocyte proliferation. *J. Biol. Chem.* 1996 271 14156-62.

[252] Prins BA, Weber MJ, Hu RM, Pedram A, Daniels M, Levin ER. Atrial natriuretic peptide inhibits mitogen-activated protein kinase through the clearance receptor. Potential role in the inhibition of astrocyte proliferation. *J. Biol. Chem.* 1996;271:14156-62.

[253] Trindade P, Hutchinson HG, Pollman MJ, Gibbons GH, Pratt RE. Atrial natriuretic peptide (ANP) and C-type natriuretic peptide (CNP) induce apoptosis in vascular smooth muscle cells (VMSC) *Circulation* 1995 92 I-696.

[254] Wu CF, Bishopric NH, Pratt RE. Atrial natriuretic peptide induces apoptosis in neonatal rat cardiac myocytes. *J. Biol. Chem.* 1997;272:14860-6.

[255] Pollman MJ, Yamada T, Horiuchi M, Gibbons GH. Vasoactive substances regulate vascular smooth muscle cell apoptosis. *Cir. Res.* 1996;79 748-56.

[256] Krajewski S, Bodrug S, Krajewski M, et al. Immunohistochemical analysis of Mcl-1 protein in human tissues. *Am. J. Pathol.* 1995;146:1309-19.

[257] Kozopas KM, Yang T, Buchan HL, Zhou P, Craig RW. MCLI, a gene expressed in programmed myeloid cell differentiation, has sequence similarity to BCL2. *Proc. Natl. Acad. Sci. USA* 1993;90:3516-20.

[258] Kiefer MC, Brauer MJ, Powers VC, et al. Modulation of apoptosis by the widely distributed Bcl-2 homologue Bak. *Nature* 1995;374:736-9.

[259] Anand-Srivastava MB. Natriuretic peptide receptor-C signaling and regulation. *Peptides* 2005;26:1044-59.

[260] Rose RA, Giles WR. Natriuretic peptide C receptor signalling in the heart and vasculature. *J. Physiol.* 2008;586:353-66.

[261] Melo LG, Veress AT, Ackermann U, et al. Chronic regulation of arterial blood pressure in ANP transgenic and knockout mice: role of cardiovascular sympathetic tone. *Cardiovasc. Res.* 1999;43:437-44.

[262] Steinhelper ME, Cochrane KL, Field LJ. Hypotension in transgenic mice expressing atrial natriuretic factor fusion genes. *Hypertension* 1990;16:301-7.

[263] Lin KF, Chao J, Chao L. Human atrial natriuretic peptide gene delivery reduces blood pressure in hypertensive rats. *Hypertension* 1995;26:847-53.

[264] Schillinger KJ, Tsai SY, Taffet GE, et al. Regulatable atrial natriuretic peptide gene therapy for hypertension. *Proc. Natl. Acad. Sci. U S A* 2005;102:13789-94.

[265] Zhao D, Das S, Pandey KN. Interactive roles of NPR1 gene-dosage and salt diets on cardiac angiotensin II, aldosterone and pro-inflammatory cytokines levels in mutant mice. *Journal of hypertension* 2013;31:134-44.

[266] Schreier B, Borner S, Volker K, et al. The heart communicates with the endothelium through the guanylyl cyclase-A receptor: acute handling of intravascular volume in response to volume expansion. *Endocrinology* 2008;149:4193-9.

[267] Felker GM, Petersen JW, Mark DB. Natriuretic peptides in the diagnosis and management of heart failure. *Cmaj* 2006;175:611-7.

[268] See R, de Lemos JA. Current status of risk stratification methods in acute coronary syndromes. *Curr. Cardiol. Rep.* 2006;8:282-8.

[269] Jaffe AS, Babuin L, Apple FS. Biomarkers in acute cardiac disease: the present and the future. *J. Am. Coll. Cardiol.* 2006;48:1-11.

[270] Ellmers LJ, Scott NJ, Piuhola J, et al. Npr1-regulated gene pathways contributing to cardiac hypertrophy and fibrosis. *Journal of molecular endocrinology* 2007;38:245-57.

[271] Vellaichamy E, Zhao D, Somanna N, Pandey KN. Genetic disruption of guanylyl cyclase/natriuretic peptide receptor-A upregulates ACE and AT1 receptor gene expression and signaling: role in cardiac hypertrophy. *Physiological genomics* 2007;31:193-202.

[272] Chen HH, Burnett JC, Jr. The natriuretic peptides in heart failure: diagnostic and therapeutic potentials. *Proc. Assoc. Am. Physicians* 1999;111:406-16.

[273] Nakao K, Itoh H, Saito Y, Mukoyama M, Ogawa Y. The natriuretic peptide family. *Curr. Opin. Nephrol. Hypertens* 1996;5:4-11.

[274] Tsutamoto T, Kanamori T, Morigami N, Sugimoto Y, Yamaoka O, Kinoshita M. Possibility of downregulation of atrial natriuretic peptide receptor coupled to guanylate cyclase in peripheral vascular beds of patients with chronic severe heart failure. *Circulation* 1993;87:70-5.

[275] Knowles JW, Esposito G, Mao L, et al. Pressure-independent enhancement of cardiac hypertrophy in natriuretic peptide receptor A-deficient mice. *J. Clin. Invest.* 2001;107:975-84.

[276] Xue H, Wang S, Wang H, et al. Atrial natriuretic peptide gene promoter polymorphism is associated with left ventricular hypertrophy in hypertension. *Clin. Sci.* (Lond) 2008;114:131-7.

[277] Doust JA, Pietrzak E, Dobson A, Glasziou P. How well does B-type natriuretic peptide predict death and cardiac events in patients with heart failure: systematic review. *BWJ* 2005:330-625.

[278] Khan IA, Fink J, Nass C, Chen H, Christenson R, deFilippi CR. N-terminal pro-B-type natriuretic peptide and B-type natriuretic peptide for identifying coronary artery disease and left ventricular hypertrophy in ambulatory chronic kidney disease patients. *Am. J. Cardiol.* 2006;97:1530-4.

[279] Girsen A, Ala-Kopsala M, Makikallio K, Vuolteenaho O, Rasanen J. Cardiovascular hemodynamics and umbilical artery N-terminal peptide of proB-type natriuretic peptide in human fetuses with growth restriction. Ultrasound in obstetrics & gynecology : *the official journal of the International Society of Ultrasound in Obstetrics and Gynecology* 2007;29:296-303.

[280] Kocylowski RD, Dubiel M, Gudmundsson S, et al. Biochemical tissue-specific injury markers of the heart and brain in postpartum cord blood. *American journal of obstetrics and gynecology* 2009;200:273 e1- e25.

[281] Zahabi A, Picard S, Fortin N, Reudelhuber TL, Deschepper CF. Expression of constitutively active guanylate cyclase in cardiomyocytes inhibits the hypertrophic effects of isoproterenol and aortic constriction on mouse hearts. *J. Biol. Chem.* 2003;278:47694-9.

[282] Deschepper CF, Masciotra S, Zahabi A, Boutin-Ganache I, Picard S, Reudelhuber TL. Function alterations of the Nppa promoter are linked to cardiac ventricular hypertrophy in WKY/WKHA rat crosses. *Circ. Res.* 2001 88 223-8.

[283] Nakanishi M, Saito Y, Kishimoto I, et al. Role of natriuretic peptide receptor guanylyl cyclase-A in myocardial infarction evaluated using genetically engineered mice. *Hypertension* 2005;46:441-7.

[284] Ellmers LJ, Knowles JW, Kim HS, Smithies O, Maeda N, Cameron VA. Ventricular expression of natriuretic peptides in Npr1(-/-) mice with cardiac hypertrophy and fibrosis. *American journal of physiology* 2002;283:H707-14.

[285] Scott NJ, Ellmers LJ, Lainchbury JG, et al. Influence of natriuretic peptide receptor-1 on survival and cardiac hypertrophy during development. *Biochim. Biophys. Acta* 2009;1792:1175-84.

[286] Parthasarathy A, Gopi V, Umadevi S, et al. Suppression of atrial natriuretic peptide/natriuretic peptide receptor-A-mediated signaling upregulates angiotensin-II-induced collagen synthesis in adult cardiac fibroblasts. *Mol. Cell. Biochem.* 2013;378:217-28.

[287] Li Y, Kishimoto I, Saito Y, et al. Guanylyl cyclase-A inhibits angiotensin II type 1A receptor-mediated cardiac remodeling, an endogenous protective mechanism in the heart. *Circulation* 2002;106:1722-8.

[288] Kilic A, Bubikat A, Gassner B, Baba HA, Kuhn M. Local actions of atrial natriuretic peptide counteract angiotensin II stimulated cardiac remodeling. *Endocrinology* 2007;148:4162-9.

[289] Sabrane K, Kruse MN, Fabritz L, et al. Vascular endothelium is critically involved in the hypotensive and hypovolemic actions of atrial natriuretic peptide. *J. Clin. Invest.* 2005;115:1666-74.

[290] Holtwick R, van Eickels M, Skryabin BV, et al. Pressure-independent cardiac hypertrophy in mice with cardiomyocyte-restricted inactivation of the atrial natriuretic peptide receptor guanylyl cyclase-A. *J. Clin. Invest.* 2003;111:1399-407.

[291] Morita E, Yasue H, Yoshimura M, et al. Increased plasma levels of brain natriuretic peptide in patients with acute myocardial infarction. *Circulation* 1993;88:82-91.

[292] Phillips PA, Sasadeus J, Hodsman GP, Horowitz J, Saltups A, Johnston CI. Plasma atrial natriuretic peptide in patients with acute myocardial infarction: effects of streptokinase. *Br. Heart J.* 1989;61:139-43.

[293] Tomoda H. Atrial natriuretic peptide in acute myocardial infarction. *Am. J. Cardiol.* 1988;62:1122-3.

[294] Yamahara K, Itoh H, Chun TH, et al. Significance and therapeutic potential of the natriuretic peptides/cGMP/cGMP-dependent protein kinase pathway in vascular regeneration. *Proc. Natl. Acad. Sci. U S A* 2003;100:3404-9.

[295] Suwa M, Seino Y, Nomachi Y, Matsuki S, Funahashi K. Multicenter prospective investigation on efficacy and safety of carperitide for acute heart failure in the 'real world' of therapy. Circulation journal : *official journal of the Japanese Circulation Society* 2005;69:283-90.

[296] Park K, Itoh H, Yamahara K, et al. Therapeutic potential of atrial natriuretic peptide administration on peripheral arterial diseases. *Endocrinology* 2008;149:483-91.

[297] Colucci WS, Elkayam U, Horton DP, et al. Intravenous nesiritide, a natriuretic peptide, in the treatment of decompensated congestive heart failure. Nesiritide Study Group. *The New England journal of medicine* 2000;343:246-53.

[298] Nishikimi T, Kuwahara K, Nakagawa Y, Kangawa K, Minamino N, Nakao K. Complexity of molecular forms of B-type natriuretic peptide in heart failure. *Heart* 2013;99:677-9.

[299] Tokudome T, Kishimoto I, Yamahara K, et al. Impaired recovery of blood flow after hind-limb ischemia in mice lacking guanylyl cyclase-A, a receptor for atrial and brain natriuretic peptides. *Arterioscler. Thromb. Vasc. Biol.* 2009;29:1516-21.

[300] Kuhn M, Volker K, Schwarz K, et al. The natriuretic peptide/guanylyl cyclase--a system functions as a stress-responsive regulator of angiogenesis in mice. *J. Clin. Invest.* 2009;119:2019-30.

[301] Newton-Cheh C, Larson MG, Vasan RS, et al. Association of common variants in NPPA and NPPB with circulating natriuretic peptides and blood pressure. *Nat. Genet.* 2009;41:348-53.

[302] Rubattu S, Bigatti G, Evangelista A, et al. Association of atrial natriuretic peptide and type a natriuretic peptide receptor gene polymorphisms with left ventricular mass in human essential hypertension. *J. Am. Coll Cardiol.* 2006;48:499-505.

[303] Webber MA, Marder SR. Better pharmacotherapy for schizophrenia: what does the future hold? *Curr. Psychiatry Rep.* 2008;10:352-8.

[304] Ji W, Foo JN, O'Roak BJ, et al. Rare independent mutations in renal salt handling genes contribute to blood pressure variation. *Nat. Genet.* 2008;40:592-9.

[305] Lifton RP, Gharavi AG, Geller DS. Molecular mechanisms of human hypertension. *Cell* 2001;104:545-56.

[306] Levy D, DeStefano AL, Larson MG, et al. Evidence for a gene influencing blood pressure on chromosome 17. Genome scan linkage results for longitudinal blood pressure phenotypes in subjects from the framingham heart study. *Hypertension* 2000;36:477-83.

[307] Nakayama T, Soma M, Takahashi Y, Rehemudula D, Kanmatsuse K, Furuya K. Functional deletion mutation of the 5'-flanking region of type A human natriuretic peptide receptor gene and its association with essential hypertension and left ventricular hypertrophy in the Japanese. *Circ. Res.* 2000;86:841-5.

[308] Knowles JW, Erickson LM, Guy VK, Sigel CS, Wilder JC, Maeda N. Common variations in noncoding regions of the human natriuretic peptide receptor A gene have quantitative effects. *Hum. Genet.* 2003;112:62-70.

In: Natriuretic Peptides
Editor: Kailash N. Pandey

ISBN: 978-1-63321-080-6

Chapter 3

MANP: A NOVEL DESIGNER NATRIURETIC PEPTIDE GC-A ACTIVATOR FOR THE TREATMENT OF HYPERTENSION AND HEART FAILURE

Giuseppe Puccia*, Valentina Cannone, Paul M. McKie and John C Burnett Jr.
Cardiorenal Research Laboratory, Division of Cardiovascular Diseases, Mayo Clinic and Foundation, Rochester, MN, US

ABSTRACT

Human arterial hypertension is the most frequent risk factor in the world for cardiovascular disease associated with increased mortality and morbidity. Atrial Natriuretic Peptide (ANP) is a hormone produced by the cardiac atria and has a pivotal role in the regulation of the cardiovascular system and body fluid homeostasis. Its well-known vasomodulatory properties and renin-angiotensin-aldosterone system (RAAS) counteractions, conferred by binding to the Guanylyl Cyclase A Receptor (CG-A) producing the second messenger cyclic GMP) make it a robust therapeutic agent for heart failure and hypertension. Currently, ANP is administered for treatment of HF by continuous intravenous infusion due to short half-life, the main limiting factor for its chronic use. Recently studies have defined a metabolic role of ANP beyond its blood pressure-lowering, natriuretic- and aldosterone-inhibiting properties, such that the ANP gene polymorphism rs5068 is associated with a low cardiovascular risk profile with protection from metabolic syndrome. Further physiological studies established that the gut peptide Glucagon-like Protein 1 (GLP-1) can stimulate the release of ANP from the atria also resulting in cardiorenal protection. Based upon these cardiorenal, humoral and metabolic properties of ANP, we engineered Mutant ANP (MANP) which represents a new and novel ANP-based peptide therapeutic consisting of 40 amino acids where the

* Corresponding Author: Giuseppe Puccia MD; Mayo Clinic; 200 First Street SW; Rochester, MN 55905; Fax: 507-266-4710; Phone: 651-592-2207; E-mail: puccia.giuseppe@mayo.edu.

unique long C-terminus confers to the peptide a greater resistance to degradation than its native form. Indeed, in vitro studies demonstrated fewer sites for degradation and a greater cGMP production than ANP. *In vivo* studies showed sustained cardiorenal actions with enhanced vasorelaxation not only in normal canines but also in experimental acute hypertension and in hypertensive acute heart failure. In summary, for these reasons, MANP represents a new therapeutic agent for the treatment of chronic and resistant hypertension that is now entering human clinical trials.

INTRODUCTION

Human hypertension is a growing health burden worldwide that in its most severe form, uncontrolled or resistant hypertension, threatens the increase in cardiovascular morbidity and mortality. Indeed, resistant hypertension is one of the major causes of heart failure, myocardial infarction, stroke and chronic kidney disease [1]. While there remains a great emphasis upon the prevention of hypertension through restriction of dietary sodium and other lifestyle modifications, increasing progress has been made in understanding fundamental mechanisms and identifying important potential molecular targets which may lead to innovative therapeutics. Here we will focus on one of the body's most fundamental blood pressure regulating systems – the cardiac hormone atrial natriuretic peptide (ANP), its guanylyl cyclase receptor A (GC-A) and its second messenger cGMP [2]. Here, too, we will introduce a major advance in ANP therapeutics which goes beyond native ANP and soon will enter human trials in hypertension. Specifically, we will discuss MANP, a designer natriuretic peptide developed at the Mayo Clinic, representing a new and novel GC-A agonist strategy targeting resistant hypertension [3].

The field of natriuretic peptides and the concept of the heart as an endocrine organ began in 1956 with the report of atrial granules similar to those in endocrine glands [4]. In 1964, Jamieson and Palade eloquently wrote: *"Large populations of spherical, electron-opaque granules are characteristically found in muscle fibers of mammalian atria. They are absent in muscle fibers of the ventricles. They form and store a population of granules presumably secretory in nature"* [5]. It was the critical research in 1981 from de Bold et al. that reported an experiment demonstrating that the intravenous infusion of atrial extract from the rat induced a lowering of blood pressure, an increase in sodium, chloride, and potassium excretion, together with prompt diuresis [6]. In 1983, Flynn and co-workers purified a 28 amino acid peptide with diuretic and natriuretic activity from rat atrial muscle, calling it "Atrial Natriuretic Factor" (ANF) [7]. Further studies demonstrated that the atrium of several mammals, including humans, contain ANF, in specific granules [8] that is released by atrial distension suggesting a role in blood volume homeostasis [9].

Following these discoveries, progress on the identification and actions of ANF now called "Atrial Natriuretic Peptide" (ANP) has been remarkable. Work led to the 1988 demonstration of the existence of two additional, structurally related but genetically distant peptide hormones called B-Type natriuretic peptide (BNP) made of 32 amino acids [10] and in 1990 C-Type natriuretic peptide (CNP) with 22 amino acids [11].

As of today, recombinant human ANP, Carperitide (HANP®, Daiichisankyo Co., Ltd, Tokyo, Japan), and recombinant human BNP, Nesiritide (Natrecor®, Scios Inc., Sunnyvale, CA, USA), are used for treatment of heart failure in Japan and the United States, respectively, for their vasodilatory and natriuretic actions together with inhibition of the renin-angiotensin-aldosterone system, [12-14]. It should also be noted that patients with acute myocardial infarction who are given ANP have lower infarct sizes, and improved outcomes. Thus, ANP has emerged as a safe and efficacious therapeutic for heart failure and myocardial infarction [15].

Carperitide and Nesiritide being peptides are administered by intravenous infusion. Importantly, recent studies have demonstrated that subcutaneous delivery of BNP is a safe and efficacious chronic protein therapeutic strategy in patients with stable heart failure resulting in reduction in myocardial mass, improved myocardial function and enhanced symptoms [16]. In addition, preclinical studies report oral administration of natriuretic peptides results in systemic GC-A activation, promising even easier modes of peptide administration [17, 18].

Biology and Receptor Actions of Atrial Natriuretic Peptide

ANP is a cardiac hormone increasingly is recognized as playing a primary role in the regulation of blood pressure homeostasis [19]. ANP is synthesized as a 151 amino acid precursor hormone [20], cleaved to produce a 126 amino acid pro-hormone and then processed to its biologically active 28 amino acid peptide form by a transmembrane cardiac serine protease called Corin which may be altered in heart failure [21-23]. ANP is secreted from the heart into the circulation mostly in response to atrial wall stretch resulting from increased intravascular volume [24] as well as by hormonal stimulation by angiotensin, endothelin, arginine-vasopressin, and in pathological and physiological conditions like acute myocardial infarction, cardiac valve disease and physical exercise [25].

The ring-shaped structure, formed by an intramolecular disulfide bridge, is necessary for binding to its receptors that carry intrinsic guanylate cyclase catalytic activity in its CG-A receptor complex and which is highly expressed with documented effects in vascular smooth muscle, heart, kidney, brain, lung, adipose, and adrenal gland [26] with auto/paracrine actions in renal tubular cells [27], pericardium [28], endothelium [29] and in the heart [30]. Most recently, Kim and colleagues have established that the gut hormone GLP-1 may also release ANP from the cardiac atria via EPAC2 signaling, thus resulting in blood pressure reductions with natriuretic and lipolytic actions [31, 32].

The actions of ANP on the vasculature are well known, supporting its role as an anti-hypertensive therapeutic. ANP relaxes renal and skeletal muscle resistance arteries in a concentration-dependent manner [33] with a direct involvement of vascular smooth muscle cells [34]. In fact, it was demonstrated that local ANP infusion in humans induced a dose-dependent vasodilatation that was significantly more than that produced by equimolar doses of BNP [35]. Relevant to this observation are binding studies showing both ANP and BNP can bind CG-A with high affinity, but ANP is 4 to 70 times more potent than BNP. This result corresponds well with the 10-fold difference in half maximal effective concentration (EC50s) of these GC-A activating hormones [36, 37]. This observation may explain the potential mechanism underlying the different vascular responses to ANP and BNP. Furthermore, the ability of one or more natriuretic peptides to chronically relax vascular smooth muscle is explained by the elevated blood pressure in the GC-A gene-disrupted mice [19,38,39] and animal studies testing these presumptions that have shown that genetically reduced ANP concentration leads to salt-sensitive hypertension; whereas genetically increased ANP concentration leads to hypotension [40,41].

GC-A is a transmembrane protein containing, to our knowledge, at least four distinct regions composed of an extracellular ligand-binding domain containing a bound chloride essential for ANP attachment [42], a single transmembrane spanning region, an intracellular protein kinase-like homology domain and a guanylate cyclase catalytic domain [43]. Because ANP has no internal symmetry, binding of ANP to CG-A is asymmetric and leads to a head-to-head reciprocal monomer dimerization [44]. With the activation of GC-A, the increase of cGMP stimulates four known cGMP effector targets including cGMP-dependent protein kinases (PKG) [45], cGMP-dependent phosphodiesterases (PDEs) [46], cGMP-dependent ion channels [47] and Na^+-K^+-ATPase pump [48]. These different effector mechanisms operate synergistically to lower intracellular calcium with vasorelaxant effects, thus underscoring the remarkable redundant cellular mechanisms contributing to vasorelaxation and reduction in blood pressure.

Of special interest is the activation of the Na^+-K^+-ATPase pump that can be directly activated by cGMP and thus possibly by ANP [49,50]. By increasing the Na+ and K+ electrochemical gradient with a consequent cell hyperpolarization, the Na+/K+ pump stimulates the Na^+-$Ca2^+$-exchanger enhancing calcium efflux and reducing calcium influx [51,52]. It is also established that these two systems are co-localized, placed nearby sarcoplasmic reticulum and coupled functionally [53-55]. Alterations of these processes and structure or an alteration in the activity of these pumps can modulate myogenic tone and, hence, blood pressure by influencing intracellular calcium concentration [56] or clearance [57]. In fact, it was demonstrated that transgenic mice overexpressing Na^+-K^+-ATPase pump (specifically alpha 2 isoform) not only have lowering of blood pressure with a concomitant faster arterial relaxation, but in addition have increased levels of the Ca2+ clearance proteins Na^+-$Ca2^+$ exchanger, plasma membrane $Ca2^+$-ATPase (PMCA) and sarcoplasmic reticulum $Ca2^+$-ATPase (SERCA) [58]. One of the vasorelaxant effects within smooth muscle cells (SMC) also correlates with cGMP-mediated activation of a protein phosphokinase G (PKG) which catalyses the phosphorylation of the target substrate protein. Three isotypes of PKG have been identified, with the Type I (PKGI) the prominent isotype in the cardiovascular system and highly expressed in vascular smooth muscle cells as well as in vascular endothelial cells [59,60].

PKGI activates the plasmalemmal calcium ATPase pump with the effect of extrusion of calcium [61, 62], thus reducing intracellular calcium levels. PKG also promotes phosphorylation-mediated opening of the large conductance, calcium-sensitive potassium channel (BK_{Ca}), increases potassium efflux hyperpolarizing cell membrane and reduces calcium influx by closing nearby voltage-gated calcium channels [63,64]. BK_{Ca} channels are also indirectly regulated by PKG via PKG activation of an associated protein phosphatase 2A [65-67].

Inositol 1,4,5 Triphosphate (IP_3) is also an important second messenger that binds its receptor (IP_3R) complex with a regulatory protein (see below) triggering SMC contraction by inducing release of calcium from intracellular stores. Importantly, PKG binds both IP_3R [68] and the important regulatory protein of IP_3-inducing calcium release called "IP_3R-Associated PKGI Substrate" (IRAG) that is co-localized as a macromolecular complex in the perinuclear endoplasmic reticulum region [69,70]. This action interferes with the IP_3 and induces SMC relaxation by decreasing efflux of calcium from endoplasmic reticulum [71].

The ANP/GC-A/cGMP effector molecule PKG regulates intracellular calcium concentration and then muscular tone through interaction with phospholamban (PLB) [72, 73]. PLB is a membrane protein that regulates, through direct physical interaction, the Sarcoplasmic Endoplasmic Reticulum Ca^{2+}-ATPase pump (SERCA). The former is a transmembrane ion transporter that when active performs the vital task of re-sequestering cytoplasmic Ca^{2+} to the sarcoendoplasmic store, terminating Ca^{2+}-induced signaling such as muscle contraction. PLB dephosphorylated suppresses SERCA activity. As PLB is phosphorylated by PKG, it dissociates from the Ca^{2+}-ATPase resulting in activation of the pump determining acceleration of the calcium transport into sarcoplasmatic reticulum [74-77].

Smooth muscle contraction is also dependent on myosin light chain phosphorylation, which in turn, is regulated by the relative activities of myosin light chain kinase and myosin phosphatase. An increase in cytosolic free $Ca2^{+}$ concentration where upon $Ca2^{+}$ saturates the four $Ca2^{+}$-binding sites of calmodulin [78]. $Ca2^{+}$-calmodulin activates myosin light chain kinase catalyzing phosphorylation of the motor protein myosin at level of its 20 kDa regulatory light chain subunits (RLC20) [79]. The phosphorylation of RLC20 induces a rotational mobility opening, activating the structure and thereby stimulating contraction of the smooth muscle [80]. The myosin light chain phosphatase (MLCP) when in the inhibited state leads to muscle contraction [81], but when activated by PKGI it dephosphorylates RLC20, leading to SMC relaxation [82, 83].

Another important system regulating vascular tone modulated by ANP and cGMP is the phosphodiesterase enzymes. These enzymes degrade cyclic nucleotides (cGMP and cAMP); hence, modifying intracellular concentrations of cGMP and cAMP. The phosphodiesterases comprise a family of 11 isoenzymes (PDE1-PDE11) differently expressed in tissues, organized according to specificity for the cAMP or cGMP and activated or inhibited by them. In particular, of interest is the PDE3 expressed together with the PDE5 in vascular smooth muscle cells [84, 85, 86]. PDE3 is directly inhibited by cGMP, resulting in a significant increase in intracellular cGMP and cAMP natriuretic peptide mediated inducing a vasorelaxant effect [87].

While we think of ANP as involving only particulate guanylyl cyclase generated cGMP, some studies suggest that the cardiorenal actions of ANP may, in part, be due to activation of cardiovascular and renal nitric oxide synthase (NOS) [88]. Furthermore, the NPR-C natriuretic receptor mediates activation of NOS by ANP in the kidney [89]. Additionally, it was also shown that ANP through GC-A in the kidney also may result in NOS activation [90, 91]. Recent studies have also demonstrated that the renal effects of ANP may be mediated, at least in part, by interaction with the NO system in a genetic model of hypertension. In fact, ANP infusion augmented renal NOS activity without modifying its expression and assuming that ANP induces renal endothelial nitric oxide synthase (eNOS) stimulation, interacting with the NPR-C receptor and activating a pathway that involves Gi1–2 protein and Ca-calmodulin. On the other hand, ANP also interacts with GC-A in the kidney, enhancing cGMP formation via particulate guanylyl cyclase, which in turn activates PKG. The impaired NO system response to ANP in hypertensive animals, involving alterations in both pathways through GC-A or NPR-C, could participate in maintenance of the high levels of arterial blood pressure in this genetic model of hypertension supporting this concept [92].

Underscoring the pleotropic blood pressure-regulating action of ANP are studies demonstrating that ANP activity can modulate cardiovascular sympathetic system activity, suggesting that the chronic vasodilator effect of ANP may, at least in part, be dependent on attenuation of vascular sympathetic tone [93]. Of note, there is evidence that vagal afferent endings in the atria reflexively change autonomic activity to the heart and blood vessels, particularly those in the kidney [94]. Recent studies have elucidated ANP's central actions [95], demonstrating the importance of the paraventricular nucleus (PVN) spinal neurones in the hypothalamus as an integrating center in the control of blood volume and then blood pressure [96,97] on renal sympathetic blood outflow by releasing vasopressin on spinal sympathetic neurons [98].

In the kidney, which plays such a key role in blood pressure homeostasis, ANP increases glomerular filtration rate (GFR) due to an elevation of glomerular capillary hydrostatic pressure from selective preglomerular afferent arterial vasodilation involving the arcuate arterioles and from slight constriction of glomerular efferent arterioles [99-103], thus increasing hydrostatic pressure within the glomerular capillaries. In addition, ANP may enhance the filtration surface area and filtration coefficient by relaxing mesangial cells [104]. An additional mechanism by which ANP may enhance renal function is related to inhibition of renin release and angiotensin II-mediated synthesis and release of aldosterone [99,105,106]. Indeed, this action on the renin-angiotensin-aldosterone system (RAAS) is a hallmark of ANP/GC-A/cGMP signaling properties. Specifically, the absence of GC-A results in greater renin and angiotensin levels and increased renin mRNA expression than observed in wild type mice [107]. Further, adrenal renin content and renin mRNA levels as well as angiotensin and aldosterone concentrations are elevated in GC-A knockout mice as compared to wild-type mice, further underscoring the key role of ANP signaling in RAAS regulation.

In the kidney, several nephron sites in the renal tubule represent targets for natriuretic actions of ANP. In the proximal tubule, ANP specifically inhibits Na^+-ATPase activity by activation of the cGMP/PKG pathway through CG-A located in luminal and basolateral membranes, increasing renal sodium excretion and decreasing the activity of the Na+/H+ exchanger and of the sodium-phosphate cotransporter [108-110]. In Henle's loop, where there is the Na^+-K^+-2Cl cotransporter, ANP inhibits vasopressin-stimulated chloride transport and decreases chloride reabsorption [111,112]. A few studies have demonstrated decreased sodium reabsorption ANP-induced in the distal tubule [113,114] and is supported by the discovery of the expression of CG-A receptor mRNA in this specific site [115].

A key target for ANP in the kidney is at the level of the inner medullary collecting duct (IMCD). ANP suppresses sodium absorption by inhibiting an amiloride-sensitive cation channel in the apical membrane by a phosphorylation-independent mechanism. Moreover, increases in urinary sodium excretion may occur through PKG activation of the cation channel [116] and by stimulating basolateral furosemide-sensitive Na^+-K^+-$2Cl^-$ cotransporter. This effect induces negativization of the lumen voltage and suggests the involvement of electrogenic Cl^- secretion through the apical Cystic Fibrosis Transmembrane Conductance Regulator (CFTR) chloride channel, which can also be directly activated by cGMP [117,118].

Studies have also demonstrated that ANP inhibits Na^+-K^+-ATPase by PKG. It seems that this effect is specific for a particular tubular isoenzyme [119,120]. The pump can also be regulated by phosphatases. In fact, endogenous phosphatase inhibitors are expressed in renal tubules [121]. In addition, the activity of the CG-A receptor and the expression of its gene have been demonstrated to be stimulated by increased extracellular tonicity through a p38 Mitogen-activated protein kinase (MAPK) [122] and it is likely exerted by reduction of cellular cGMP, that, in turn, is caused by hypertonic-induced inhibition of the nitric oxide signaling system [123].

Thus, the integration of the cardiovascular system with renal function and the RAAS following activation by ANP clearly plays a key role in the control of blood pressure. Such widespread cardiorenal and adrenal actions of ANP/GC-A/cGMP signaling underscore the primacy of ANP in both blood pressure regulation and in hypertensive disease. Thus, ANP and novel analogues of ANP represent an opportunity for the treatment of human hypertension, especially the huge unmet need in the area of resistant hypertension.

MANP, a Novel ANP-Like Designer Natriuretic Peptide for Resistant Hypertension

The therapeutic natriuretic peptide human recombinant ANP (Carperitide) is currently approved in Japan for the treatment of acute heart failure. Due to the rapid degradation of ANP, it is currently administered by continuous intravenous infusion. The short half-life of ANP limits its utility in the outpatient setting and in disease states requiring chronic administration, such as hypertension. Thus, we and others have pursued the molecular design of chimeric natriuretic peptides that combine selected amino acid sequences from the native natriuretic peptides to produce novel designer peptides whose biologic actions go beyond those of the native peptides. This structural modification of the natriuretic peptides with the purpose of increasing potency represents a novel and rapidly developing therapeutic direction. The most advanced designer natriuretic peptide is CD-NP (Cenderitide), which integrates the mature CNP 22 amino acid structure with the C-terminus of Dendroaspis natriuretic peptide (DNP) [124]. CD-NP co-activates both GC-A and GC-B and is in clinical trials for acute heart failure.

MANP is the most recently engineered novel natriuretic peptide designed on the basis of information from an *ANP* gene frameshift mutation found in a single caucasian family of Northern European ancestry [125]. MANP was engineered based upon translation of the mutant gene resulting in a fusion protein consisting of the 28 amino acids of native ANP with a 12 amino acid extension to the C-terminus, resulting in a 40 amino acid peptide (Figure 1). The development of MANP is based on the understanding that the C-terminus of natriuretic peptides plays an important role in their degradation profile, particularly by neprilysin (NEP) that is highly expressed in the endothelium and the kidney. Specifically, recent kinetic studies demonstrate that the rank order for hydrolysis by NEP is ANP>CNP>BNP. Indeed, studies of DNP, which is isolated from the venom of the green mamba and possesses a C-terminus of 15 amino acids, show that it is highly resistant to degradation by NEP. Importantly, the C-terminus of MANP is 17 amino acids, and the therapeutic potential of MANP is based upon the hypothesis that the fusion protein would resist degradation by NEP, resulting in sustained biologic activity bases. Below, we discuss the most up-to-date in vitro and *in vivo* studies of MANP, as well as future directions.

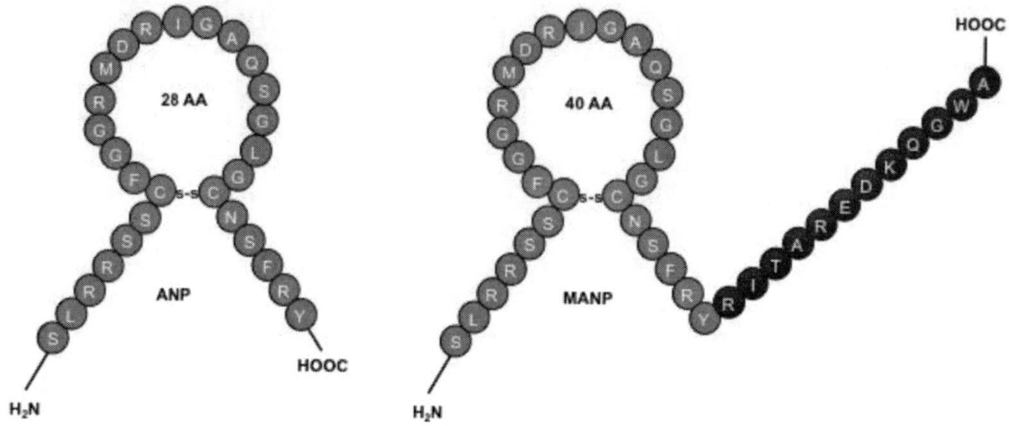

Figure 1. Amino acid structures for atrial natriuretic peptide and MANP.

In Vitro Studies of MANP

Early studies of MANP assessed the ability of MANP to activate GC-A receptors linked to cGMP generation in cultured human cardiac fibroblasts that are known to express GC-A [3]. In these studies, increasing MANP concentrations resulted in incremental increases in cGMP generation and suggested that the extended C-terminus of MANP does not render it unable to bind to GC-A. There was no demonstrated difference in GC-A cGMP generation between MANP and ANP.

Importantly, Dickey and colleagues assessed the binding of MANP to the GC-A receptor, demonstrated equivalent binding of native ANP and MANP to GC-A, including to the NPR-C (Figure 2) [126]. Importantly, in assessing guanylate cyclase activity and cGMP generation, these investigators reported that the highest concentration of MANP was more potent than ANP in activating cGMP in vitro.

They also assessed the degradation profile of MANP compared with native ANP to the key natriuretic peptide degrading enzyme NEP (Figure 2). As noted previously, studies have demonstrated that the C-terminus of the NPs is a key determinant of the degradation profile by NEP.

Indeed, MANP revealed greater resistance to NEP degradation than native ANP. They also noted there was no significant difference between MANP and native ANP in binding to the natriuretic peptide clearance receptor NPR-C. Further, Ralat et al., which employed mass spectrometry, demonstrated that MANP possessed fewer sites than any other known natriuretic peptides for degradation by a second endopeptidase, insulin-degrading enzyme (IDE) [127].

Thus, these studies suggest that the extended C-terminus of MANP, when compared with native ANP, provides greater resistance to NEP and IDE degradation and does not impair binding to GC-A and cGMP generation. Such studies underscore the attractive properties of MANP as a favored natriuretic peptide for therapeutic use in hypertension and its potential for use as a chronic treatment.

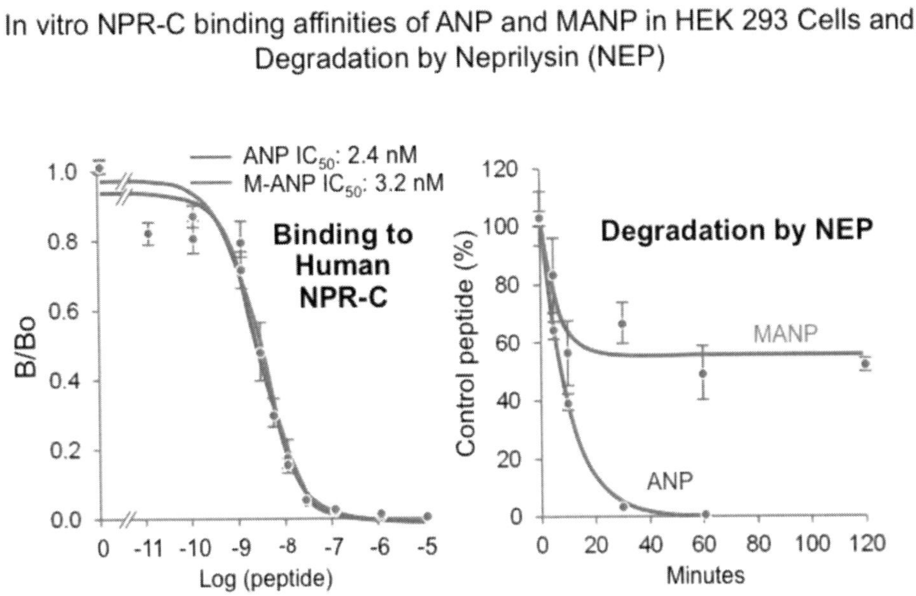

Figure 2. ANP and MANP receptor binding in HEK293 cells at increasing molar concentrations to the natriuretic peptide C-receptor (NPR-C). Studies assessing the degradation of ANP and MANP in vitro by neprilysin.

In Vivo Studies of MANP

Based on in vitro studies of MANP which demonstrated high resistance to NEP and IDE degradation, studies were undertaken to assess whether this degradation profile resulted in enhanced *in vivo* actions. This also was based on the intriguing observation of Dickey et al., that the highest molar concentrations of MANP compared to ANP resulted in significantly greater cGMP activation [126].

The *in vivo* hemodynamic and neurohumoral actions of MANP were first compared with native ANP in normal canines [3]. Equimolar concentrations of MANP and ANP were administered by continuous intravenous infusion and the responses of each peptide were determined. We observed that MANP had greater and more sustained hemodynamic properties than native ANP (Figure 3). Importantly, following MANP infusion, there was a sustained reduction in blood pressure, whereas native ANP resulted in only a transient reduction in blood pressure. In addition to these important hemodynamic actions and despite a reduction in blood pressure, there was significant inhibition of the RAAS, with marked inhibition of angiotensin and aldosterone activation following MANP infusion, compared with native ANP infusion.

Importantly, there was greater enhancement of renal function with MANP, despite a marked reduction in blood pressure, including greater increase in renal blood flow and GFR, compared with native ANP. MANP also possessed greater natriuretic and diuretic actions that were localized to the distal nephron where there was significant reduction in sodium reabsorption. Indeed, MANP resulted in greater urinary ANP immunoreactivity excretion compared with native ANP. This finding suggests that after it is filtered into the nephron, MANP is resistant to renal degradation by NEP and IDE. This resistance to degradation in the nephron where there are high NEP and IDE concentrations likely represents a key mechanism for the greater renal and systemic biologic actions of MANP. An alternative explanation is that the C-terminus of ANP enhances ANP interactions with the GC-A receptor, resulting in greater cGMP activation, enhanced vasorelaxation, and augmented renal responses.

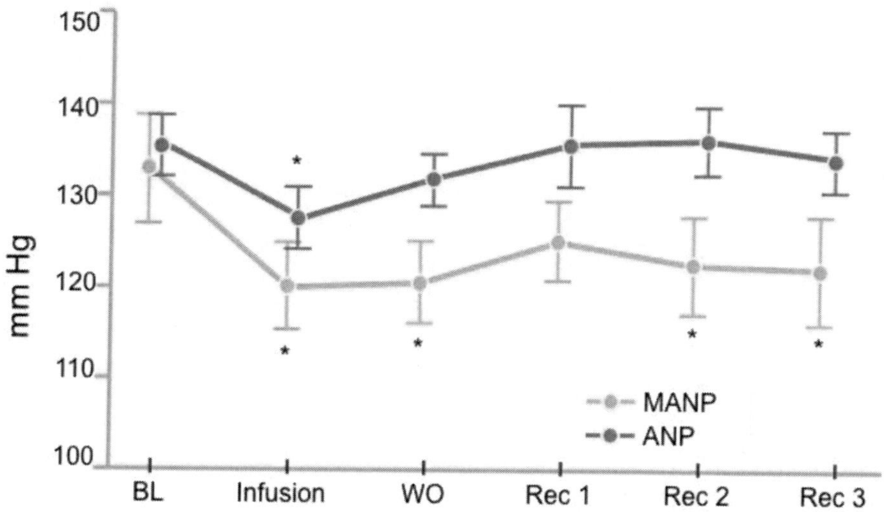

Figure 3. Mean arterial pressure during and following ANP or MANP infusion in normal canines infused over 45 minutes then discontinued with a 120 minute period of observation. *$P < 0.05$ vs baseline.

In Vivo Studies of MANP in Experimental Acute Hypertension

While MANP possesses potent cardiorenal actions under physiological conditions, it was not known if such properties would be observed in the setting of hypertension. Therefore, the cardiorenal and neurohumoral actions of MANP were also investigated in a canine model of acute hypertension induced by infusion of angiotensin II [128]. Here MANP was compared with vehicle and recombinant human BNP. A single intravenous infusion of MANP lowered blood pressure despite concomitant infusion of angiotensin II. The significant reduction in blood pressure was associated with a significant decrease in systemic vascular resistance. Despite a reduction in systemic blood pressure, MANP significantly increased renal blood flow and GFR. In addition, there was a marked increase in diuresis and natriuresis after MANP infusion. Despite the continuous infusion of angiotensin II and an increase in plasma angiotensin levels, aldosterone levels were suppressed after MANP infusion.

For the first time, the actions of MANP in acute hypertension were also compared with those of recombinant human BNP (Nesiritide). Importantly, in normal humans, equimolar BNP is two to three times more potent than ANP, a difference that may be secondary to the much greater metabolic clearance of ANP compared with BNP. Further, in experimental heart failure, BNP has more potent cardiorenal enhancing actions than ANP at equimolar doses [129]. Although BNP is known to be log-fold more potent than ANP with a clinical dose 1/10 that of native ANP, the biologic actions of MANP are similar to those of BNP. Indeed and most importantly, there was greater natriuresis with MANP than with BNP. These findings suggest that the extended C-terminus of MANP renders ANP more biologically active and, with regard to natriuresis, more potent than BNP.

In Vivo Studies of MANP in Experimental Hypertensive Acute Heart Failure

Uncontrolled hypertension or a sudden, severe elevation of blood pressure may result in acute heart failure. Here, too, there is an unmet need for new and effective as well as safe novel therapeutic agents. Therefore, we developed a new model of acute hypertensive heart failure and defined for the first time the actions of intravenously administered MANP compared to placebo and nitroglycerin (NTG) [130]. The NTG comparison was motivated by the widespread use of NTG in hypertensive heart failure and also because NTG represents a nitrovasodilator that functions through cGMP like MANP but via soluble guanylyl cyclase. Indeed, with the reported compartmentalization of cGMP in the cell following stimulation of particulate guanylyl cyclase by natriuretic peptides as compared to soluble guanylyl cyclase activation by nitrovasodilators, we hypothesized that MANP would go beyond NTG in enhancing and improving cardiorenal and neurohumoral function [131,132].

The model was produced by ten days of rapid ventricular pacing to produce a model of dilated cardiomyopathy followed by an acute infusion of angiotensin II to produce a rise in blood pressure. While both NTG and MANP reduced blood pressure and pulmonary capillary wedge pressure (Figure 4), the actions of MANP were sustained while the responses to NTG were transient. MANP had superior actions on the kidney compared to NTG with greater renal vasodilatation, increases in GFR and greater rises in diuresis and natriuresis (Figure 5). Importantly, only MANP suppressed aldosterone during the infusion (Figure 6).

These findings, taken together, point to MANP possessing attractive cardiorenal protective actions in the setting of experimental hypertension and hypertensive heart failure. These preclinical studies lay the foundation for further studies in humans to define efficacy and safety.

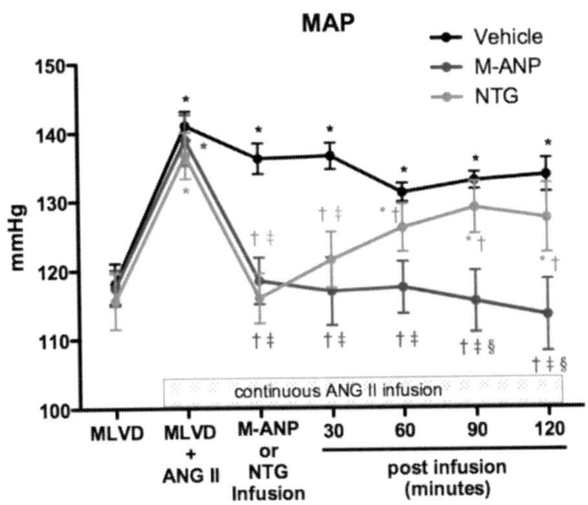

Figure 4. (Continued).

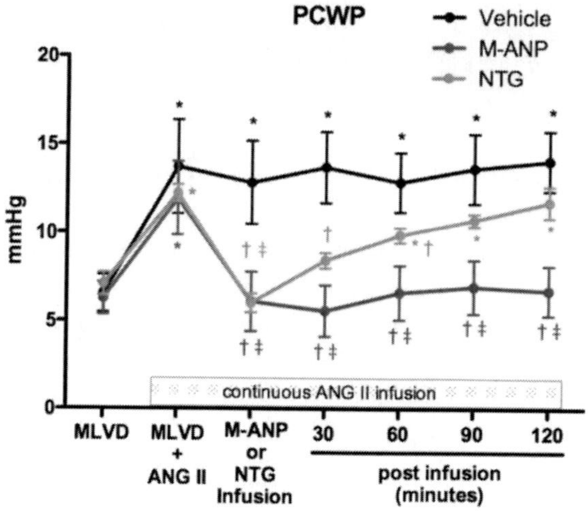

Figure 4. Mean arterial pressure and mean pulmonary capillary wedge pressure during infusion of MANP, NTG or placebo in mild left ventricular function and in mild left ventricular dysfunction with angiotensin induced hypertension. *P <0.05 vs MLVD and +P<0.05 vs MLVD +HTN.

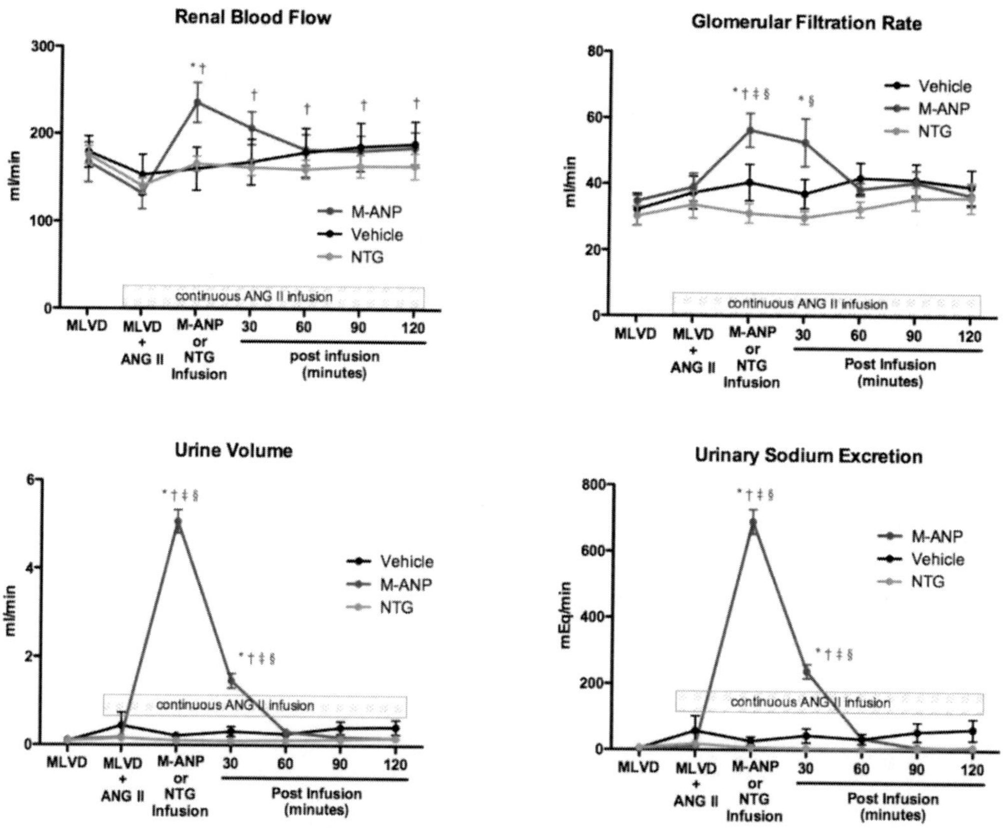

Figure 5. Renal blood flow, glomerular filtration rate, urine volume, and urinary sodium excretion during infusion of MANP, NTG or placebo in mild left ventricular function and in mild left ventricular dysfunction with angiotensin induced hypertension. *P<0.05 vs MLVD and +P <0.05 vs MLVD + HTN.

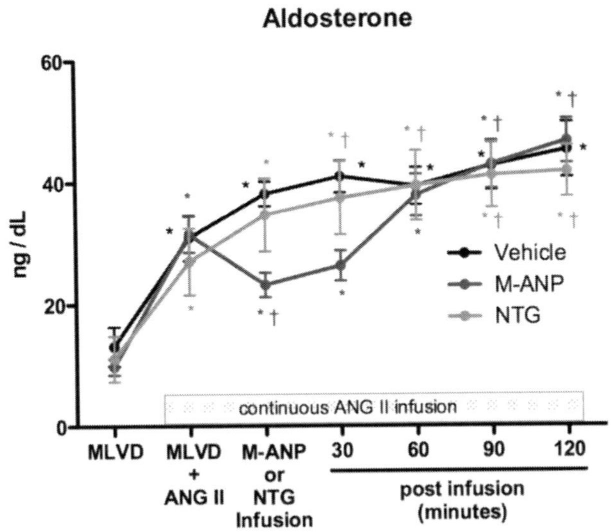

Figure 6. Aldosterone during infusion of MANP, NTG or placebo in mild left ventricular function and in mild left ventricular dysfunction with angiotensin induced hypertension. *P<0.05 vs MLVD and +P<0.05 vs MLVD + HTN.

Going Beyond Hypertension: Metabolic Syndrome As a Therapeutic Target for MANP

The natriuretic peptide precursor A gene (*NPPA*) localized to chromosome 1 encodes the preprohormone from which ANP and the biologically inactive N-terminal-proANP (NT-proANP) are derived in equimolar amount after cleavage [133]. The genetic variant rs5068 is located in the 3' untranslated region of *NPPA*. And several studies have investigated the clinical phenotype associated with this single nucleotide polymorphism [134]. In a recent genome-wide association study, the G minor allele of rs5068 is associated with higher circulating levels of ANP, lower systolic and diastolic blood pressures and odds of hypertension in community-based cohorts from Framingham, MA, Sweden and Finland [134].

In an analysis conducted on a randomly selected cohort (n=1608) of white American residents of Olmsted County, MN, Cannone et al. showed that genotype frequencies of rs5068 were AA: 89.9%, AG: 9.7%, and GG: 0.4% and confirmed the association between the G allele of rs5068 and higher plasma levels of NT-proANP, which is an estimator of ANP secretion, (median values, AA genotype: 2584 versus AG+GG genotypes: 2188 pg/mL, p= 0.006) and lower systolic blood pressure (delta = -4.28 mmHg, p=0.051) in the age, sex and body mass index (BMI) adjusted analysis [135]. Interestingly, this study also showed that rs5068 minor allele is associated with lower body mass index (BMI) (delta= -1.22 kg/m^2, p=0.006) and waist circumference (delta= -2.45 cm, p=0.021), as well as reduced rate of obesity (odds ratio=0.54, p=0.002) and metabolic syndrome (odds ratio= 0.58, p=0.025). Carriers of the G allele also have higher levels of the "protective" high-density lipoprotein cholesterol (delta= 2.46 mg/dL, p=0.019).

Importantly, such associations were replicated in a random sample of the general population of Ventimiglia di Sicilia, Italy [136]. The carriers of the G allele have lower systolic blood pressure (estimate (SE): -5.2 (2.4) mmHg, p= 0.03) and BMI (estimate (SE): -1.7 (0.8) kg/m^2, p=0.04); prevalence of hypertension (odds ratio: 0.35; 95% CI 0.18 to 0.70; p= 0.003) and metabolic syndrome (odds ratio: 0.42; 95% CI: 0.20-0.89, p=0.02) is lower between the carriers of the minor allele when compared to the homozygotes for the major allele. Obesity tends to be less frequent in the AG and GG genotypes (odds ratio: 0.52; 95% CI 0.26 to 1.04; p= 0.07). The phenotype associated with rs5068 genotypes has also been investigated in a cohort of coronary artery disease patients, findings show carriers of the G allele have higher ANP and NT-proANP plasma levels while homozygotes for the G allele are less likely to have a history of hypertension.

In consideration of the natriuretic, vasodilating and renin-angiotensin-aldosterone inhibiting properties of ANP (ref), the association between the genetic variant rs5068 and lower blood pressure values and prevalence of hypertension is probably mediated by the increase in circulating levels of ANP. The lipolytic action of ANP provides a physiological explanation for the lower BMI and rate of obesity observed in the carriers of the G allele [137]. The lower prevalence of metabolic syndrome might be a consequence of both the cardiovascular and metabolic effects of this natriuretic peptide. An interesting study conducted by Arora et al. provides insights regarding the genetic mechanism that may lead to increased circulating levels of ANP in the carriers of the G allele. Indeed, the microRNA miR-425 might function as a regulator of ANP production that inhibits *NPPA* gene expression, whereas the G minor allele may confer resistance against miR-425, causing chronically higher circulating levels of ANP [138].

Based on these fascinating findings of a role for ANP/GC-A/cGMP signaling in metabolic tissues, studies are ongoing to define the role of MANP as a potential therapeutic in obesity and metabolic syndrome in models of metabolic disease.

Chronic Natriuretic Peptide Delivery: The Next Frontier

Leader et al. recently wrote that we are entering into an era of protein therapeutics [139]. This declaration is based upon the great selectivity and safety of peptides that target specific receptors and lack the nonspecific actions that characterize small molecules. MANP emerges as the most advanced protein therapeutic agent ever developed for hypertension. As such, human hypertension may be one of the most important molecular targets for protein therapeutics, based not only upon the beneficial renal, vascular, and humoral actions of MANP, but also upon the emerging evidence from large-scale human genetic studies that endogenous ANP from the heart may play a fundamental role in blood pressure regulation and in mechanisms of human hypertension extending now into metabolic disease.

As natriuretic peptide therapeutics move forward, it is important to understand the broad spectrum of delivery systems to make chronic protein therapeutics possible. Whether administered by simple subcutaneous injection, use of a pump for subcutaneous delivery, inhalation of peptide, oral administration, or even gene therapy, MANP may revolutionize our ability to target hypertension. Indeed, combining a genetic test to determine whether a hypertensive patient lacks a protective ANP genotype together with chronic MANP treatment may represent an example of the new field of theragnostics specifically applied to human hypertension.

Such an approach, indeed, has been realized in both preclinical and clinical research. Subcutaneous injection of the native natriuretic peptide BNP together with technologies for its delivery by oral pegylation strategies as well as gene delivery platforms have been reported [16, 17, 18, 140]. Thus, chronic natriuretic peptide therapeutics is both safe and efficacious, having demonstrated reductions in cardiac hypertrophy and improvement in myocardial function in both experimental hypertension and in human heart failure.

CONCLUSION

ANP is increasingly recognized as having a fundamental role in blood pressure, cardiorenal homeostasis and metabolic regulation, and these findings support the development of ANP-based therapeutics for hypertension such as MANP. Indeed, ANP (Carperitide) is approved in Japan as an intravenous agent for heart failure, and investigation has also established beneficial renoprotective actions after bypass surgery and prevention of contrast-induced nephropathy [141,142]. MANP is an innovative ANP-based peptide that is resistant to proteolytic degradation, with greater cGMP-activating properties and possessing greater ability to lower blood pressure, enhance renal function, and suppress the RAAS than native ANP. In animal models of hypertension, MANP reduces blood pressure via multiple mechanisms that include vasodilatation, natriuresis, and suppression of RAAS. As recently emphasized by De Bold, the significance of the natriuretic peptides as therapeutic agents is their diverse pharmacologic properties, which may potentially replace the need for complex and expensive polypharmacy [143]. From this perspective, ANP exerts beneficial cardiorenal effects without activating the RAAS, as often occurs with other agents used to treat hypertension or heart failure. A blood pressure-lowering and natriuretic therapeutic that also enhances renal function is highly attractive when compared with other conventional diuretic agents and vasodilators, which often reduce renal perfusion and activate the RAAS. The importance of therapies that enhance renal function in hypertension is underscored by the Studying the Treatment of Acute Hypertension (STAT) registry, which establishes that acute and chronic kidney diseases are common comorbidities in patients with acute hypertension and are associated with greater risk for heart failure, cardiac arrest, and death [144]. These findings again emphasize the need for therapeutics aimed at preventing or mitigating the severity of kidney injury to reduce subsequent increased morbidity and mortality in human hypertension.

In conclusion, MANP is a novel designer natriuretic peptide that has greater resistance to proteolytic degradation than native ANP and greater cGMP-activating properties. In animal models of hypertension and heart failure, the biologic properties of MANP were greater than those of native natriuretic peptides as well as cGMP nitrovasodilators such as nitroglycerin. Importantly, MANP possesses properties that enhance renal function while inhibiting the RAAS and lowering blood pressure, thus underscoring its therapeutic potential in cardiovascular disease states, including resistant hypertension. Additional animal studies are ongoing and a First-in-Human Phase 1 clinical trial in human hypertension is planned for 2014.

Financial disclosure: Mayo Clinic has licensed MANP to Zumbro Discovery. Dr. Burnett is co-founder of Zumbro Discovery and Chief Scientific Officer

Grant Support: PO1 HL76611

References

[1] Daugherty S, Powers J, Magid D, Tavel H, Masoudi F, Margolis F, O'Connor P, Selby J, Ho P. Incidence and prognosis of resistant Hypertension in hypertensive patients. *Circulation* 2012;125:1635-1642.

[2] Pandey K. Guanyly cyclase/atrial natriuretic peptide receptor-A: role in the pathophysiology of cardiovascular regulation. *Can. J. Physiol. Pharmacol.* 2011;89:557-573.

[3] McKie PM, Cataliotti A, Huntley BK, Martin FL, Olson TM, Burnett JC Jr. A human atrial natriuretic peptide gene mutation reveals a novel peptide with enhanced blood pressure-lowering, renal-enhancing, and aldosterone-suppressing actions. *J. Am. Coll. Cardiol.* 2009;54(11):1024-32.

[4] Kirsch B. Electron microscopy of the atrium of the heart. I. Guinea pig. *Exp. Med. Surg.* 1956;14:99–112.

[5] J. D. Jamieson, M.D., G. E. Palade, M.D. Specific Granules In Atrial Muscle Cells. *J. Cell Biol.* 1964;23:151-172.

[6] De Bold AJ, Borenstein HB, Veress AT, Sonnenberg H. A rapid and potent natriuretic response to intravenous injection of atrial myocardial extract in rats. *Life Sci.* 1981;28(1):89-94.

[7] Flynn TG, de Bold ML, de Bold AJ. The amino acid sequence of an atrial peptide with potent diuretic and natriuretic properties. *Biochem. Biophys. Res. Commun.* 1983;117(3):859-65.

[8] De Bold AJ, Salerno TA. Natriuretic activity of extracts obtained from hearts of different species and from various rat tissues. *Can. J. Physiol. Pharmacol.* 1983;61(2):127-30.

[9] Bolli P, Müller FB, Linder L, et al. The vasodilator potency of atrial natriuretic peptide in man. *Circulation* 1987;75(1):221-8.

[10] Maekawa K, Sudoh T, Furusawa M, et al. Cloning and sequence analysis of cDNA encoding a precursor for porcine brain natriuretic peptide. *Biochem. Biophys. Res. Commun.* 1988;157(1):410-6.

[11] Sudoh T, Minamino N, Kangawa K, Matsuo H. C-type natriuretic peptide (CNP): a new member of natriuretic peptide family identified in porcine brain. *Biochem. Biophys. Res. Commun.* 1990;168(2):863-70.

[12] Colucci WS, Elkayam U, Horton DP, et al. Intravenous nesiritide, a natriuretic peptide, in the treatment of decompensated congestive heart failure. Nesiritide Study Group. *N. Engl. J. Med.* 2000;343(4):246-53.

[13] Publication Committee for the VMAC Investigators (Vasodilatation in the Management of Acute CHF). Intravenous nesiritide vs nitroglycerin for treatment of decompensated congestive heart failure: a randomized controlled trial. *JAMA* 2002;287(12):1531-40.

[14] Suwa M, Seino Y, Nomachi Y, Matsuki S, Funahashi K. Multicenter prospective investigation on efficacy and safety of carperitide for acute heart failure in the 'real world' of therapy. *Circ. J.* 2005;69(3):283-90.

[15] Kitakaze M, Asakura M, Kim J, et al.; J-WIND investigators. Human atrial natriuretic peptide and nicorandil as adjuncts to reperfusion treatment for acute myocardial infarction (J-WIND): two randomised trials. *Lancet* 2007;370(9597):1483-93.

[16] Chen HH, Glockner JF, Schirger JA, Cataliotti A, Redfield MM, Burnett JC Jr. Novel protein therapeutics for systolic heart failure: chronic subcutaneous B-type natriuretic peptide. *J. Am. Coll. Cardiol.* 2012;60(22):2305-12.

[17] Cataliotti A, Schirger JA, Martin FL, et al. Oral human brain natriuretic peptide activates cyclic guanosine 3',5'-monophosphate and decreases mean arterial pressure. *Circulation* 2005;112(6):836-40.

[18] Cataliotti A, Chen HH, Schirger JA. Et al. Chronic actions of a novel oral B-type natriuretic peptide conjugate in normal dogs and acute actions in angiotensin II-mediated hypertension. *Circulation* 2008;118(17):1729-36.

[19] Pandey P. The functional genomics of guanylyl cyclase/natriuretic peptide receptor-A: perspectives and paradigms. *FEBS Journal*. 2011;278:1792-1807.

[20] Maki M, Takayanagi R, Misono KS, Pandey KN, Tibbetts C, Inagami T. Structure of rat atrial natriuretic factor precursor deduced from cDNA sequence. *Nature* 1984;309(5970):722-4.

[21] Yan W, Wu F, Morser J, Wu Q. Corin, a transmembrane cardiac serine protease, acts as a pro-atrial natriuretic peptide-converting enzyme. *Proc. Natl. Acad. Sci. USA* 2000;97(15):8525-9.

[22] Dries DL, Victor RG, Rame JE, Cooper RS, Wu X, Zhu X, Leonard D, Ho SI, Wu Q, Post W, Drazner MH. Corin gene minor allele defined by 2 missense mutations is common in blacks and associated with high blood pressure and hypertension. *Circulation* 2005;112: 2403–2410.

[23] Ichiki T, Boerrigter G, Huntley BK, Sangaralingham SJ, McKie PM, Harty G, Harders EG, Burnett JC, Jr. Differential expression of the pro-natriuretic peptide convertases corin and furin in experimental heart failure and atrial fibrosis. *Am. J. Physiol. Regul. Integr. Com. Physiol.* 2013;304:R102-R109.

[24] Edwards BS, Zimmerman RS, Schwab TR, Heublein DM, Burnett JC Jr. Atrial stretch, not pressure, is the principal determinant controlling the acute release of atrial natriuretic factor. *Circ. Res.* 1988;62(2):191-5.

[25] Radziszewski W, Surdacki A, Vuorinen P, et al. Plasma ANP and cyclic GMP after physical exercise in patients with mitral valve disease and in healthy subjects. *Int. J. Cardiol.* 1993;40(3):203-9.

[26] Wilcox JN, Augustine A, Goeddel DV, Lowe DG. Differential regional expression of three natriuretic peptide receptor genes within primate tissues. *Mol. Cell Biol.* 1991;11(7):3454-62.

[27] Lo CS, Chen CH, Hsieh TJ, Lin KD, Hsiao PJ, Shin SJ. Local action of endogenous renal tubular atrial natriuretic peptide. *J. Cell Physiol.* 2009;219(3):776-86.

[28] Mebazaa A, Wetzel RC, Dodd-o JM, et al. Potential paracrine role of the pericardium in the regulation of cardiac function. *Cardiovasc. Res.* 1998;40(2):332-42.

[29] Kuhn M. Endothelial actions of atrial and B-type natriuretic peptides. *Br. J. Pharmacol.* 2012;166(2):522-31.

[30] D'Souza SP, Davis M, Baxter GF. Autocrine and paracrine actions of natriuretic peptides in the heart. *Pharmacol. Ther.* 2004;101(2):113-29.

[31] Kim M, Platt MJ, Shibasaki T, et al. GLP-1 receptor activation and Epac2 link atrial natriuretic peptide secretion to control of blood pressure. *Nat. Med.* 2013;19(5):567-75.

[32] Buglioni A, Burnett JC Jr. A gut-heart connection in cardiometabolic regulation. *Nat. Med.* 2013;19(5):534-6.

[33] Hughes AD, Nielsen H, Sever PS. The effect of atrial natriuretic peptide on human isolated resistance arteries. *Br. J. Pharmacol.* 1989;97(4):1027-30.

[34] Winquist RJ, Faison EP, Waldman SA, Schwartz K, Murad F, Rapoport RM. Atrial natriuretic factor elicits an endothelium-independent relaxation and activates particulate guanylate cyclase in vascular smooth muscle. *Proc. Natl. Acad. Sci. USA* 1984;81(23):7661-4.

[35] Van der Zander K, Houben AJ, Kroon AA, de Leeuw PW. Effects of brain natriuretic peptide on forearm vasculature: comparison with atrial natriuretic peptide. *Cardiovasc. Res.* 1999;44(3):595-600.

[36] Bennett BD, Bennett GL, Vitangcol RV, et al. Extracellular domain-IgG fusion proteins for three human natriuretic peptide receptors. Hormone pharmacology and application to solid phase screening of synthetic peptide antisera. *J. Biol. Chem.* 1991;266(34):23060-7.

[37] Koller KJ, Goeddel DV. Molecular biology of the natriuretic peptides and their receptors. *Circulation* 1992;86(4):1081-8.

[38] Lopez MJ, Wong SK, Kishimoto I, et al. Salt-resistant hypertension in mice lacking the guanylyl cyclase-A receptor for atrial natriuretic peptide. *Nature* 1995;378(6552):65-8.

[39] Lopez MJ, Garbers DL, Kuhn M. The guanylyl cyclase-deficient mouse defines differential pathways of natriuretic peptide signaling. *J. Biol. Chem.* 1997;272(37):23064-8.

[40] John SW, Krege JH, Oliver PM, et al. Genetic decreases in atrial natriuretic peptide and salt-sensitive hypertension. Science 1995;267:679–681.

[41] Steinhelper ME, Cochrane KL, Field LJ. Hypotension in transgenic mice expressing atrial natriuretic factor fusion genes. *Hypertension* 1990;16(3):301-7.

[42] Misono KS. Atrial natriuretic factor binding to its receptor is dependent on chloride concentration: A possible feedback-control mechanism in renal salt regulation. *Circ. Res.* 2000;86(11):1135-9.

[43] Ogawa H, Qiu Y, Ogata CM, Misono KS. Crystal structure of hormone-bound atrial natriuretic peptide receptor extracellular domain: rotation mechanism for transmembrane signal transduction. *J. Biol. Chem.* 2004;279(27):28625-31.

[44] Misono KS, Ogawa H, Qiu Y, Ogata CM. Structural studies of the natriuretic peptide receptor: a novel hormone-induced rotation mechanism for transmembrane signal transduction. *Peptides* 2005;26(6):957-68.

[45] Schlossmann J, Feil R, Hofmann F. Insights into cGMP signalling derived from cGMP kinase knockout mice. *Front Biosci.* 2005;10:1279-89.

[46] Rybalkin SD, Yan C, Bornfeldt KE, Beavo JA. Cyclic GMP phosphodiesterases and regulation of smooth muscle function. *Circ. Res.*. 2003 Aug 22;93(4):280-91.

[47] Kaupp UB, Seifert R. Cyclic nucleotide-gated ion channels. *Physiol. Rev.* 2002;82(3):769-824.

[48] Tamaoki J, Tagaya E, Nishimura K, Isono K, Nagai A. Role of Na(+)-K+ ATPase in cyclic GMP-mediated relaxation of canine pulmonary artery smooth muscle cells. *Br. J. Pharmacol.* 1997;122(1):112-6.

[49] Rapoport RM, Schwartz K, Murad F. Effect of sodium-potassium pump inhibitors and membrane-depolarizing agents on sodium nitroprusside-induced relaxation and cyclic guanosine monophosphate accumulation in rat aorta. *Circ. Res.* 1985;57(1):164-70.

[50] De Mey JG, Vanhoutte PM. Interaction between Na+,K+ exchanges and the direct inhibitory effect of acetylcholine on canine femoral arteries. *Circ. Res.* 1980;46(6):826-36.

[51] Blanco G, Mercer RW. Isozymes of the Na-K-ATPase: heterogeneity in structure, diversity in function. *Am. J. Physiol.* 1998;275(5 Pt 2):F633-50.

[52] Clausen T, Nielsen OB. The Na+,K(+)-pump and muscle contractility. *Acta Physiol. Scand.* 1994;152(4):365-73.

[53] Juhaszova M, Blaustein MP. Distinct distribution of different Na+ pump alpha subunit isoforms in plasmalemma. Physiological implications. *Ann. NY Acad. Sci.* 1997;834:524-36.

[54] Lynch RM, Weber CS, Nullmeyer KD, Moore ED, Paul RJ. Clearance of store-released Ca2+ by the Na+-Ca2+ exchanger is diminished in aortic smooth muscle from Na+-K+-ATPase alpha 2-isoform gene-ablated mice. *Am. J. Physiol. Heart Circ. Physiol.* 2008;294(3):H1407-16.

[55] Moore ED, Etter EF, Philipson KD, et al. Coupling of the Na+/Ca2+ exchanger, Na+/K+ pump and sarcoplasmic reticulum in smooth muscle. *Nature* 1993;365(6447):657-60.

[56] Blaustein MP, Zhang J, Chen L, Hamilton BP. How does salt retention raise blood pressure? *Am. J. Physiol. Regul. Integr. Comp. Physiol.* 2006;290(3):R514-23.

[57] Pritchard TJ, Bowman PS, Jefferson A, Tosun M, Lynch RM, Paul RJ. Na(+)-K(+)-ATPase and Ca(2+) clearance proteins in smooth muscle: a functional unit. *Am. J. Physiol. Heart Circ. Physiol.* 2010;299(2):H548-56.

[58] Pritchard TJ, Parvatiyar M, Bullard DP, Lynch RM, Lorenz JN, Paul RJ. Transgenic mice expressing Na+-K+-ATPase in smooth muscle decreases blood pressure. *Am. J. Physiol. Heart Circ. Physiol.* 2007;293(2):H1172-82.

[59] Keilbach A, Ruth P, Hofmann F. Detection of cGMP dependent protein kinase isozymes by specific antibodies. *Eur. J. Biochem.* 1992;208(2):467-73.

[60] Vaandrager AB, de Jonge HR. Signalling by cGMP-dependent protein kinases. *Mol. Cell Biochem.* 1996;157(1-2):23-30.

[61] Eggermont JA, Vrolix M, Wuytack F, Raeymaekers L, Casteels R. The (Ca2+-Mg2+)-ATPases of the plasma membrane and of the endoplasmic reticulum in smooth muscle cells and their regulation. *J. Cardiovasc. Pharmacol.* 1988;12 Suppl 5:S51-5.

[62] Eggermont JA, Vrolix M, Raeymaekers L, Wuytack F, Casteels R. Ca2+-transport ATPases of vascular smooth muscle. *Circ. Res.* 1988;62(2):266-78.

[63] Peng W, Hoidal JR, Farrukh IS. Regulation of Ca(2+)-activated K+ channels in pulmonary vascular smooth muscle cells: role of nitric oxide. *J. Appl. Physiol.* 1996;81(3):1264-72.

[64] Alioua A, Tanaka Y, Wallner M, et al. The large conductance, voltage-dependent, and calcium-sensitive K+ channel, Hslo, is a target of cGMP-dependent protein kinase phosphorylation *in vivo*. *J. Biol. Chem.* 1998;273(49):32950-6.

[65] Swayze RD, Braun AP. A catalytically inactive mutant of type I cGMP-dependent protein kinase prevents enhancement of large conductance, calcium-sensitive K+ channels by sodium nitroprusside and cGMP. *J. Biol. Chem.* 2001;276(23):19729-37.

[66] White RE, Lee AB, Shcherbatko AD, Lincoln TM, Schonbrunn A, Armstrong DL. Potassium channel stimulation by natriuretic peptides through cGMP-dependent dephosphorylation. *Nature* 1993;361(6409):263-6.

[67] Zhou XB, Ruth P, Schlossmann J, Hofmann F, Korth M. Protein phosphatase 2A is essential for the activation of Ca2+-activated K+ currents by cGMP-dependent protein kinase in tracheal smooth muscle and Chinese hamster ovary cells. *J. Biol. Chem.* 1996;271(33):19760-7.

[68] Tertyshnikova S, Yan X, Fein A. cGMP inhibits IP3-induced Ca2+ release in intact rat megakaryocytes via cGMP- and cAMP-dependent protein kinases. *J. Physiol.* 1998;512 (Pt 1):89-96.

[69] Schlossmann J, Ammendola A, Ashman K, et al. Regulation of intracellular calcium by a signalling complex of IRAG, IP3 receptor and cGMP kinase Ibeta. *Nature* 2000;404(6774):197-201.

[70] Ammendola A, Geiselhöringer A, Hofmann F, Schlossmann J. Molecular determinants of the interaction between the inositol 1,4,5-trisphosphate receptor-associated cGMP kinase substrate (IRAG) and cGMP kinase Ibeta. *J. Biol. Chem.* 2001;276(26):24153-9.

[71] Ruth P, Wang GX, Boekhoff I, et al. Transfected cGMP-dependent protein kinase suppresses calcium transients by inhibition of inositol 1,4,5-trisphosphate production. *Proc. Natl. Acad. Sci. USA* 1993;90(7):2623-7.

[72] Raeymaekers L, Jones LR. Evidence for the presence of phospholamban in the endoplasmic reticulum of smooth muscle. *Biochim. Biophys. Acta* 1986;882(2):258-65.

[73] Raeymaekers L, Hofmann F, Casteels R. Cyclic GMP-dependent protein kinase phosphorylates phospholamban in isolated sarcoplasmic reticulum from cardiac and smooth muscle. *Biochem. J.* 1988;252(1):269-73.

[74] Colyer J. Phosphorylation states of phospholamban. *Ann. NY Acad. Sci.* 1998;853:79-91.

[75] Toyoshima C, Nakasako M, Nomura H, Ogawa H. Crystal structure of the calcium pump of sarcoplasmic reticulum at 2.6 A resolution. *Nature* 2000;405(6787):647-55.

[76] Zhang Z, Lewis D, Strock C, et al. Detailed characterization of the cooperative mechanism of Ca(2+) binding and catalytic activation in the Ca(2+) transport (SERCA) ATPase. *Biochemistry* 2000;39(30):8758-67.

[77] Cornwell TL, Pryzwansky KB, Wyatt TA, Lincoln TM. Regulation of sarcoplasmic reticulum protein phosphorylation by localized cyclic GMP-dependent protein kinase in vascular smooth muscle cells. *Mol. Pharmacol.* 1991;40(6):923-31.

[78] Berridge MJ. Smooth muscle cell calcium activation mechanisms. *J. Physiol.* 2008;586(Pt 21):5047-61.

[79] Ihara E, MacDonald JA. The regulation of smooth muscle contractility by zipper-interacting protein kinase. *Can. J. Physiol. Pharmacol.* 2007;85(1):79-87.

[80] Kast D, Espinoza-Fonseca LM, Yi C, Thomas DD. Phosphorylation-induced structural changes in smooth muscle myosin regulatory light chain. *Proc. Natl. Acad. Sci. USA* 2010;107(18):8207-12.

[81] Hartshorne DJ, Ito M, Erdödi F. Role of protein phosphatase type 1 in contractile functions: myosin phosphatase. *J. Biol. Chem.* 2004;279(36):37211-4.

[82] Lincoln TM, Dey N, Sellak H. Invited review: cGMP-dependent protein kinase signaling mechanisms in smooth muscle: from the regulation of tone to gene expression. *J. Appl. Physiol.* 2001;91(3):1421-30.

[83] Surks HK, Mendelsohn ME. Dimerization of cGMP-dependent protein kinase 1alpha and the myosin-binding subunit of myosin phosphatase: role of leucine zipper domains. *Cell Signal* 2003;15(10):937-44.

[84] Rybalkin SD, Yan C, Bornfeldt KE, Beavo JA. Cyclic GMP phosphodiesterases and regulation of smooth muscle function. *Circ. Res.* 2003;93(4):280-91.

[85] Beavo JA. Cyclic nucleotide phosphodiesterases: functional implications of multiple isoforms. *Physiol. Rev.* 1995;75(4):725-48.

[86] Maurice DH. Cyclic nucleotide phosphodiesterase-mediated integration of cGMP and cAMP signaling in cells of the cardiovascular system. *Front Biosci.* 2005;10:1221-8.

[87] Pauvert O, Salvail D, Rousseau E, Lugnier C, Marthan R, Savineau JP. Characterisation of cyclic nucleotide phosphodiesterase isoforms in the media layer of the main pulmonary artery. *Biochem. Pharmacol.* 2002;63(9):1763-72.

[88] Costa MA, Elesgaray R, Loria A, Balaszczuk AM, Arranz C. Atrial natriuretic peptide influence on nitric oxide system in kidney and heart. *Regul. Pept.* 2004;118:151–157.

[89] Costa MA, Gonzalez Bosc LV, Majowicz MP, Vidal NA, Balaszczuk AM, Arranz CT. Atrial natriuretic peptide modifies arterial blood pressure through nitric oxide pathway in rats. *Hypertension* 2000;35: 1119–1123.

[90] Costa MA, Elesgaray R, Balaszczuk AM, Arranz C. Role of NPR-C natriuretic receptor in nitric oxide system activation induced by atrial natriuretic peptide. *Regul. Pept.* 2006;135: 63–68.

[91] Elesgaray R, Caniffi C, Rodríguez Ierase D, et al. Signaling cascade that mediates endothelial nitric oxide synthase activation induced by atrial natriuretic peptide. *Regul. Pept.* 2008;151: 130–134.

[92] Fernández AP, Serrano J, Castro S, Salazar FJ, López JC, Nava E. Distribution of nitric oxide synthases and nitrotyrosine in the kidney of spontaneously hypertensive rats. *J. Hypertens* 2003;21: 2375–2388.

[93] Melo LG, Veress AT, Ackermann U, et al. Chronic regulation of arterial blood pressure in ANP transgenic and knockout mice: role of cardiovascular sympathetic tone. *Cardiovasc. Res. 1999*;43(2):437-44.

[94] Thomas CJ, Rankin AJ, Head GA, Woods RL. ANP enhances bradycardic reflexes in normotensive but not spontaneously hypertensive rats. *Hypertension* 1997;29(5):1126-32.

[95] Schultz HD, Steele MK, Gardner DG. Central administration of atrial peptide decreases sympathetic outflow in rats. *Am. J. Physiol.* 1990;258(5 Pt 2):R1250-6.

[96] Coote JH. A role for the paraventricular nucleus of the hypothalamus in the autonomic control of heart and kidney. *Exp. Physiol.* 2005;90(2):169-73.

[97] Kenney MJ, Weiss ML, Haywood JR. The paraventricular nucleus: an important component of the central neurocircuitry regulating sympathetic nerve outflow. *Acta Physiol. Scand.* 2003;177(1):7-15.

[98] Malpas SC, Coote JH. Role of vasopressin in sympathetic response to paraventricular nucleus stimulation in anesthetized rats. *Am. J. Physiol.* 1994;266(1 Pt 2):R228-36.

[99] Burnett JC Jr., Granger JP, Opgenorth TF. Effects of synthetic atrial natriuretic factor on renal function and renin release. *Am. J. Physiol.*. 247:F863-F866, 1984.

[100] Cogan MG. Atrial natriuretic factor can increase renal solute excretion primarily by raising glomerular filtration. *Am. J. Physiol.* 1986;250(4 Pt 2):F710-4.

[101] Marin-Grez M, Fleming JT, Steinhausen M. Atrial natriuretic peptide causes pre-glomerular vasodilatation and post-glomerular vasoconstriction in rat kidney. *Nature* 1986;324(6096):473-6.

[102] Veldkamp PJ, Carmines PK, Inscho EW, Navar LG. Direct evaluation of the microvascular actions of ANP in juxtamedullary nephrons. *Am. J. Physiol.* 1988;254(3 Pt 2):F440-4.

[103] Aalkjaer C, Mulvany MJ, Nyborg NC. Atrial natriuretic factor causes specific relaxation of rat renal arcuate arteries. *Br. J. Pharmacol* 1985;86(2):447-53.

[104] Appel RG, Wang J, Simonson MS, Dunn MJ. A mechanism by which atrial natriuretic factor mediates its glomerular actions. *Am. J. Physiol.* 1986;251(6 Pt 2):F1036-42.

[105] Kudo T, Baird A. Inhibition of aldosterone production in the adrenal glomerulosa by atrial natriuretic factor. *Nature* 1984-1985;312(5996):756-7.

[106] Chartier L, Schiffrin E, Thibault G, Garcia R. Atrial natriuretic factor inhibits the stimulation of aldosterone secretion by angiotensin II, ACTH and potassium in vitro and angiotensin II-induced steroidogenesis *in vivo*. *Endocrinology* 1984;115(5):2026-8.

[107] Shi SJ, Nguyen HT, Sharma GD, Navar LG, Pandey KN. Genetic disruption of atrial natriuretic peptide receptor-A alters renin and angiotensin II levels. *Am. J. Physiol.* 2001; 281(4):F665– F673.

[108] Caruso-Neves C, Vives D, Dantas C, et al. Ouabain-insensitive Na+-ATPase of proximal tubules is an effector for urodilatin and atrial natriuretic peptide. *Biochim. Biophys. Acta* 2004;1660(1-2):93-8.

[109] Vives D, Farage S, Motta R, Lopes AG, Caruso-Neves C. Atrial natriuretic peptides and urodilatin modulate proximal tubule Na(+)-ATPase activity through activation of the NPR-A/cGMP/PKG pathway. *Peptides* 2010;31(5):903-8.

[110] Winaver J, Burnett JC, Tyce GM, Dousa TP. ANP inhibits Na(+)-H+ antiport in proximal tubular brush border membrane: role of dopamine. *Kidney Int.* 1990;38(6):1133-40.

[111] Nonoguchi H, Tomita K, Marumo F. Effects of atrial natriuretic peptide and vasopressin on chloride transport in long- and short-looped medullary thick ascending limbs. *J. Clin. Invest.* 1992;90(2):349-57.

[112] Bailly C. Effect of luminal atrial natriuretic peptide on chloride reabsorption in mouse cortical thick ascending limb: inhibition by endothelin. *J. Am. Soc. Nephrol.* 2000;11(10):1791-7.

[113] Rabelink TJ, Koomans HA, van de Stolpe A, Bijlsma JA, Dorhout Mees EJ. Effects of atrial natriuretic peptide on distal tubule function in humans. *Kidney Int.* 1990;37(3):996-1001.

[114] Ader JL, Tran-Van T, Praddaude F. Nephron sites of the natriuretic effect of intrarenal infusion of atrial natriuretic peptide in the rat. *Am. J. Hypertens* 1990;3(11):870-2.

[115] Terada Y, Moriyama T, Martin BM, Knepper MA, Garcia-Perez A. RT-PCR microlocalization of mRNA for guanylyl cyclase-coupled ANF receptor in rat kidney. *Am. J. Physiol.* 1991;261(6 Pt 2):F1080-7.

[116] Light DB, Corbin JD, Stanton BA. Dual ion-channel regulation by cyclic GMP and cyclic GMP-dependent protein kinase. *Nature 1*990;344(6264):336-9.

[117] Rocha AS, Kudo LH. Atrial peptide and cGMP effects on NaCl transport in inner medullary collecting duct. *Am. J. Physiol.* 1990;259(2 Pt 2):F258-68.

[118] Forte LR, London RM, Krause WJ, Freeman RH. Mechanisms of guanylin action via cyclic GMP in the kidney. *Annu. Rev. Physiol.* 2000;62:673-95.

[119] Scavone C, Scanlon C, McKee M, Nathanson JA. Atrial natriuretic peptide modulates sodium and potassium-activated adenosine triphosphatase through a mechanism involving cyclic GMP and cyclic GMP-dependent protein kinase. *J. Pharmacol. Exp. Ther* 1995;272(3):1036-43.

[120] Bełtowski J, Górny D, Marciniak A. The mechanism of Na+, K+-ATPase inhibition by atrial natriuretic factor in rat renal medulla. *J. Physiol. Pharmacol.* 1998;49(2):271-83.

[121] Li D, Aperia A, Celsi G, da Cruz e Silva EF, Greengard P, Meister B. Protein phosphatase-1 in the kidney: evidence for a role in the regulation of medullary Na(+)-K(+)-ATPase. *Am. J. Physiol.* 1995;269(5 Pt 2):F673-80.

[122] Chen S, Gardner DG. Osmoregulation of natriuretic peptide receptor signaling in inner medullary collecting duct. A requirement for p38 MAPK. *J. Biol. Chem.* 2002;277(8):6037-43.

[123] Chen S, Cao L, Intengan HD, Humphreys M, Gardner DG. Osmoregulation of endothelial nitric-oxide synthase gene expression in inner medullary collecting duct cells. Role in activation of the type A natriuretic peptide receptor. *J. Biol. Chem.* 2002;277(36):32498-504

[124] Lisy O, Huntley B, McCormick D, Kurlansky P, Burnett JC Jr. Design, synthesis, and actions of a novel chimeric natriuretic peptide: CD-NP. *J. Am. Coll. Card* 2008;52:60-68, 2008.

[125] Hodgson-Zingman DM, Karst ML, Zingman LV, et al. Atrial natriuretic peptide frameshift mutation in familial atrial fibrillation. *N. Engl. J. Med.* 2008 Jul 10;359(2):158-65.

[126] Dickey DM, Yoder AR, Potter LR. A familial mutation renders atrial natriuretic Peptide resistant to proteolytic degradation. *J. Biol. Chem.* 2009 Jul 17;284(29):19196-20010.

[127] Ralat LA, Guo Q, Ren M, et al. Insulin-degrading enzyme modulates the natriuretic peptide-mediated signaling response. *J. Biol. Chem.* 2011 Feb 11;286(6):4670-9.

[128] McKie PM, Cataliotti A, Boerrigter G, et al. A novel atrial natriuretic peptide based therapeutic in experimental angiotensin II mediated acute hypertension. . Hypertension. 2010 Dec;56(6):1152-9.

[129] Chen HH, Cataliotti A, Schirger JA, Martin FL, Burnett Jr JC. Equimolar doses of atrial and brain natriuretic peptides and urodi- latin have differential renal actions in overt experimental heart failure. *Am. J. Physiol.*. 2005;288(5):R1093–7.

[130] McKie P, Cataliotti A, Ichiki T, Sangaralingham SJ, Chen H, and Burnett Jr JC. M-atrial natriuretic peptide and nitroglycerin in a canine model of experimental acute

hypertensive heart failure: differential actions of two cGMP activating therapeutics. *Circulation* in press 2014.

[131] Castro LR, Verde I, Cooper DM, Fischmeister R. Cyclic guanosine monophosphate compartmentation in rat cardiac myocytes. *Circulation* 2006; 113: 2221–2228.

[132] Martin F, Supaporn T, Chen H, Sandberg S, Matsuda Y, Jougasaki M, Burnett JC Jr. Distinct roles for renal particulate and soluble guanylyl cyclases in preserving renal function in experimental acute heart failure. *Am. J. Physiol.*. 2007;293:R1580-R1585.

[133] Nemer M, Chamberland M, Sirois D, et al. Gene structure of human cardiac hormone precursor, pronatriodilatin. *Nature.* 1984 Dec 13-19;312(5995):654-661.

[134] Newton-Cheh C, Larson MG, Vasan RS, et al. Association of common variants in NPPA and NPPB with circulating natriuretic peptides and blood pressure. *Nat. Genet.* 2009;41:348–53.

[135] Cannone V, Boerrigter G, Cataliotti A, Costello-Boerrigter LC, Olson TM, McKie PM, Heublein DM, Lahr BD, Bailey KR, Averna M, Redfield MM, Rodeheffer RJ, Burnett JC Jr. A genetic variant of the atrial natriuretic peptide gene is associated with cardiometabolic protection in the general community. *J. Am. Coll. Cardiol.* 58 (6):629-636, 2011.

[136] Cannone V, Cefalu'AB, Noto D, Scott CG, Bailey KR, Cavera G, Pagano M, Sapienza M, Averna M, Burnett Jr. JC. The atrial natriuretic peptide genetic variant rs5068 is associated with a favorable cardiometabolic phenotype in a mediterranean population. *Diabetes Care* 2013;56:2850-2856.

[137] Sengenes C, Bouloumie A, Hauner H, et al. Involvement of a cGMP-dependent pathway in the natriuretic peptide-mediated hormone-sensitive lipase phosphorylation in human adipocytes. *J. Biol. Chem.* 2003;278:48617–26.

[138] Arora P, Wu C, Khan AM, Bloch DB, Davis-Dusenbery BN, Ghorbani A, Spagnolli E, Martinez A, Ryan A, Tainsh LT, Kim S, Rong J, Huan T, Freedman JE, Levy D, Miller KK, Hata A, Del Monte F, Vandenwijngaert S, Swinnen M, Janssens S, Holmes TM, Buys ES, Bloch KD, Newton-Cheh C, Wang TJ. Atrial natriuretic peptide is negatively regulated by microrna-425. *J. Clin. Invest.*. 2013;123:3378-3382.

[139] Leader B, Baca QJ, Golan DE. Protein therapeutics: a summary and pharmacological classification. *Nat. Rev. Drug Discov.* 2008 Jan;7(1):21-39.

[140] Cataliotti A, Tonne JM, Bellavia D, Martin FL, Oehler EA, Harders GE, Campbell JM, Peng KW, Russell SJ, Malatino LS, Burnett JC Jr, Ikeda Y. Long-term cardiac pro-B-type natriuretic peptide gene delivery prevents the development of hypertensive heart disease in spontaneously hypertensive rats. *Circulation* 2011 Mar 29;123(12):1297-305.

[141] Hata N, Seino Y, Tsutamoto T, et al. Effects of carperitide on the long-term prognosis of patients with acute decompensated chronic heart failure: the PROTECT multicenter randomized controlled study. *Circ. J.* 2008 Nov;72(11):1787-93.

[142] Morikawa S, Sone T, Tsuboi H, et al. Renal protective effects and the prevention of contrast-induced nephropathy by atrial natriuretic peptide. *J. Am. Coll. Cardiol.* 2009 Mar 24;53(12):1040-6.

[143] de Bold AJ. Cardiac natriuretic peptides: gaining further insights into structure-function relationships. *J. Am. Coll. Card.* 2009 Sep 8;54(11):1033-4.

[144] Szczech LA, Granger CB, Dasta JF, Amin A, Peacock WF, McCullough PA, et al. Acute kidney injury and cardiovascular outcomes in acute severe hypertension. *Circulation* 2010;121 (20):2183–91.

In: Natriuretic Peptides
Editor: Kailash N. Pandey

ISBN: 978-1-63321-080-6

Chapter 4

ATRIAL NATRIURETIC FACTOR RECEPTOR GUANYLATE CYCLASE: THE JOURNEY

Rameshwar K. Sharma and Teresa Duda
Research Divisions of Biochemistry and Molecular Biology,
Salus University, Elkins Park, PA, US

ABSTRACT

Discovery of the ANF-RGC (Atrial Natriuretic Factor Receptor Guanylate Cyclase) is an exceptional historical event. First, it seeded in generating the field of cyclic GMP as an intracellular signaling agent. Second, it terminated an over two decades-long debate over the independent existence of a membrane guanylate cyclase, the enzyme that catalyzes the formation of cyclic GMP from GTP. Third, unlike the three cellular signaling components composition of its predecessors, first adenylate cyclase and then G-protein-coupled-receptors, ANF-RGC was a single transduction component, being both a receptor and the signal transducer. Third, in contrast to the seven- transmembrane-spanning structure of G-proteins, it was a single-transmembrane-spanning protein. Fourth, being a physiological receptor of the two most hypotensive hormones, ANF and BNP, it was a potential modulator of the cardiovascular processes, including blood pressure, natriuresis and fluid secretion. The objective of this brief review of 5,000 words is to focus on the chronological development, recognize major contributions of the original investigators, correct misplaced facts, and speculate on the present and future trend of the field of ANF-RGC signal transduction. For comprehensive discussion of the general field of membrane guanylate cyclases the reader is referred to four earlier reviews of the author, the subject matters of these have been freely borrowed with the appropriate citations in this presentation [origin to the year 2002 (Sharma in Mol. Cell Biochem. 230:3–30, 2002), to the year 2004 (Duda et al., in Peptides 26:969–984, 2005); to the year 2010 (Sharma in Mol. Cell Biochem.istry 334: 3-36, 2010) and of the Ca^{2+}-modulated membrane guanylate cyclase: to the year 1997 (Pugh et al., in Biosci Rep 17:429–473, 1997); to the year 2012 (Sharma and Duda in Front Mol Neurosci 5: 42, 2012)].

1. INTRODUCTION

Prior to the purification and characterization of the ANF-RGC (Atrial Natriuretic Factor Receptor Guanylate Cyclase), the development of the membrane guanylate cyclase transduction field passed through JUBILATION, DISAPPOINTMENT AND DENIAL. Finally its CONSOLIDATION resulted through a COMPETING CONCEPT where this author (RKS) was privileged to make a major contribution [reviewed in: 1]. The efforts of this author involved building on small projects, brick by brick, and emerged from the chasm of disbelief through slow steady continuous work. The first ''Historical'' part of this article is an encapsulated narration on these historical periods of this field. To make the presentation simple and chronologically continuous, the subject matters of the earlier reviews and publications of the author have been freely borrowed with the appropriate citations.

2. HISTORY: YEARS 1963–1987

JUBILATION. Following the discovery of cyclic AMP, cyclic GMP was first identified in rat urine [2]. This raised the possibility that, besides cyclic AMP, an additional second messenger hormonal system was present in the vertebrates. Boosting this possibility, six years later cyclic GMP [3, 4] and guanylate cyclase, the enzyme catalyzing its biosynthesis from GTP were detected in all animal tissues [4, 5].

Mimicking the conceptual framework formulated for the second messenger role of the cyclic AMP system, the course for the construction of the cyclic GMP signaling model began [5]. Great excitement was generated with the "Ying Yang" hypothesis. This hypothesis postulated that the biological regulation of the cell is governed by the opposing biological activities of cyclic AMP and cyclic GMP [6]. Both cyclic nucleotides serve as second messengers in cellular regulation.

DISAPPOINTMENT. "Ying Yang" hypothesis and the second messenger concept of cyclic GMP was seriously compromised with (1) the failed attempts to demonstrate a hormonally dependent guanylate cyclase in every tested system [7]; (2) the guanylate cyclase activity was nonspecifically stimulated by a variety of agents such as polyunsaturated fatty acids, peroxides, hydroperoxides, free radicals, ascorbic acid, sodium nitroprusside and other agents, including cigarette smoke, that presumably affected the oxidation-reduction potential of the biochemical reactions [8]; and (3) there was a consensus that cyclic GMP-dependent protein kinase, at the time the only known effector enzyme of cyclic GMP signal transduction, did not phosphorylate a specific protein (substrate) clearly distinct from that of the cyclic AMP-dependent protein kinase. It acted sub-servant to the cyclic AMP system, lacked specificity and phosphorylated the substrate through cyclic AMP-dependent protein kinase [9]. No hormonally dependent cyclic GMP signaling component existed in any of the biological systems [9].

DENIAL. These conflicting observations resulted in the turnaround of thinking of the original investigators who were the proponents for the second messenger role of cyclic GMP in cellular signaling.

They then reinterpreted their original results to be artifactually attributable to the "assay conditions, in crude preparations, or possibly other factors" [8]. Their view was transformed into an aggressive belief with the discovery of a nonspecific form of guanylate cyclase, and their previous observations related to the stimulus-dependent rises in cyclic GMP were linked to the indirect activation of soluble guanylate cyclases [7, 8]. These conclusions are summed up in their two, of many, high-impact reviews:

> "While the metabolism of cyclic GMP has many parallelism to that of cyclic AMP, it is becoming quite clear that the two systems have distinct differences and that the functions of one system may not necessarily provide clues to the function of the other. In retrospect, the apparent parallelisms of the two systems hindered the originality of approaches to the cyclic GMP system and may, in fact, have impeded some of the developments in the area. Certainly, many early experiments with guanylate cyclase and cyclic GMP conducted in this laboratory and others were done based on previous experience and biases with the cyclic AMP system..., the properties of the soluble and particulate activities in tissues are quite different.... Minor differences have also been noted between particulate activities and in plasma and endoplasmic reticulum preparations. In spite of these significant physical and/or kinetic differences that suggest two or more forms of guanylate cyclase may exist in tissues, it is not yet known whether or not the proteins or their subunits are different. The apparent differences may be attributable to assay conditions, other constituents in crude preparations, or possibly other factors" [8]. (Emphasis added).
>
> "Considerable effort has been put forth to demonstrate an activation of the enzyme by hormonal or other physiological agents that promote the accumulation of cGMP in intact cells. To date, no specific direct effect of such an agent on guanylate cyclase activity in vitro has been convincingly demonstrated. This is a characteristic of the cGMP system that contrasts sharply with its cAMP counterpart; adenylate cyclase from disrupted cells as a general rule is activable by the agents that stimulate cellular cAMP accumulation....It is becoming evident that the greatest dissimilarity between guanylate cyclase and adenylate cyclase may be that the activity of the former is modulated indirectly with respect to the cellular action of hormones (and other cell stimuli) compared with the direct effect that hormones seem to have on the activity of adenylate cyclase" [7] (Emphasis added).

CONSOLIDATION. Against heavy odds modeled by the above-intimidating concept, which lasted almost to the mid1980s and at the time considered cyclic AMP as the solo hormonal second messenger system, slowly, yet steadily, a parallel COMPETING CONCEPT, held primarily by this author's group and a few other laboratories, persisted and evolved [reviewed in: 1] . The work based on the model systems of the isolated adrenal fasciculate and the adrenocortical carcinoma cells established that the membrane guanylate cyclase was, indeed, a bona fide component of the hormonal signal transduction system and, importantly, it was a distinct enzyme from that of the soluble guanylate cyclase enzyme [1, 10, 11]. These studies are briefly recounted below.

Both cell models of the isolated adrenal fasciculate and the adrenocortical carcinoma cells had no phosphodiesterase activity. For this reason the fasciculata cells were highly sensitive to the physiological levels of ACTH, responding to as little as 1 μU of ACTH in the synthesis of corticosterone [10]. The physiological concentrations of the ACTH (1 to 10 μU) did not raise the level of cyclic AMP in the fasciculata cells although they significantly stimulated steroidogenesis [12].

A significant increase in the cyclic AMP level occurred only at the supra-physiological (>10 IU) concentrations of ACTH [12]. An added important observation was that beyond the maximal steroidogenic concentrations of ACTH there was a rise in cyclic AMP levels and yet no further elevation of steroidogenesis [12, 13]. These findings demonstrated that cyclic AMP was not the physiological second messenger of ACTH. And, cyclic AMP was not the sole hormonal second messenger.

In contrast, (a) under physiological concentrations of the hormone, an excellent temporal correlation existed between cyclic GMP formation [12, 14, 15], phosphorylation, and corticosterone synthesis [13, 16, 17]; (b) exogenous cyclic GMP stimulated the protein kinase activity followed by the steroidogenic activity [13]; and (c) exogenously supplied cyclic GMP mimicked the exogenously supplied ACTH in stimulating the direct transformation of cholesterol to corticosterone [18]. Consistent with the earlier findings [19, 20, 21], these studies also found that Ca^{2+} was an obligatory factor in the process of steroidogenesis, but it was unable to promote steroidogenesis by itself [22]. The conclusions were that cyclic GMP and Ca^{2+} were the two complementary physiological second messengers of the hormone.

Because the cells contained no phosphodiesterase activity [23], the findings also indicated that the hormone-dependent cyclic GMP formation reflected the presence of the hormonally-dependent membrane guanylate cyclase in these cells. RECALL, at the time there was a general denial of the independent existence of the hormonally-dependent membrane guanylate cyclase transduction system, including that of cyclic GMP-dependent protein kinase, in any biological system (*vide supra*). The studies were, therefore, channeled toward the direct characterization of these two components of the cyclic GMP pathway in the adrenal gland.

The cyclic-GMP-dependent protein kinase was purified to homogeneity and characterized from this tissue and its specificity for the cyclic GMP analogues was assessed [24]. The 8-bromo cyclic GMP was established as the most potent stimulator of the protein kinase. Significantly, compared to this, dibutyryl cyclic GMP was 10,000-fold less active.

This is an important point, because the corresponding dibutyryl cyclic AMP analogue was ~ 10-fold more potent than the parent cyclic AMP in its biological activity; and it had been argued that if, indeed, cyclic GMP were to be the real hormonal second messenger dibutyryl cyclic GMP should have also been more potent than the parent cyclic GMP in stimulating steroidogenesis. But it did not. The potency of dibutyryl cyclic GMP for the cyclic GMP-dependent protein kinase provided the possible answer and the results also linked the cyclic-GMP-dependent protein kinase with the ACTH-dependent process of steroidogenesis [24].

Finally, the direct presence of the ACTH/Ca^{2+}-dependent guanylate cyclase activity was demonstrated in the adrenal cortex and in the adrenocortical carcinoma [25 - 27]. And, the distinct features between the membrane and the soluble form of guanylate cyclases were established (Table 1 of [25]). Most vitally, the membrane guanylate cyclase was not stimulated by the free radical- and nitric-oxide-generating agents, including hemin. The soluble guanylate cyclase was.

Subsequent studies by other groups showed that particulate guanylate cyclase activity could indeed be stimulated by ACTH in cultured neurons [28] and by ANF in rat tissues [29, 30]. For the first time, the existence of a hormonally dependent membrane guanylate cyclase in several different cell types that was clearly functionally and kinetically distinct from the soluble form was established.

The stage was now set to directly purify the membrane guanylate cyclase from the rat adrenocortical carcinoma cells, and study its functional and physical characteristics. This achievement resulted in a remarkably surprising finding.

3. ANF-RGC DISCOVERY

First purified and functionally characterized vertebrate membrane guanylate cyclase was the atrial natriuretic factor receptor, ANF-RGC [31 - 33]. A plasma membrane guanylate cyclase was purified and its absolute purity was established by (1) protein staining; (2) isoelectric focusing of the native protein and its iodinated form yielded a symmetrical protein peak which superimposed over the guanylate cyclase activity peak; (3) Western blot analysis using a monospecific polyclonal antibody raised against the native protein; and (4) the protein antibody blocked almost all its guanylate cyclase activity. Thus, the protein was pure and it was a membrane guanylate cyclase. Further analysis demonstrated that it bound ANF in a stoichiometric ratio, proving that the guanylate cyclase was an ANF receptor. It was named Atrial Natriuretic Factor Receptor Guanylate Cyclase, ANF-RGC.

Contradicting the early conclusions on the membrane guanylate cyclase properties [7, 8], ANF-RGC activity was independent of the nitric-oxide-generating agents—catalase, hemin, dithiothreitol, arachidonic acid, and tufstin, including cigarette smoke. ANF-RGC discovery contributed in solving the prior confusion on the nonspecific stimulation of guanylate cyclase in diverse biological systems. Only the soluble form was stimulated by the nitric oxide-generating agents. Importantly, these agents functioned through heme, an important factor that through the below noted commentary settled the accuracy of this historical discovery [33].

In a parallel study and reported in about the same time frame, the tentative conclusion was drawn that in the rat lung an ANF-binding and guanylate cyclase activity may reside on a single protein [34]. These authors raised the issue of priority in their discovery and this matter was debated and settled in a commentary [33], discussed below.

The rebuttal pointed out that, in contrast to the homogeneous form of the adrenocortical ANF-RGC [31 – 33], by the authors own admission, the lung guanylate cyclase preparation was crude, in addition to the 120-kD protein it contained at least 13 minor bands and the authors stated that it was only about 95% pure [34]. The possibility existed that the 5% contaminant contained the separate ligand-binding or the cyclase activity. There were two other serious problems with the claim that the lung enzyme was the real ANF-RGC. (1) In contrast to the 1:1 stoichiometry, the lung enzyme bound only 14.5% of ANF at the noted theoretical value [34]; (2) the lung enzyme was stimulated by hemin [34], a characteristic of the soluble guanylate cyclase, not possessed by the membrane guanylate cyclase.

Subsequently, two other laboratories reported purification of the ANF-RGC from the adrenal cortex, consolidating the concept that this membrane guanylate cyclase is, indeed, the receptor for the hormone ANF [35, 36].

The discovery of ANF-RGC had a powerful impact on the cellular signal transduction field. In describing this new paradigm it was stated "coexistence of the ANF receptor and guanylate cyclase activities on a single polypeptide chain indicates that the mechanism of transmembrane signal transduction involving mediation by second messenger, cyclic GMP, is different from the well-established adenylate cyclase system. In the hormone-dependent adenylate cyclase, there is an assemblage of individual components—receptor, GTP-binding protein, and catalytic moiety—for signal transduction. In contrast, the presence of dual activities—receptor-binding and enzymic—on a single polypeptide chain indicates that this transmembrane protein contains both the information for signal recognition and its translation into a 'second messenger" [32].

The two attributes of ANF-RGC, being a surface receptor as well as a guanylate cyclase, were pictorially shown in a model [37 - 39; modified version of Figure 1 in 39]. It depicted ANF-RGC as a single-transmembrane-spanning protein, the ANF-receptor domain residing on the outside, the catalytic domain protruding inside the cell and a transmembrane segment separating its two portions (Figure 1 in [39]).

Does the adrenal ANF-RGC reside in other tissues? The Western blot and immunocytochemical analysis showed its presence in the rat and mouse testes [40], in the rat neuronal cells located in the ventral horn region of the spinal cord, and in the cerebellum Purkinje cell somata and Purkinje cell dendrites [37]. With the same antibody probe, ANF-RGC was directly purified to homogeneity and its functional linkage with ANF was established in the rat and mouse testes [40], and in the renal glomerulus cells [41]. These studies established ubiquity of ANF-RGC in the selected mammalian tissues.

4. ANF Meets ANF-RGC

In the time frame the trustworthiness of the membrane guanylate cyclase was being established, cotemporaneous findings from the de Bold's [42] and then other groups demonstrated that ANF regulates sodium excretion, water balance, and blood pressure [43 - 45].

With the evidence that ANF stimulates guanylate cyclase activity in crude particulate fractions of the rat [29] and urinary cyclic GMP excretion parallel to the physiological processes of natriuresis and diuresis [30], the discovery of ANF-RGC immediately suggested that being present in the adrenal gland [32, 35, 36] and kidney [41], it is a modulator of the processes of natriuresis and diuresis. In accordance with this clue and together with the earlier studies that ANF raises the levels of cyclic GMP and down-regulates the aldosterone formation in the glomerulosa cells of the adrenal gland [46 - 49], the concept arose that ANF-RGC via cyclic GMP modulates the production of aldosterone, which, in turn, through the kidney controls natriuresis and diuresis.

5. Three-Member Surface Receptor ANF-RGC Family

Almost 3 years after its purification, ANF-RGC mRNA was cloned from the rat brain [50, 51], adrenal gland [52] and the mouse Leydig tumor cells [53].

Its predicted protein sequence validated that it was a single-transmembrane-spanning protein, the transmembrane region dividing it in two roughly equal portions: extracellular and intracellular. Molecular cloning of the two other membrane guanylate cyclases, receptors for the respective natriuretic peptide hormones CNP [54, 55] and enterotoxin [56 - 61], termed

CNP-RGC and STa-RGC, established a three-member surface receptor family of membrane guanylate cyclases.

The present status of the number of the natriuretic factor surface receptor guanylate cyclases and their ligands—ANF, BNP and CNP—remain the same. ANF and BNP are the ligands of ANF-RGC, CNP of CNP-RGC and Star-GC of enterotoxin and the endogenously produced guanylin from the intestine [reviewed in: 1].

6. The Family Expands

Inheriting a common trait of being the hormone receptors, the belief arose that the membrane guanylate cyclase is exclusively the hormone receptor family. This belief was not sustained with the subsequent discoveries of the photoreceptor ROS-GCs (Rod Outer Segment membrane guanylate cyclases) 1 and 2; and of ONE-GC (Olfactory receptor Neuroepithelial membrane guanylate cyclase (This topic is out of the scope for the present review, for a comprehensive discussion the reader is referred to [62]).

All these guanylate cyclases were modulated by the intracellular changes of $[Ca^{2+}]_i$, and primarily were the vital components of sensory neurons transduction machineries [reviewed in: 62]. The ROS-GC represented the second and ONE-GC, the third subfamily of membrane guanylate cyclases.

Their differences from each other being that the former was exclusively $[Ca^{2+}]_i$-modulated and the latter by the odorant guanylin with $[Ca^{2+}]_i$. Some groups to this time refer to these two subfamilies as orphan receptor guanylate cyclases [reviewed in: 63].

7. Common Family Traits

All members of the membrane guanylate cyclase family (1) are single-transmembrane-spanning proteins. (2) A helical TMD (transmembrane domain) divides the protein roughly into two equal regions: N-terminal, the extracellular (ECD) and C-terminal, the intracellular (ICD). (3) ICD is composed of multiple modular blocks [50 – 52, 64 - 67], it contains the core catalytic domain (CCD) which is the common translation center for all signals into the generation of cyclic GMP. (4) The structures of the family members are homologous to varying degrees ranging from ~29% between ANF-RGC and ROS-GCs to ~59% with CNP-RGC.

The CCD structure is largely conserved, ranging from ~63% conservation between ANF-RGC and Star-GC to 94% with CNP-RGC and about 68% with ROS-GCs and ONE-GC; there is 93% conservation between ROS-GC1 and ROS-GC2, and ~ 86% between ROS-GC1 and ONE-GC. (5) Juxtaposed to the N-terminal side of CCD resides the SHD (Signaling Helix Domain) with about ~50 t0 75% structural conservation among all guanylate cyclases. It is a 5-heptad coil to coil structure. Its signaling function differs depending on the nature of the guanylate cyclase.

Because it was previously believed that it keeps CCD in the dimeric state, it was named as the dimerization domain (DD) [68]. But this no longer being the case [69], it has been renamed as SHD [69]. (6) After transmission of the signal to the core catalytic domain it is translated into the accelerated production of cyclic GMP by a structural change in the homodimeric antiparallel configuration of CCD [69].

To better understand this discussion on the structural topology of the guanylate cyclases, the reader may find figure 1 of the ANF-RGC model as a useful general guide.

8. Topography: Experimental Validation

ANF and CNP signaling sites. To define the domains where ANF and CNP signaling sites reside, ANF-RGC was constructed from the recombinant GCα, a guanylate cyclase which had two changes in its residues, Gln^{338}His and Leu^{364}Pro [52]. GCα did not bind and was not stimulated by ANF, yet it had the basal guanylate cyclase activity. Restoration of these residues to their original forms and their site directed analyses established that (1) Leu^{364} is critical for ANF-binding and ANF-dependent guanylate cyclase activity of ANF-RGC [52].

Thus, the ANF signaling site resided in the ECD segment of ANF-RGC. (2) The identical approach proved that the Glu^{332} residue resided in the ECD of CNP-RGC and it was critical for binding CNP and activation of CNP-RGC [70]. (3) Transmigration studies of these residues demonstrated that the folding patterns of the binding domains of ANF-RGC for ANF and of CNP-RGC for CNP were identical or very similar [71].

9. ATP Modulation

Early studies with crude membrane preparations suggested that ATP accelerates ANF-dependent ANF-RGC activity. Two mechanisms were proposed; one, direct [72] and the other indirect via an accessory protein [73]. Because the preparations were crude the correct molecular mechanism between the two could not be settled.

Subsequently, two groups in parallel demonstrated that ATP is obligatory for ANF dependent ANF-RGC transduction activity [74, 75]. Since the non-hydrolyzable analogs, ATPγS and AMP-PNP, mimicked ATP, both groups proposed that ATP acts directly by allosteric regulation of the cyclase [74, 75]. Subsequent study with CNP-RGC showed that ATP is also obligatory for its transduction activity [76].

These two groups, however, proposed contrasting models of ANF signal transduction, and they played an important role in shaping the future field of ANF-RGC. MODEL 1, Garbers group stated "...binding of ANP [ANF] to the extracellular domain of its receptor initiates a conformational change in the protein kinase-like domain, resulting in derepression of guanylate cyclase activity" [77]. Its central idea was that KHD in native ANF-RGC suppresses its catalytic module activity; ANF functions by relieving this suppression".

MODEL 2, Sharma's group challenged MODEL 1 on several grounds [78 – 81; reviewed in 39], most compelling of which was that partial or complete deletion of the KHD does not cause any significant elevation in the basal activity of the guanylate cyclase [78, 79].

Therefore, ANF signaling does not involve overcoming the KHD suppression of the guanylate cyclase activity, a central theme of the MODEL 1. In contrast, MODEL 2, proposed that the repressor of the CCD is the disulfide structural motif of ANF-RGC [79].

This motif is juxtaposed to the N-terminus of the TM (mistakenly referred as N-terminus in ref. 1). The motif represses the ARM domain, which, in turn, inhibits the CCD. ANF overcomes the disulfide structural motif-dependent inhibition of ARM (ATP Regulated Module) (discussed below), resulting in activation of the CCD of ANF-RGC (refer Figure 1. for simplicity in domain abbreviations).

ARM domain. To address the issue of whether ATP regulates, directly or indirectly, the transduction activity of ANF-RGC, ICDs of these guanylate cyclases were targeted [meticulously reviewed in: 1, 82]. Briefly, 3D-structure of the ICD: of ANF-RGC was modeled, its structure based features were experimentally validated, it underwent step-by-step dissection and biochemical analysis of its dissected components via mass spectroscopy, examination through steady state time-resolved tryptophan fluorescence and Forster Resonance Energy Transfer and site directed and reconstitution.

The findings were (1) ICD at its N-terminus side contains a glycine rich cluster (Grc) region--Gly^{503}-Arg-Gly^{505}-Ser^{506}-Asn-Tyr-Gly^{509} (Figure 1); this cluster is essential for ATP-modulation of ANF-dependent ANF-RGC signaling [78], hence it was named ARM. (2) The counterpart of the Grc region in CNP-RGC is Leu^{497}-Arg-Gly^{499}-Ser-Ser-Tyr-Gly^{503}. Since changing the ANF-RGC-Grc to the CNP-RGC-Grc region did not affect the ANF/ATP signaling [83], it was concluded that the two regions are interchangeable, are functionally identical, are essential in the ATP-dependent transduction activities of both ANF-RGC and CNP-RGC, and the ATP-dependent transduction mechanisms of these two guanylate cyclases are identical. (3) The Grc-residues, *Gly^{505} of ANF-RGC and Gly^{499} of CNPRGC represent the respective ATP bimodal switches of ANF and CNP signal transductions* [84]; and the core Grc sequence critical in formation of the functional ATP-binding pocket in both ANF-RGC and CNPRGC is common and is represented by the Gly-X-X-X-Gly motif. (4) ICD binds ATP directly [78, 80]. (5) It stretches from D^{481} to K^{771}. (6) GEOMETRY [80, 81, 85]. It is composed of two interconnected lobes. The smaller, N-terminal lobe is composed of five β strands, four of which are arranged in a twisted β-sheet, and one helix. The larger, C-terminal lobe is predominantly α-helical, composed of 8 α-helices and 2β-strands. (7) ATP folds in its binding pocket composed of Leu^{511}, Thr^{513}, Thr^{514}, Glu^{515}, Gln^{517}, Ala^{533}, Lys^{535}, Glu^{551}, Thr^{580}, Cys^{583}, Gly^{586}, Ser^{587}, Val^{635}, Lys^{630}, Ser^{632}, Asn^{633}, Thr^{645}, Asp^{646}. The Grc motif, Gly^{503}-Gly^{505}-Gly^{509}, spans the tip of the β1 strand and part of the connecting loop between β1 and β2 strands. It constitutes the floor of the pocket and produces a design that is conducive to fit the ATP molecule. (8) ATP interaction with its pocket induces three significant changes in the ARM's configuration: (i) it moves and rotates the floor of the pocket comprising β1, β2 strands, and the loop between the strands. The movement is of 3–4 A°; the rotation is by 15°. Gly^{505} is a critical PIVOT for both the movement and the rotation; (ii) It moves by 2–7 A°, but does not rotate, the β4 and β5 strands, and the loop; (iii) It moves the αEF helix by 2–5 A°. The movement exposes a hydrophobic sequence motif, ^{669}WTAPPELL^{675}, which involves interaction between the ARM and the catalytic domains and results in the ATP-dependent activation of ANF-RGC. (9) ATP through its ARM also causes distal hydrophobic and electrostatic changes in ANF-RGC. One of these relates to the surface hydrophobicity in the ^{669}WTAPPELL^{675} motif, which is distal to the ATP-binding pocket [39, 80, 81].

This change causes interaction between the ARM and the CCD domains of ANF-RGC. In addition, in another electrostatic potential change there is a loss of the negative charge in the WTAPPELL motif. Thus, it depicted this motif as a critical signaling component of the ARM.

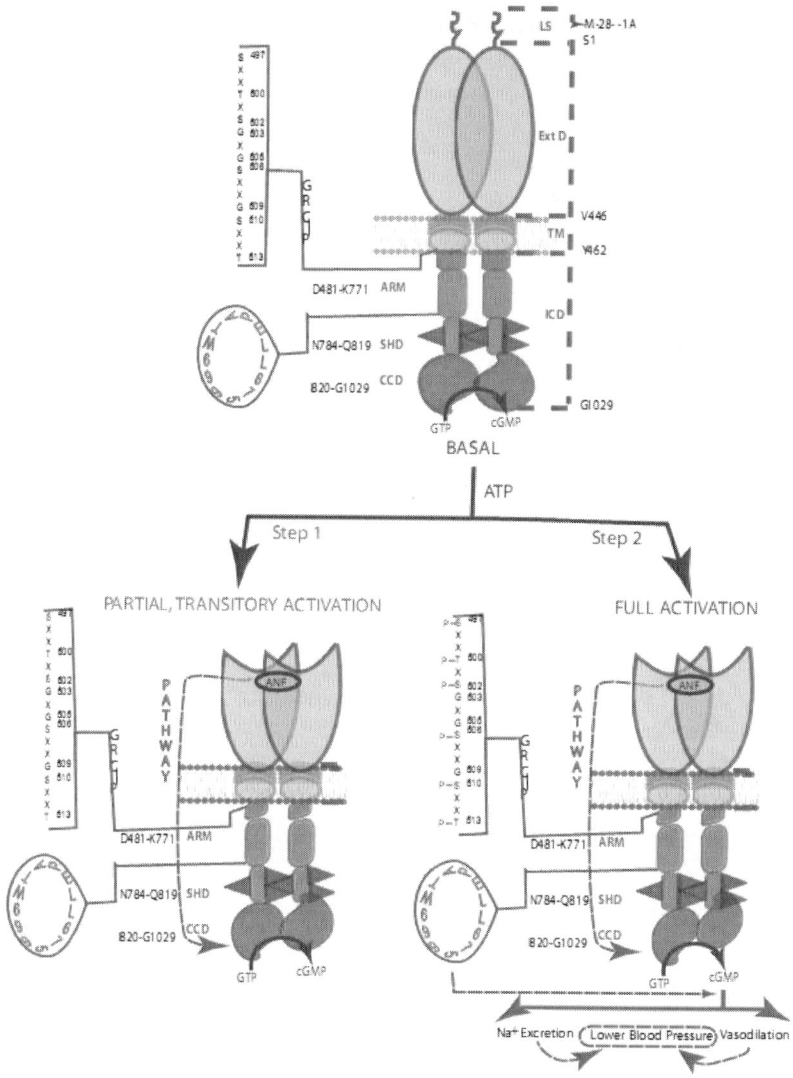

Figure 1. Two-Step ATP-modulated ANF signaling ANF-RGC model. *Upper panel*: ANF-RGC is a single transmembrane spanning homodimer protein. The dashed lines on the right show the defined boundaries of its segments: LS, leader sequence; Ext D, extracellular domain; TM, transmembrane domain; ICD, intracellular domain. The functional domains housed in the intracellular domain (ICD), their designated names and the amino acid residues constituting their boundaries are indicated at the left: ARM, the ATP regulated module; SHD-signaling helix domain; CCD-core catalytic domain. The CCD exists as an antiparallel homodimer. The sequence of the ARM domain's region of the glycine rich cluster and the phosphorylation sites are shown at the left-hand side as GRC~P; the helical ^{669}WTAPELL^{675} motif is shown in blue. In the absence of ANF, ANF-RGC is in its BASAL state, generating minimal cyclic GMP. Step1-*Lower left-hand panel*: The signal originates by the binding of one molecule of ANF to the ECD of ANF-RGC dimer. The binding modifies the juxtamembrane region, twists the trans-membrane domain , and induces a structural change in the ARM domain and prepares it for the ATP binding. ARM domain binds ATP what leads to a cascade of temporal and spatial changes and exposure of the hydrophobic motif, ^{669}WTAPELL^{675}, which facilitates its direct (or indirect) interaction with the catalytic module resulting in its partial, about 50%, activation. Step2: *Lower right-hand panel*: The six phosphorylation sites are brought from their buried to the exposed states (GRC~P domain). ATP through a hypothetical protein kinase phosphorylates the residues and the full activation (additional 50%) of ANF-RGC is achieved. ANF-dependent cyclic GMP is generated and it functions as the ANF second messenger of blood pressure regulation. The red dashed arrow indicates the physiological targets of cyclic GMP.

10. Two-Step ATP-Modulated ANF Signaling ANF-RGC Model Locked in with Blood Pressure Regulation

ANF signal transduction (Figure 1). In the absence of ANF, ANF-RGC is in its BASAL state (Top Panel), generating minimal cyclic GMP. The signal originates by the binding of one molecule of ANF to the ECD of ANF-RGC dimer [52, 86] (Left Panel). The binding modifies the juxtamembrane region where the disulfide ^{423}C-C^{432} structural motif is a key element [79, 87]. The signal twists the trans-membrane domain [88], induces a structural change in the ARM domain and prepares it for the ATP binding [89].

Step1. ARM domain binds ATP what leads to a cascade of temporal and spatial changes [80]. They involve (1) shift in ATP binding pocket position by 3-4 Å and rotation of its floor by 15°; G^{505} acts as a critical PIVOT for both the shift and the rotation; (2) movement by 2-7 Å but not the rotation of its β4 and β5 strands and its loop; (3) movement of its αEF helix by 2-5 Å. This movement exposes the hydrophobic motif, ^{669}WTAPELL^{675}, which facilitates its direct (or indirect) interaction with the catalytic module resulting in its partial, about 50%, activation [90] (Left Panel).

Step2: The six phosphorylation sites are brought from their buried to the exposed states [91] (Right Panel, GRCP domain). ATP through a hypothetical protein kinase phosphorylates the residues and the full activation (additional 50%) of ANF-RGC is achieved. ANF-dependent cyclic GMP is generated and it functions as the ANF second messenger of blood pressure regulation. Concomitantly, phosphorylation converts the ATP binding site from high to low affinity, ATP dissociates, and ANF-RGC returns to its ground state [91]. *(ii) Post cyclic GMP production events.*

Cyclic GMP generated suppresses the RAS-dependent secretion of aldosterone through a yet not fully understood mechanism. It is known, however, that it affects a number of effectors of aldosterone synthesis. They include cyclic GMP-gated channels, cyclic GMP-dependent protein kinases, and cyclic GMP-regulated phosphodiesterases [92 – 95]. Lowered aldosterone level results in the increase of sodium excretion and diuresis in the kidney and in maintaining physiological blood pressure.

With the exception of one group, there is a general consensus for the validity of the "Two-Step ATP-modulated ANF signaling ANF-RGC model". In the dissenting group's model ATP has no direct role in the ANF signaling of ANF-RGC; its only role is to phosphorylate ANF-RGC through a hypothetical protein kinase [96, 97].

Conclusion

Followed by its turbulent history where even its existence was questioned the discovery of ANF-RGC awakened the field of membrane guanylate cyclase transduction. It demonstrated that this membrane guanylate represents a new surface receptor family. The family sprouted and generated branches. Each branch was unique to itself in its biochemistry and physiology. The ROS-GC branch was linked with phototransduction, was $[Ca^{2+}]_i$-modulated and unlike all others was $[Ca^{2+}]$-inhibited. It was a 2-component signal transduction system, composed of a Ca^{2+}-sensor component and the transducer ROS-GC component. Intriguingly, this branch changed its behavior when accompanied by its Ca^{2+}

sensor component S100B protein. It was stimulated by the Ca^{2+} spikes. This behavior occurred in the photoreceptor cones [98], were absent in the rods and was also present in the photoreceptor bipolar synapse region [99]. In both cones and in the bipolar-synapse region it functioned as a $[Ca^{2+}]$ bimodal switch. In its third branch, ONE-GC (the olfactory receptor neuroepithelial guanylate cyclase), it became a surface receptor and was intracellularly modulated by the $[Ca^{2+}]$ signals generated in the ONE-GC neurons; the $[Ca^{2+}]$ signals amplifying the receptor signals. In this ever-growing saga, the latest findings of these authors demonstrate that ANF-RGC is a bimodal transduction switch [100]. It is modulated both by the hormonal signals of ANF and BNP and also by the intracellular $[Ca^{2+}]$ signal sensed by its sensor neurocalcin δ [100]. In the bimodal mode, it becomes a blood pressure regulator present in the glands of the adrenal and kidney. All indications are that the field will continue to grow. At the present time cyclic GMP, its generated second messenger, has been established as an omnipresent intracellular messenger of prokaryotes and eukaryotes. It plays a critical role in the control of physiological processes of cardiovasculature, smooth muscle relaxation, blood volume, cellular growth, sensory transduction, neural plasticity, learning and memory. The authors feel privileged to contribute in and watch its growth.

ACKNOWLEDGMENTS

The author (RKS) appreciates the past continuous support for over three-and– a half decade research from the National Institutes of health and the initial awards from the National Science Foundation and the Damon Runyon Walter Winchell Cancer Fund. The present work is supported by the National Heart, Lung and Blood Institute awards HL084584 and S82701 (TD).

REFERENCES

[1] Sharma RK. Membrane guanylate cyclase is a beautiful signal transduction machine: overview. *Mol. Cell Biochem.* 2010;334:3-36.

[2] Ashman DF, Lipton R, Melicow MM, Price TD. Isolation of adenosine 3', 5'-monophosphate and guanosine 3', 5'-monophosphate from rat urine. *Biochem. Biophys. Res. Commun.* 1963;11:330-334.

[3] Goldberg ND, Dietz SB, O'Toole AG. Cyclic guanosine 3',5'-monophosphate in mammalian tissues and urine. *J. Biol. Chem.* 1969;244:4458-4466.

[4] Ishikawa E, Ishikawa S, Davis JW, Sutherland EW. Determination of guanosine 3',5'-monophosphate in tissues and of guanyl cyclase in rat intestine. *J. Biol. Chem.* 1969;244:6371-6376.

[5] Goldberg ND, O'Dea RF, Haddox MK. Cyclic GMP. *Adv. Cyclic Nucleotide Res.* 1973;3:155-223.

[6] Goldberg ND, Haddox MK, Nicol SE, Glass DB, Sanford CH, Kuehl FA Jr., Estensen R. Biologic regulation through opposing influences of cyclic GMP and cyclic AMP: the Yin Yang hypothesis. *Adv. Cyclic Nucleotide Res.* 1975;5:307-330.

[7] Goldberg ND, Haddox MK. Cyclic GMP metabolism and involvement in biological regulation. *Annu. Rev. Biochem.* 1977;46:823-896.

[8] Murad F, Arnold WP, Mittal CK, Braughler JM. Properties and regulation of guanylate cyclase and some proposed functions for cyclic GMP. *Adv. Cyclic Nucleotide Res.* 1979;11:175-204.

[9] Gill GN, McCune RW. Guanosine 3',5'-monophosphate-dependent protein kinase. *Curr. Top Cell Regul* 1979;15:1-45.

[10] Kitabchi AE, Sharma RK. Corticosteroidogenesis in isolated adrenal cells of rats. I. Effect of corticotropins and 3',5'-cyclic nucleotides on corticosterone production. *Endocrinology* 1971;88:1109-1116.

[11] Sharma RK, Hashimoto K. Ultrastructural studies and metabolic regulation of isolated adrenocortical carcinoma cells of rat. *Cancer Res.* 1972;32:666-674.

[12] Sharma RK, Ahmed NK, Sutliff LS, Brush JS. Metabolic regulation of steroidogenesis in isolated adrenal cells of the rat. ACTH regulation of cGMP and cAMP levels and steroidogenesis. *FEBS Lett.* 1974;45:107-110.

[13] Sharma RK, Ahmed NK, Shanker G. Metabolic regulation of steroidogenesis in isolated adrenal cells of rat. Relationship of adrenocorticotropin-, adenosine 3':5'-monophosphate-and guanosine 3':5'-monophosphate-stimulated steroidogenesis with the activation of protein kinase. *Eur. J. Biochem.* 1976;70:427-433.

[14] Harrington CA, Fenimore DC, Farmer RW. Regulation of adrenocortical steroidogenesis by cyclic 3'-5'-guanosine monophosphate in isolated rat adrenal cells. *Biochem. Biophys. Res. Commun.* 1978;85:55-61.

[15] Neri G, Gambino AM, Mazzocchi G, Nussdorfer GG. Effects of chronic treatment with ACTH on the intracellular levels of cyclic-AMP and cyclic-GMP in the rat adrenal cortex. *Experientia* 1978;34:815-817.

[16] Perchellet JP, Shanker G, Sharma R. Regulatory role of guanosine 3',5'-monophosphate in adrenocorticotropin hormone-induced steroidogenesis. *Science* 1978;199:311-312.

[17] Sharma RK. Metabolic regulation of steroidogenesis in adrenocortical carcinoma cells of rat. Effect of adrenocorticotropin and adenosine cyclic 3':5'-monophosphate on corticosteroidogenesis. *Eur. J. Biochem.* 1973;32:506-512.

[18] Sharma RK, Sawhney RS. Metabolic regulation of steroidogenesis in isolated adrenal cell. Investigation of the adrenocorticotropic hormone, guanosine 3',5'-monophosphate, and adenosine 3',5'-monophosphate control step. *Biochemistry* 1978;17:316-321.

[19] Sayers G, Beall RJ, Seelig S. Isolated adrenal cells: adrenocorticotropic hormone, calcium, steroidogenesis, and cyclic adenosine monophosphate. *Science* 1972;175:1131-1133.

[20] Haksar A, Péron FG. The role of calcium in the steroidogenic response of rat adrenal cells to adrenocorticotropic hormone. *Biochim. Biophys. Acta* 1973;313:363-371.

[21] Bowyer F, Kitabchi AE. Dual role of calcium in steroidogenesis in the isolated adrenal cell of rat. *Biochem. Biophys. Res. Commun.* 1974;57:100-105.

[22] Perchellet JP, Sharma RK. Mediatory role of calcium and guanosine 3', 5'-monophosphate in adrenocorticotropin-induced steroidogenesis by adrenal cells. *Science* 1979;203:1259-1261.

[23] Kitabchi AE, Wilson DB, Sharma RK. Steroidogenesis in isolated adrenal cells of rat. II. Effect of caffeine on ACTH and cyclic nucleotide-induced steroidogenesis and its

relation to cyclic nucleotide phosphodiesterase (PDE). *Biochem. Biophys. Res. Commun.* 1971;44:898-904.

[24] Ahrens H, Paul AK, Kuroda Y, Sharma RK. Adrenocortical cyclic GMP-dependent protein kinase: purification, characterization, and modification of its activity by calmodulin, and its relationship with steroidogenesis. *Arch. Biochem. Biophys.* 1982;215:597-609.

[25] Nambi P, Aiyar NV, Sharma RK. Adrenocorticotropin-dependent particulate guanylate cyclase in rat adrenal and adrenocortical carcinoma: comparison of its properties with soluble guanylate cyclas and its relationship with ACTH-induced steroidogenesis. *Arch. Biochem. Biophys.* 1982;217:638-646.

[26] Nambi P, Sharma RK. Adrenocorticotropic hormone-responsive guanylate cyclase in the particulate fraction of rat adrenal glands. *Endocrinology* 1981;108:2025-2027.

[27] Nambi P, Sharma RK. Demonstration of ACTH-sensitive particulate guanylate cyclase in adrenocortical carcinoma. *Biochem. Biophys. Res. Commun.* 1981;100:508-514.

[28] Anglard P, Zwiller J, Vincendon G, Louis JC. Regulation of cyclic AMP and cyclic GMP levels by adrenocorticotropic hormone in cultured neurons. *Biochem. Biophys. Res. Commun.* 1985;133:286-292.

[29] Waldman SA, Rapoport RM, Murad F. Atrial natriuretic factor selectively activates particulate guanylate cyclase and elevates cyclic GMP in rat tissues. *J. Biol. Chem.* 1984;259:14332-14334.

[30] Hamet P, Tremblay J, Pang SC, Garcia R, Thibault G, Gutkowska J, Cantin M, Genest J. Effect of native and synthetic atrial natriuretic factor on cyclic GMP. *Biochem. Biophys. Res. Commun.* 1984;123:515-527.

[31] Paul AK. Particulate guanylate cyclase from adrenocortical carcinoma 494. Purification, biochemical and immunological characterization. Doctoral Thesis, University of Tennessee, 1986.

[32] Paul AK, Marala RB, Jaiswal RK, Sharma RK. Coexistence of guanylate cyclase and atrial natriuretic factor receptor in a 180-kD protein. *Science* 1987;235:1224-1226.

[33] Sharma RK. Guanylate cyclase and the atrial natriuretic factor receptor. Response to Waldman SA, Leitman DC, Anderson J, Murad F. *Science* 1988;240:805-806.

[34] Kuno T, Andresen JW, Kamisaki Y, Waldman SA, Chang LY, Saheki S, Leitman DC, Nakane M, Murad F. Co-purification of an atrial natriuretic factor receptor and particulate guanylate cyclase from rat lung. *J. Biol. Chem.* 1986;261:5817-5823.

[35] Takayanagi R, Inagami T, Snajdar RM, Imada T, Tamura M, Misono KS. Two distinct forms of receptors for atrial natriuretic factor in bovine adrenocortical cells. Purification, ligand binding, and peptide mapping. *J. Biol. Chem.* 1987;262:12104-12113.

[36] Meloche S, McNicoll N, Liu B, Ong H, De Léan A. Atrial natriuretic factor R1 receptor from bovine adrenal zona glomerulosa: purification, characterization, and modulation by amyloidal. *Biochemistry* 1988;27:8151-8158.

[37] Sharma RK, Jaiswal RK, Duda T. Second messenger role of cyclic GMP in atrial natriuretic factor receptor mediated signal transduction: 180-kDa membrane guanylate cyclase, its coupling with atrial natriuretic factor receptor and its regulation by protein kinase C. In: Biological and Molecular Aspects of Atrial Factors. Alan R. Liss, Inc., 1988;77-96.

[38] Sharma RK, Marala RB, Paul AK. Mediatory role of of cyclic GMP in receptor mediated signal transduction: Membrane guanylate cyclase and its coupling with atrial natriuretic factor receptor. In: B. M. Brenner, J.H. Laragh (eds). Advances in peptide research, vol. II. American Society of Hypertension Symposium Series. Raven Press, New York, 1988;198;61-77.

[39] Sharma RK. Evolution of the membrane guanylate cyclase transduction system. *Mol. Cell Biochem.* 2002 Jan;230(1-2):3-30.

[40] Marala RB, Sharma RK. Characterization of atrial-natriuretic-factor-receptor-coupled membrane guanylate cyclase from rat and mouse testes. *Biochem J* 1990;251:301-304.

[41] Ballermann BJ, Marala RB, Sharma RK. Characterization and regulation by protein kinase C of renal glomerular atrial natriuretic peptide receptor-coupled guanylate cyclase. *Biochem. Biophys. Res. Commun.* 1988;157:755-761.

[42] de Bold AJ. Atrial natriuretic factor of the rat heart. Studies on isolation and properties. *Proc. Soc. Exp. Biol. Med.* 1982;170:133-138.

[43] Cantin M, Genest J. The heart, an endocrine gland. *Ann. Endocrinol.* (Paris) 1985;46:219-228.

[44] Schwartz D, Geller DM, Manning PT, Siegel NR, Fok KF, Smith CE, Needleman P. Ser-Leu-Arg-Arg-atriopeptin III: the major circulating form of atrial peptide. *Science* 1985;229:397-400.

[45] Atlas SA, Laragh JH. Atrial natriuretic peptide: a new factor in hormonal control of blood pressure and electrolyte homeostasis. *Annu. Rev. Med.* 1986;37:97-414.

[46] Atarashi K, Mulrow PJ, Franco-Saenz R, Snajdar R, Rapp J. Inhibition of aldosterone production by an atrial extract. *Science* 1984;224:992-994.

[47] Chartier L, Schiffrin E, Thibault G, Garcia R. Atrial natriuretic factor inhibits the stimulation of aldosterone secretion by angiotensin II, ACTH and potassium in vitro and angiotensin II-induced steroidogenesis *in vivo*. *Endocrinology* 1984;115:2026-2028.

[48] De Léan A, Racz K, Gutkowska J, Nguyen TT, Cantin M, Genest J. Specific receptor-mediated inhibition by synthetic atrial natriuretic factor of hormone-stimulated steroidogenesis in cultured bovine adrenal cells. *Endocrinology* 1984;115:1636-1638.

[49] Kudo T, Baird A. Inhibition of aldosterone production in the adrenal glomerulosa by atrial natriuretic factor. *Nature* 1984;312:756-757.

[50] Chinkers M, Garbers DL, Chang MS, Lowe DG, Chin HM, Goeddel DV, Schulz S. A membrane form of guanylate cyclase is an atrial natriuretic peptide receptor. *Nature* 1989;338:78-83.

[51] Lowe DG, Chang MS, Hellmiss R, Chen E, Singh S, Garbers DL, Goeddel DV. Human atrial natriuretic peptide receptor defines a new paradigm for second messenger signal transduction. *EMBO J.* 1989;8:1377-1384.

[52] Duda T, Goraczniak RM, Sharma RK. Site-directed mutational analysis of a membrane guanylate cyclase cDNA reveals the atrial natriuretic factor signaling site. *Proc. Natl. Acad. Sci. USA* 1991; 88:7882-7886.

[53] Pandey KN, Singh S. Molecular cloning and expression of murine guanylate cyclase/atrial natriuretic factor receptor cDNA. *J. Biol. Chem.* 1990;265:12342-12348.

[54] Schulz S, Singh S, Bellet RA, Singh G, Tubb DJ, Chin H, Garbers DL. The primary structure of a plasma membrane guanylate cyclase demonstrates diversity within this new receptor family. *Cell* 1989;58:1155-1162.

[55] Chang MS, Lowe DG, Lewis M, Hellmiss R, Chen E, Goeddel DV. Differential activation by atrial and brain natriuretic peptides of two different receptor guanylate cyclases. *Nature* 1989;341:68-72.

[56] Currie MG, Fok KF, Kato J, Moore RJ, Hamra FK, Duffin KL, Smith CE. Guanylin: an endogenous activator of intestinal guanylate cyclase. *Proc. Natl. Acad. Sci. USA* 1992;89:947-951.

[57] Wiegand RC, Kato J, Currie MG. Rat guanylin cDNA: characterization of the precursor of an endogenous activator of intestinal guanylate cyclase. *Biochem. Biophys. Res. Commun.* 1992;185:812-817.

[58] de Sauvage FJ, Camerato TR, Goeddel DV. Primary structure and functional expression of the human receptor for Escherichia coli heat-stable enterotoxin. *J. Biol. Chem.* 1991;266:17912-17918.

[59] Singh S, Singh G, Heim JM, Gerzer R. Isolation and expression of a guanylate cyclase-coupled heat stable enterotoxin receptor cDNA from a human colonic cell line. *Biochem. Biophys. Res. Commun.* 1991;179:1455-1463.

[60] Hamra FK, Forte LR, Eber SL, Pidhorodeckyj NV, Krause WJ, Freeman RH, Chin DT, Tompkins JA, Fok KF, Smith CE, Duffin KL, Siegel NR, Currie MG. Uroguanylin: structure and activity of a second endogenous peptide that stimulates intestinal guanylate cyclase. *Proc. Natl. Acad. Sci. USA* 1993;90:10464-10468.

[61] Khare S, Wilson D, Wali RK, Tien XY, Bissonnette M, Niedziela SM, Bolt MJ, Sitrin MD, Brasitus TA. Guanylin activates rat colonic particulate guanylate cyclase. *Biochem. Biophys. Res. Commun.* 1994;203:1432-1437.

[62] Sharma RK, Duda T. Ca(2+)-sensors and ROS-GC: interlocked sensory transduction elements: a review. *Front Mol. Neurosci.* 2012;5:42.

[63] Pandey KN. The functional genomics of guanylyl cyclase/natriuretic peptide receptor-A: perspectives and paradigms. *FEBS J.* 2011;278:1792-1807.

[64] Sharma RK, Duda T, Sitaramayya A. Plasma membrane guanylate cyclase is a multimodule transduction system: Minireview. *Amino Acids* 1994;7:117-127.

[65] Sharma RK, Duda T. Plasma membrane guanylate cyclase. A multimodule transduction system. *Adv. Exp. Med. Biol.* 1997;407:271-279.

[66] Drewett JG, Garbers DL. The family of guanylyl cyclase receptors and their ligands. *Endocr. Rev.* 1994;15:135-162.

[67] Garbers DL, Koesling D, Schultz G. Guanylyl cyclase receptors. *Mol. Biol. Cell* 1994;5:1-5.

[68] Wilson EM, Chinkers M. Identification of sequences mediating guanylyl cyclase dimerization. *Biochemistry.* 1995;34:4696-4701.

[69] Duda T, Pertzev A, Sharma RK. Differential Ca(2+) sensor guanylate cyclase activating protein modes of photoreceptor rod outer segment membrane guanylate cyclase signaling. *Biochemistry* 2012;51:4650-4657.

[70] Duda T, Goraczniak RM, Sharma RK. Glutamic acid-332 residue of the type C natriuretic peptide receptor guanylate cyclase is important for signaling. *Biochemistry* 1994;33:7430-3743.

[71] Duda T, Goraczniak RM, Sharma RK. Single amino acid residue-linked signaling shifts in the transduction activities of atrial and type C natriuretic factor receptor guanylate cyclases. *Biochem. Biophys. Res. Commun.* 1995;212:1046-1053.

[72] Kurose H, Inagami T, Ui M. Participation of adenosine 5'-triphosphate in the activation of membrane-bound guanylate cyclase by the atrial natriuretic factor. *FEBS Lett.* 1987;219:375-379.

[73] Chang CH, Kohse KP, Chang B, Hirata M, Jiang B, Douglas JE, Murad F. Characterization of ATP-stimulated guanylate cyclase activation in rat lung membranes. *Biochim. Biophys. Acta* 1990;1052:159-165.

[74] Marala RB, Sitaramayya A, Sharma RK. Dual regulation of atrial natriuretic factor-dependent guanylate cyclase activity by ATP. *FEBS Lett.* 1991;281:73-76.

[75] Chinkers M, Singh S, Garbers DL. Adenine nucleotides are required for activation of rat atrial natriuretic peptide receptor/guanylyl cyclase expressed in a baculovirus system. *J. Biol. Chem.* 1991;266:4088-4093.

[76] Duda T, Goraczniak RM, Sitaramayya A, Sharma RK. Cloning and expression of an ATP-regulated human retina C-type natriuretic factor receptor guanylate cyclase. *Biochemistry* 1993;32:1391-1395.

[77] Chinkers M, Garbers DL. The protein kinase domain of the ANP receptor is required for signaling. *Science* 1989;245:1392-1394.

[78] Goraczniak RM, Duda T, Sharma RK. A structural motif that defines the ATP-regulatory module of guanylate cyclase in atrial natriuretic factor signalling. *Biochem. J.* 1992;282:533-537.

[79] Duda T, Sharma RK. Two membrane juxtaposed signaling modules in ANF-RGC are interlocked. *Biochem. Biophys. Res. Commun.* 2005;332:149-156.

[80] 80Duda T, Venkataraman V, Ravichandran S, Sharma RK. ATP-regulated module (ARM) of the atrial natriuretic factor receptor guanylate cyclase. *Peptides* 2005;26:969-984.

[81] Sharma RK, Yadav P, Duda T. Allosteric regulatory step and configuration of the ATP-binding pocket in atrial natriuretic factor receptor guanylate cyclase transduction mechanism. *Can. J. Physiol. Pharmacol.* 2001;79:682-691.

[82] Duda T. Atrial natriuretic factor-receptor guanylate cyclase signal transduction mechanism. *Mol. Cell Biochem.* 2010;334:37-51.

[83] Duda T, Goraczniak RM, Sharma RK. Core sequence of ATP regulatory module in receptor guanylate cyclases. *FEBS Lett.* 1993;315:143-148.

[84] Duda T, Goraczniak RM, Sharma RK. The glycine residue of ATP regulatory module in receptor guanylate cyclases that is essential in natriuretic factor signaling. *FEBS Lett.* 1993;335:309-314A.

[85] Duda T, Yadav P, Jankowska A, Venkataraman V, Sharma RK.Three dimensional atomic model and experimental validation for the ATP-Regulated Module (ARM) of the atrial natriuretic factor receptor guanylate cyclase. *Mol. Cell Biochem.* 2001;217:165-172.

[86] van den Akker F, Zhang X, Miyagi M, Huo X, Misono KS, Yee VC. Structure of the dimerized hormone-binding domain of a guanylyl-cyclase-coupled receptor. *Nature* 2000;406:101-104.

[87] Ogawa H, Qiu Y, Ogata CM, Misono KS. Crystal structure of hormone-bound atrial natriuretic peptide receptor extracellular domain: rotation mechanism for transmembrane signal transduction. *J. Biol. Chem.* 2004;279: 28625-28631.

[88] Parat M, Blanchet J, De Léan A. Role of juxtamembrane and transmembrane domains in the mechanism of natriuretic peptide receptor A activation. *Biochemistry* 2010;49:4601-4189.

[89] Burczynska B, Duda T, Sharma RK. ATP signaling site in the ARM domain of atrial natriuretic factor receptor guanylate cyclase. *Mol. Cell Biochem.* 2007;301:93-107.

[90] Duda T, Bharill S, Wojtas I, Yadav P, Gryczynski I, Gryczynski Z, Sharma RK. Atrial natriuretic factor receptor guanylate cyclase signaling: new ATP-regulated transduction motif. *Mol. Cell Biochem.* 2009;324:39-53.

[91] Duda T, Yadav P, Sharma RK. Allosteric modification, the primary ATP activation mechanism of atrial natriuretic factor receptor guanylate cyclase. *Biochemistry* 2011;50:1213-1225.

[92] Lohmann SM, Vaandrager AB, Smolenski A, Walter U, De Jonge HR. () Distinct and specific functions of cGMP-dependent protein kinases. *Trends Biochem. Sci.* 1997;22:307-312.

[93] Pfeifer A, Ruth P, Dostmann W, Sausbier M, Klatt P, Hofmann F. Structure and function of cGMP-dependent protein kinases. *Rev. Physiol. Biochem. Pharmacol.* 1999;135:105-149.

[94] MacFarland RT, Zelus BD, Beavo JA. High concentrations of a cGMP-stimulated phosphodiesterase mediate ANP-induced decreases in cAMP and steroidogenesis in adrenal glomerulosa cells. *J. Biol. Chem.* 1991;266:136-142.

[95] Côté M, Payet MD, Rousseau E, Guillon, G, Gallo-Payet N. Comparative involvement of cyclic nucleotide phosphodiesterases and adenylyl cyclase on adrenocorticotropin-induced increase of cyclic adenosine monophosphate in rat and human glomerulosa cells. *Endocrinology* 1999;140:3594-3601.

[96] Antos LK, Abbey-Hosch SE, Flora DR, Potter LR. ATP-independent activation of natriuretic peptide receptors. *J. Biol. Chem.* 2005;280:26928-26932.

[97] Potter LR, Abbey-Hosch S, Dickey DM. Natriuretic peptides, their receptors, and cyclic guanosine monophosphate-dependent signaling functions. *Endocr. Rev.* 2006;27:47-72.

[98] Wen XH, Duda T, Pertzev A, Venkataraman V, Makino CL, Sharma RK. S100B serves as a Ca(2+) sensor for ROS-GC1 guanylate cyclase in cones but not in rods of the murine retina. *Cell Physiol. Biochem.* 2012;29:417-430.

[99] Duda T, Koch KW, Venkataraman V, Lange C, Beyermann M, Sharma RK. Ca(2+) sensor S100beta-modulated sites of membrane guanylate cyclase in the photoreceptor-bipolar synapse. *EMBO J.* 2002;21:2547-2556.

[100] Duda T, Pertzev A, Sharma RK. Ca(2+) modulation of ANF-RGC: new signaling paradigm interlocked with blood pressure regulation. *Biochemistry* 2012;51:9394-9405.

In: Natriuretic Peptides
Editor: Kailash N. Pandey

ISBN: 978-1-63321-080-6

Chapter 5

NATRIURETIC PEPTIDE RECEPTOR: STRUCTURE AND SIGNALING MECHANISM

Kunio Misono*
Aki Bioscience Incorporated, Reno, Nevada

ABSTRACT

Atrial natriuretic peptide (ANP) and homologous B-type natriuretic peptide (BNP) are cardiac hormones that dilate blood vessels and stimulate natriuresis and diuresis, thereby lowering blood pressure and blood volume. ANP and BNP counterbalance the actions of the renin-angiotensin-aldosterone and neurohormonal systems, and play a central role in blood pressure and blood volume regulation. ANP and BNP also suppress cardiomyocyte growth and, hence, regulate cardiac remodeling. These activities are mediated by the A-type natriuretic peptide receptor (NPRA), a single transmembrane segment, guanylate cyclase (GC) linked receptor that functions as a homodimer. Our group has expressed the ligand-binding extracellular domain (ECD) of NPRA and performed structural and functional studies. ECD had the ligand binding affinity and selectivity indistinguishable from the full-length NPRA. The X-ray crystal structures of ECD with and without bound ANP showed that ECD occurs as a preformed dimer and that ANP binding causes 24° counter-clockwise rotation of each ECD monomer. This led us to propose that the rotation of the ECD triggers transmembrane signaling. Recently, the crystal structures of the GC catalytic domains (GCD) of green algae GC Cyg12 and cyanobacter GC Cya2 were reported. These structures resemble the structure of adenylate cyclase (AC) catalytic core which consists of C1 and C2 subdomain heterodimer. AC is activated by binding of $G_s\alpha$ to C2 and ensuing 7° rotation of C1 domain. It has been suggested that a similar domain rotation may occur in the GCD dimer and activate the GC activity.

In NPRA, the ANP-induced rotation of the ECD monomers, transmitted across the cell membrane, may induce a similar rotation in each of the dimerized GC catalytic domains. This rotation brings the active site residues into the catalytically competent geometry, thereby manifesting the GC enzymatic activity.

* E-mail: misono.ks@gmail.com.

Keywords: Natriuretic peptides, atrial natriuretic peptide receptor, guanylate cyclase, adenylate cyclase, single transmembrane segment receptor, transmembrane signal transduction, X-ray crystallography, single particle electron microscopy, hormone binding, structural motif

Enzyme: guanylate cyclase, EC 4.6.1.2

Subdivision: Structural biology

ABBREVIATIONS

ANP	atrial natriuretic peptide;
BNP	B-type natriuretic peptide;
CNP	C-type natriuretic peptide;
NPRA	ANP receptor (A-type natriuretic peptide receptor);
NPR	natriuretic peptide receptor;
RAA	renin-angiotensin-aldosterone;
GC	guanylate cyclase;
ECD	extracellular domain;
ICD	intracellular domain;
PKLD	protein kinase-like domain;
GCD	guanylate cyclase domain;
AC	adenylate cyclase

1. NATRIURETIC PEPTIDES AND PHYSIOLOGICAL ACTIVITIES

Atrial natriuretic factor, now more widely referred to as atrial natriuretic peptide (ANP), was discovered by deBold in 1981 as a principle in rat atrial extract that stimulates natriuresis [1]. ANP was subsequently shown also to relax aortic smooth muscle *in vitro* [2]. ANP was soon isolated and sequenced independently by several groups [3-8]. ANP has the structure shown in Figure 1. It has an intra-chain disulfide bond which is essential for its biological activity [7].

The residues essential for the activity were identified by amino acid residue substitutions and deletions via peptide synthesis approach. The critical residues for the activity include: Arg14, Phe8, the C-terminal Asn-Ser-Phe-Arg sequence, and the 17-residue ring structure (a comprehensive review can be found in [9]).

ANP stimulates salt excretion by increasing renal glomerular filtration rate and inhibiting Na^+-reabsorption at the collecting duct. ANP also suppresses aldosterone, vasopressin, and renin secretion. In the vasculature, ANP relaxes smooth muscle contraction [2, 10], causing an immediate hypotensive effect. By these activities, ANP plays a central role in blood pressure and blood volume regulation (reviews [11-13]). ANP also suppress cardiac cell proliferation, thereby playing a role in the maintenance and remodeling of the heart [14-17].

```
ANP        SLRRSSCFGGRIDRIGAQSGLGCNSFRY
BNP    NSKMAHSSSCFGQKIDRIGAVSRLGCDGLRLF
CNP          GLSKGCFGLKLDRIGSMSGLGC
```

Figure 1. Amino acid sequences of ANP, BNP and CNP from rat. Two Cys residues in each peptide form an intra-molecular disulfide-bond, which is essential for the activity [7]. Conserved residues are boxed.

Transgenic studies have shown that mice lacking ANP gene develop salt-sensitive hypertension [18]. Those lacking the ANP receptor gene develop salt-insensitive essential hypertension accompanied with severe cardiac hypertrophy, fibrosis, and dilatation, reminiscent of hypertensive heart disease in humans [19]. The data suggest that ANP plays important regulatory roles in the cardiovascular system, and aberrations in ANP and ANP receptor systems may lead to hypertension and heart disease. Peptides that are analogous to ANP have been identified, including B-type natriuretic peptide (BNP) and C-type natriuretic peptide (CNP) (Figure 1). BNP, produced mainly by the heart ventricle, has the activities essentially identical to ANP [20]. On the other hand, CNP has weak natriuretic and vasorelaxant activities [21]. It is thought to mediate fluid regulation via the central nervous system [20, 22-24] and also play local regulatory roles in variety of tissues [25-29]. ANP, BNP and CNP share common structural features such as the 17-residue peptide ring closed by a disulfide-bond and several conserved residues within the peptide ring (Figure 1).

2. Natriuretic Peptide Receptors — Molecular Topology

The hormonal activities of ANP and BNP are mediated by the A-type natriuretic peptide receptor (NPRA). It has a ligand selectivity in the order ANP>BNP>>CNP. The closely related B-type natriuretic peptide receptor (NPRB) has a reverse selectivity in the order CNP>>ANP>BNP [22, 24, 30, 31]. Both NPRA and NPRB are coupled to guanylate cyclase (GC), generating cGMP in response to ligand binding. cGMP, in turn, mediates hormonal activities via cGMP-regulated ion-channels, protein kinases, phosphodiesterases and possibly other effector molecules. NPRA was first purified from rat lung [32] and bovine adrenal cortex [33], where the ANP binding activity and the GC activity were co-purified. This finding lead to a series of cloning studies utilizing then the known sequence homology among GC enzyme domains. These studies have identified series of particulate GC's including: GC-A (identical to NPRA) [34], GC-B (NPRB) [30], GC-C (guanylin or enterotoxin receptor) [35] as well as the orphan receptors: retinal GC [36], retinal GC-2 [37], GC-E [38] and GC-F [38] in the eye; GC-D in the olfactory sensory neurons [39]; and GC-G in peripheral organs [40]. All were found to be single-span transmembrane proteins carrying an intracellular protein kinase-like domain (PKLD) and a GC catalytic domain (GCD).

Yet another related receptor, the natriuretic peptide clearance-receptor (NPR-C), has been found. NPR-C lacks the intracellular domain and is not GC-coupled [41]. It binds ANP, BNP, and CNP as well as some hormonally inactive fragments of those peptides with equally high affinities. NPR-C is thought to bind and remove excess peptides and their fragments from circulation [42, 43]. No hormonal activities have been linked to NPR-C.

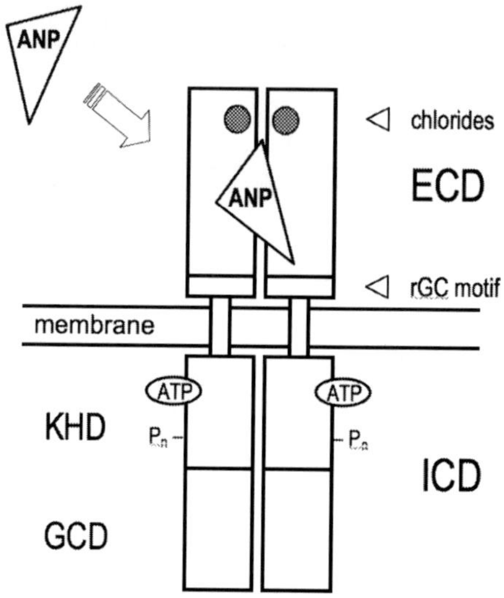

Figure 2. The molecular topology of the NPRA. The NPRA occurs as a pre-formed homodimer. Each monomer contains an extracellular domain (ECD) that binds ANP, a transmembrane segment, and an intracellular domain (ICD) consisting of a protein kinase-like domain (PKLD) and a GC catalytic domain (GCD). ANP binding to the dimerized ECD is asymmetric. The ECD contains a highly conserved chloride binding site [45, 64, 65] and a juxtamembrane GC-signature motif [46]. Bound chloride (cyan ball) is essential for ANP binding [44]. The juxtamembrane GC-signature motif plays a critical role in transmembrane signal transduction. The PKLD binds the positive allosteric effector ATP [48, 49] and is phosphorylated at multiple sites [51].

The over all molecular topology of NPRA is depicted in Figure 2. NPRA consists of an extracellular hormone-binding domain (ECD), a transmembrane segment, and an intracellular domain (ICD) consisting of PKLD and GCD. NPRA occurs as a homodimer. One molecule of ANP binds to the dimerized ECD. Each ECD contains a bound chloride near the ANP binding site, which is essential for ANP-binding [44, 45]. ECD also contains a unique highly conserved structure, termed receptor-GC motif, near the membrane, which consists of Pro417, disulfide linked Cys423-Cys432 and Phe425 [46]. The structure apparently is critical in signaling since single-residue mutations in this structure produce either uncoupled or constitutively active NPRA mutants [47].

PKLD binds ATP which augments GC activation by ANP [48, 49]. PKLD is also phosphorylated. Dephosphorylation in PKLD causes receptor desensitization [50, 51].

3. Expression of ECD, and Biochemical and Biophysical Characterization

To investigate the signaling mechanism of NPRA, our group has expressed and purified the ECD in a soluble form, and studied its structure and function [52]. The ECD was a logical starting point because it is the site where ANP binds and initiates signal transduction. The extracellular domain is also an obvious target for drug development.

ECD was expressed in COS cells and CHO cells after truncation of the transmembrane and intracellular domains [52]. ECD, which was secreted from the cells, was purified in a single step by ANP-affinity chromatography. The purified ECD bound ANP at an affinity (K_d at about 1 nM) comparable to that of the native full-length receptor. By LC/MS and Edman sequencing, we determined the disulfide-bond structure [53], and the sites of glycosylation and type of attached oligosaccharides [54] as schematically summarized in Figure 3. No other post-translational modifications were found.

Deglycosylation of ECD with endoglycosidases F2 and H did not affect ANP binding activity [54, 55]. On the other hand, glycosylation was found to be essential for the expression of active NPRA [56-58]. Together these findings suggest that glycosylation contributes to correct folding of the polypeptide and/or its translocation to the cell surface. However, the glycosy structure itself is not involved in or required for ANP binding.

We found that ANP binding to the receptor required the presence of chloride and was chloride concentration-dependent [44, 45]. This finding suggests that a change in chloride concentration at a target site, more specifically at the renal collecting tubule, may modulate ANP binding to the receptor and, hence, natriuretic responses [44, 45, 59].

ANP increases sodium excretion in urine by inhibiting Na^+ reabsorption at the collecting. This mechanism involves: binding of ANP to NPRA, resulting increase in GMP, and closure of Na^+ channels by cGMP (Figure 4; more discussions in the legend). In salt depletion or during intravenous diuretics treatments, sodium and chloride levels in the lumen at the collecting tubule decreases to low mM to sub-mM levels. At such low chloride levels, ANP cannot bind NPRA on the luminal surface of the collecting tubule and hence is unable to inhibit the Na^+ channels. This mechanism prevents undesirable loss of sodium. This autonomous, chloride-dependent regulatory mechanism operating on NPRA on the luminal surface of the collecting tubule may explain the attenuated natriuretic responses to ANP, such as those observed in animals in low-salt diet [60] and in late-stage heart failure patients [61-63]. X-ray crystallographic studies of ECD indeed have found a protein bound chloride in the ECD structure (see below) [45, 64]. Sedimentation equilibrium studies showed that the purified ECD in the absence of ANP was in a monomer-dimer equilibrium with a dissociation constant of ~500 nM. In the presence of ANP, the dissociation constant was reduced to 10 nM, suggesting strong stabilization of the ANP-bound complex [45]. Analysis of ANP binding by isothermal titration calorimetry gave a receptor-to-ANP binding stoichiometry at 2:1. The same analysis gave a large ∂H value at -42.5 kcal/mol, again suggesting strong stabilization in the complex (unpublished data).

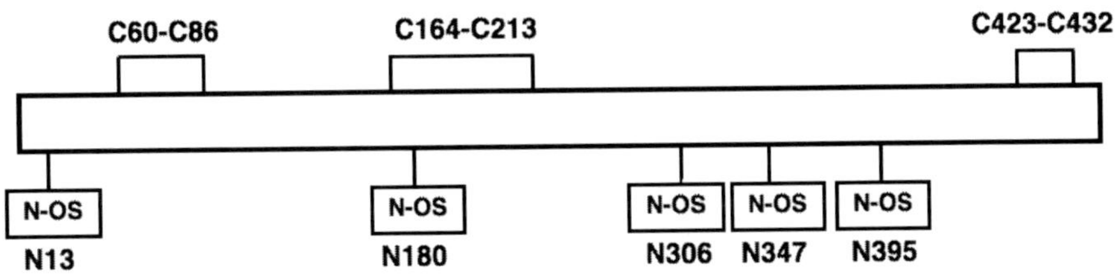

Figure 3. Diagram illustrating the covalent structure of the ECD. The ECD contains five N-linked oligosaccharides (N-OS, in boxes) [85] and three disulfide-bonds (orange lines) [86]. The glycosylated Asn residues and disulfide-bonded Cys residues are indicated. No free Cys residue is present in the ECD. No other post-translational modifications are found.

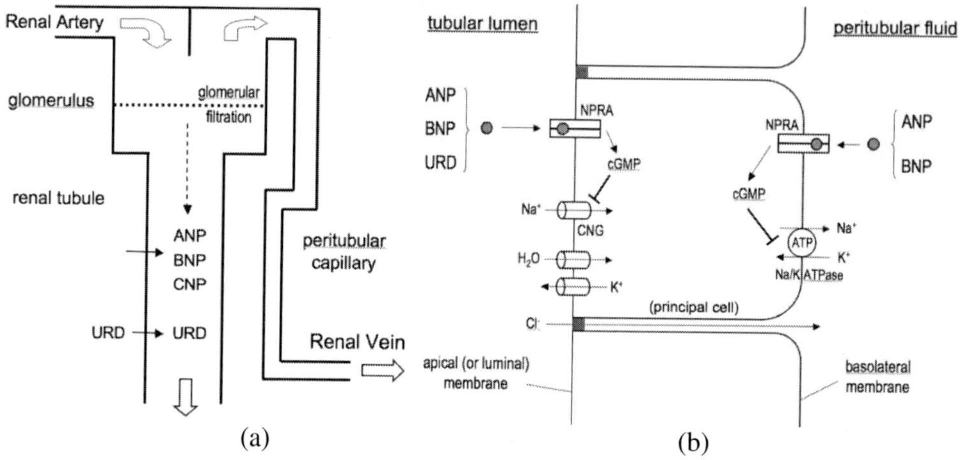

Figure 4. (a) Renal glomerular filtration. Approximately, 25% of the cardiac output pass through the kidney (renal blood flow, roughly 1.3 liter per min), of which about 10% is filtered through the glomerulus. Most of the filtered electrolytes and water (>99%) must be reabsorbed and returned to circulation. This reabsorption is essential for survival. Small molecules with the molecular weight less than ~20,000 are also filtered through the glomerulus [87, 88], which include natriuretic peptides. Natriuretic peptides survive brush border proteases at least in part and are excreted in urine [89-94]. Natriuretic peptides and urodilatin (URD) [95] are also synthesized by the tubular cells and secreted into the lumen [96]. Natriuretic peptide levels in urine are known to increase for example in heart failure [97, 98]. (b) Roles of natriuretic peptides in the regulation of sodium transport. ANP, BNP and URD in luminal fluid bind to NPRA localized on the luminal membrane of collecting tubule cells [95, 99, 100], stimulating cGMP production. cGMP, in turn, inhibits the cGMP-gated channel (CNG) and blocks Na^+ reabsorption [101], thus increasing Na^+ excretion (or natriuresis). On the basolateral side, ANP and BNP in peritubular circulation bind to and stimulate NPRA on the basolateral membrane. Increased cGMP inhibits Na^+/K^+ ATPase on the basolateral membrane and prevent transport of Na^+ back into circulation [102]. The former luminal mechanism is believed to function in rapid and direct control of sodium transport, while the latter basolateral mechanism functions in a slower regulation. Chloride-dependent control of NPRA occurs on the luminal membrane. Low level of Cl^- in the luminal fluid, which occurs for example in salt depletion or diuretics treatment, makes NPRA unable to bind its ligand and produce cGMP. This makes the sodium channels remain open irrespective of the presence of natriuretic peptides, allowing Na^+ reabsorption to occur. This chloride-dependent regulatory mechanism on NPRA may be responsible for the renal insensitivity to ANP and BNP found in certain physiological and disease conditions [44, 45, 59].

4. X-Ray Crystallographic Studies of the ECD Structure

Our lab has determined the crystal structures of both the apo ECD [64] and hormone-bound ANP-ECD complex [65]. These structures have revealed the interactions involved in hormone binding and a hormoneinduced structural change in the receptor that apparently triggers signal transduction.

4.1. Crystal Structure of the Apo ECD

ECD crystallizes as a dimer, consistent with its tendency to spontaneously dimerize at high concentrations [45, 66].

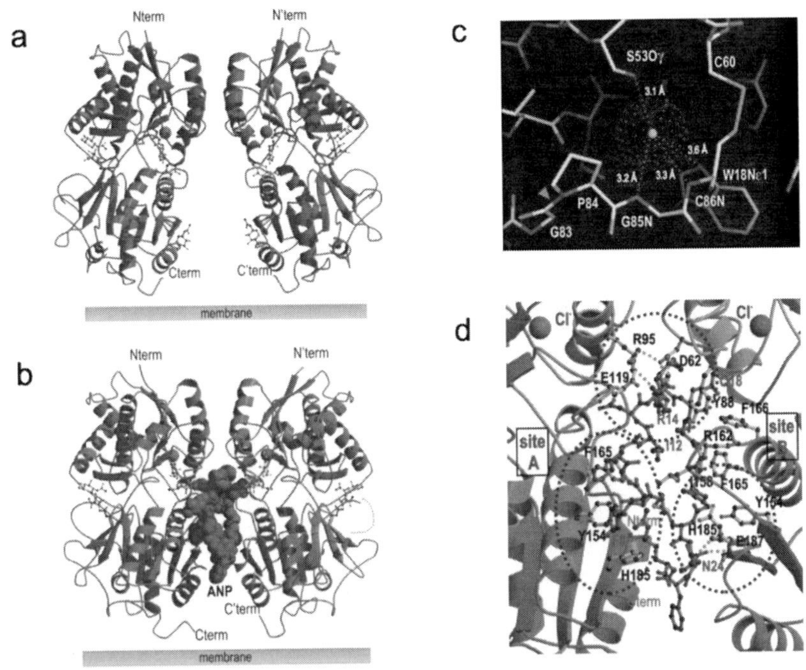

Figure 5. (a, b) Crystal structures of apo ECD dimer (PDB: 1DP4) and ANP-ECD complex (PDB: 1T34) [64, 65]. ANP is shown in green. Protein-bound chloride atoms are shown by magenta balls. (c) Close-up view of chloride and the chloride binding site. (d) Close-up view of ANP binding interactions. Major interactions are shown circled.

Each ECD molecule contains two subdomains: the membrane-distal and membrane-proximal subdomains (Figure 5a). ECD occurs dimerized through the membrane-distal subdomain. A non-covalently associated chloride ion is located near the dimerization interface and also close to the ANP-binding site (see below). The highly conserved receptor-GC motif [47] is anchored to the base of the membrane-proximal subdomain near the membrane via residues Pro417 and Phe425 [64, 65].

4.2. Crystal Structure of the ANP-ECD Complex

Two ECD molecules binds one molecule of ANP, forming a 2:1 complex (Figure 5b) [65]. Because the ANP molecule has no internal symmetry, its binding is asymmetric. Two ECD monomers are related by non-crystallographic two-fold symmetry. A non-covalently bound chloride, required for binding, is localized near the ANP binding site (Figure 5c). The strategically located chloride site close to the ANP binding site and to the dimerization interface is consistent with its critical role in NPRA regulation.

Binding of ANP to the receptor involves three major interactions: i) Arg14 of ANP (Arg14(ANP)) hydrogen bonding to receptor residues Asp62, Arg95, and Asp119; ii) Phe8(ANP) making aromatic contacts with Tyr154, Phe165, and Tyr172; and iii) the C-terminal sequence of ANP (residues Asn24-Arg27) forming a partial parallel β-sheet structure with the receptor sequence Gln186- Phe188 (Figure 5d). These binding interactions in the crystal structure are consistent with the structure-activity relationship described for ANP [9].

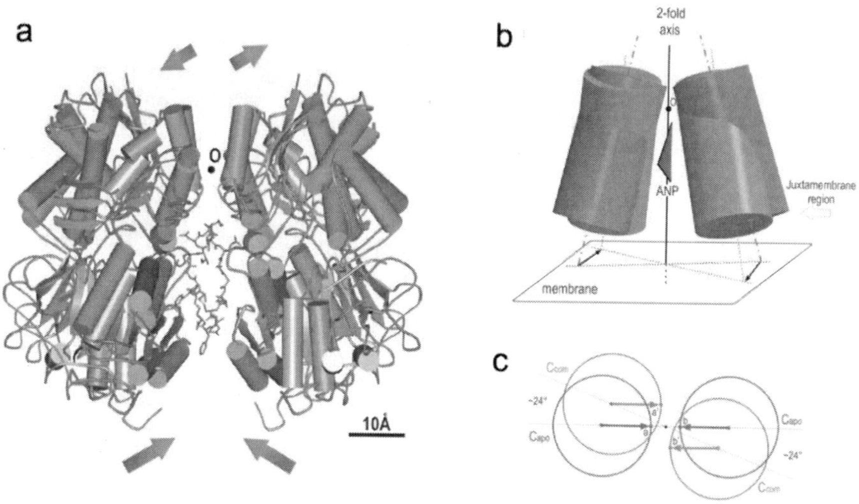

Figure 6. (a) ANP binding induced conformational change in the ECD dimer. ANP binding causes the dimer-interface in the apo ECD dimer (blue) to open and the two monomers to undergo a twisting motion (arrows) to form an ANP-bound complex (orange). (b) Schematic illustration of ANP-induced change in the ECD dimer structure. ANP binding causes a twisting motion of the two ECD monomers from the apo position (blue) to the complex position (orange) [65, 67]. (c) Viewed toward the membrane, the juxtamembrane domains in the apo form (blue circles) translate to the complex position (orange circles) with essentially no change in the inter-domain distance. The arrows depict parallel translocation. This motion causes a change in the angular relationship between the two domains equivalent to rotating each domain by 24° counter-clockwise. Because the dimerized receptor is free to spin or move about in the cell membrane, this rotation motion occurring in the juxtamembrane domains would be the only structural change that is "recognized" by the receptor upon ANP binding.

Namely, the ANP residues essential for its biological activities are Arg-14, Phe8 and the C-terminal Asn-Ser-Phe-Arg sequence. These residues contributing to binding are all conserved in BNP (Figure 1), being consistent with the fact that ANP and BNP have essentially the same activities and potencies.

4.3. ANP-Induced Conformational Change in the Dimerized ECD

Binding of ANP does not induce a significant change in each monomer structure (root mean square deviation of 426 Cα atoms, 0.64 Å), but causes a large shift in the relative positions of the two ECD monomers [65].

Upon ANP binding, the two monomers undergo a twist motion, and their membrane-proximal subdomains close onto the ANP molecule (Figure 6a, 6b). The motion causes the two juxtamembrane domains of the dimer to translate In opposite directions in a Ferris wheel-like motion (Figure 6c). This alters the relative angular relationship between the two juxtamembrane domains, that is equivalent to rotating each domain by 24° counter-clockwise (looking toward the cell membrane). This is the only significant structural change that occurs in response to ANP binding. This ligand-induced rotation of the ECD monomers likely triggers transmembrane signal transduction [65, 67].

The structures of the apo ECD dimer and the ANP-ECD complex were also observed by single particle electron microscopy [68] (not shown). This method allows determination of the native structures in the absence of crystal contacts and in closer to native buffer solution conditions. The three-dimensional reconstructions of the apo ECD dimer and the ANP-ECD complex are in close agreement with their respective x-ray structures. The data further confirm that the ECD occurs as a pre-formed homodimer in the head-to-head configuration and undergoes the large and distinct quaternary structural change, as seen in X-ray structures, in response to ANP binding.

5. Studies of the ICD

To date there is no direct structure information available on the ICD. Our current knowledge of the ICD is based on studies done by mutagenesis techniques such as site-directed mutagenesis, deletion mutagenesis, and creation of chimeric proteins. These studies have shown the ICD consists of the regulatory PKLD, the catalytic GCD, and possibly a linker segment connecting the two domains and playing a regulatory role.

5.1. PKLD

The PKLD is thought to be the site of ATP binding. ATP is a positive allosteric effector for NPRA, which augments GC activation by ANP [48, 49]. Contrary to this model, others have reported the absence of such stimulatory effects by ATP [69]. The PKLD is phosphorylated at multiple sites [51, 70]. Desensitization of NPRA in cultured cells upon extended exposure to ANP is accompanied with dephosphorylation [70, 71]. The PKLD structure has been modeled based on sequence homology to protein tyrosine kinases [72]. This model has found some support by site-directed mutagenesis data. The actual structure of the PKLD is yet to be determined. Thus, the structure and the regulatory role of this domain remain largely unsolved.

5.2. Putative Linker Segment

The PKLD is connected to the C-terminal GCD by a putative ~50 residue linker region. Deletion mutagenesis studies have suggested that this region is necessary for receptor dimerization and the GC activity [73]. By modeling based on its amino acid sequence, this region has been proposed to form an amphipathic helix in the NPRA monomer and a coiled-coil structure in the dimer.

On the other hand, more recent studies by systematic site-directed mutagenesis in the guanylin receptor (or GC-C) have suggested that this region does not contain a coiled coil structure nor is necessary for dimerization [74]. Thus, the conformation or the role of this linker region remains uncertain.

5.3. GCD

The structure of the NPRA GCD has not been determined. The GC catalytic cores of NO-binding soluble GCs as well as those of membrane bound GCs all have high sequence similarity to adenylate cyclase (AC) catalytic core. Hence, it has been long speculated that the structure of GCD may be similar to that of the AC catalytic core. It is only recently that the crystal structures of GC have been determined: One structure is that of a 188-residue GC catalytic core of Cyg12 [75] (a GC from eukaryotic unicellular green algae which is homologous to mammalian soluble GCs); and the other is the structure of a 202-residue catalytic core of Cya2 [76] (a GC from prokaryote cyanobacter, which is homologous to mammalian membrane-bound GCs). These are the first structures for any GCs that have been determined more than 10 years after the first reports of the AC structures [77-79].

Both Cyg12 GCD and Cya2 GCD were expressed in *E. coli,* and without the putative linker region discussed above. Yet, both formed and crystallized as dimers, casting doubts on the proposed role of the linker region in dimerization and GC activity [73]. Cyg12 GCD had a specific activity of 2.8 μmol/min/mg in the presence of Mn^{2+} but much lower activity (less than 1%) in the presence of Mg^{2+} [75], as generally found for mammalian GCs [80] and ACs [81]. On the other hand, the activity of Gya2 GCD was significantly lower at 1.5 nmol/min/mg [76]. Yet, the GC activity showed a similar metal ion-dependence exhibiting a significantly higher specific activity in the presence of Mn^{2+} than in the presence of Mg^{2+}, and even higher when both metal ions were present. In eukaryotes, manganese is a trace element, and magnesium ions are assumed to be the physiological active-site ions. The enhancement of the catalytic activity by Mn^{2+} ions is considered unlikely to have any physiological relevance [81]. Nevertheless, the observed GCD homodimerization and the metal ion-dependence of the catalytic activity support the integrity of the expressed proteins.

As expected from the high sequence homology with ACs, both Cyg12 GCD and Gya2 GCD monomers have basically the same protein fold as the mammalian AC catalytic core. Each GCD monomer contains a 7-stranded β-sheet surrounded by several α-helices. In the dimer, two GCD monomers are related by a two-fold symmetry axis that runs through the central dimer cleft, and form a wreath-like structure (Figure 7a, Cyg12 GCD dimer). The central cavity between the two monomers contains two symmetric active sites. The catalytic residues in each active site are supplied jointly by both monomers. The active-site residues in each monomer are located at positions homologous to their counterparts in ACs. Such conserved active site residues include: two metal-binding Asp residues, ribose-orienting Asn residue, transition state-stabilizing Arg residue, triphosphate-binding Arg and Lys residues [75]. The guanine base-recognition residues Glu and Cys in Cyg12 [75], and Glu and Gly residues in Cya2 [76] are similarly conserved at the positions close to the locations of adenine base-recognizing Lys and Asp in ACs [82].

In AC, the catalytic core consists of a heterodimer of C1 and C2 subdomains [77-79]. C1 and C2 domains are related by a pseudo two-fold symmetry and form a heterodimeric wreath-like structure. By structural comparison, the dimer structure of Cyg12 GCD (Figure 7a) is similar to the open, inactive conformation of the AC catalytic core, which must close to be catalytically active [75].

On the other hand, the Cya2 GCD dimer is in a closed conformation that must open in order to bind the substrate GTP for catalysis [76]. This closed structure of the Cya2 GCD dimer may explain its low specific GC activity.

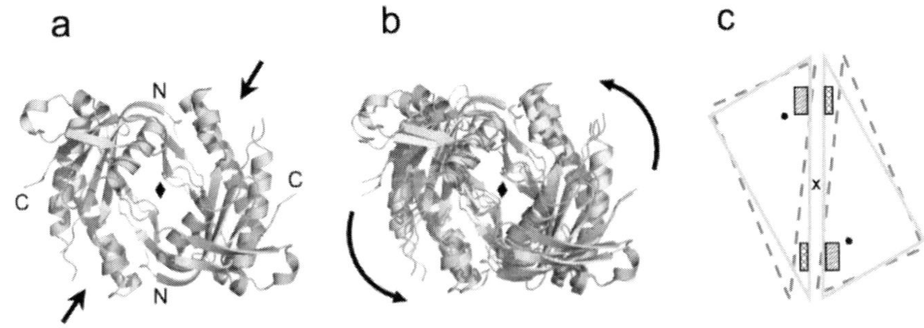

Figure 7. (a) Structure of Cyg12 GCD dimer (PDB: 3ET6), which is an open inactive conformation [75]. The arrows show the surface grooves in the GCD that correspond to the $G_s\alpha$ binding site in AC C1 domain [78]. The N- and C-terminal ends of each monomer are labeled. The 2-fold symmetry axis in the dimer runs perpendicular to the plane of the page. The dimer structure is seen from the C-terminal end. (b) Model of the closed active GCD dimer conformation (yellow) overlaid onto the open inactive GCD dimer (cyan). The rotation of each of the two domains (around each own axis) leads to the closed and active conformation (arrows). (c) Schematic presentation of a possible activation mechanism of GC. GCD monomers are represented by triangles. The open inactive GCD dimer conformation is depicted by two broken line triangles. Upon receptor activation, rotation of each GCD monomer around the axis (estimate center of rotation shown by the dot in the triangle) to the active closed position (solid line triangles) brings the active site residues (shown schematically by hatched and cross-hatched boxes) into the catalytically active positions, thereby manifesting the GC activity.

Interestingly, the specific activity of Cyg12 GCD at 2.8 μmol/min/mg is roughly comparable to the GC activities of the full-length NPRA purified from various tissues and species, which range from 1.8 μmol/min/mg to 23 μmol/min/mg [32, 33, 83, 84]. Together, these data seem to suggest that the structure of Cyg12 GCD dimer may reflect the structure of the GCD in the dimerized full-length NPRA in its basal state.

6. Possible Mechanism of NPRA Signaling

Signaling by G-protein coupled receptors involves stimulation of AC by $G_s\alpha$ released from the heterotrimeric G-protein in response to receptor activation by a ligand. A possible mechanism for this AC activation by $G_s\alpha$ has been proposed based on the crystal structures [77-79]. In the proposed mechanism, $G_s\alpha$ binds to the C2 domain of the AC C1/C2 heterodimer. This binding causes a 7° rotation of the C1 domain around an axis that runs through the C1 domain and roughly parallel to the central cleft axis. This movement brings the catalytic residues on the C1 domain closer to the catalytic residues on the C2 domain, thereby forming a catalytically competent active site [78].

In the Cyg12 GCD crystal structure, two GCD monomers are in an inactive, open conformation (Figure 7a) [75]. It has been suggested that activation of Cyg12 GCD may be mediated by a domain rotation similar to that of the AC C1 domain rotation induced by $G_s\alpha$ binding to C2. Figure 7b shows an overlay of the closed, activated GCD dimer configuration modeled based on such rotations of the two monomers over the open inactive conformation determined for Cyg12 GCD dimer. Closing of the dimer conformation brings the active site residues into proper proximity and geometry, thereby enabling enzyme catalysis. This movement is schematically shown in Figure 7c.

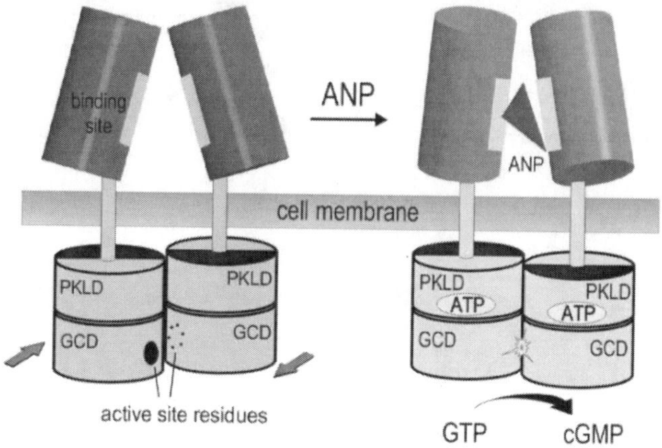

Figure 8. Rotation mechanism proposed for transmembrane signaling by the NPRA [65, 103]. ANP binding causes a 24° counter-clockwise rotation of each of the two ECD monomers. This rotation, transduced across the membrane, reorients the two ICDs. The reorientation of the ICD and its subdomain GCD brings two active sites of the GCD dimer to optimal proximity and orientation, thereby manifesting the GC catalytic activity. In the figure, only one of the two symmetric active sites in the GCD dimer is illustrated.

The NPRA and other receptor-GCs exist as homodimers. Their GCDs are similarly expected to form homodimer structures. We speculate that the ANP-induced rotation of the two ECD monomers and their juxtamembrane regions may be transduced across the transmembrane helices and, through the PKLD, causes the two GCD domains to rotate [65, 67] (Figure 8). This rotation may bring the GCD dimer from an open inactive conformation to a closed and active conformation (modeled in Figure 7b), thereby enabling GC catalysis. In this signaling process, the PKLD may play a regulatory role by binding with the allosteric effector ATP or by its phosphorylation state. A segment in PKLD (or the putative linker region) may also interact with the groves on the GCD corresponding to the $G_s\alpha$ binding site in AC catalytic core and participate in GCD activation.. The actual and detailed mechanism of GC activation by ANP, namely the mechanism of signal transduction, must await determination of the NPRA's GCD structure and ultimately the structure of the full-length NPRA with and without bound ANP.

Acknowledgments

I thank my research associates, research assistants, students and my collaborators who worked with me in this study. Without their contributions, this work surely would not have been possible. This research was supported by grants from the National Institute of Health and the American Heart Association.

REFERENCES

[1] de Bold AJ, Borenstein HB, Veress AT, et al. A rapid and potent natriuretic response to intravenous injection of atrial myocardial extract in rats. *Life Science* 1981; 28:89-94.

[2] Currie MG, Geller DM, Cole BR, et al. Bioactive cardiac substances: potent vasorelaxant activity in mammalian atria. *Science* 1983; 221:71-73.

[3] Currie MG, Geller DM, Cole BR, et al. Purification and sequence analysis of bioactive atrial peptides (atriopeptins). *Science* 1984; 223:67-69.

[4] Flynn TG, de Bold ML, de Bold AJ. The amino acid sequence of an atrial peptide with potent diuretic and natriuretic properties. *Biochem. Biophys. Res. Commun.* 1983; 117: 859-865.

[5] Atlas SA, Kleinert HD, Camargo MJ, et al. Purification, sequencing and synthesis of natriuretic and vasoactive rat atrial peptide. *Nature* 1984; 309: 717-719.

[6] Kangawa K, Matsuo H. Purification and complete amino acid sequence of alpha-human atrial natriuretic polypeptide (alpha-hANP). *Biochem. Biophys. Res. Commun.* 1984; 118: 131-139.

[7] Misono KS, Fukumi H, Grammer RT, et al. Rat atrial natriuretic factor: complete amino acid sequence and disulfide linkage essential for biological activity. *Biochem. Biophys. Res. Commun.* 1984; 119:524-529.

[8] Misono KS, Grammer RT, Fukumi H, et al. Rat atrial natriuretic factor: isolation, structure and biological activities of four major peptides. *Biochem. Biophys. Res. Commun.* 1984; 123:444-451.

[9] Bovy PR. Structure activity in the atrial natriuretic peptide family. *Med. Res. Rev.* 1990; 10:115-142.

[10] Grammer RT, Fukumi H, Inagami T, et al. Rat atrial natriuretic factor. Purification and vasorelaxant activity. *Biochem. Biophys. Res. Commun.* 1983; 116: 696-703.

[11] Cantin M, Gutkowska J, Thibault G, et al. The heart as an endocrine gland. *J. Hypertens. Suppl.* 1984; 2:S329-331.

[12] Sonnenberg H. The physiology of atrial natriuretic factor. *Can. J. Physiol. Pharmacol.* 1987; 65:2021-2023.

[13] Gerzer R, Heim JM, Schutte B, et al. Cellular mechanisms of action of atrial natriuretic factor. *Klin. Wochenschr.* 1987; 65:109-114.

[14] Itoh H, Pratt RE, Dzau VJ. Atrial natriuretic polypeptide inhibits hypertrophy of vascular smooth muscle cells. *J. Clin. Invest.* 1990; 86: 1690-1697.

[15] Chrisman TD, Garbers DL. Reciprocal antagonism coordinates C-type natriuretic peptide and mitogen-signaling pathways in fibroblasts. *J. Biol. Chem.* 1999; 274: 4293-4299.

[16] Suenobu N, Shichiri M, Iwashina M, et al. Natriuretic peptides and nitric oxide induce endothelial apoptosis via a cGMP-dependent mechanism. *Arterioscler. Thromb. Vasc. Biol.* 1999; 19:140-146.

[17] Wu CF, Bishopric NH, Pratt RE. Atrial natriuretic peptide induces apoptosis in neonatal rat cardiac myocytes. *J. Biol. Chem.* 1997; 272:14860-14866.

[18] John SW, Krege JH, Oliver PM, et al. Genetic decreases in atrial natriuretic peptide and salt-sensitive hypertension [published erratum appears in Science 1995 Mar 24; 267 (5205): 1753]. *Science* 1995; 267:679-681.

[19] Oliver PM, Fox JE, Kim R, et al. Hypertension, cardiac hypertrophy, and sudden death in mice lacking natriuretic peptide receptor A. *Proc. Natl. Acad. Sci. U S A* 1997; 94: 14730-14735.

[20] Nakao K, Ogawa Y, Suga S, et al. Molecular biology and biochemistry of the natriuretic peptide system. I: Natriuretic peptides. *J. Hypertens.* 1992; 10:907-912.

[21] Sudoh T, Minamino N, Kangawa K, et al. C-type natriuretic peptide (CNP): a new member of natriuretic peptide family identified in porcine brain. *Biochem. Biophys. Res. Commun.* 1990; 168:863-870.

[22] Koller KJ, Lowe DG, Bennett GL, et al. Selective activation of the B natriuretic peptide receptor by C-type natriuretic peptide (CNP). *Science* 1991; 252:120-123.

[23] Nakao K, Ogawa Y, Suga S, et al. Molecular biology and biochemistry of the natriuretic peptide system. II: Natriuretic peptide receptors. *J. Hypertens.* 1992; 10: 1111-1114.

[24] Imura H, Nakao K, Itoh H. The natriuretic peptide system in the brain: implications in the central control of cardiovascular and neuroendocrine functions. *Front. Neuroendocrinol.* 1992; 13:217-249.

[25] Agui T, Yamada T, Legros G, et al. Expression of receptors for atrial natriuretic peptide on the murine bone marrow-derived stromal cells. *Endocrinology* 1992; 130: 2487-2494.

[26] Suga S, Nakao K, Itoh H, et al. Endothelial production of C-type natriuretic peptide and its marked augmentation by transforming growth factor-beta. Possible existence of "vascular natriuretic peptide system". *J. Clin. Invest.* 1992; 90: 1145-1149.

[27] Hagiwara H, Sakaguchi H, Itakura M, et al. Autocrine regulation of rat chondrocyte proliferation by natriuretic peptide C and its receptor, natriuretic peptide receptor-B. *J. Biol. Chem.* 1994; 269:10729-10733.

[28] Yasoda A, Ogawa Y, Suda M, et al. Natriuretic peptide regulation of endochondral ossification. Evidence for possible roles of the C-type natriuretic peptide/guanylyl cyclase-B pathway. *J. Biol. Chem.* 1998; 273:11695-11700.

[29] Suda M, Komatsu Y, Tanaka K, et al. C-Type natriuretic peptide/guanylate cyclase B system in rat osteogenic ROB-C26 cells and its down-regulation by dexamethazone. *Calcif. Tissue Int.* 1999; 65:472-478.

[30] Schulz S, Singh S, Bellet RA, et al. The primary structure of a plasma membrane guanylate cyclase demonstrates diversity within this new receptor family. *Cell* 1989; 58:1155-1162.

[31] Hashimoto Y, Nakao K, Hama N, et al. Pharmacokinetics of brain natriuretic peptide in rats. *Chem. Pharm. Bull.* (Tokyo) 1992;40:1650-1652.

[32] Kuno T, Andresen JW, Kamisaki Y, et al. Co-purification of an atrial natriuretic factor receptor and particulate guanylate cyclase from rat lung. *J. Biol. Chem.* 1986; 261: 5817-5823.

[33] Takayanagi R, Inagami T, Snajdar RM, et al. Two distinct forms of receptors for atrial natriuretic factor in bovine adrenocortical cells. Purification, ligand binding, and peptide mapping. *J. Biol. Chem.* 1987; 262:12104-12113.

[34] Chinkers M, Garbers DL, Chang MS, et al. A membrane form of guanylate cyclase is an atrial natriuretic peptide receptor. *Nature* 1989; 338:78-83.

[35] Schulz S, Green CK, Yuen PS, et al. Guanylyl cyclase is a heat-stable enterotoxin receptor. *Cell* 1990; 63:941-948.

[36] Shyjan AW, de Sauvage FJ, Gillett NA, et al. Molecular cloning of a retina-specific membrane guanylyl cyclase. *Neuron* 1992; 9:727-737.
[37] Lowe DG, Dizhoor AM, Liu K, et al. Cloning and expression of a second photoreceptor-specific membrane retina guanylyl cyclase (RetGC), RetGC-2. *Proc. Natl. Acad. Sci. U S A* 1995; 92:5535-5539.
[38] Yang RB, Foster DC, Garbers DL, et al. Two membrane forms of guanylyl cyclase found in the eye. *Proc. Natl. Acad. Sci. U S A* 1995; 92:602-606.
[39] Fulle HJ, Vassar R, Foster DC, et al. A receptor guanylyl cyclase expressed specifically in olfactory sensory neurons. *Proc. Natl. Acad. Sci. U S A* 1995; 92:3571-3575.
[40] Schulz S, Wedel BJ, Matthews A, et al. The cloning and expression of a new guanylyl cyclase orphan receptor. *J. Biol. Chem.* 1998; 273:1032-1037.
[41] Fuller F, Porter JG, Arfsten AE, et al. Atrial natriuretic peptide clearance receptor. Complete sequence and functional expression of cDNA clones. *J. Biol. Chem.* 1988; 263:9395-9401.
[42] Maack T, Suzuki M, Almeida FA, et al. Physiological role of silent receptors of atrial natriuretic factor. *Science* 1987; 238:675-678.
[43] Cohen D, Koh GY, Nikonova LN, et al. Molecular determinants of the clearance function of type C receptors of natriuretic peptides. *J. Biol. Chem.* 1996; 271: 9863-9869.
[44] Misono KS. Atrial natriuretic factor binding to its receptor is dependent on chloride concentration: A possible feedback-control mechanism in renal salt regulation. *Circ. Res.* 2000; 86:1135-1139.
[45] Ogawa H, Qiu Y, Philo JS, et al. Reversibly bound chloride in the atrial natriuretic peptide receptor hormone-binding domain: possible allosteric regulation and a conserved structural motif for the chloride-binding site. *Protein Sci.* 2010; 19:544-557.
[46] Huo X, Abe T, Misono KS. Ligand binding-dependent limited proteolysis of the atrial natriuretic peptide receptor: juxtamembrane hinge structure essential for transmembrane signal transduction. *Biochemistry* 1999; 38:16941-16951.
[47] Qiu Y, Ogawa H, Miyagi M, et al. Constitutive activation and uncoupling of the atrial natriuretic peptide receptor by mutations at the dimer interface. Role of the dimer structure in signalling. *J. Biol. Chem.* 2004; 279: 6115-6123.
[48] Kurose H, Inagami T, Ui M. Participation of adenosine 5'-triphosphate in the activation of membrane-bound guanylate cyclase by the atrial natriuretic factor. *FEBS Lett.* 1987; 219:375-379.
[49] Chinkers M, Singh S, Garbers DL. Adenine nucleotides are required for activation of rat atrial natriuretic peptide receptor/guanylyl cyclase expressed in a baculovirus system. *J. Biol. Chem.* 1991; 266:4088-4093.
[50] Potter LR. Phosphorylation-dependent regulation of the guanylyl cyclase-linked natriuretic peptide receptor B: dephosphorylation is a mechanism of desensitization. *Biochemistry* 1998; 37:2422-2429.
[51] Potter LR, Hunter T. Phosphorylation of the kinase homology domain is essential for activation of the A-type natriuretic peptide receptor. *Mol. Cell Biol.* 1998; 18:2164-2172.
[52] Misono KS, Sivasubramanian N, Berkner K, et al. Expression and purification of the extracellular ligand-binding domain of the atrial natriuretic peptide (ANP) receptor. *Biochemistry* 1999; 38:516-523.

[53] Miyagi M, Misono KS. Disulfide bond structure of the atrial natriuretic peptide receptor extracellular domain: conserved disulfide bonds among guanylate cyclase-coupled receptors. *Biochim. Biophys. Acta* 2000; 1478:30-38.
[54] Miyagi M, Zhang X, Misono KS. Glycosylation sites in the atrial natriuretic peptide receptor: oligosaccharide structures are not required for hormone binding. *Eur. J. Biochem.* 2000; 267:5758-5768.
[55] Abe T, Nishiyama K, Snajdar R, et al. Aortic smooth muscle contains guanylate-cyclase-coupled 130-kDa atrial natriuretic factor receptor as predominant receptor form. Spontaneous switching to 60-kDa C-receptor upon cell culturing. *Eur. J. Biochem.* 1993; 217:295-304.
[56] Lowe DG, Fendly BM. Human natriuretic peptide receptor-A guanylyl cyclase. Hormone cross- linking and antibody reactivity distinguish receptor glycoforms. *J. Biol. Chem.* 1992; 267:21691-21697.
[57] Koller KJ, Lipari MT, Goeddel DV. Proper glycosylation and phosphorylation of the type A natriuretic peptide receptor are required for hormone-stimulated guanylyl cyclase activity. *J. Biol. Chem.* 1993; 268:5997-6003.
[58] Fenrick R, Bouchard N, McNicoll N, et al. Glycosylation of asparagine 24 of the natriuretic peptide receptor-B is crucial for the formation of a competent ligand binding domain. *Mol. Cell Biochem.* 1997; 173:25-32.
[59] Misono KS, Philo JS, Arakawa T, et al. Structure, signaling mechanism and regulation of natriuretic peptide receptor-guanylate cyclase. *FEBS J.* 2011; 278:1818-1829.
[60] Veress AT, Honrath U, Chong CK, et al. Renal resistance to ANF in salt-depleted rats is independent of sympathetic or ANG-aldosterone systems. *Am. J. Physiol.* 1997; 272: F545-550.
[61] Warner L, Skorecki K, Blendis LM, et al. Atrial natriuretic factor and liver disease [published erratum appears in Hepatology 1993 Jun;17(6):1174]. *Hepatology* 1993; 17:500-513.
[62] Koepke JP, DiBona GF. Blunted natriuresis to atrial natriuretic peptide in chronic sodium- retaining disorders. *Am. J. Physiol.* 1987; 252:F865-871.
[63] Burnett JC, Jr., Kao PC, Hu DC, et al. Atrial natriuretic peptide elevation in congestive heart failure in the human. *Science* 1986; 231:1145-1147.
[64] van den Akker F, Zhang X, Miyagi M, et al. Structure of the dimerized hormone-binding domain of a guanylyl-cyclase-coupled receptor. *Nature* 2000; 406:101-104.
[65] Ogawa H, Qiu Y, Ogata CM, et al. Crystal structure of hormone-bound atrial natriuretic peptide receptor extracellular domain: rotation mechanism for transmembrane signal transduction. *J. Biol. Chem.* 2004; 279:28625-28631.
[66] Misono KS, Sivasubramanian N, Berkner K, et al. Expression and purification of the extracellular ligand-binding domain of the atrial natriuretic peptide (ANP) receptor: monovalent binding with ANP induces 2:2 complexes. *Biochemistry* 1999; 38:516-523.
[67] Misono KS, Ogawa H, Qiu Y, et al. Structural studies of the natriuretic peptide receptor: a novel hormone-induced rotation mechanism for transmembrane signal transduction. *Peptides* 2005; 26:957-968.
[68] Ogawa H, Qiu Y, Huang L, et al. Structure of the atrial natriuretic peptide receptor extracellular domain in the unbound and hormone-bound states by single-particle electron microscopy. *FEBS J.* 2009; 276:1347-1355.

[69] Antos LK, Abbey-Hosch SE, Flora DR, et al. ATP-independent activation of natriuretic peptide receptors. *J. Biol. Chem.* 2005; 280:26928-26932.
[70] Schroter J, Zahedi RP, Hartmann M, et al. Homologous desensitization of guanylyl cyclase A, the receptor for atrial natriuretic peptide, is associated with a complex phosphorylation pattern. *Febs. J.* 2010; 277:2440-2453.
[71] Potter LR, Garbers DL. Dephosphorylation of the guanylyl cyclase-A receptor causes desensitization. *J. Biol. Chem.* 1992; 267:14531-14534.
[72] Duda T, Venkataraman V, Ravichandran S, et al. ATP-regulated module (ARM) of the atrial natriuretic factor receptor guanylate cyclase. *Peptides* 2005; 26:969-984.
[73] Wilson EM, Chinkers M. Identification of sequences mediating guanylyl cyclase dimerization. *Biochemistry* 1995; 34:4696-4701.
[74] Saha S, Biswas KH, Kondapalli C, et al. The linker region in receptor guanylyl cyclases is a key regulatory module: mutational analysis of guanylyl cyclase C. *J. Biol. Chem.* 2009; 284:27135-27145.
[75] Winger JA, Derbyshire ER, Lamers MH, et al. The crystal structure of the catalytic domain of a eukaryotic guanylate cyclase. *BMC Struct. Biol.* 2008; 8:42.
[76] Rauch A, Leipelt M, Russwurm M, et al. Crystal structure of the guanylyl cyclase Cya2. *Proc. Natl. Acad. Sci. U S A* 2008; 105:15720-15725.
[77] Zhang G, Liu Y, Ruoho AE, et al. Structure of the adenylyl cyclase catalytic core. *Nature* 1997; 386:247-253.
[78] Tesmer JJ, Sunahara RK, Gilman AG, et al. Crystal structure of the catalytic domains of adenylyl cyclase in a complex with Gsalpha.GTPgammaS. *Science* 1997; 278:1907-1916.
[79] Tesmer JJ, Sunahara RK, Johnson RA, et al. Two-metal-Ion catalysis in adenylyl cyclase. *Science* 1999; 285:756-760.
[80] Winger JA, Marletta MA. Expression and characterization of the catalytic domains of soluble guanylate cyclase: interaction with the heme domain. *Biochemistry* 2005; 44: 4083-4090.
[81] Hurley JH. Structure, mechanism, and regulation of mammalian adenylyl cyclase. *J. Biol. Chem.* 1999; 274:7599-7602.
[82] Liu Y, Ruoho AE, Rao VD, et al. Catalytic mechanism of the adenylyl and guanylyl cyclases: modeling and mutational analysis. *Proc. Natl. Acad. Sci. U S A* 1997; 94: 13414-13419.
[83] Paul AK, Marala RB, Jaiswal RK, et al. Coexistence of guanylate cyclase and atrial natriuretic factor receptor in a 180-kD protein. *Science* 1987; 235:1224-1226.
[84] Radany EW, Gerzer R, Garbers DL. Purification and characterization of particulate guanylate cyclase from sea urchin spermatozoa. *J. Biol. Chem.* 1983; 258:8346-8351.
[85] Miyagi M, Zhang X, Misono KS. Glycosylation sites in the atrial natriuretic peptide receptor oligosaccharide structures are not required for hormone binding. *Eur. J. Biochem.* 2000; 267:5758-5768.
[86] Miyagi M, Misono KS. Disulfide bond structure of the atrial natriuretic peptide receptor extracellular domain: conserved disulfide bonds among guanylate cyclase- coupled receptors. *Biochim. Biophys. Acta* 2000; 1478:30-38.
[87] Maack T, Johnson V, Kau ST, et al. Renal filtration, transport, and metabolism of low-molecular-weight proteins: a review. *Kidney Int.* 1979; 16:251-270.

[88] Carone FA, Peterson DR, Oparil S, et al. Renal tubular transport and catabolism of proteins and peptides. *Kidney Int.* 1979; 16:271-278.

[89] Totsune K, Takahashi K, Satoh F, et al. Urinary immunoreactive brain natriuretic peptide in patients with renal disease. *Regul. Pept.* 1996; 63:141-147.

[90] Marumo F, Sakamoto H, Ando K, et al. Concentrations of atrial natriuretic peptide in plasma and urine of kidney disease patients. *Clin. Chem.* 1990; 36:1650-1653.

[91] Suzuki Y, Suzuki H, Ohtake R, et al. Changes in the plasma and urine alpha human atrial natriuretic peptide (alpha hANP) concentration in patients with thyroid disorders. *Endocrinol. Jpn.* 1988; 35:907-913.

[92] Bentzen H, Pedersen RS, Nyvad O, et al. Effect of exercise on natriuretic peptides in plasma and urine in chronic heart failure. *Int. J. Cardiol.* 2004; 93:121-130.

[93] Suzuki Y, Suzuki H, Ohtake R, et al. Plasma and urine concentrations of atrial natriuretic peptide in patients with diabetes mellitus. *Pancreas* 1988; 3:404-408.

[94] Ando K, Umetani N, Kurosawa T, et al. Atrial natriuretic peptide in human urine. *Klin. Wochenschr.* 1988; 66:768-772.

[95] Forssmann WG, Richter R, Meyer M. The endocrine heart and natriuretic peptides: histochemistry, cell biology, and functional aspects of the renal urodilatin system. *Histochem. Cell Biol.* 1998; 110:335-357.

[96] Mattingly MT, Brandt RR, Heublein DM, et al. Presence of C-type natriuretic peptide in human kidney and urine. *Kidney Int.* 1994; 46:744-747.

[97] Cortes R, Rivera M, Salvador A, et al. Urinary B-type natriuretic peptide levels in the diagnosis and prognosis of heart failure. *J. Card. Fail.* 2007; 13:549-555.

[98] Ng LL, Geeranavar S, Jennings SC, et al. Diagnosis of heart failure using urinary natriuretic peptides. *Clin. Sci.* (Lond) 2004; 106:129-133.

[99] Forssmann W, Meyer M, Forssmann K. The renal urodilatin system: clinical implications. *Cardiovasc. Res.* 2001; 51:450-462.

[100] Hirsch JR, Kruhoffer M, Adermann K, et al. Cellular localization, membrane distribution, and possible function of guanylyl cyclases A and 1 in collecting ducts of rat. *Cardiovasc. Res.* 2001; 51:553-561.

[101] Light DB, Corbin JD, Stanton BA. Dual ion-channel regulation by cyclic GMP and cyclic GMP-dependent protein kinase. *Nature* 1990; 344:336-339.

[102] Scavone C, Scanlon C, McKee M, et al. Atrial natriuretic peptide modulates sodium and potassium-activated adenosine triphosphatase through a mechanism involving cyclic GMP and cyclic GMP-dependent protein kinase. *J. Pharmacol. Exp. Ther.* 1995; 272: 1036-1043.

[103] Garrett RH, Grisham CM. *Biochemistry. 5th edition* ed. Boston, MA: Brooks/Cole; 2013.

In: Natriuretic Peptides
Editor: Kailash N. Pandey

ISBN: 978-1-63321-080-6

Chapter 6

Physiology and Patho-Physiology of Signal Transduction via Cardiac Natriuretic Peptides and Guanylyl Cyclase-A

Ichiro Kishimoto*

Department of Endocrinology and Metabolism,
National Cerebral and Cardiovascular Center, Japan

Abstract

After discovery of atrial natriuretic peptide (ANP) and brain natriuretic peptide (BNP), which are cardiac hormones synthesized in, processed in, and secreted from cardiac myocytes, nearly 30 years have passed, whilst the diagnosis of clinical heart failure adapting natriuretic peptides has been established. These cardiac natriuretic peptides act through their cognate receptor, guanylyl cyclase (GC)-A. From the research using the developmental engineering technique in recent years, the diverse physiological/pathophysiological roles of natriuretic peptides are becoming clear. In fact, it is revealed that the ANP/BNP/GC-A signaling pathway is tightly involved in cardiovascular, renal and metabolic health and diseases. Moreover, human studies imply the relationship between the natriuretic peptide/receptor system and hypertension/cardiac hypertrophy. In the present article, I would like to focus the roles of cardiac natriuretic peptides, i.e., ANP and BNP, and their cognate receptor GC-A.

The Natriuretic Peptide Family

In the early 80's, the isolation and purification of a biologically active atrial factor with potent diuretic and natriuretic activities and the determination of its amino acid sequences were reported in rats and humans [1-7]. The discovery of atrial natriuretic peptide (ANP)

* 5-7-1 Fujishiro-dai, Suita City, Osaka, 565-8565, Japan. Phone +81-6-6833-5012. Fax +81-6-6835-5402. e-mail; kishimot@hsp.ncvc.go.jp.

revealed that the heart is not only a mechanical pump driving the blood circulation, but also an endocrine organ. When blood volume is expanded in situations such as heart failure or renal failure, ANP secreted from the heart dilates vessels, increases renal glomerular filtration and salt/water excretion, and inhibits the release of aldosterone from the adrenal gland, which collectively results in a reduction in blood pressure and body fluid volume. In 1988, a second peptide was identified [8] from the porcine brain and named as brain natriuretic peptide (BNP), but it was soon revealed that the BNP content of the brain is negligible in species other than pig, and that BNP is mainly synthesized in and secreted from cardiac ventricles [9]. Although the two peptides share many common biochemical properties, BNP differs from ANP in its synthesis and secretion [10], and the plasma concentration of BNP reflects the severity of heart failure more closely than ANP [11]. Another member of the natriuretic peptide family was identified from the porcine brain, in 1990, and termed C-type natriuretic peptide (CNP) [12]. Although this third congener was initially thought to function only in the brain, CNP was subsequently found to be produced in peripheral tissues such as vascular endothelial cells [13], smooth muscle cells and macrophages [14]. Because plasma levels of CNP are much lower than those of the other two natriuretic peptides, it is thought that CNP most likely acts locally as a paracrine factor rather than as a circulating hormone.

Discovery of GC-A As a Receptor for ANP

One of the active research fields for the communication-of-information system has stemmed from "the second messenger theory", which U.S. pharmacologist Earl W. Sutherland advocated [15]. Some of the disciples of Professor Sutherland in Vanderbilt University, such as Dr. David L Garbers, considered that cyclic guanosine monophosphate (cyclic GMP) would work as a secondary messenger, and continued research with interest to the physiological function of the cyclic GMP generating enzymes (guanylyl cyclase, GC). It had been postulated that the factor which activates respiration and motility of a sperm existed in a sea urchin egg. In process of the research, it was found that cyclic GMP level of the sea urchin sperm was remarkably elevated by the extract of a sea urchin egg jelly layer [16]. In 1981, the structural properties of the egg peptides, resact and speract, which cause various physiological and biochemical changes to the sperm including the increases of respiration, cell motility, intracellular pH, Ca2+, cyclic AMP level and cyclic GMP level, were first reported by Japanese researcher, Professor Norio Suzuki [17]. In 1988, Professor Garbers cloned a receptor for resact from the sea urchin sperm, which was a membrane form of guanylyl cyclase [18].

One of the earliest events following the binding of ANP to its receptor on target cells is an increase in the cytosolic levels of cyclic GMP. This suggested that cyclic GMP might act as a mediator of the physiological effects of the hormone and that the biologically active ANP receptor is coupled to the enzyme guanylyl cyclase (GC), which catalyzes the formation of the second messenger cyclic GMP. A group led by Dr. Garbers used a part of the cloned sea urchin guanylyl cyclase cDNA as a probe to screen various cDNA libraries, which enabled them to clone the first mammalian GC (thus called GC-A) from rat and human [19]. Expression of the cloned enzyme showed its ability to generate cyclic GMP upon binding to

ANP, thus confirmed that GC-A was an ANP receptor. Subsequent studies revealed that three ligands and three receptors compose the natriuretic peptide system.

Atrial natriuretic peptide (ANP), and brain natriuretic peptide (BNP) exert biological functions through their cognate receptor, guanylyl cyclase (GC)-A, while C-type natriuretic peptide (CNP) acts through the receptor called GC-B. Clearance (C) receptor mediates the metabolism of all natriuretic peptides from blood circulation (Figure 1).

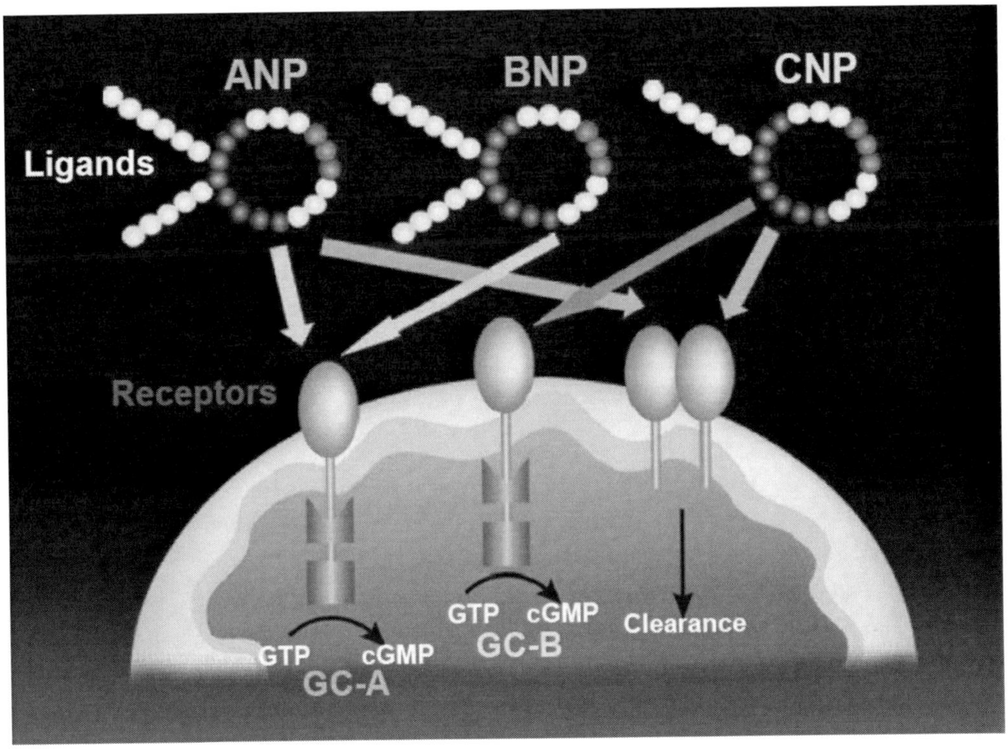

Figure 1. The natriuretic peptide system in the mammals. The natriuretic peptide system is composed of at least three ligands, atrial (ANP), brain (BNP), and C-type (CNP) natriuretic peptides, and three receptors, guanylyl cyclase (GC)-A, GC-B and Clearance (C) receptor.

Blood Pressure Regulation

In recent years, the research using genetically engineered animals revealed various physiological and patho-physiological roles of the signal transduction system through natriuretic peptides and their receptors [20]. Transgenic animals, which constitutively express fusion genes consisting of the transthyretin promoter/ANP gene or of the serum amyloid P component/ BNP gene, were generated so that the hormone expression is targeted to the liver [21, 22]. The animals showed 10- to 100-fold increase in plasma natriuretic peptide concentration, and had significantly lower blood pressure than their non-transgenic littermates. The chronic reduction in blood pressure was mainly due to a reduction in total peripheral resistance, suggesting that cardiac natriuretic peptides involve blood pressure regulation by relaxing resistant arteries, when expressed abundantly.

Since GC-A is a common receptor for cardiac natriuretic peptides, i.e., ANP and BNP, the analysis of the mice deficient for GC-A gene reveals functional significance of the cardiac hormonal pathway. In 1995, the disruption of the GC-A gene is reported to result in chronically elevated blood pressure by about 25 mmHg in systolic pressure in mice on a standard salt diet [23]. Unlike mice heterozygous for the ANP gene which display salt-sensitive hypertension, blood pressures of GC-A heterozygotes remained elevated and unchanged despite increasing dietary salt intake. In 1997, another group also reported that the mice lacking GC-A, display elevated blood pressures and cardiac hypertrophy with interstitial fibrosis resembling that seen in human hypertensive heart disease [24].

To examine the tissue(s) responsible for the hypertensive phenotype of systemic GC-A null mice, endothelium-specific deletion of GC-A was achieved by crossing of mice with floxed GC-A with transgenic mice expressing Cre recombinase under the control of the Tie2 promoter/enhancer. The mice with endothelium-restricted disruption of the GC-A gene displayed significantly increased systolic and diastolic blood pressure compared to those of their floxed GC-A littermates [25]. In contrast, smooth muscle cell-restricted deletion of GC-A did not significantly affect blood pressure [26], indicating that GC-A not in vascular smooth muscle cells but in endothelial cells is indispensable for chronic regulation of blood pressure.

Regulation of Blood Volume

In wild-type mice, after expansion of the plasma volume, urine output as well as urinary sodium and cyclic GMP excretion increased rapidly and markedly. In systemic GC-A-deficient animals, although there were comparable or even more increase in plasma ANP levels, the blood volume expansion-induced rapid urinary responses were abolished [27]. On the contrary, the knock-in overexpression of GC-A mice showed augmented responses to volume expansion in urinary flow and sodium excretion along with rises in both glomerular filtration rate and renal plasma flow, compared with wild-type mice after volume expansion [28]. It is, therefore, indicated that GC-A is the key of the natriuretic, diuretic, and renal hemodynamic responses to acute blood volume expansion. In fact, plasma volumes of animals with systemic GC-A deletion are expanded by 30%. Interestingly, the mice lacking GC-A specifically in vascular endothelium are volume expanded by 11–13% [25], suggesting that GC-A in the endothelium at least partly accounts for chronic blood volume regulatory effects. Since previous experiments indicated that ANP increased capillary permeability of the endothelium to macromolecules like albumin [28], these data imply that the endogenous ANP/GC-A pathway regulates chronic transvascular fluid balance by increasing microvascular permeability [29].

Cardiac Hypertrophy

Cardiac synthesis and secretion of ANP and BNP are known to be increased according to the severity of cardiac remodeling in humans as well as in animals [30]. Because the two cardiac natriuretic peptides share a common receptor (i.e., GC-A), we analyzed the cardiac phenotype of the mice lacking GC-A to reveal the full effects of cardiac natriuretic peptide signaling. Notably, targeted deletion of the GC-A gene resulted in marked cardiac hypertrophy and fibrosis [20].

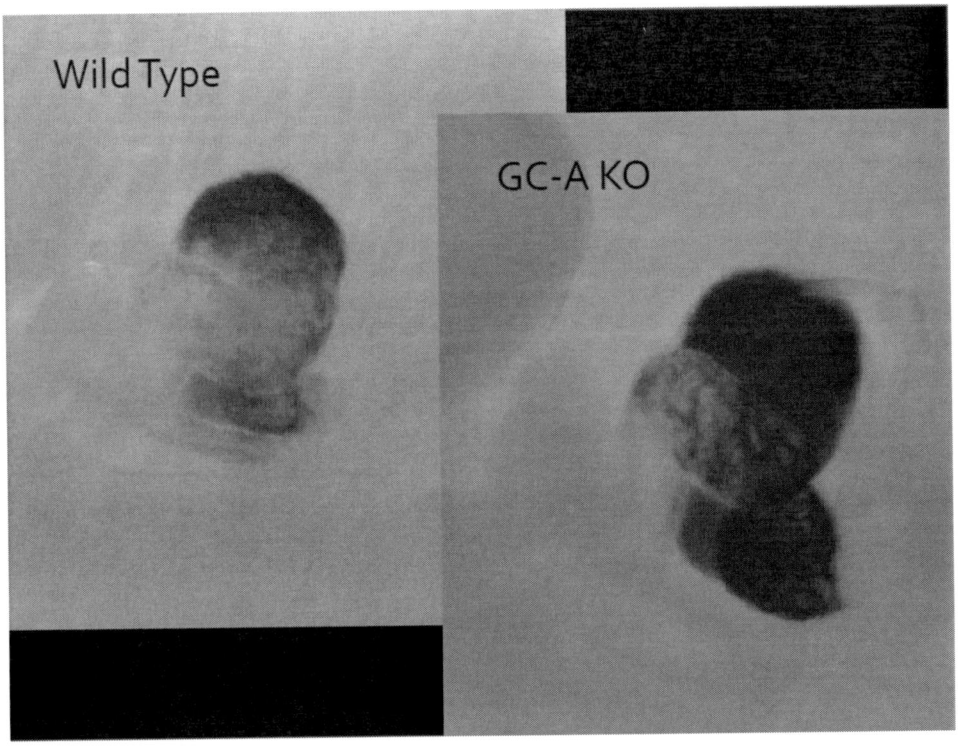

Figure 2. Detection of ANP mRNA in mouse embryo using whole mount in situ hybridization. As early as E9.5, the cardiac expression of atrial natriuretic peptide mRNA was augmented in the heart of the mice deficient for GC-A compared to that of the wild type mice.

In addition, the cardiac hypertrophy was disproportionately severe [24, 31], given the modest rise in blood pressure [23, 24]. Moreover, the expression of ANP mRNA is augmented as early as 9.5 dpc when looping of the heart tube morphogenesis occurs (Figure 2). These findings suggest cardioprotective actions of endogenous ANP/BNP/GC-A signaling pathway within the heart.

More direct evidences of local anti-hypertrophic GC-A signaling were obtained from the animals in which the GC-A gene was conditionally targeted. The effects of selectively overexpressing GC-A gene in the cardiac myocytes of wild-type or of GC-A null animals were analyzed and it is found that, while introduction of the GC-A transgene did not alter blood pressure or heart rate, it reduced cardiac myocyte size in both wild-type and null mice [32]. Coincident with the reduction in myocyte size, the cellular content of both ANP mRNA and ANP was significantly reduced, which suggests the existence of a local regulatory mechanism that governs cardiac myocyte size and gene expression via a GC-A-mediated pathway [32].

Conversely, Professor Kuhn's group inactivated the GC-A gene selectively in cardiac myocytes by homologous loxP/Cre-mediated recombination, which enabled them to circumvent the systemic hypertensive phenotype associated with germline disruption of GC-

A gene [33]. Mice with cardiac myocyte-restricted GC-A deletion exhibited mild cardiac hypertrophy, markedly increased transcription of cardiac hypertrophy markers such as ANP, which is consistent with the idea that one local function of the ANP/GC-A system is to moderate the molecular program of cardiac hypertrophy [34, 35].

THE ROLE OF ANGIOTENSIN II IN CARDIAC HYPERTROPHY OF GC-A DEFICIENT MICE

Because it is well known that angiotensin II receptor blockers have beneficial effects in patients and animals with cardiac remodeling, whether GC-A interacts with AT1A, a major murine type 1 receptor for angiontensin II, signaling in the heart is studied. The effect of targeted deletion of angiontensin II receptor gene on cardiac hypertrophy of GC-A null mice was analyzed by generating mice deficient for both GC-A and AT1A [36]. Interestingly, the cardiac hypertrophy and fibrosis normally seen in GC-A null mice were greatly diminished by deletion of AT1A gene. In addition, pharmacological blockade of AT1 using a selective antagonist, CS-866, also improved the cardiac phenotype of GC-A null mice. Furthermore, stimulation of AT1 receptor by exogenous infusion of a sub-pressor dose angiontensin II (i.e., a dose that did not affect blood pressure) significantly exacerbated cardiac hypertrophy and interstitial fibrosis in GC-A null mice but not in wild-type animals [36]. It is, therefore, suggested that endogenous GC-A inhibits angiotensin II-mediated excessive cardiac remodeling and that cardiac hypertrophy and fibrosis of GC-A deficient mice are attributable, at least in part, to augmented cardiac angiontensin II signaling [34].

THE MOLECULAR MECHANISM OF GC-A-MEDIATED INHIBITION OF CARDIAC HYPERTROPHY

To identify the molecular mechanism underlying the cardiac hypertrophy seen in the mice deficient for GC-A, DNA microarrays were performed for the identification of genes upregulated in the hypertrophied heart [37]. Among several genes known to be upregulated in cardiac hypertrophy (e.g., α-skeletal actin, ANP and BNP), expression of a gene called MCIP1 was increased. Because MCIP1 gene is reportedly regulated by calcineurin, which is known to be a critical regulator of cardiac hypertrophy, it is hypothesized that activity in the calcineurin pathway is enhanced in the heart of GC-A deficient mice. To test the idea, cultured neonatal cardiac myocytes is used to determine whether pharmacological inhibition of GC-A would increase calcineurin activity, and it was found that it did. In the opposite direction, stimulation of GC-A by ANP inhibited calcineurin activity, suggesting that it is by inhibiting the calcineurin pathway that cardiac GC-A signaling activated by locally secreted natriuretic peptides exerts its anti-hypertrophic effect. In fact, chronic treatment with FK506, which, in combination with FKBP (FK506 binding protein), inhibits the phosphatase activity of calcineurin, significantly reduced heart weight to body weight ratios, cardiac myocyte size, and the collagen volume fraction in GC-A deficient mice [37]. In wild-type mice, FK506 had no effect on these parameters.

RGS-4 Mediates ANP/GC-A Anti-Hypertrophic Action

A group of Professor Mendelsohn reported that cyclic GMP-dependent protein kinase (PKG) Iα attenuates signaling by the thrombin receptor protease-activated receptor-1 (PAR-1) through direct activation of regulator of G-protein signaling-2 (RGS-2) [38]. PKGIα binds directly to and phosphorylates RGS-2, which significantly increases the GTPase activity of Gαq, thereby terminating PAR-1 signaling. Given that cyclic GMP is an intracellular second messenger for natriuretic peptides, it is hypothesized that RGS might mediate the cardioprotective effect of GC-A signaling. In cultured cardiac myocytes, ANP stimulated the binding of PKGIα to RGS4, the dominant RGS in cardiac myocytes in a physiological condition, which lead to the phosphorylation of RGS4 and its subsequent association with Gαq [39]. In addition, cardiac myocyte-specific overexpression of RGS4 in GC-A null mice significantly rescued the cardiac phenotype of these mice. Oppositely, overexpression of a dominant-negative form of RGS4 blocked the inhibitory effects of ANP on cardiac hypertrophy [39]. It is, therefore, suggested that GC-A activates cardiac RGS4, which then inhibits the activity of Gαq and its downstream hypertrophic effectors.

On the other hand, Drs. Takimoto and Kass demonstrated that PKG activation reflecting chronic inhibition of cyclic GMP-selective phosphodiesterase 5 (PDE5) suppressed maladaptive cardiac hypertrophy by inhibiting Gαq-coupled stimulation, though this effect was not observed in mice lacking RGS2 [40]. This suggests that RGS2 also mediates cardioprotective actions of PKG in pathological conditions such as pressure overload or excessive Gαq activation due to hypertrophic stimuli.

The Role of GC-A in Myocardial Infarction and in Peripheral Arterial Disease

It is well known that plasma levels of ANP and BNP are dramatically elevated early after myocardial infarction [41]. To examine the significance of this upregulation of natriuretic peptides, experimental myocardial infarction was induced by ligating the left coronary artery in mice lacking GC-A [42]. GC-A deficient mice showed significantly higher mortality than wild type mice, reflecting a higher incidence of acute heart failure, which was associated with diminished water and sodium excretion. By 4 weeks after infarction, left ventricular remodeling, including myocardial hypertrophy and fibrosis, and impairment of left ventricular systolic function were significantly more severe in mice lacking GC-A than wild type mice [42]. It is thus suggested that GC-A activation by endogenous cardiac natriuretic peptides protects against acute heart failure and attenuates chronic cardiac remodeling after acute myocardial infarction.

The role of the natriuretic peptide system in peripheral arterial diseases has also been suggested by a group led by Drs. Itoh and Nakao. Yamahara et al. found an activation of the natriuretic peptides/cyclic GMP/ PKG pathway accelerated vascular regeneration and blood flow recovery in a murine model of peripheral arterial diseases, in which leg ischemia was induced by a femoral arterial ligation [43]. Park et al. reported that intraperitoneal injection of carperitide, a recombinant human ANP, accelerated blood flow recovery with increasing capillary density in the ischemic legs [44]. In fact, carperitide administration improves the ankle-brachial pressure index, and mitigates intermittent claudication, rest pain, and ulceration in patients with peripheral arterial diseases [44], which is possibly due to the role of the exogenous ANP and BNP in angiogenesis. The contributions of endogenous ANP and BNP on vascular remodeling are also suggested by animal experiments. When the hind-limb ischemia model was applied to GC-A deficient mice, autoamputation or severe ulcers were noted in 60% of mice during the 28-day observation period [45]. In wild type mice, by contrast, partial amputation or mild ulcers were detected in only 20% of mice (2/10). Laser Doppler perfusion imaging revealed that the recovery of blood flow in the ischemic limb was significantly inhibited in GC-A null mice, as compared with wild type mice [45]. Kuhn et al. reported that vascular regeneration in response to critical hind limb ischemia was severely impaired in mice with systemic deletion of the GC-A gene [46]. Similar attenuation of ischemic angiogenesis was observed in mice with conditional, endothelial cell-restricted GC-A deletion, whereas smooth muscle cell-restricted GC-A ablation did not affect ischemic neovascularization [46], which suggests it is the endothelial GC-A that stimulates endothelial regeneration after induction of ischemia. Taken together, these evidences suggest that the natriuretic peptide pathway significantly contributes peripheral vascular remodeling during ischemia.

Kidney Podocytes

The ANP/BNP/GC-A pathway is demonstrated to antagonize the renin-angiotensin-aldosterone system, but whether this mechanism contributes to their renoprotective effect is poorly understood. Recently, a group led by Dr. Mukoyama examined kidney functions of mice lacking GC-A under conditions of high aldosterone and high dietary salt [47]. After 4 weeks of administering aldosterone and a high-salt diet, GC-A deficient mice, but not wild-type mice, exhibited accelerated hypertension with massive proteinuria. Aldosterone-treated GC-A deficient mice had marked mesangial expansion, segmental sclerosis, severe podocyte injury, and increased oxidative stress. Reduction of blood pressure with hydralazine failed to lessen such changes. In clear contrast, blockade of the renin-angiotensin-aldosterone system markedly reduced albuminuria, podocyte injury, and oxidative stress. In cultured podocytes, natriuretic peptides inhibited aldosterone-induced phosphorylation of mitogen-activated protein kinase [47]. Taken together, the authors concluded that renoprotective properties of the endogenous natriuretic peptide/GC-A system may result from the local inhibition of the renin-angiotensin-aldosterone system and of the oxidative stress in podocytes .

Obesity and Plasma BNP

Overweight is increasing globally and is a highly dangerous group of a cardiovascular disease or diabetes. Furthermore, overweight brings about many metabolic disorders used as the base of the coronary-arteries risk factor. Clinical and epidemiology study revealed a negative correlation between plasma concentration of BNP and a body mass index (BMI). In 3,389 participants of the U.S. Framingham cohort study without heart failure or renal insufficiency, BNP concentration was negatively correlation with BMI [48]. Moreover, a negative correlation between BNP level or N-terminal BNP level and BMI was similarly reported in 2,707 Dallas Heart Study participants [49].

Furthermore, in patients with heart failure of various extents, it has been reported that overweight patients have relatively low BNP concentration compared to that in normal weight or in lean person after adjustment of cardiac function parameters. A similar result was obtained from the about 2,000 participants of the Suita cohort study, a cohort of general population in urban area [50]. Furthermore, in 608 type 2 diabetes, without chronic atrial fibrillation or cardiomyopathy, when the relations between BNP level and BMI or various obesity indices were analyzed in the multiple linear regression analysis, similar negative correlation was found [51]. The negative correlation was independent of age, sex, systolic arterial pressure, a pulse, a serum creatinine level, a regional wall motion abnormality, a left atrial dimension, left ventricular diameter, fractional shortening, or left ventricular weight. That is, the factors (for example, cardiac or renal function) which are known to influence BNP now cannot explain the negative correlation of a corpulence degree and BNP level, but existence of a certain new mechanism is implied.

Insulin Resistance and Adiponectin

From Framingham study, it has been reported that high insulin resistance seen by HOMA-IR (homeostasis model assessment index) is associated with decreased plasma BNP level in blood, even after adjustments with BMI [52]. Adiponectin is one of the adipocytokines secreted from adipose cells and it is well known that the blood concentration of adiponectin is decreased with overweight. Moreover, a decrease in the serum adiponectin level is accompanied by insulin resistance and arteriosclerosis. From the analysis of Japanese general population cohort of 1,538 person, a group led by Dr. Kitakaze have reported that the adiponectin concentration blood and BNP concentration in blood is positively correlated independently of other confounding factors. Furthermore, ANP and BNP increase the adiponectin secretion from cultured human adipose cells. Moreover, it is shown clearly that the adiponectin concentration is increase in the patients with heart failure treated with ANP (25 ng/kg/min) four days compared with the placebo-treated group [53]. It is, therefore, suggested that adiponectin might be involved the negative correlation between cardiac natriuretic peptides and obesity at least partly.

The Inhibitory Role of Endogenous GC-A in Fat Accumulation

It is reported that heterozygous deficiency of GC-A increased susceptibility to high-fat induced obesity, so that the body weight of high-fat-fed GCA+/- mice increased significantly compared to that of wild-type mice, even though food intake was similar between genotypes [54]. The blood glucose levels after intraperitoneal administration of glucose were significantly higher in GCA+/- mice on the high-fat diet. The total fat weight estimated from the computerized tomography showed significant increases in high-fat fed GCA+/- mice, in parallel with the increase in body weight [54]. Thus, the role of endogenous GC-A-mediated signal transduction in lipid metabolism is suggested. Galitzky and others [56] prepared adipose cells from the human subcutaneous fat after cosmetic surgery and demonstrated that, in these cells, 1) GC-A exists, 2) ANP at a physiological concentration (10 pM) promotes the lipid degradation, 3) the lipolytic action is accompanied by the rise of intracellular cyclic GMP concentration and activation of cyclic GMP-dependent protein kinase without affecting cAMP concentration. Moreover, in the latest biochemical analysis, ANP-induced activation of cyclic GMP-dependent kinase stimulates hormone sensitive lipase (HSL) through phosphorylation of its serine residue [56]. Although HSL activation is promoted by the rise of the cAMP concentration induced by β adrenergic stimulation and decreased by the fall of the cAMP concentration induced by insulin, ANP could act on adipose cells directly, and contribute to lipolysis through cyclic GMP independently of cAMP. Birkenfeld and others [57] infused synthetic human ANP at a rate of 25 ng/kg/min for 2 hours in 12 healthy adult men and examined the lipolytic action after a high fat test meal. Accompanying decreases in blood pressure by about 4 mmHg, ANP simultaneously increased blood glycerol and free fatty acid concentrations. From these data, it is suggested that ANP could promote lipolysis both in animals and in humans.

Energy Metabolism

When Miyashita and others [54] loaded high fat diet on the genetically engineered animals of BNP overexpression, the fewer weight gain was observed compared with the wild-type control group. Moreover, the adipose weight increment and fatty liver after high-fat-diet load were also suppressed. In the animal with overexpression of PKG, which is a downstream signaling molecule of ANP/BNP/GC-A, the similar but more strong phenotype was observed. Expiration gas analysis revealed that, in the mice, the increase in oxygen consumption and the decrease in the respiratory exchange ratio were observed under standard diets. Moreover, from the analysis of PPARdelta in the skeletal muscle and of UCP-3/UCP-1 in brown adipose tissue, the authors concluded that the activation of BNP/PKG cascade promotes mitochondria biosynthesis and fatty acid oxidation in mouse skeletal muscle, and promotes a heat production in brown adipose tissue, and, as a result of increased energy consumption, the amount of adipose tissue decreases. These results show the close relation between the signal transduction system of cardiac natriuretic peptide and energy metabolism, and foresee the progress of the field.

Human Genetics

The importance of natriuretic peptide signaling in cardiovascular (patho-)physiology suggested by the results from the animal studies summarized above makes it likely that human subjects with defective natriuretic peptide signaling are prone to cardiovascular diseases. In fact, there have been several studies looking at the polymorphisms in each component of the natriuretic peptide system and their association with cardiovascular diseases. Of note, the animal experiments indicating that genetically determined quantitative changes in the expression of GC-A gene affect blood pressure and heart weight in a gene dose-dependent fashion [58] suggests GC-A alleles could significantly affect genetic risks for essential hypertension and cardiac hypertrophy in humans. Human GC-A gene is located on chromosome 1q21 and is comprised of 22 exons spanning 16 kb. Several polymorphisms within the 5'-franking region of GC-A gene were determined recently. These include an 8 base pairs insertion/deletion polymorphism at position -60 and CT dinucleotide repeat at position -293 (Figure 3). It has been demonstrated that these polymorphisms in the 5'-franking region decrease the transcriptional activity of GC-A gene and were significantly associated with human hypertension and cardiac hypertrophy.

Nakayama et al. examined the promoter region of the human GC-A gene and identified an 8 nucleotides insertion/deletion mutation at position -60 (60 base pairs upstream of the ATG codon) [59]. After genotyping 200 subjects with essential hypertension and 200 normotensive control subjects, they found that 9 individuals heterozygous for the mutated allele exhibited either hypertension, left ventricular hypertrophy or both. In vitro reporter gene analysis revealed that the deletion mutation reduced the transcriptional activity of the GC-A promoter to less than 30% of the wild-type allele, suggesting that this deletion reduces receptor expression and increases susceptibility to hypertension or left ventricular hypertrophy in those carrying the mutation [59].

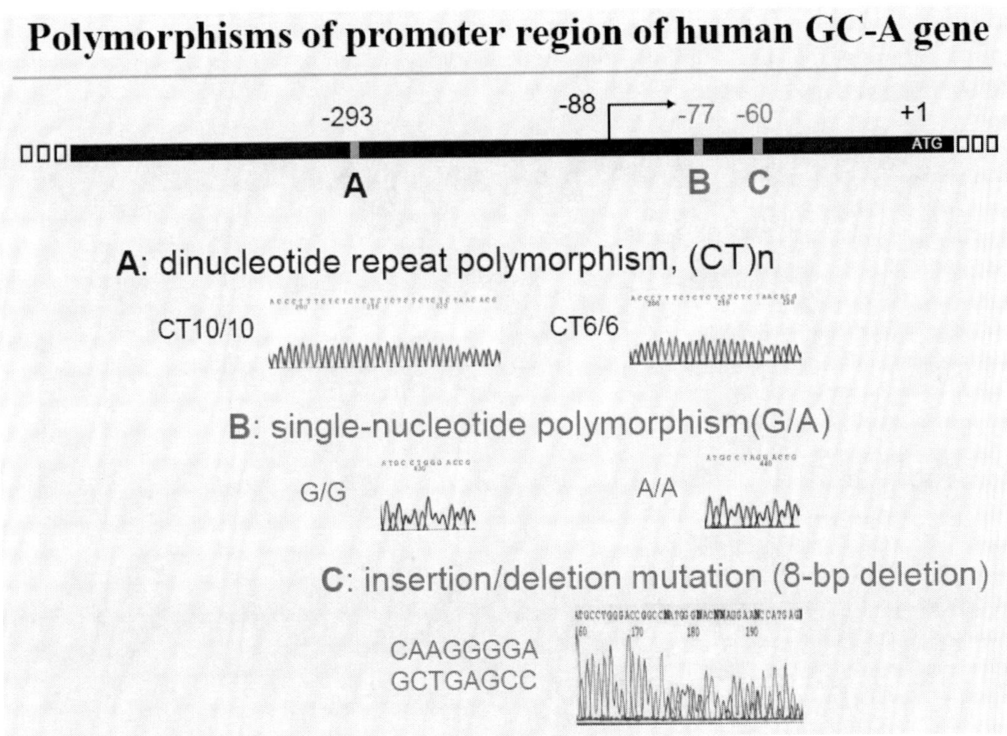

Figure 3. The polymorphisms in the 5'-franking region of human GC-A gene. The decreases in transcriptional activities of GC-A gene were significantly associated with hypertension and cardiac hypertrophy.

In addition, Rubattu et al. investigated the relationships between GC-A polymorphisms and left ventricular structure in approximately 200 subjects with essential hypertension [60]. Those investigators identified GC-A gene variants that were significantly associated with left ventricular mass index and left ventricular septal thickness. Based on their findings, they concluded that endogenous GC-A contributed significantly to ventricular remodeling in human essential hypertension [60]. It is also examined the association between polymorphisms within the GC-A promoter and essential hypertension in a group of Japanese subjects (177 hypertensive and 170 normotensive) and identified five allele types in which 6, 9, 10, 11 or12 CT dinucleotide repeats around position -293 upstream of the ATG codon [61].

The frequency of the (CT)n=6 allele was significantly higher among hypertensive subjects than among normotensives, while the frequencies of the other four allele types did not differ between the two groups. Promoter-reporter analyses carried out in cultured human aortic smooth muscle cells using a luciferase gene fused to the 5'-flanking region of the GC-A gene revealed that, like the promoter containing an 8 bp deletion at position -60, the promoter containing (CT)n=6 at position -293 drove less transcriptional activity than the promoter containing (CT)n=10 (control) [61].

Collectively, the findings summarized in this section suggest that people carrying certain variants of the human GC-A gene are more susceptible to cardiovascular diseases such as hypertension and cardiac hypertrophy. It is, therefore, proposed that the functional polymorphisms of GC-A gene in human are potent and novel susceptibility markers for genetical risks of the cardiovascular disease phenotypes, which have been demonstrated using GC-A deficient mice.

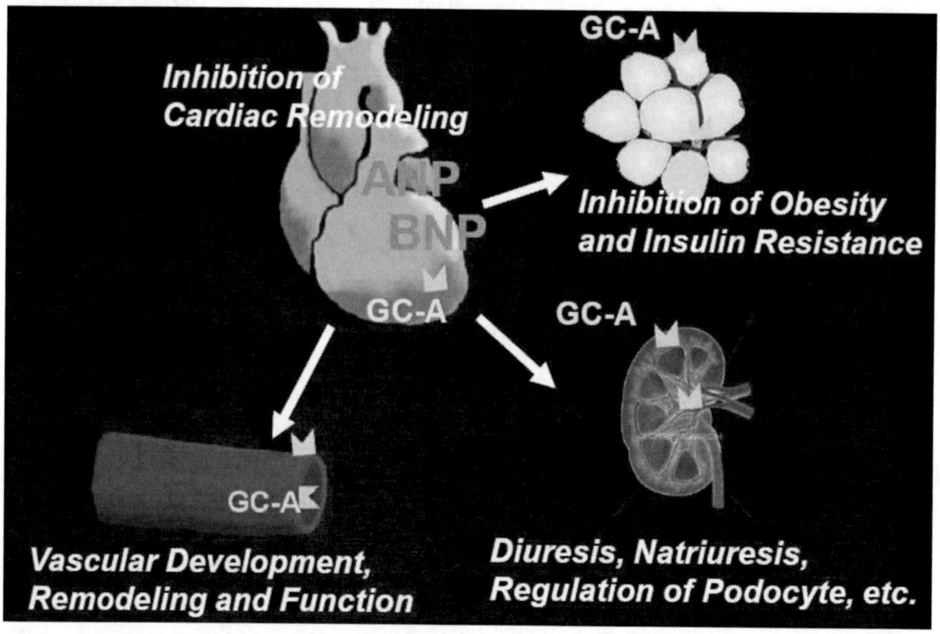

Figure 4. A schema showing the target tissues of cardiac hormones in the body. Atrial and Brain natriuretic peptides are cardiac hormones synthesized in, processed in and secreted from cardiac myocytes, which exert potent physiological and path-physiological effects on blood vessels, kidney and fat tissues through their cognate receptor, guanylyl cyclase(GC)-A.

CONCLUSION

As mentioned above, the role of cardiac natriuretic peptides/GC-A signal transduction pathway is involved broad areas of the body and physiological/patho-physiological functions as schematically depicted in Figure 4. The suggested roles of the pathway as the causes or modifiers of various cardio/vascular/renal/metabolic morbidities all require further verification and, as for these hypotheses, research progress of a related field is strongly desired.

ACKNOWLEDGMENTS

The author has nothing to disclose. I would like to acknowledge the extensive productive works of Dr. Takeshi Tokudome, the continuous supports from Dr. Kenji Kangawa and Dr. Kazuwa Nakao, and excellent secretarial work of Ms. Yuko Yasokawa. This work has been supported by the research grants from Japanese Ministry of Education, Culture, Sports, Science and Technology, and Ministry of Health, Labour and Welfare, and by the Intramural Research Fund (22-3-2) for Cardiovascular Diseases of National Cerebral and Cardiovascular Center.

As a person who had performed joint research on GC-A, and had continued a friendship with late Professor David L Garbers, who has passed away in the youth of 62 years old in 2006, the collaborative works with him were summarized praying for the bliss of the dead.

REFERENCES

[1] de Bold, A.J., Borenstein, H.B., Veress, A.T. and Sonnenberg, H. (1981). A rapid and potent natriuretic response to intravenous injection of atrial myocardial extract in rats. *Life Sci.* 28, 89-94.

[2] Flynn, T.G., de Bold, M.L. and de Bold, A.J. (1983). The amino acid sequence of an atrial peptide with potent diuretic and natriuretic properties. *Biochem. Biophys. Res. Commun.* 117, 859-65.

[3] Thibault G, Garcia R, Seidah NG, Lazure C, Cantin M, Chrétien M, Genest J. (1983) Purification of three rat atrial natriuretic factors and their amino acid composition. *FEBS Lett.* 164(2):286-90

[4] Kangawa, K. and Matsuo, H. (1984). Purification and complete amino acid sequence of alpha-human atrial natriuretic polypeptide (alpha-hANP). *Biochem. Biophys. Res. Commun.* 118, 131-9.

[5] Misono KS, Fukumi H, Grammer RT, Inagami T. (1984). Rat atrial natriuretic factor: complete amino acid sequence and disulfide linkage essential for biological activity. *Biochem. Biophys. Res. Commun.* 119(2):524-9

[6] Currie MG, Geller DM, Cole BR, Siegel NR, Fok KF, Adams SP, Eubanks SR, Galluppi GR, Needleman P. (1984). Purification and sequence analysis of bioactive atrial peptides (atriopeptins). *Science* 223:67–69.

[7] Kangawa K, Tawaragi Y, Oikawa S, Mizuno A, Sakuragawa Y, Nakazato H, Fukuda A, Minamino N, Matsuo H 1984 Identification of rat atrial natriuretic polypeptide and characterization of the cDNA encoding its precursor. *Nature* 312:152–155.

[8] Sudoh, T., Kangawa, K., Minamino, N. and Matsuo, H. (1988). A new natriuretic peptide in porcine brain. *Nature* 332, 78-81.

[9] Mukoyama, M. et al. (1991). Brain natriuretic peptide as a novel cardiac hormone in humans. Evidence for an exquisite dual natriuretic peptide system, atrial natriuretic peptide and brain natriuretic peptide. *J. Clin. Invest.* 87, 1402-12.

[10] Nakagawa, O., Ogawa, Y., Itoh, H., Suga, S., Komatsu, Y., Kishimoto, I., Nishino, K., Yoshimasa, T., Nakao, K. (1995). Rapid transcriptional activation and early mRNA turnover of brain natriuretic peptide in cardiocyte hypertrophy. Evidence for brain natriuretic peptide as an "emergency" cardiac hormone against ventricular overload. *J. Clin. Invest.* 96, 1280-7.

[11] Mukoyama, M., Nakao, K., Saito, Y., Ogawa, Y., Hosoda, K., Suga, S., Shirakami, G., Jougasaki, M., Imura, H. (1990). Increased human brain natriuretic peptide in congestive heart failure. *N. Engl. J. Med.* 323, 757-8.

[12] Sudoh, T., Minamino, N., Kangawa, K. and Matsuo, H. (1990). C-type natriuretic peptide (CNP): a new member of natriuretic peptide family identified in porcine brain. *Biochem. Biophys. Res. Commun.* 168, 863-70.

[13] Suga, S., Itoh, H., Komatsu, Y., Ishida, H., Igaki, T., Yamashita, J., Doi, K., Chun, T. H., Yoshimasa, T., Tanaka, I., Nakao, K. (1998). Regulation of endothelial production of C-type natriuretic peptide by interaction between endothelial cells and macrophages. *Endocrinology* 139, 1920-6.

[14] Naruko, T., Ueda, M., van der Wal, A.C., van der Loos, C.M., Itoh, H., Nakao, K. and Becker, A.E. (1996). C-type natriuretic peptide in human coronary atherosclerotic lesions. *Circulation* 94, 3103-8.

[15] Sutherland EW Studies on the mechanism of hormone action. *Science* 1972 177(4047):401-8.

[16] Garbers, D. L. Hardman, J. G. (1975) Factors released from sea urchin eggs affect cyclic nucleotide metabolism. *Nature* (London), 257:677-678.

[17] Suzuki, N., Nomura, K., Ohtake, H.,Isaka, S. (1981) Purification and the primary structure of sperm-activating peptides from the jelly coat of sea urchin eggs. *Biochem. Biophys. Res. Commun.*, 99:1238-1244.

[18] Singh, S., Lowe, D. G., Thorpe, D. S., Rodriguez, H, H., Kuang, W. J., Dangott, L. J., Chinkers, M., Goeddel D. V., Garbers, D. L. (1988) Membrane guanylate cyclase is a cell-surface receptor with homology to protein kinases. *Nature,* 334:708-712.

[19] Chinkers, M., Garbers, D.L., Chang, M.S., Lowe, D.G., Chin, H.M., Goeddel, D.V. and Schulz, S. (1989). A membrane form of guanylate cyclase is an atrial natriuretic peptide receptor. *Nature* 338, 78-83.

[20] Kishimoto I, Tokudome T, Nakao K, Kangawa K. Natriuretic peptide system: anoverview of studies using genetically engineered animal models. *FEBS J.* 2011 Jun;278(11):1830-41.

[21] Steinhelper ME, Cochrane KL, Field LJ. (1990). Hypotension in transgenic mice expressing atrial natriuretic factor fusion genes. *Hypertension* 16(3):301-7.

[22] Ogawa Y, Itoh H, Tamura N, Suga S, Yoshimasa T, Uehira M, Matsuda S, Shiono S, Nishimoto H, Nakao K. (1994). Molecular cloning of the complementary DNA and

gene that encode mouse brain natriuretic peptide and generation of transgenic mice that overexpress the brain natriuretic peptide gene. *J. Clin. Invest.* 93(5):1911-21.

[23] Lopez, M.J., Wong, S.K., Kishimoto, I., Dubois, S., Mach, V., Friesen, J., Garbers, D.L. and Beuve, A. (1995). Salt-resistant hypertension in mice lacking the guanylyl cyclase-A receptor for atrial natriuretic peptide. *Nature* 378, 65-8.

[24] Oliver PM, Fox JE, Kim R, Rockman HA, Kim HS, Reddick RL, Pandey KN, Milgram SL, Smithies O, Maeda N. (1997) Hypertension, cardiac hypertrophy, and sudden death in mice lacking natriuretic peptide receptor A. *Proc. Natl. Acad. Sci. U S A.* 94(26): 14730-5.

[25] Sabrane K, Kruse MN, Fabritz L, Zetsche B, Mitko D, Skryabin BV, Zwiener M, Baba HA, Yanagisawa M, Kuhn M. (2005). Vascular endothelium is critically involved in the hypotensive and hypovolemic actions of atrial natriuretic peptide. *J. Clin. Invest.* 115(6): 1666-74.

[26] Holtwick R, Gotthardt M, Skryabin B, Steinmetz M, Potthast R, Zetsche B, Hammer RE, Herz J, Kuhn M. (2002). Smooth muscle-selective deletion of guanylyl cyclase-A prevents the acute but not chronic effects of ANP on blood pressure. *Proc. Natl. Acad. Sci. U S A.* 99(10): 7142-7.

[27] Kishimoto I, Dubois SK, Garbers DL. (1996). The heart communicates with the kidney exclusively through the guanylyl cyclase-A receptor: acute handling of sodium and water in response to volume expansion. *Proc. Natl. Acad. Sci. U S A.* 93(12): 6215-9.

[28] Shi SJ, Vellaichamy E, Chin SY, Smithies O, Navar LG, Pandey KN. (2003). Natriuretic peptide receptor A mediates renal sodium excretory responses to blood volume expansion. *Am. J. Physiol. Renal. Physiol.* 285(4): F694-702.

[29] Curry FR. (2005). Atrial natriuretic peptide: an essential physiological regulator of transvascular fluid, protein transport, and plasma volume. J. Clin. Invest. 115(6): 1458-61.

[30] Nakao, K., Itoh, H., Saito, Y., Mukoyama, M. and Ogawa, Y. (1996). The natriuretic peptide family. *Curr. Opin. Nephrol. Hypertens.* 5, 4-11.

[31] Knowles, J.W., Esposito, G., Mao, L., Hagaman, J.R., Fox, J.E., Smithies, O., Rockman, H.A. and Maeda, N. (2001). Pressure-independent enhancement of cardiac hypertrophy in natriuretic peptide receptor A-deficient mice. *J. Clin. Invest.* 107, 975-84.

[32] Kishimoto, I., Rossi, K. and Garbers, D.L. (2001). A genetic model provides evidence that the receptor for atrial natriuretic peptide (guanylyl cyclase-A) inhibits cardiac ventricular myocyte hypertrophy. *Proc. Natl. Acad. Sci. U S A* 98, 2703-6.

[33] Kuhn M, Holtwick R, Baba HA, Perriard JC, Schmitz W, Ehler E. (2002) Progressive cardiac hypertrophy and dysfunction in atrial natriuretic peptide receptor (GC-A) deficient mice. *Heart.* 87(4): 368-74

[34] Kishimoto I, Tokudome T, Horio T, Garbers DL, Nakao K, Kangawa K. Natriuretic Peptide Signaling via Guanylyl Cyclase (GC)-A: An Endogenous Protective Mechanism of the Heart. *Curr. Cardiol. Rev.* 2009 Jan; 5(1): 45-51.

[35] Molkentin JD. A friend within the heart: natriuretic peptide receptor signaling. *J. Clin. Invest.* 2003; 111: 1275-7.

[36] Li Y, Kishimoto I, Saito Y, et al. Guanylyl cyclase-A inhibits angiotensin II type 1A receptor-mediated cardiac remodeling, an endogenous protective mechanism in the heart. *Circulation* 2002; 106: 1722-8.

[37] Tokudome, T., Horio, T., Kishimoto, I., Soeki, T., Mori, K., Kawano, Y., Kohno, M., Garbers, D. L., Nakao, K., Kangawa, K. (2005). Calcineurin-nuclear factor of activated T cells pathway-dependent cardiac remodeling in mice deficient in guanylyl cyclase A, a receptor for atrial and brain natriuretic peptides. *Circulation* 111, 3095-104.

[38] Tang KM, Wang GR, Lu P, Karas RH, Aronovitz M, Heximer SP, Kaltenbronn KM, Blumer KJ, Siderovski DP, Zhu Y, Mendelsohn ME. Regulator of G-protein signaling-2 mediates vascular smooth muscle relaxation and blood pressure. *Nat. Med.* 2003; 9: 1506-12.

[39] Tokudome T, Kishimoto I, Horio T, Arai Y, Daryl O. Schwenke, Hino J, Okano I, Kawano Y, Kohno M, Nakao K, Kangawa K. RGS4 mediates anti-hypertrophic effect of locally secreted natriuretic peptides in the heart. *Circulation* 2008; 117: 2329-39.

[40] Takimoto E, Koitabashi N, Hsu S, Ketner EA, Zhang M, Nagayama T, Bedja D, Gabrielson KL, Blanton R, Siderovski DP, Mendelsohn ME, Kass DA. Regulator of G protein signaling 2 mediates cardiac compensation to pressure overload and antihypertrophic effects of PDE5 inhibition in mice. *J. Clin. Invest.* 2009 Feb; 119(2): 408-20.

[41] Morita E, Yasue H, Yoshimura M, Ogawa H, Jougasaki M, Matsumura M, Mukoyama M, Nakao K. Increased plasma levels of brain natriuretic peptide in patients with acute myocardial infarction. *Circulation.* 1993; 88: 82–91.

[42] Nakanishi, M., Saito, Y., Kishimoto, I., Harada, M., Kuwahara, K., Takahashi, N., Kawakami, R., Nakagawa, Y., Tanimoto, K., Yasuno, S., Usami, S., Li, Y., Adachi, Y., Fukamizu, A., Garbers, D. L., Nakao, K. (2005). Role of natriuretic peptide receptor guanylyl cyclase-A in myocardial infarction evaluated using genetically engineered mice. *Hypertension* 46, 441-7.

[43] Yamahara, K., Itoh, H., Chun, T. H., Ogawa, Y., Yamashita, J., Sawada, N., Fukunaga, Y., Sone, M., Yurugi-Kobayashi, T., Miyashita, K., Tsujimoto, H., Kook, H., Feil, R., Garbers, D. L., Hofmann, F., Nakao, K. (2003). Significance and therapeutic potential of the natriuretic peptides/cGMP/cGMP-dependent protein kinase pathway in vascular regeneration. *Proc. Natl. Acad. Sci. U S A* 100, 3404-9.

[44] Park K, Itoh H, Yamahara K, Sone M, Miyashita K, Oyamada N, Sawada N, Taura D, Inuzuka M, Sonoyama T, Tsujimoto H, Fukunaga Y, Tamura N, Nakao K. (2008). Therapeutic potential of atrial natriuretic peptide administration on peripheral arterial diseases. *Endocrinology* 149, 483-91.

[45] Tokudome T, Kishimoto I, Yamahara K, Osaki T, Minamino N, Horio T, Sawai K, Kawano Y, Miyazato M, Sata M, Kohno M, Nakao K, Kangawa K. (2009). Impaired Recovery of Blood Flow After Hind-Limb Ischemia in Mice Lacking Guanylyl Cyclase-A, a Receptor for Atrial and Brain Natriuretic Peptides. *Arterioscler. Thromb. Vasc. Biol.* 29, 1516-21.

[46] Kuhn M, Völker K, Schwarz K, Carbajo-Lozoya J, Flögel U, Jacoby C, Stypmann J, van Eickels M, Gambaryan S, Hartmann M, Werner M, Wieland T, Schrader J, Baba HA. (2009). The natriuretic peptide/guanylyl cyclase--a system functions as a stress-responsive regulator of angiogenesis in mice. *J. Clin. Invest.* 119, 2019-30.

[47] Ogawa Y, Mukoyama M, Yokoi H, Kasahara M, Mori K, Kato Y, Kuwabara T, Imamaki H, Kawanishi T, Koga K, Ishii A, Tokudome T, Kishimoto I, Sugawara A, Nakao K. Natriuretic peptide receptor guanylyl cyclase-A protects podocytes from aldosterone-induced glomerular injury. *J. Am. Soc. Nephrol.* 2012 Jul; 23(7): 1198-209.

[48] Wang TJ, Larson MG, Levy D, Benjamin EJ, Leip EP, Wilson PW, Vasan RS. Impact of obesity on plasma natriuretic peptide levels. *Circulation*. 2004 Feb 10; 109(5): 594-600.

[49] Das SR, Drazner MH, Dries DL, Vega GL, Stanek HG, Abdullah SM, Canham RM, Chung AK, Leonard D, Wians FH Jr, de Lemos JA. Impact of body mass and body composition on circulating levels of natriuretic peptides: results from the Dallas Heart Study. Circulation. 2005 Oct 4; 112(14): 2163-8.

[50] Sugisawa T, Kishimoto I, Kokubo Y, Makino H, Miyamoto Y, Yoshimasa Y. Association of plasma B-type natriuretic peptide levels with obesity in a general urban Japanese population: the Suita Study. *Endocr. J.* 2010; 57(8): 727-33.

[51] Sugisawa T, Kishimoto I, Kokubo Y, Nagumo A, Makino H, Miyamoto Y, Yoshimasa Y. Visceral fat is negatively associated with B-type natriuretic peptide levels in patients with advanced type 2 diabetes. *Diabetes Res. Clin. Pract.* 2010 Aug; 89(2):174-80.

[52] Wang TJ, Larson MG, Keyes MJ, Levy D, Benjamin EJ, Vasan RS. Association of plasma natriuretic peptide levels with metabolic risk factors in ambulatory individuals. *Circulation.* 2007 Mar 20; 115(11): 1345-53.

[53] Tsukamoto O, Fujita M, Kato M, Yamazaki S, Asano Y, Ogai A, Okazaki H, Asai M, Nagamachi Y, Maeda N, Shintani Y, Minamino T, Asakura M, Kishimoto I, Funahashi T, Tomoike H, Kitakaze M. Natriuretic peptides enhance the production of adiponectin in human adipocytes and in patients with chronic heart failure. *J. Am. Coll. Cardiol.* 2009 Jun 2; 53(22): 2070-7.

[54] Miyashita K, Itoh H, Tsujimoto H, Tamura N, Fukunaga Y, Sone M, Yamahara K, Taura D, Inuzuka M, Sonoyama T, Nakao K. Natriuretic peptides/cGMP/cGMP-dependent protein kinase cascades promote muscle mitochondrial biogenesis and prevent obesity. *Diabetes.* 2009 Dec;58(12):2880-92.

[55] Sengenès C, Berlan M, De Glisezinski I, Lafontan M, Galitzky J. Natriuretic peptides: a new lipolytic pathway in human adipocytes. *FASEB J.* 2000 Jul; 14(10):1345-51.

[56] Sengenes C, Bouloumie A, Hauner H, Berlan M, Busse R, Lafontan M, Galitzky J. Involvement of a cGMP-dependent pathway in the natriuretic peptide-mediated hormone-sensitive lipase phosphorylation in human adipocytes. *J. Biol. Chem.* 2003 Dec 5; 278(49): 48617-26.

[57] Birkenfeld AL, Boschmann M, Moro C, Adams F, Heusser K, Franke G, Berlan M, Luft FC, Lafontan M, Jordan J. Lipid mobilization with physiological atrial natriuretic peptide concentrations in humans. *J. Clin. Endocrinol. Metab*. 2005 Jun; 90(6): 3622-8.

[58] Oliver PM, John SW, Purdy KE, Kim R, Maeda N, Goy MF, Smithies O. Natriuretic peptide receptor 1 expression influences blood pressures of mice in a dose-dependent manner. *Proc. Natl. Acad. Sci. U S A.* 1998 Mar 3;95(5):2547-51.

[59] Nakayama T, Soma M, Takahashi Y, Rehemudula D, Kanmatsuse K, Furuya K. Functional deletion mutation of the 5'-flanking region of type A human natriuretic peptide receptor gene and its association with essential hypertension and left ventricular hypertrophy in the Japanese. *Circ. Res.* 2000; 86: 841-5.

[60] Rubattu S, Bigatti G, Evangelista A, Lanzani C, Stanzione R, Zagato L, Manunta P, Marchitti S, Venturelli V, Bianchi G, Volpe M, Stella P. Association of atrial natriuretic peptide and type a natriuretic peptide receptor gene polymorphisms with left ventricular mass in human essential hypertension. *J. Am. Coll. Cardiol.* 2006; 48: 499-505.

[61] Usami S, Kishimoto I, Saito Y, Harada M, Kuwahara K, Nakagawa Y, Nakanishi M, Yasuno S, Kangawa K, Nakao K. Association of CT dinucleotide repeat polymorphism in the 5'-flanking region of the guanylyl cyclase (GC)-A gene with essential hypertension in the Japanese. *Hypertens. Res.* 2008; 31: 89-96.

In: Natriuretic Peptides
Editor: Kailash N. Pandey

ISBN: 978-1-63321-080-6

Chapter 7

NATRIURETIC PEPTIDE RECEPTOR-C IN PHYSIOLOGY AND PATHOPHYSIOLOGY

Madhu B. Anand-Srivastava*

Department of Molecular and Integrative Physiology, Université de Montréal, Quebec, Canada

ABSTRACT

The discovery of atrial natriuretic peptide (ANP) has led to considerable effort to unravel its implication in health and disease and the mechanisms evoked by ANP and related signaling aberrancies are believed to be implicated in the pathogenesis of diverse cardiovascular diseases, such as hypertension, atherosclerosis, hypertrophy and diabetes. The natriuretic peptides (NP) are a family of three polypeptide hormones termed atrial natriuretic peptide (ANP), brain natriuretic peptide (BNP), and C-type natriuretic peptide (CNP). All of these peptides are produced in mammalian hearts, including humans. ANP regulates a variety of physiological functions by interacting with receptors on the plasma membrane. NPR-A and NPR-B are guanylyl cyclase receptors, whereas NPR-C are coupled to adenylyl cyclase inhibition or phospholipase C activation through inhibitory guanine nucleotide-regulating protein. Unlike other G protein-coupled receptors, NPR-C have a single transmembrane domain and a short cytoplasmic domain of 37 amino acids, the cytoplasmic domain has a structural specificity like those of other single-transmembrane-domain receptors and 37 amino-acid cytoplasmic domain peptide is able to exert is inhibitory effect on adenylyl cyclase. The activation of NPR-C by C-ANP_{4-23} (a ring-deleted peptide of ANP) and C-type natriuretic peptide inhibits the mitogen-activated protein kinase activity stimulated by endothelin-3, platelet-derived growth factor and phorbol-12 myristate 13-acetate. C-ANP also inhibits mitogen-induced stimulation of DNA synthesis, indicating that the NPR-C plays a role in cell proliferation through an inhibition of mitogen-activated protein kinase and suggesting that the NPR-C might also be coupled to other signal transduction mechanism(s) or that there might be an

* Correspondence and request for printing: Dr. Madhu B. Anand-Srivastava, Department of Molecular and Integrative Physiology, Faculty of Medecine, Université de Montréal, C.P. 6128, Succ. Centre-Ville, Montréal (Québec) H2X 1H1, Canada. Tel: +1 514-343-2091, Fax: +1 514-343-2111. E-mail: madhu.anand-srivastava@umontreal.ca.

interaction of the NPR-C with some other signaling pathways. The levels of NP receptors as well as the signalling pathways activated by these receptors are altered in several cardiovascular diseases including hypertension, hypertrophy, atherosclerosis, diabetes etc. ANP receptor binding is decreased in most organs in hypertensive subjects and hypertensive animals. Platelet NPR-C are decreased in number in hypertensive patients and spontaneously hypertensive rats. NPR-A, NPR-B and NPR-C are decreased in number in deoxycorticosterone acetate–salt-treated kidneys and vasculature. These alterations in NP receptor subtypes may be related to the pathophysiology of hypertension. Several hormones such as angiotensin II, ANP and catecholamines, whose levels are increased in hypertension, downregulate or upregulate NPR-C and NPR-C receptor-mediated inhibition of adenylyl cyclase. It can be suggested that the antihypertensive action of several types of drugs such as angiotensin converting enzyme inhibitors, angiotensin type 1 receptor antagonists and [beta]$_2$-adrenergic antagonists may partly be attributed to their ability to modulate the expression and function of the NPR-C. In this review, we provide an overview of the NPR-C and signaling in both physiological and pathological conditions.

1. Introduction

Natriuretic peptides (NP) are a group of structurally but genetically distinct peptides termed atrial natriuretic peptide (ANP), brain natriuretic peptide (BNP) and C-type natriuretic peptide (CNP) [1-3]. All of these peptides are produced in mammalian hearts, including humans [4, 5] and exert variety of physiological functions including blood pressure, progesterone secretion, renin release, vasopressin release, endothelin release etc [6-8] by interacting with receptors on the plasma membrane either to modulate the levels of second messengers such as cAMP [9-14], cGMP [15-17], inositol turnover [8, 18, 19] or to affect ion channels. BNP and CNP; other members of natriuretic family act as antagonists to vasopressin, endothelin, and renin–angiotensin–aldosterone system [1] and [20] . CNP has also been shown to be a vasodilator produced by endothelial cells [3, 21] Compared with ANP, BNP has an additional six amino acid sequences at its amino terminal end [2, 22], whereas the carboxy terminal extension is absent in CNP [3].

2. Natriuretic Peptide Receptors and Cell Signaling

Natriuretic peptide receptors are divided into two major categories, those that activate guanylyl cyclase referred to as NPR-A [23, 24] and NPR-B [25, 26] and those that do not, termed as NPR-C [27]. NPR-A and NPR-B are membrane guanylyl cyclase receptors and have a relative molecular mass of approximately 120 kDa. NPR-A preferentially binds ANP and BNP, while NPR-B is more selective for CNP. On the other hand, NPR-C is a disulfide-bridged homodimer of ~ 60 kDa units with a broad range of ligands, including ANP, BNP, CNP and the synthetic des[Gln^{18}, Ser^{19}, Gly^{20}, Leu^{21}, Gly^{22}] ANP$_{4-23}$-amide C-ANP [28] and shows no affinity for NPR-A or NPR-B [28, 29]. Two NPR-C like proteins of 67 and 77 kDa have been identified in rat glomerular membranes and their electrophoretic mobilities with and without dithiothreitol suggest that both form disulfide bridged dimers [30]. The 77 kDa protein is involved in ligand internalization as a clearance receptor [31] and has a low affinity for CNP whereas 67 kDa protein shows a high affinity for CNP and is coupled to adenylyl cyclase inhibition through inhibitory guanine nucleotide regulatory protein Gi [30, 32, 33] and is devoid of guanylyl cyclase and kinase activities. The coupling of NPR-C to adenylyl cyclase signal transduction was further supported by the studies using antisense oligonucleotide of NPR-C [34].

NPR-A and NPR-B contain an extracellular ligand binding domain (ECD), a single transmembrane domain and a cytoplasmic domain containing protein kinase-like homology domain (KHD), dimerization domain and guanylyl cyclase (GC) catalytic domain [35]. On the other hand, NPR-C is homologous to NPR-A and NPR-B in its extracellular and transmembrane domain but contains a short 37 amino acid cytoplasmic domain [36] (Figure 1). The cytoplasmic domain of NPR-C contains several specific Gi activator sequences of 12 or 17 amino acids that have been shown to inhibit adenylyl cyclase activity through a pertussis toxin-sensitive Giα protein [37, 38]. The coupling of NPR-C to the activation of phospholipase C through Giα protein has also been shown [39, 40] (Figure 2), However, Mouawad et.al have shown that the NPR-C activation by C-ANP_{4-23} and resultant decrease in cAMP levels contributes to the activation of PI turnover signaling and suggested a cross-talk between NPR-C-mediated adenylyl cyclase inhibition and PLC signaling pathways [41] (Figure 3).

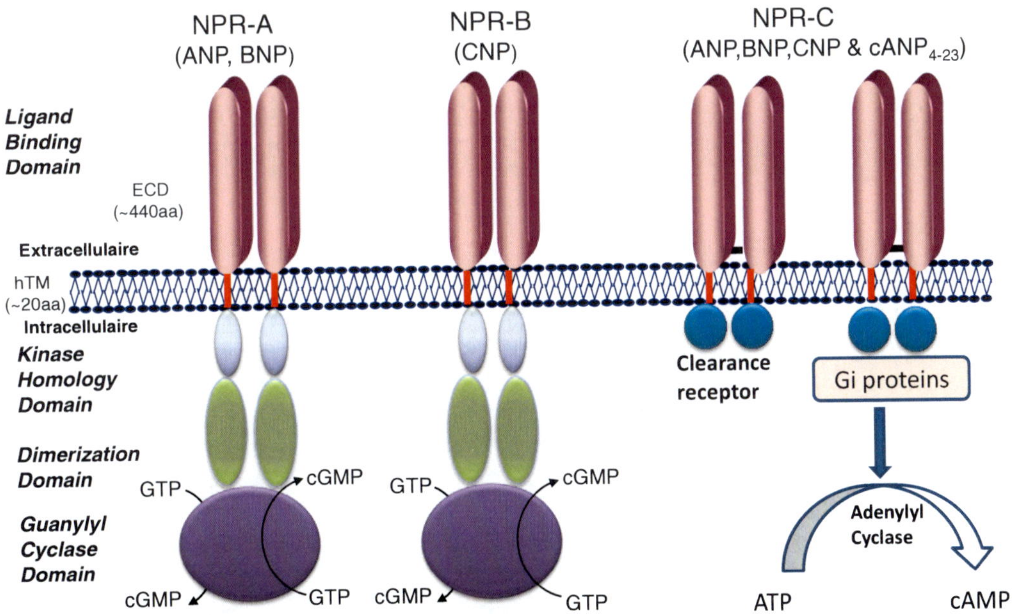

Figure 1. Subtypes of natriuretic peptide (NP) receptors. There are at least three distinct NP receptors, NPR-A, NPR-B and NPR-C. NPR-A and NPR-B are guanylyl cyclase receptors whereas NPR-C is coupled to inhibition of adenylyl cyclase. NPR-A and NPR-B contain an extracellular ligand binding domain (ECD), a single transmembrane domain and a cytoplasmic domain containing protein kinase-like homology domain (KHD), dimerization domain and guanylyl cyclase (GC) catalytic domain. On the other hand, NPR-C is homologous to NPR-A and NPR-B in its extracellular and transmembrane domain but contains a short 37 amino acid cytoplasmic domain. Two NPR-C like proteins of 67 and 77 kDa have been identified. The 77 kDa protein is involved in ligand internalization as a clearance receptor and has a low affinity for CNP whereas 67 kDa protein shows a high affinity for CNP and is coupled to adenylyl cyclase inhibition through inhibitory guanine nucleotide regulatory protein Gi.

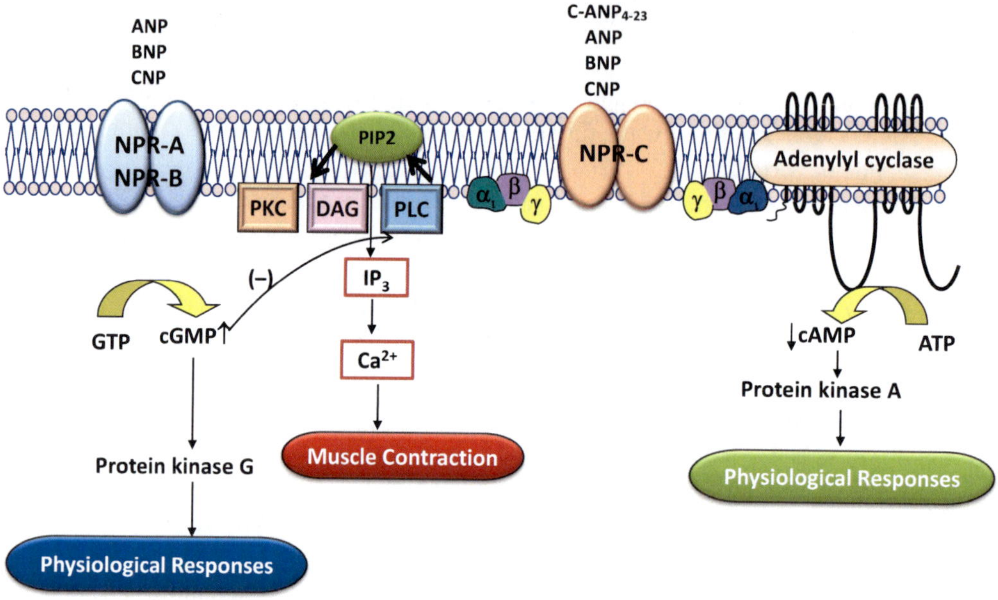

Figure 2. Natriuretic peptide receptors and associated signal transduction mechanisms activated by different natriuretic peptides. NPR-A and NPR-B are activated by ANP, BNP and CNP and result in the stimulation of cyclic GMP formation. The latter inhibits phospholipase C (PLC) activity and results in decreased production of diacylglycerol (DAG) and inositol triphosplate (IP3). The other major type of receptor is NPR-C, which is coupled to adenylyl cyclase through Gi protein. C-ANP$_{4-23}$ (a ring deleted peptide of ANP) ANP, BNP and CNP interact with receptor and inhibit adenylyl cyclase activity and suppress cAMP concentrations. Adapted from (Anand-Srivastava 2005).

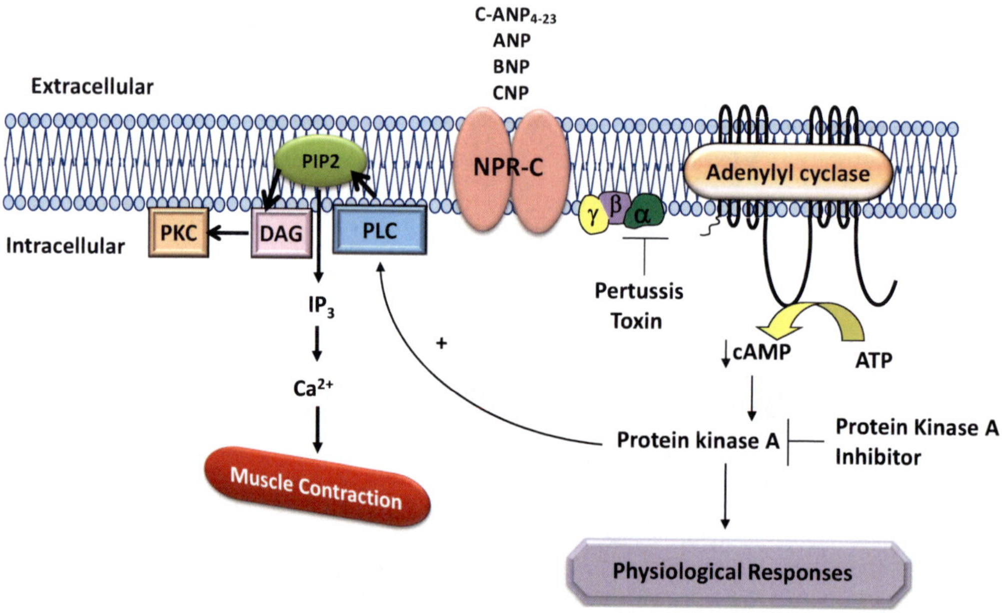

Figure 3. Summary of the cross-talk between NPR-C-mediated inhibition of adenylyl cyclase and activation of PI turnover. NPR-C activation by C-ANP$_{4-23}$ inhibits adenylyl cyclase activity through PT-sensitive Gi protein and the resultant decreased levels od cAMP stimulates PI turnover in VSMCs. Adapted from (Anand-Srivastava. 2005).

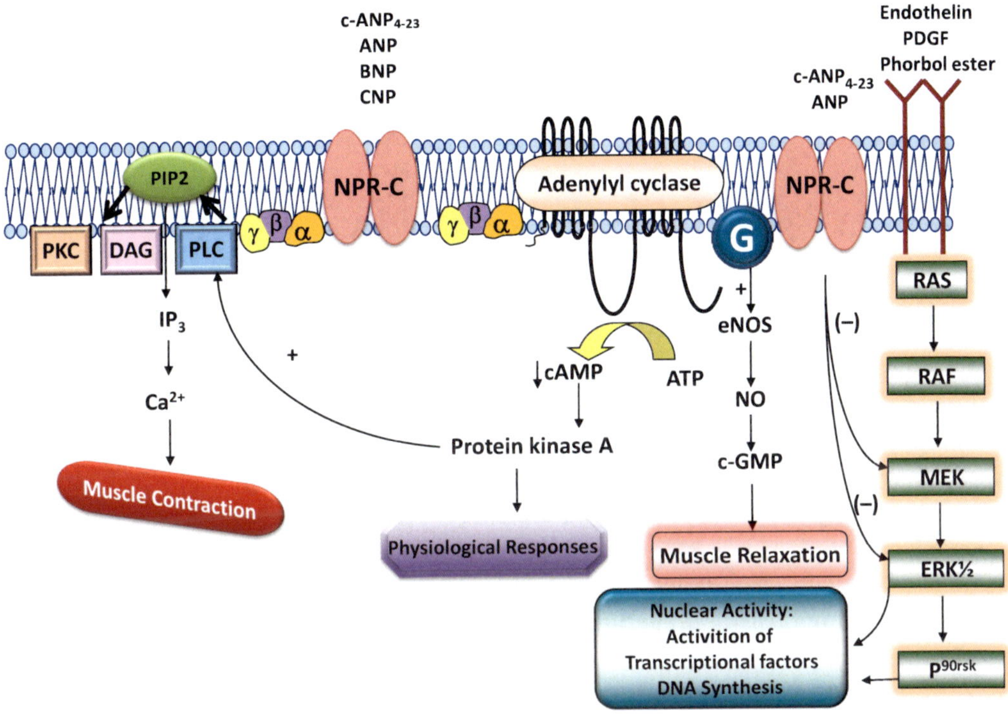

Figure 4. Summary of NPR-C-coupled signaling systems. NPR-C is coupled to adenylyl cyclase transduction system through pertussis toxin-sensitive Gi protein. The activation of NPR-C by C-ANP$_{4-23}$ and other NP results in the inhibition of adenylyl cyclase activity and decreased cAMP levels. The NPR-C-induced decreased cAMP levels stimulate PI turnover signaling. NPR-C activation also increased endothelial nitric oxide synthase (eNOS) activity and results in the increased formation of nitric oxide (NO). NO stimulates soluble guanylyl cyclase activity and increases the formation of cGMP, a second messger implicated in vasorelaxation. NPR-C activation by CNP and C-ANP$_{4-23}$ attenuates MAP kinase activity stimulated by mitogenic factor such as endothelin (ET-1), platelet-derived growth factor (PDGF) and phorbol ester and inhibits cell proliferation. Adapted from (Anand-Srivastava. 2005).

The modulation of other signaling pathways by NPR-C activation has also been suggested. Trachte [42] has shown that NPR-C activation by CNP and 15 amino acid peptide fragment of cytoplasmic domain of NPR-C inhibited protein kinase C (PKC) activity as well as evoked dopamine efflux in pheochromocytoma (PC12) cells. NPR-C activation has also been shown to activate nitric oxide synthase in various tissues including smooth muscle [43, 44]. In addition, NPR-C was also shown to inhibit MAP kinase activity in astrocyte [45] (Figure 4).

NPR-C, both non-liganded and bound to CNP, has been crystallized [46]. The non-liganded structure was determined to 2.9 A with one receptor ECD, or a half-dimer in the asymmetric unit, with the second half of the dimer related by a crystallographic two-fold axis. However, NPR-C structure bound with CNP was determined to be 2.0 A. A single CNP molecule binds in the center of the two-fold symmetric dimeric receptors with the stoichiometry of 1:2. The NPR-C gene has been shown to resides on human chromosome 5p14-p13 [47].

NPR-C are distributed widely in several tissues and cells including platelets, vascular smooth muscle cells (VSMC), glomeruli, collecting ducts, pituitary glands, adrenal glands, the cerebral cortex, the brain striatum, the ciliary process of the eye, purkinje fibers of the cardiac conduction system, and Leydig tumor cells [8, 10-14, 48-53]. The density of these receptors in most tissues is higher than that of NPR-A/B. For example, in endothelial cells, the NPR-C comprise about 94% of the total ANP receptor population [54].

3. Cross-Talk between NPR-A and NPR-C

An interaction between guanylyl cyclase receptor (NPR-A, NPR-B) and NPR-C has been demonstrated [55]. Our earlier studies showed that knocking down of NPR-C by NPR-C antisense enhanced the responsiveness of ANP to stimulate cGMP levels in A10 VSMC [34]. We further showed that the knockdown of NPR-C by antisense resulted in the upregulation of NPR-A (unpublished observations). The knock down of NPR-A by antisense or SiRNA has also been shown to upregulate the expression of NPR-C and Giα proteins and increased inhibition of adenylyl cyclase resulting in decreased levels of cAMP in VSMC [56]. The re-expression of NPR-A in antisense-treated cells reversed the enhanced expression of NPR-C to control levels. The mechanism by which knockdown of NPR-A upregulates the expression of NPR-C and Giα proteins was attributed to the decreased levels of intracellular cGMP, because the treatment of antisense-treated cells with 8Br-cGMP was shown to reverse the enhanced levels of NPR-C as well as of Giα proteins to control levels. Furthermore, 8Br-cGMP and PD98059, a MEK inhibitor was also shown to reverse the antisense-induced enhanced levels of NPR-C and Giα proteins to control levels [56]. These results suggest that a cross-talk exists between the two receptors. Thus, it may be suggested that the interplay of NPR-C and NPR-A may play an important role in the regulation of physiological functions including cell proliferation and blood pressure.

4. NPR-C: A Regulator of Physiological Functions

NPR-C initially was proposed as a clearance receptor [57], the main function of this receptor was considered to remove the bound ligand by internalization and degradation [58]. However, accumulating evidence has suggested that this receptor does not only function as a clearance receptor but plays an important role in the regulation of variety of physiological functions. In this regard, 77 kDa subunit of NPR-C protein is implicated in ligand internalization as a clearance receptor [31] , whereas 67 kDa protein of NPR-C that is coupled to adenylyl cyclase inhibiton through Gi regulatory protein and exerts physiological functions including the regulation of fluid homeostasis [30, 42, 59-70].

NPR-C activation was shown to exert antiproliferative effects in three (SY-5Y, SH-IN, Neuro2a) neuroblastoma cell lines [71]. The implication of NPR-C in ANP-induced inhibition of rat astroglial, endothelial and vascular smooth muscle cell proliferation has also been shown [4, 60, 72-74]. NPR-C activation-mediated inhibition of vascular smooth muscle cell proliferation was shown to involve MAPK and PI3-kinase pathways [72]. The contribution of different cyclins of D type, A and E and cyclin-dependent kinases (CDK) in ANP-mediated inhibition of ET-3-stimulated cell proliferation in astrocytes has also been reported [75]. We have also shown that C-ANP_{4-23} and small peptide fragments of cytoplasmic domain of NPR-C with Gi activator sequences inhibited vasoactive peptide-induced cell proliferation [72] through Giα/MAPK/P13K/AKT signalling pathways in VSMC. We have further shown that C-ANP_{4-23} attenuated the enhanced expression of cell cycle proteins, cyclin D1/cdk2/cd4 (G1-S phase) through MAP kinase/Gi signalling and resulted in the attenuation of the hyperproliferation of VSMC from SHR [76]. Furthermore, BNP and CNP have been shown to exert antiproliferative actions in cardiac fibroblasts through a non-cGMP-mediated mechanism involving the NPR-C [77].

In addition, several effects of CNP appear to be mediated by NPR-C signaling. In the heart, CNP/NPR-C signalling has been linked to electrophysiological activities [7]. The role of NPR-C in mediating the negative chronotropic effect of CNP on heart rate has also been shown in cardiac myocytes, including the sinoatrial node myocytes [67]. In rat hypothalamus, CNP inhibits the calcium current in magnocellular neurosecretory cells through NPR-C [78]. The implication of NPR-C and NO-dependent pathway in NPs-mediated stimulation of the $Na^{(+)}$-$K^{(+)}$ pump has also been demonstrated [43]. Furthermore, CNP/NPR-C signalling has also been demonstrated as a fundamental pathway underlying endothelium-derived hyperpolarizing factor-dependent regulation of vascular tone in the rat mesenteric resistance arteries and in the coronary vasculature [79]. The activation of NPR-C by CNP has also been shown to regulate coronary blood flow and prevents myocardial ischemia/reperfusion injury by reducing the infarct size in the rat model [80] and may thus be involved in cardiovascular homeostasis.

NPR-C activation has also been shown to inhibit electrically induced purinergic and adrenergic contractile force generation in rabbit isolated vasa deferentia [70]. In addition, NPR-C-mediated inhibition of adenylyl cyclase and neurotransmission in nerve growth factor-treated PC12 cells has also been demonstrated [18, 70]. Furthermore, NPR-C activation was also shown to inhibit *in vivo* translation of the endothelin message and the secretion of endothelin from cultured bovine aortic endothelial cells [62] and basal and luteinizing hormone-stimulated progesterone secretion in Leydig tumor cells [33]. In addition, NPR-C activation has also been shown to mediate the ANP-induced attenuation of COX-2 expression [81].

NPR-C and cAMP inhibition has also been implicated in modulating endothelial permeability in coronary endothelial cells [63]. The knockout mice lacking NPR-C gene Npr3 showed reduced blood pressure and skeletal deformities associated with a considerable increase in bone turnover [82]. In addition, the elimination of NPR-C by antisense has also been reported to attenuate inhibitory effects of CNP on evoked neurotransmitter efflux in PC12 cells [83]. The role of NPR-C as a modulator of vascular endothelial growth factor (VEGF) production has also been documented [64]. C-ANP_{4-23} was shown to inhibit VEGF production and endothelin (ET) or hypoxia-stimulated VEGF transcription in VSMC and suggested the implication of NPR-C in vascular remodeling and angiogenesis. The activation of NPR-C has also been shown to stimulate pancreatic fluid and protein secretion in the rats [65]. In addition, CNP has also been shown to increase the secretion of amylase through the activation of NPR-C coupled to PLC pathway and downstream effectors in the exocrine pancreas and suggest a role of NPR-C in the regulation of the digestive functions [84]. Furthermore, the involvement of NPR-C in the hepatocyte protection against hypoxic injury has also been shown [66]. Activation of NPR-C with C-ANP_{4-23} increased hepatocyte tolerance to hypoxia, which was linked to the stimulation of protein kinase C delta through Gi and phospholipase C signaling. NPR-C has been shown to play a role in maintaining the retinal functions in dopaminergic and cholinergic amacrine cells of retina [69].

Natriuretic peptides have also been shown to inhibit cardiac hypertrophy in pressure and volume overload hypertrophy rat models [85, 86]. Genetic ablation of the NPR-A gene in cardiac myocytes of mice results in cardiac hypertrophy as well as fibrosis from birth [87]. In addition, targeted overexpression of a dominant-negative NPR-A receptor in the suprarenal aorta-banded mouse heart has been shown to result in increased cardiac hypertrophy and fibrosis [88]. The ascending aorta-banded rat heart, characterized by extensive hypertrophy and fibrosis [89] has also been shown to exhibit an increased expression of NPR-A and NPR-C [85]. On the other hand, C-$ANP_{4\text{-}23}$ and small peptide fragments of cytoplasmic domain of NPR-C with Gi activator sequences were shown to inhibit vasoactive peptide-induced vascular hypertrophy through Gqα/MAPK/P13K/AKT signaling pathways in VSMC [90]. The role of Gqα protein and associated signaling pathway including PKC in the development and progression of cardiac hypertrophy has also been reported [91-95].We also show that NPR-C activation by C-$ANP_{4\text{-}23}$ inhibited the hypertrophy of VSMC from SHR through the inhibition of the enhanced expression of Gqα proteins (unpublished observations).

NPR-C activation by C-$ANP_{4\text{-}23}$ has also been reported to inhibit the expression of Giα proteins in A10 VSMC [96]. In addition, the enhanced expression of Giα proteins, enhanced superoxide anion production, enhanced NADPH oxidase activity and enhanced expression of different subunits of NADPH oxidase such as Nox4 and $p47^{phox}$ proteins exhibited by VSMC from SHR was also shown to be attenuated by NPR-C activation by C-$ANP_{4\text{-}23}$ [97]. These studies suggest a role of C-$ANP_{4\text{-}23}$, an NPR-C agonist, as an antioxidant. Thus, it may be possible that C-$ANP_{4\text{-}23}$ due to its antioxidant property and its ability to suppress the heightened expression of Giα proteins may have a potential to serve as a therapeutic agent for the treatment of hypertension. This is supported by our recent study showing that treatment of SHR with C-$ANP_{4\text{-}23}$ attenuated the development of hypertension in SHR which was associated with its ability to decrease the enhanced expression of Giα proteins and oxidative stress (unpublished observations). Furthermore, genetic variants in *NPR3* gene encoding NPR-C have been reported to be associated with hypertension, a conventional risk factor for stroke, in the general populations of Caucasians [98, 99], Japanese [100] and Chinese [101]. Thus, from the foregoing discussion, it is now clear that NPR-C may not be solely clearance receptor as has been reported earlier [57, 58] but plays an important role in the regulation of various physiological functions.

5. Hormonal Regulation of NPR-C Expression and Signaling

The levels of NPR-C have been reported to be regulated by several hormonal factors that regulate blood pressure. Exposure of vascular smooth muscle cells to angiotensin II (Ang II) has been shown to decrease NP receptor whereas the production of cGMP in response of ANP was augmented suggesting that guanylyl cyclase-coupled NP receptors were upregulated and uncoupled sites were downregulated [102]. On the other hand, a downregulation of the NPR-A and cGMP production by Ang II treatment of VSMC has also been reported [103, 104]. In addition, Ang II infusion in rats decreased the ANP binding sites of both NPR-A and NPR-C [105]. On the other hand, Ang II treatment of VSMC from mesenteric arteries exaggerated the ANP-induced production of cGMP without affecting the binding [106]. In accord with these data, Palaparti and Anand-Srivastava have also shown [107] that Ang II treatment of VSMC did not downregulate NPR-C; however, NPR-C-mediated inhibition of adenylyl cyclase was completely abolished. This could be attributed to the uncoupling of NPR-C from adenylyl cyclase due to the phosphorylation of Giα-2 proteins by the protein kinase C pathway [107]. On the other hand, Yoshimoto et al. [108] have reported that Ang II treatment of cultured aortic smooth muscle cells decreased the NPR-C mRNA levels, which was reversed by AT1 antagonist CV-11974, however, NPR-A or NPR-B expression was not altered by this treatment.

Endothelin (ET-1) and arginine-vasopressin (AVP) have also been shown to modulate NP receptor and signaling. AVP has been shown to decrease ANP receptor binding and ANP-induced cGMP accumulation in VSMC [103, 104] whereas treatment of rat cultured VSMC from mesenteric arteries with ET-1 and AVP did not alter the density of NP binding sites and cGMP production [106]. However, we showed that pre-treatment of A10 VSMC with AVP decreased the ANP receptor binding activity as well as the expression of NPR-C protein [109] which was prevented by GF109203X, an inhibitor of protein kinase C. The decreased expression of NPR-C was reflected in an attenuation of NPR-C-mediated inhibition of adenylyl cyclase suggesting a role of PKC signaling in AVP-induced downregulation of NPR-C protein and NPR-C- mediated inhibition of adenylyl cyclase activity [109].

In addition, ET-1, a regulator of blood pressure has also been reported to attenuate the expression of NPR-C protein NPR-C-mediated inhibition of adenylyl cyclase [110]. In addition, ANP has also been reported to regulate its own receptors. Exposure of cultured VSMC to ANP was shown to downregulate ANP receptors [103, 111] whereas glomerular ANP receptors did not undergo downregulation [112]. On the other hand, Anand-Srivastava [96] has shown that VSMC exposed to C-ANP_{4-23} resulted in the reduction in ANP binding activity and the attenuation of NPR-C-mediated inhibition of adenylyl cyclase by C-ANP_{4-23} as well as 17 amino acid fragment peptide of cytoplasmic domain [96]. C-ANP_{4-23} treatment also decreased the expression of the Giα regulatory protein in these cells [113], which may be partly responsible for the attenuated inhibition of adenylyl cyclase induced by C-ANP_{4-23} and cytoplasmic peptide fragment of NPR-C.

A differential regulation of NP receptor subtype by transforming growth factor-β1 (TGF-β1) has been reported. TGF-β1 augmented the expression of NPR-C at the transcriptional level in murine thymic stromal cell line (MRL 104.8a), whereas NPR-A and NPR-B were downregulated [114]. On the other hand, fibroblast growth factors (FGF-1 and FGF-2) and platelet-derived growth factor BB (PDGF-BB) that activate tyrosine kinase receptors were shown to reduce the expression of NPR-C mRNA in pulmonary arterial smooth muscle cells [115]. Down regulation of NPR-C expression by FGF-1, FGF-2, and PDGF-BB was inhibited by the selective FGF-1 receptor tyrosine kinase inhibitor PD166866 and mitogen-activated protein/extracellular signal-regulated kinase inhibitors U-0126 and PD98059 [115]. In addition, epidermal growth factor (EGF) and the apparented peptide HB-EGF, which shares the same receptor with EGF were reported to downregulate NPR-C mRNA in cultured mesengial cells, whereas PDGF and insulin-like growth factor-1 (IGF-1) were ineffective in reducing the levels of NPR-C mRNA [116].

The regulation of NP receptors by catecholamines in cultured VSMC has also been reported. Noradrenaline was shown to downregulate NPR-C in a time- and dose-dependent manner and this effect was antagonized by a β2-selective adrenergic antagonist (ICI 118,551) but not by α1-, α2-, and β1-adrenergic antagonists [117]. In addition, isoproterenol, 8-Br-cAMP, forskolin, and sodium fluoride were also reported to decrease the steady-state level of NPR-C mRNA. On the other hand, NPR-A and NPR-B receptor mRNA levels were not affected [117].

The modulation of NPR-C by salt has also been documented. Dietary salt supplementation has been shown to downregulate steady-state NPR-C mRNA levels in kidney independently of ANP, whereas the levels of NPR-A and NPR-B were not altered [115]. In addition, treatment of vascular endothelial cells with NaCl resulted in a marked reduction in the number of NPR-C, whereas the density of NPR-A was not altered. The reduction in number of NPR-C occurred at a post-translational level according to a RNAase protection assay [118].

Nitric oxide (NO) that plays a role in the regulation of variety of physiological functions including vasorelaxation has also been shown to decrease the expression of NPR-C, Giα proteins and NPR-C-mediated adenylyl cyclase signaling in VSMC through MAPK pathway [119, 120]. In addition, peroxynitrite, ($ONOO^-$), a powerful oxidant and exerts a prolonged vasorelaxant action under physiological conditions [121, 122] also attenuates the expression of NPR-C as well as Giα proteins in VSMC [119]. The downregulation of NPR-C and Giα protein and resultant attenuation of adenylyl cyclase inhibition and increased levels of cAMP may be an important mechanism through which NO and $ONOO^-$, exert vasorelaxation and regulates blood pressure. In addition, cGMP, a second messenger of NO as well as of NPR-A/NPR-B activation has also been reported to attenuate the expression of NPR-C in VSMC [123].

6. NPR-C in Pathophysiology

Abnormalities in NPs levels have been implicated in pathogenesis of cardiocerebral diseases such as hypertension [124, 125], coronary artery disease [126], and heart failure [127, 128]. A decreased or increased NP receptor density has also been reported in cultured VSMC from spontaneously hypertensive rats (SHR) and other models of experimental hypertension [6]. The levels of NPR-C mRNA were down regulated in aorta from stroke-prone SHR (SHR/SD/Izm) whereas NPR-A were upregulated as compared to WKY control rats [129]. The downregulation of vascular NPR-C was attributed to Ang II-mediated mechanism because the treatment with AT1 receptor antagonist TCV-116 reversed the attenuated NPR-C mRNA to the WKY Izm level [129]. NPR-C was also downregulated in several tissues from DOCA-salt hypertensive rats (HR).

A significant decrease in NPR-C and NPR-C-mediated adenylyl cyclase signaling has been reported in platelets from SHR [130] whereas platelets from hypertensive patients did not exhibit any alteration in NP receptor density [131, 132]. In addition, a reduction in number of NP receptor binding sites in various neuronal tissues, such as the subfornical organ, the area postrema, the nucleus of the solitary tract [65] and cereberal microvessels [133] from SHR, has been reported. A decrease in NP receptor binding sites without any change in ANP-mediated increases in cGMP levels has also been reported in stellate ganglion tissue [134] from SHR, which suggests that the receptor, which is downregulated could be the NPR-C subtype rather than the guanylyl cyclase-coupled receptor.

In addition, aorta from Nw-nitro-L-arginine-methyl ester (L-NAME) hypertensive rats exhibited decreased expression of NPR-C and NPR-A mRNA, whereas AT1 and ACE mRNA expression was enhanced [135]. We also showed that NPR-C-mediated inhibition of adenylyl cyclase in hearts was completely abolished in L-NAME hypertensive rats [136], which may be attributed to the downregulation of NPR-C. The downregulation of NPR-C in these HR may be due to the increased levels of Ang II, because losartan treatment of these rats restored the attenuated inhibition of C-ANP_{4-23}-mediated adenylyl cyclase towards control levels [137]. A marked reduction of NPR-C receptor number has also been observed in salt-loaded Dahl salt-sensitive hypertensive rats, suggesting a role of this mechanism in the pathophysiology of hypertension in this animal model [138].

Thus taken together, it may be suggested that the downregulation of NPR-C and upregulation of NPR-A may not be related to the regulation of blood pressure in hypertensive rats because the ablation of NPR-C in NPR-C-knockout mice has been shown to result in decreased blood pressure [82] whereas the ablation of NPR-A in NPR-A knockout mice showed increased blood pressure [139]. However, the underlying mechanisms responsible for the upregulation of NPR-A and downregulation of NPR-C in SHR and whether the altered expression of these receptors is the cause or effect of hypertension is not clear and needs to be investigated. In addition, other factors including Ang II, endothelin and growth factor and their receptors that are upregulated in hypertension may be responsible for the altered expression of NP receptors and contribute to the enhanced blood pressure in hypertensive rats.

Increased expressions of NPR-A and NPR-B have also been reported in aorta from obese/hyperglycaemic Wistar fatty rats, whereas no difference in renal levels of mRNA for both receptors was detected [129]. Plasma ANP levels and CNP mRNA levels were also shown to be significantly increased in aorta from these rats. Administration of HS-142-1 an antagonist of NPR-A and NPR-B resulted in a significant augmentation in the systolic blood pressure and a larger decrease in cGMP as compared to Wistar lean rats [129]. However, the levels of NPR-C in these rats need to be investigated.

CONCLUSION

There are at least three distinct NP receptors, NPR-A, NPR-B and NPR-C. NPR-A and NPR-B are guanylyl cyclase receptors and their activation results in increased cGMP levels, whereas activation of the NPR-C results in inhibition of adenylyl cyclase activity, augmentation of phospholipase C activity, stimulation of eNOS or inhibition of MAPK activity. NPR-C regulates a variety of physiological functions which suggests that NPR-C is not merely a clearance receptor as has been initially viewed.

A cross-talk between the two receptors NPR-A and NPR-C has been shown that may play an important role in pathphysiology. In this regard, it may be possible that the decreased blood pressure in NPR-C knockout mice may not be due to the downregulation of NPR-C only but the upregulation of NPR-A and resultant increased levels of cGMP as well as the decreased expression of Giα proteins may also be the contributing factors. Similarly, the increased blood pressure in NPR-A knockout mice that results in the decreased levels of cGMP may also be attributed to the upregulation of NPR-C, Giα proteins and resultant decreased levels of cAMP. These possibilities need to be investigated.

The alteration in number of the NPR-C and of the signaling mechanisms by several humoral factors and other interventions may play important roles in the regulation of the blood pressure. In addition, NPR-C -induced decreased expression of Giα proteins and nitroxidative stress in VSMC from SHR may be the important mechanisms for the regulation of physiological functions including blood pressure. Furthermore, the inhibition of endothelin secretion by ANP via the NPR-C could be another mechanism involved in the regulation of the blood pressure. Similarly, the NPR-C-mediated inhibition of MAPK activity and protein synthesis stimulated by endothelin and other mitogenic factors could also contribute to the inhibition of cell proliferation and hypertrophy associated with hypertension.

REFERENCES

[1] Brenner BM, Ballermann BJ, Gunning ME, Zeidel ML. Diverse biological actions of atrial natriuretic peptide. *Physiol. Rev.* 1990;70:665-99.

[2] Sudoh T, Kangawa K, Minamino N, Matsuo H. A new natriuretic peptide in porcine brain. *Nature* 1988;332:78-81.

[3] Sudoh T, Minamino N, Kangawa K, Matsuo H. C-type natriuretic peptide (CNP): a new member of natriuretic peptide family identified in porcine brain. *Biochem. Biophys. Res. Commun.* 1990;168:863-70.

[4] Levin ER, Gardner DG, Samson WK. Natriuretic peptides. *N. Engl. J. Med.* 1998; 339:321-8.

[5] Yandle TG. Biochemistry of natriuretic peptides. *J. Intern. Med.* 1994;235:561-76.

[6] Anand-Srivastava MB. Atrial natriuretic peptide-C receptor and membrane signalling in hypertension. *J. Hypertens.* 1997;15:815-26.

[7] Anand-Srivastava MB. Natriuretic peptide receptor-C signaling and regulation. *Peptides* 2005;26:1044-59.

[8] Anand-Srivastava MB, Trachte GJ. Atrial natriuretic factor receptors and signal transduction mechanisms. *Pharmacol. Rev.* 1993;45:455-97.

[9] Anand-Srivastava MB, Franks DJ, Cantin M, Genest J. Atrial natriuretic factor inhibits adenylate cyclase activity. *Biochem. Biophys. Res. Commun.* 1984;121:855-62.

[10] Anand-Srivastava MB, Cantin M, Genest J. Inhibition of pituitary adenylate cyclase by atrial natriuretic factor. *Life Sci.* 1985;36:1873-9.

[11] Anand-Srivastava MB, Genest J, Cantin M. Inhibitory effect of atrial natriuretic factor on adenylate cyclase activity in adrenal cortical membranes. *FEBS Lett.* 1985;181:199-202.

[12] Anand-Srivastava MB, Cantin M. Atrial natriuretic factor receptors are negatively coupled to adenylate cyclase in cultured atrial and ventricular cardiocytes. *Biochem. Biophys. Res. Commun.* 1986;138:427-36.

[13] Anand-Srivastava MB, Vinay P, Genest J, Cantin M. Effect of atrial natriuretic factor on adenylate cyclase in various nephron segments. *Am. J. Physiol.* 1986;251:F417-23.

[14] Bianchi C, Anand-Srivastava MB, De Lean A, et al. Localization and characterization of specific receptors for atrial natriuretic factor in the ciliary processes of the eye. *Curr. Eye Res.* 1986;5:283-93.

[15] Hamet P, Tremblay J, Pang SC, et al. Effect of native and synthetic atrial natriuretic factor on cyclic GMP. *Biochem. Biophys. Res. Commun.* 1984;123:515-27.

[16] Waldman SA, Rapoport RM, Murad F. Atrial natriuretic factor selectively activates particulate guanylate cyclase and elevates cyclic GMP in rat tissues. *J. Biol. Chem.* 1984;259:14332-4.

[17] Winquist RJ, Faison EP, Waldman SA, Schwartz K, Murad F, Rapoport RM. Atrial natriuretic factor elicits an endothelium-independent relaxation and activates particulate guanylate cyclase in vascular smooth muscle. *Proc. Natl. Acad. Sci. U. S. A.* 1984; 81:7661-4.

[18] Fujii K, Ishimatsu T, Kuriyama H. Mechanism of vasodilation induced by alpha-human atrial natriuretic polypeptide in rabbit and guinea-pig renal arteries. *J. Physiol.* 1986;377:315-32.

[19] Rapoport RM. Cyclic guanosine monophosphate inhibition of contraction may be mediated through inhibition of phosphatidylinositol hydrolysis in rat aorta. *Circ. Res.* 1986;58:407-10.

[20] Ruskoaho H. Atrial natriuretic peptide: synthesis, release, and metabolism. *Pharmacol. Rev.* 1992;44:479-602.

[21] Suga S, Nakao K, Itoh H, et al. Endothelial production of C-type natriuretic peptide and its marked augmentation by transforming growth factor-beta. Possible existence of "vascular natriuretic peptide system". *J. Clin. Invest.* 1992;90:1145-9.

[22] Kambayashi Y, Nakao K, Mukoyama M, et al. Isolation and sequence determination of human brain natriuretic peptide in human atrium. *FEBS Lett.* 1990;259:341-5.

[23] Chinkers M, Garbers DL, Chang MS, et al. A membrane form of guanylate cyclase is an atrial natriuretic peptide receptor. *Nature* 1989;338:78-83.

[24] Lowe DG, Chang MS, Hellmiss R, et al. Human atrial natriuretic peptide receptor defines a new paradigm for second messenger signal transduction. *EMBO J.* 1989;8:1377-84.

[25] Chang MS, Lowe DG, Lewis M, Hellmiss R, Chen E, Goeddel DV. Differential activation by atrial and brain natriuretic peptides of two different receptor guanylate cyclases. *Nature* 1989;341:68-72.

[26] Schulz S, Singh S, Bellet RA, et al. The primary structure of a plasma membrane guanylate cyclase demonstrates diversity within this new receptor family. *Cell* 1989;58:1155-62.

[27] Fuller F, Porter JG, Arfsten AE, et al. Atrial natriuretic peptide clearance receptor. Complete sequence and functional expression of cDNA clones. *J. Biol. Chem.* 1988;263:9395-401.

[28] Koller KJ, Lowe DG, Bennett GL, et al. Selective activation of the B natriuretic peptide receptor by C-type natriuretic peptide (CNP). *Science* 1991;252:120-3.

[29] Brown J, Chen Q. Regional expression of natriuretic peptide receptors during the formation of arterial neointima in the rabbit. *Circ. Res.* 1995;77:906-18.

[30] Woodard GE, Li X, Rosado JA. Water deprivation enhances the inhibitory effect of natriuretic peptides on cAMP synthesis in rat renal glomeruli. *Am. J. Physiol. Renal Physiol.* 2004;287:F418-26.

[31] Brown J, Zuo Z. Receptor proteins and biological effects of C-type natriuretic peptides in the renal glomerulus of the rat. *Am. J. Physiol.* 1994;266:R1383-94.

[32] Anand-Srivastava MB, Srivastava AK, Cantin M. Pertussis toxin attenuates atrial natriuretic factor-mediated inhibition of adenylate cyclase. Involvement of inhibitory guanine nucleotide regulatory protein. *J. Biol. Chem.* 1987;262:4931-4.

[33] Anand-Srivastava MB, Sairam MR, Cantin M. Ring-deleted analogs of atrial natriuretic factor inhibit adenylate cyclase/cAMP system. Possible coupling of clearance atrial natriuretic factor receptors to adenylate cyclase/cAMP signal transduction system. *J. Biol. Chem.* 1990;265:8566-72.

[34] Palaparti A, Li Y, Anand-Srivastava MB. Inhibition of atrial natriuretic peptide (ANP) C receptor expression by antisense oligodeoxynucleotides in A10 vascular smooth-muscle cells is associated with attenuation of ANP-C-receptor-mediated inhibition of adenylyl cyclase. *Biochem. J.* 2000;346 Pt 2:313-20.

[35] Pandey KN. Stoichiometric analysis of internalization, recycling, and redistribution of photoaffinity-labeled guanylate cyclase/atrial natriuretic factor receptors in cultured murine Leydig tumor cells. *J. Biol. Chem.* 1993;268:4382-90.

[36] Porter JG, Arfsten A, Fuller F, Miller JA, Gregory LC, Lewicki JA. Isolation and functional expression of the human atrial natriuretic peptide clearance receptor cDNA. *Biochem. Biophys. Res. Commun.* 1990;171:796-803.

[37] Anand-Srivastava MB, Sehl PD, Lowe DG. Cytoplasmic domain of natriuretic peptide receptor-C inhibits adenylyl cyclase. Involvement of a pertussis toxin-sensitive G protein. *J. Biol. Chem.* 1996;271:19324-9.

[38] Pagano M, Anand-Srivastava MB. Cytoplasmic domain of natriuretic peptide receptor C constitutes Gi activator sequences that inhibit adenylyl cyclase activity. *J. Biol. Chem.* 2001;276:22064-70.

[39] Murthy KS, Makhlouf GM. Identification of the G protein-activating domain of the natriuretic peptide clearance receptor (NPR-C). *J. Biol. Chem.* 1999;274:17587-92.

[40] Hirata M, Chang CH, Murad F. Stimulatory effects of atrial natriuretic factor on phosphoinositide hydrolysis in cultured bovine aortic smooth muscle cells. *Biochim. Biophys. Acta* 1989;1010:346-51.

[41] Mouawad R, Li Y, Anand-Srivastava MB. Atrial natriuretic peptide-C receptor-induced attenuation of adenylyl cyclase signaling activates phosphatidylinositol turnover in A10 vascular smooth muscle cells. *Mol. Pharmacol.* 2004;65:917-24.

[42] Trachte GJ. Natriuretic peptides suppress protein kinase C activity to reduce evoked dopamine efflux from pheochromocytoma (PC12) cells. *Endocrinology* 2003;144:94-100.

[43] William M, Hamilton EJ, Garcia A, et al. Natriuretic peptides stimulate the cardiac sodium pump via NPR-C-coupled NOS activation. *Am. J. Physiol. Cell Physiol.* 2008;294:C1067-73.

[44] Murthy KS, Teng B, Jin J, Makhlouf GM. G protein-dependent activation of smooth muscle eNOS via natriuretic peptide clearance receptor. *Am. J. Physiol.* 1998; 275: C1409-16.

[45] Prins BA, Weber MJ, Hu RM, Pedram A, Daniels M, Levin ER. Atrial natriuretic peptide inhibits mitogen-activated protein kinase through the clearance receptor. Potential role in the inhibition of astrocyte proliferation. *J. Biol. Chem.* 1996; 271:14156-62.

[46] He X, Chow D, Martick MM, Garcia KC. Allosteric activation of a spring-loaded natriuretic peptide receptor dimer by hormone. *Science* 2001;293:1657-62.

[47] Kone BC. Molecular biology of natriuretic peptides and nitric oxide synthases. *Cardiovasc. Res.* 2001;51:429-41.

[48] Anand-Srivastava MB, Gutkowska J, Cantin M. The presence of atrial-natriuretic-factor receptors of ANF-R2 subtype in rat platelets. Coupling to adenylate cyclase/cyclic AMP signal-transduction system. *Biochem. J.* 1991;278 (Pt 1):211-7.

[49] Bianchi C, Thibault G, De Lean A, Genest J, Cantin M. Atrial natriuretic factor binding sites in the jejunum. *Am. J. Physiol.* 1989;256:G436-41.

[50] De Lean A, Gutkowska J, McNicoll N, Schiller PW, Cantin M, Genest J. Characterization of specific receptors for atrial natriuretic factor in bovine adrenal zona glomerulosa. *Life Sci.* 1984;35:2311-8.

[51] Konrad EM, Bianchi C, Thibault G, et al. Localization and characterization of binding sites for circulating and cerebroventricular atrial natriuretic factor in rat choroid plexus. *Neuroendocrinology* 1990;51:304-14.

[52] Ohashi M, Fujio N, Nawata H, Matsuo H, Kato K. Alpha-human atrial natriuretic polypeptide binding sites in human adrenal membrane fractions. *Regul. Pept.* 1988;21:271-8.

[53] Schenk DB, Phelps MN, Porter JG, Scarborough RM, McEnroe GA, Lewicki JA. Identification of the receptor for atrial natriuretic factor on cultured vascular cells. *J. Biol. Chem.* 1985;260:14887-90.

[54] Leitman DC, Andresen JW, Kuno T, Kamisaki Y, Chang JK, Murad F. Identification of multiple binding sites for atrial natriuretic factor by affinity cross-linking in cultured endothelial cells. *J. Biol. Chem.* 1986;261:11650-5.

[55] Kishimoto I, Nakao K, Suga S, et al. Downregulation of C-receptor by natriuretic peptides via ANP-B receptor in vascular smooth muscle cells. *Am. J. Physiol.* 1993;265:H1373-9.

[56] Li Y, Madiraju P, Anand-Srivastava MB. Knockdown of natriuretic peptide receptor-A enhances receptor C expression and signalling in vascular smooth muscle cells. *Cardiovasc. Res.* 2012;93:350-9.

[57] Maack T, Suzuki M, Almeida FA, et al. Physiological role of silent receptors of atrial natriuretic factor. *Science* 1987;238:675-8.

[58] Nussenzveig DR, Lewicki JA, Maack T. Cellular mechanisms of the clearance function of type C receptors of atrial natriuretic factor. *J. Biol. Chem.* 1990;265:20952-8.

[59] Tseng YC, Lahiri S, Sellitti DF, Burman KD, D'Avis JC, Wartofsky L. Characterization by affinity cross-linking of a receptor for atrial natriuretic peptide in cultured human thyroid cells associated with reductions in both adenosine 3',5'-monophosphate production and thyroglobulin secretion. *J. Clin. Endocrinol. Metab.* 1990;70:528-33.

[60] Levin ER, Frank HJ. Natriuretic peptides inhibit rat astroglial proliferation: mediation by C receptor. *Am. J. Physiol.* 1991;261:R453-7.

[61] Drewett JG, Ziegler RJ, Trachte GJ. Neuromodulatory effects of atrial natriuretic peptides correlate with an inhibition of adenylate cyclase but not an activation of guanylate cyclase. *J. Pharmacol. Exp. Ther.* 1992;260:689-96.

[62] Hu RM, Levin ER, Pedram A, Frank HJ. Atrial natriuretic peptide inhibits the production and secretion of endothelin from cultured endothelial cells. Mediation through the C receptor. *J. Biol. Chem.* 1992;267:17384-9.

[63] Hempel A, Noll T, Bach C, et al. Atrial natriuretic peptide clearance receptor participates in modulating endothelial permeability. *Am. J. Physiol.* 1998;275:H1818-25.

[64] Pedram A, Razandi M, Hu RM, Levin ER. Vasoactive peptides modulate vascular endothelial cell growth factor production and endothelial cell proliferation and invasion. *J. Biol. Chem.* 1997;272:17097-103.

[65] Sabbatini ME, Villagra A, Davio CA, Vatta MS, Fernandez BE, Bianciotti LG. Atrial natriuretic factor stimulates exocrine pancreatic secretion in the rat through NPR-C receptors. *Am. J. Physiol. Gastrointest. Liver Physiol.* 2003;285:G929-37.

[66] Carini R, De Cesaris MG, Splendore R, et al. Mechanisms of hepatocyte protection against hypoxic injury by atrial natriuretic peptide. *Hepatology* 2003;37:277-85.

[67] Rose RA, Lomax AE, Kondo CS, Anand-Srivastava MB, Giles WR. Effects of C-type natriuretic peptide on ionic currents in mouse sinoatrial node: a role for the NPR-C receptor. *Am. J. Physiol. Heart Circ. Physiol.* 2004;286:H1970-7.

[68] Woodard GE, Li X, Rosado JA. Characteristics of the renal C-type natriuretic peptide receptor in hypertrophied and developing rat kidney. *J. Mol. Endocrinol.* 2005;35:519-30.

[69] Abdelalim EM, Tooyama I. NPR-C is expressed in the cholinergic and dopaminergic amacrine cells in the rat retina. *Peptides* 2010;31:180-3.

[70] Johnson BG, Trachte GJ, Drewett JG. Neuromodulatory effect of the atrial natriuretic factor clearance receptor binding peptide, cANF(4-23)-NH2 in rabbit isolated vasa deferentia. *J. Pharmacol. Exp. Ther.* 1991;257:720-6.

[71] Lelievre V, Pineau N, Hu Z, et al. Proliferative actions of natriuretic peptides on neuroblastoma cells. Involvement of guanylyl cyclase and non-guanylyl cyclase pathways. *J. Biol. Chem.* 2001;276:43668-76.

[72] Hashim S, Li Y, Anand-Srivastava MB. Small cytoplasmic domain peptides of natriuretic peptide receptor-C attenuate cell proliferation through Gialpha protein/MAP kinase/PI3-kinase/AKT pathways. *Am. J. Physiol. Heart Circ. Physiol.* 2006;291: H3144-53.

[73] Cahill PA, Hassid A. Clearance receptor-binding atrial natriuretic peptides inhibit mitogenesis and proliferation of rat aortic smooth muscle cells. *Biochem. Biophys. Res. Commun.* 1991;179:1606-13.

[74] Itoh H, Pratt RE, Ohno M, Dzau VJ. Atrial natriuretic polypeptide as a novel antigrowth factor of endothelial cells. *Hypertension* 1992;19:758-61.

[75] Pedram A, Razandi M, Hu RM, Levin ER. Astrocyte progression from G1 to S phase of the cell cycle depends upon multiple protein interaction. *J. Biol. Chem.* 1998;273: 13966-72.

[76] El Andalousi J, Li Y, Anand-Srivastava MB. Natriuretic Peptide Receptor-C Agonist Attenuates the Expression of Cell Cycle Proteins and Proliferation of Vascular Smooth Muscle Cells From Spontaneously Hypertensive Rats: role of Gi proteins and MAPkinase/PI3kinase signaling *PLOS one* (under revision)

[77] Huntley BK, Sandberg SM, Noser JA, et al. BNP-induced activation of cGMP in human cardiac fibroblasts: interactions with fibronectin and natriuretic peptide receptors. *J. Cell Physiol.* 2006;209:943-9.

[78] Rose RA, Anand-Srivastava MB, Giles WR, Bains JS. C-type natriuretic peptide inhibits L-type Ca2+ current in rat magnocellular neurosecretory cells by activating the NPR-C receptor. *J. Neurophysiol.* 2005;94:612-21.

[79] Villar IC, Panayiotou CM, Sheraz A, et al. Definitive role for natriuretic peptide receptor-C in mediating the vasorelaxant activity of C-type natriuretic peptide and endothelium-derived hyperpolarising factor. *Cardiovasc. Res.* 2007;74:515-25.

[80] Hobbs A, Foster P, Prescott C, Scotland R, Ahluwalia A. Natriuretic peptide receptor-C regulates coronary blood flow and prevents myocardial ischemia/reperfusion injury: novel cardioprotective role for endothelium-derived C-type natriuretic peptide. *Circulation* 2004;110:1231-5.

[81] Kiemer AK, Lehner MD, Hartung T, Vollmar AM. Inhibition of cyclooxygenase-2 by natriuretic peptides. *Endocrinology* 2002;143:846-52.

[82] Matsukawa N, Grzesik WJ, Takahashi N, et al. The natriuretic peptide clearance receptor locally modulates the physiological effects of the natriuretic peptide system. *Proc. Natl. Acad. Sci. U. S. A.* 1999;96:7403-8.

[83] Trachte GJ. Depletion of natriuretic peptide C receptors eliminates inhibitory effects of C-type natriuretic peptide on evoked neurotransmitter efflux. *J. Pharmacol. Exp. Ther.* 2000;294:210-5.

[84] Sabbatini ME, Vatta MS, Davio CA, Bianciotti LG. Atrial natriuretic factor negatively modulates secretin intracellular signaling in the exocrine pancreas. *Am. J. Physiol. Gastrointest. Liver Physiol.* 2007;292:G349-57.

[85] Christoffersen TE, Aplin M, Strom CC, et al. Increased natriuretic peptide receptor A and C gene expression in rats with pressure-overload cardiac hypertrophy. *Am. J. Physiol. Heart Circ. Physiol.* 2006;290:H1635-41.

[86] Brown LA, Rutherford RA, Nunez DJ, Wharton J, Lowe DG, Wilkins MR. Downregulation of natriuretic peptide C-receptor protein in the hypertrophied ventricle of the aortovenocaval fistula rat. *Cardiovasc. Res.* 1997;36:363-71.

[87] Ellmers LJ, Knowles JW, Kim HS, Smithies O, Maeda N, Cameron VA. Ventricular expression of natriuretic peptides in Npr1(-/-) mice with cardiac hypertrophy and fibrosis. *Am. J. Physiol. Heart Circ. Physiol.* 2002;283:H707-14.

[88] Patel JB, Valencik ML, Pritchett AM, Burnett JC, Jr., McDonald JA, Redfield MM. Cardiac-specific attenuation of natriuretic peptide A receptor activity accentuates adverse cardiac remodeling and mortality in response to pressure overload. *Am. J. Physiol. Heart Circ. Physiol.* 2005;289:H777-84.

[89] Weinberg EO, Schoen FJ, George D, et al. Angiotensin-converting enzyme inhibition prolongs survival and modifies the transition to heart failure in rats with pressure overload hypertrophy due to ascending aortic stenosis. *Circulation* 1994;90:1410-22.

[90] Li Y, Hashim S, Anand-Srivastava MB. Intracellular peptides of natriuretic peptide receptor-C inhibit vascular hypertrophy via Gqalpha/MAP kinase signaling pathways. *Cardiovasc. Res.* 2006;72:464-72.

[91] Milano CA, Dolber PC, Rockman HA, et al. Myocardial expression of a constitutively active alpha 1B-adrenergic receptor in transgenic mice induces cardiac hypertrophy. *Proc. Natl. Acad. Sci. U. S. A.* 1994;91:10109-13.

[92] Bogoyevitch MA, Andersson MB, Gillespie-Brown J, et al. Adrenergic receptor stimulation of the mitogen-activated protein kinase cascade and cardiac hypertrophy. *Biochem. J.* 1996;314 (Pt 1):115-21.

[93] Akhter SA, Milano CA, Shotwell KF, et al. Transgenic mice with cardiac overexpression of alpha1B-adrenergic receptors. *In vivo* alpha1-adrenergic receptor-mediated regulation of beta-adrenergic signaling. *J. Biol. Chem.* 1997;272:21253-9.

[94] Hein L, Stevens ME, Barsh GS, Pratt RE, Kobilka BK, Dzau VJ. Overexpression of angiotensin AT1 receptor transgene in the mouse myocardium produces a lethal phenotype associated with myocyte hyperplasia and heart block. *Proc. Natl. Acad. Sci. U. S. A.* 1997;94:6391-6.

[95] Dorn GW, 2nd, Tepe NM, Wu G, Yatani A, Liggett SB. Mechanisms of impaired beta-adrenergic receptor signaling in G(alphaq)-mediated cardiac hypertrophy and ventricular dysfunction. *Mol. Pharmacol.* 2000;57:278-87.

[96] Anand-Srivastava MB. Downregulation of atrial natriuretic peptide ANP-C receptor is associated with alterations in G-protein expression in A10 smooth muscle cells. *Biochemistry* 2000;39:6503-13.

[97] Saha S, Li Y, Lappas G, Anand-Srivastava MB. Activation of natriuretic peptide receptor-C attenuates the enhanced oxidative stress in vascular smooth muscle cells from spontaneously hypertensive rats: implication of Gialpha protein. *J. Mol. Cell Cardiol.* 2008;44:336-44.

[98] Pitzalis MV, Sarzani R, Dessi-Fulgheri P, et al. Allelic variants of natriuretic peptide receptor genes are associated with family history of hypertension and cardiovascular phenotype. *J. Hypertens.* 2003;21:1491-6.

[99] Sarzani R, Dessi-Fulgheri P, Salvi F, et al. A novel promoter variant of the natriuretic peptide clearance receptor gene is associated with lower atrial natriuretic peptide and higher blood pressure in obese hypertensives. *J. Hypertens.* 1999;17:1301-5.

[100] Aoi N, Soma M, Nakayama T, et al. Variable number of tandem repeat of the 5'-flanking region of type-C human natriuretic peptide receptor gene influences blood pressure levels in obesity-associated hypertension. *Hypertens. Res.* 2004;27:711-6.

[101] Liu Y, Li S, Chen X, et al. Association of natriuretic peptide receptor-C gene with ischemic stroke and hypertension in Chinese Han population. *Clin. Exp. Hypertens.* 2012;34:504-9.

[102] Chabrier PE, Roubert P, Lonchampt MO, Plas P, Braquet P. Regulation of atrial natriuretic factor receptors by angiotensin II in rat vascular smooth muscle cells. *J. Biol. Chem.* 1988;263:13199-202.

[103] Hirata Y, Takata S, Takagi Y, Matsubara H, Omae T. Regulation of atrial natriuretic peptide receptors in cultured vascular smooth muscle cells of rat. *Biochem. Biophys. Res. Commun.* 1986;138:405-12.

[104] Smith JB, Lincoln TM. Angiotensin decreases cyclic GMP accumulation produced by atrial natriuretic factor. *Am. J. Physiol.* 1987;253:C147-50.

[105] Garcia R, Gauquelin G, Cantin M, Schiffrin EL. Renal glomerular atrial natriuretic factor receptors in one-kidney, one clip rats. *Hypertension* 1988;11:185-90.

[106] Schiffrin EL, Turgeon A, Tremblay J, Deslongchamps M. Effects of ANP, angiotensin, vasopressin, and endothelin on ANP receptors in cultured rat vascular smooth muscle cells. *Am. J. Physiol.* 1991;260:H58-65.

[107] Palaparti A, Anand-Srivastava MB. Angiotensin II modulates ANP-R2/ANP-C receptor-mediated inhibition of adenylyl cyclase in vascular smooth muscle cells: role of protein kinase C. *J. Mol. Cell Cardiol.* 1998;30:1471-82.

[108] Yoshimoto T, Naruse M, Naruse K, et al. Angiotensin II-dependent down-regulation of vascular natriuretic peptide type C receptor gene expression in hypertensive rats. *Endocrinology* 1996;137:1102-7.

[109] Bray P, Carter A, Simons C, et al. Human cDNA clones for four species of G alpha s signal transduction protein. *Proc. Natl. Acad. Sci. U. S. A.* 1986;83:8893-7.

[110] Boumati M, Li Y, Anand-Srivastava MB. Modulation of ANP-C receptor signaling by arginine-vasopressin in A-10 vascular smooth muscle cells: role of protein kinase C. *Arch. Biochem. Biophys.* 2003;415:193-202.

[111] Schiffrin EL, Poissant L, Cantin M, Thibault G. Receptors for atrial natriuretic factor in cultured vascular smooth muscle cells. *Life Sci.* 1986;38:817-26.

[112] Martin ER, Lewicki JA, Scarborough RM, Ballermann BJ. Expression and regulation of ANP receptor subtypes in rat renal glomeruli and papillae. *Am. J. Physiol.* 1989; 257:F649-57.

[113] Anand-Srivastava MB, Palaparti A, Pion J. Differential regulation of G-protein expression by vasoactive peptides. *Mol. Cell Biochem.* 1997;176:21-7.

[114] Agui T, Xin X, Cai Y, et al. Opposite actions of transforming growth factor-beta 1 on the gene expression of atrial natriuretic peptide biological and clearance receptors in a murine thymic stromal cell line. *J. Biochem.* 1995;118:500-7.

[115] Sun JZ, Oparil S, Lucchesi P, Thompson JA, Chen YF. Tyrosine kinase receptor activation inhibits NPR-C in lung arterial smooth muscle cells. *Am. J. Physiol. Lung Cell Mol. Physiol.* 2001;281:L155-63.

[116] Placier S, Bretot X, Ardaillou N, Dussaule JC, Ardaillou R. Regulation of ANP clearance receptors by EGF in mesangial cells from NOD mice. *Am. J. Physiol. Renal Physiol.* 2001;281:F244-54.

[117] Kishimoto I, Yoshimasa T, Suga S, et al. Natriuretic peptide clearance receptor is transcriptionally down-regulated by beta 2-adrenergic stimulation in vascular smooth muscle cells. *J. Biol. Chem.* 1994;269:28300-8.

[118] Katafuchi T, Mizuno T, Hagiwara H, Itakura M, Ito T, Hirose S. Modulation by NaCl of atrial natriuretic peptide receptor levels and cyclic GMP responsiveness to atrial natriuretic peptide of cultured vascular endothelial cells. *J. Biol. Chem.* 1992;267:7624-9.

[119] Bassil M, Li Y, Anand-Srivastava MB. Peroxynitrite inhibits the expression of G(i)alpha protein and adenylyl cyclase signaling in vascular smooth muscle cells. *Am. J. Physiol. Heart Circ. Physiol.* 2008;294:H775-84.

[120] Arejian M, Li Y, Anand-Srivastava MB. Nitric oxide attenuates the expression of natriuretic peptide receptor C and associated adenylyl cyclase signaling in aortic vascular smooth muscle cells: role of MAPK. *Am. J. Physiol. Heart Circ. Physiol.* 2009;296:H1859-67.

[121] Beckman JS, Beckman TW, Chen J, Marshall PA, Freeman BA. Apparent hydroxyl radical production by peroxynitrite: implications for endothelial injury from nitric oxide and superoxide. *Proc. Natl. Acad. Sci. U. S. A.* 1990;87:1620-4.

[122] Dowell FJ, Martin W. The effects of peroxynitrite on rat aorta: interaction with glucose and related substances. *Eur. J. Pharmacol.* 1997;338:43-53.

[123] Bassil M, Anand-Srivastava MB. Cyclic GMP modulates the expression of Gi protein and adenylyl cyclase signaling in vascular smooth muscle cells. *Cell Biochem. Biophys.* 2007;47:99-108.

[124] Lang RE, Unger T, Ganten D. Atrial natriuretic peptide: a new factor in blood pressure control. *J. Hypertens.* 1987;5:255-71.

[125] Nicholls MG, Espiner EA, Ikram H, Crozier IG, Richards AM. Atrial natriuretic peptide in human hypertension. *Eur. Heart J.* 1987;8 Suppl B:123-8.

[126] Jernberg T, James S, Lindahl B, et al. Natriuretic peptides in unstable coronary artery disease. *Eur. Heart J.* 2004;25:1486-93.

[127] Clerico A, Fontana M, Zyw L, Passino C, Emdin M. Comparison of the diagnostic accuracy of brain natriuretic peptide (BNP) and the N-terminal part of the propeptide of BNP immunoassays in chronic and acute heart failure: a systematic review. *Clin. Chem.* 2007;53:813-22.

[128] Doust JA, Pietrzak E, Dobson A, Glasziou P. How well does B-type natriuretic peptide predict death and cardiac events in patients with heart failure: systematic review. *BMJ* 2005;330:625.

[129] Yoshimoto T, Naruse M, Naruse K, et al. Vascular action of circulating and local natriuretic peptide systems is potentiated in obese/hyperglycemic and hypertensive rats. *Endocrinology* 1996;137:5552-7.

[130] Anand-Srivastava MB. Platelets from spontaneously hypertensive rats exhibit decreased expression of inhibitory guanine nucleotide regulatory protein. Relation with adenylyl cyclase activity. *Circ. Res.* 1993;73:1032-9.

[131] Marcil J, Thibault C, Anand-Srivastava MB. Enhanced expression of Gi-protein precedes the development of blood pressure in spontaneously hypertensive rats. *J. Mol. Cell Cardiol.* 1997;29:1009-22.

[132] Schiffrin EL, Deslongchamps M, Thibault G. Platelet binding sites for atrial natriuretic factor in humans. Characterization and effects of sodium intake. *Hypertension* 1986;8:II6-10.

[133] Okazaki M, Kobayashi H, Kuroiwa A, Izumi F. Atrial natriuretic peptide receptors in cerebral microvessels and choroid plexus of spontaneously hypertensive rats. *Brain Res.* 1990;518:292-4.

[134] Gutkind JS, Kurihara M, Castren E, Saavedra JM. Atrial natriuretic peptide receptors in sympathetic ganglia: biochemical response and alterations in genetically hypertensive rats. *Biochem. Biophys. Res. Commun.* 1987;149:65-72.

[135] Lee J, Kim S, Jung M, Oh Y, Kim SW. Altered expression of vascular natriuretic peptide receptors in experimental hypertensive rats. *Clin. Exp. Pharmacol. Physiol.* 2002;29:299-303.

[136] Di Fusco F, Anand-Srivastava MB. Enhanced expression of Gi proteins in non-hypertrophic hearts from rats with hypertension-induced by L-NAME treatment. *J. Hypertens.* 2000;18:1081-90.

[137] Hashim S, Anand-Srivastava MB. Losartan-induced attenuation of blood pressure in L-NAME hypertensive rats is associated with reversal of the enhanced expression of Gi alpha proteins. *J. Hypertens.* 2004;22:181-90.

[138] Nagase M, Ando K, Katafuchi T, Kato A, Hirose S, Fujita T. Role of natriuretic peptide receptor type C in Dahl salt-sensitive hypertensive rats. *Hypertension* 1997;30:177-83.

[139] Oliver PM, Fox JE, Kim R, et al. Hypertension, cardiac hypertrophy, and sudden death in mice lacking natriuretic peptide receptor A. *Proc. Natl. Acad. Sci. U. S. A.* 1997;94: 14730-5.

In: Natriuretic Peptides
Editor: Kailash N. Pandey

ISBN: 978-1-63321-080-6

Chapter 8

ARM, THE ATP-REGULATED MODULE OF ANF-RGC GUANYLATE CYCLASE: STRUCTURE AND FUNCTION

Teresa Duda*, Alexandre Pertzev and Rameshwar K. Sharma

The Unit of Regulatory and Molecular Biology, Research Divisions of Biochemistry and Molecular Biology, Salus University, Elkins Park, PA, US

ABSTRACT

Atrial natriuretic factor (ANF) receptor guanylate cyclase (ANF-RGC), like the other members of the membrane guanylate cyclase family, is a single transmembrane-spanning protein. The transmembrane domain separates the protein into two regions, extracellular and intracellular.

The extracellular region contains the ANF-binding domain and the intracellular region the catalytic domain located at the C-terminus of the protein. Preceding the catalytic domain, the intracellular region is composed of the following functional domains: juxtaposed 40 amino acids to the transmembrane domain is the ATP-Regulated Module (ARM) domain [also termed the kinase homology domain (KHD)], and the putative dimerization domain.

The ANF-RGC signaling is initiated by hormone, ANF, binding to its extracellular binding site. The binding signal is transmitted through the transmembrane domain to the intracellular portion where binding of ATP to the ARM domain induces a cascade of temporal and spatial changes which finally result in activation of the catalytic domain. This chapter reviews the signaling role of the ARM domain and discusses the biochemical and physiological evidence on the critical role of its transduction motif ^{669}WTAPELL^{675}.

* Corresponding authors: Phone: 215-780-3112; Fax: 215-780-3125; e-mail tduda@salus.edu.

Keywords: ANF, ANF receptor, membrane guanylate cyclase, cyclic GMP, signal transduction

1. Introduction

The discovery of a novel type of a membrane protein that was concurrently a receptor of a hormone and the transducer of its signal turned around the field of cellular signal transduction. The hormone was the atrial natriuretic factor (ANF) and the membrane protein was both its receptor and guanylate cyclase enzyme, termed therefore atrial natriuretic factor receptor guanylate cyclase, ANF-RGC [1-5]. Other terms used to describe the protein are GC-A and NPR-A. ANF-RGC responded to ANF binding with accelerated synthesis of its second messenger cyclic GMP. The cyclic GMP signaling system was entirely different from the known paradigms of the cyclic AMP and IP_3 signaling systems where an assemblage of three components -- receptor, GTP binding protein, and the transducer catalyst -- was necessary for signal transduction. In the new transduction system, a single protein, ANF-RGC, contained both the ability to recognize the ANF hormonal signal and the activity to translate the hormonal information into the production of its second messenger, cyclic GMP [detailed review of the historic aspect of the ANF-RGC discovery is provided in: ref. 6 and in another chapter in this edition].

The concept that the ANF receptor guanylate cyclase is indeed both hormone receptor and a catalyst was further supported by cloning studies [7-11]. Homology cloning technique made possible the finding of other receptor membrane guanylate cyclases: C-type natriuretic factor receptor guanylate cyclase, CNP-RGC (also known as GC-B) [12-14] and heat-stable enterotoxin (also guanylin and uroguanylin) receptor guanylate cyclase, STa-RGC (GC-C) [15-19].

Presently, there are three known members of the ANF family. They are termed ANF (alternatively called ANP), brain natriuretic peptide (BNP) and type C natriuretic peptide (CNP) [reviewed in: 20, 21]. The main sites of ANF and BNP synthesis are the atrial and the ventricular granules of the heart. With each atrial stretch, defined doses of these peptides are pulsated into the blood stream, and carried to their target tissues where they exhibit their biological activities. This mode of action was first demonstrated for ANF by deBold's group, which classified the heart as an exocrine gland, and, thereby, showed that it is not merely a blood pump [22].

In contrast to ANF and BNP, CNP is not synthesized in the heart but in the vascular endothelial cells [23-25], and may also be in the localized regions of the central nervous system. As their name indicates, to varying degrees, all three natriuretic factors induce natriuresis and exhibit vasodilatory activities through which they lower blood pressure [26-28]. Gene knock-out studies link ANF with salt-sensitive hypertension [29]. And ANF-RGC has been linked with the salt-resistant form [30]. Some studies link BNP with skeletal growth [31] and CNP with bone cartilage growth [32]. It is, therefore, likely that besides being common vasodilatory agents, the individual natriuretic peptides also possess biological activities, which are unique to them.

Identification of three structurally and functionally related membrane guanylate cyclases that were receptors for hormonally active peptides resulted in a notion that all of them, even those not discovered yet, are receptors to some extracellular ligands.

The notion was short-lived, however. The reason being that the next membrane guanylate cyclases identified by their direct purification or homology cloning, the rod outer segment guanylate cyclases, ROS-GC1 (also known as RetGC1 or GC-E) [33-36] and ROS-GC2 (RetGC2 or GC-F) [37-39], the olfactory neuroepithelium guanylate cyclase, ONE-GC (also termed as GC-D) [40, 41] and GC-G [42] did not have their specific extracellular ligands to modulate their guanylate cyclase activities. They were therefore termed the "orphan receptors" [39-43].

Soon thereafter it became evident that the photoreceptor guanylate cyclases ROS-GC1 and ROS-GC2 are not orphans. Because their physiological function is to return the photoreceptors from illuminated to the dark, resting state, regulation of their activities is designed to fulfill this function. Photoreceptors' illumination leads to depletion of cyclic GMP, closure of the cyclic GMP gated channels, and lowering the free Ca^{2+} concentration [reviewed in: 44, 45]. The ROS-GCs task is to restore the dark-level of cyclic GMP and through it opening of the cyclic GMP gated channels and increase of Ca^{2+} influx. Thus, Ca^{2+} concentration is the signal to which ROS-GCs respond. Guanylate cyclase activating proteins (GCAPs) sense the post-illumination fall in Ca^{2+} and stimulate ROS-GCs to synthesize cyclic GMP at faster rate [reviewed in:45, 46]. In this manner, ROS-GCs do not respond to an extracellular ligand but to their intracellular Ca^{2+} sensing ligands GCAPs [47-49]. GCAPs are not the only ROS-GC Ca^{2+} sensing modulators. While, conferring to the phototransduction mechanism in photoreceptor outer segments, increasing Ca^{2+} concentrations inhibit ROS-GCs activity through GCAPs, two other Ca^{2+} sensors, S100B and neurocalcin δ, stimulate ROS-GC1 in a Ca^{2+}-dependent fashion [50-53]. This stimulation, however, is not linked with phototransduction but rather with transmission of neural signals to cone ON-bipolar cells in case of S100B stimulation [54] and with retinal ganglion cells signaling in case of neurocalcin δ [55]. Evidently, the two ROS-GCs are not hormone receptors, are not orphans, but form a subfamily of membrane guanylate cyclases, the Ca^{2+} modulated subfamily.

Olfactory neuroepithelial guanylate cyclase, ONE-GC (also known as GC-D), which operates in a subset of olfactory sensory neurons [40, 41, 56] was initially classified as another member of the Ca^{2+} modulated subfamily [40]. It is modulated in a Ca^{2+}-dependent manner by neurocalcin δ [40, 57] and by GCAP1 [56, 58, 59]. Remarkably, the Ca^{2+}-GCAP1 fashion of ONE-GC modulation is opposite to that of ROS-GC modulation. Sensing increasing concentrations of Ca^{2+} GCAP1 stimulates ONE-GC while it inhibits ROS-GC [58, 59]. Thus, ONE-GC was *bona fide* Ca^{2+}-modulated cyclase. At that point it appeared that the two models of membrane guanylate cyclase operation, the hormone-dependent and Ca^{2+}-dependent, are established and that they properly describe the respective signal transduction processes. Recent findings, however, enforced a revision. After the Ca^{2+}-dependent modulation of ONE-GC activity was demonstrated, an extracellular ligand of the cyclase was found. The ligand was odorant uroguanylin [60, 61]. Binding of the odorant molecule to the extracellular domain of ONE-GC results in partial activation of the cyclase and readies it for binding of Ca^{2+} sensor neurocalcin δ, which then fully activates ONE-GC [62]. This model of ONE-GC activation is fundamentally different from the generally accepted operational model of phototransduction-linked ROS-GCs and the hormone receptor signal transduction model.

Since the cyclase unites and transduces two types of signal, the extracellular odorant and the intracellular Ca^{2+} with common outcome, it is a hybrid of the two subfamilies.

The results of the latest studies from our laboratory show that ONE-GC is not the only membrane guanylate cyclase dually modulated. Surprisingly, the founding member of the hormone receptor subfamily, the ANF-RGC, is also responsive to Ca^{2+} signal [59, 63]. The Ca^{2+} signaling is additive to the hormonal signaling, it functions in the adrenocortical zona-glomerulosa, and is physiologically linked with blood pressure regulation through the regulation of aldosterone synthesis [63].

The mechanism of GC-G regulation is not fully understood. It was suggested that the mouse form of GC-G is selectively expressed in the sperm and may be involved in the process of capacitation [64]. Other reports suggest that the cyclase is expressed in Grueneberg ganglion olfactory subsystem where it is responding to CO_2 [65] or to cool ambient temperature indicating its role in thermo-sensation [66].

1.1. Topography of Membrane Guanylate Cyclases

Primary amino acid sequence comparison of all cloned membrane guanylate cyclases suggests their identical topography. They are single transmembrane spanning proteins and the helical transmembrane domain divides each protein into two portions of more or less equal length. The N-terminal portion is extracellular (or intradiscal in case of ROS-GCs); the C-terminal portion is intracellular (interdiscal for ROS-GCs). Their functional forms are homo-dimeric [67-73]. Functional distinctiveness to each member of the guanylate cyclase family is provided by specific modular blocks within the extra- and intracellular portion of the protein. Each modular block within the extracellular region of the receptor guanylate cyclases uniquely senses its peptide hormone signal and the modular blocks within the intracellular portion of a Ca^{2+}-modulated guanylate cyclases sense Ca^{2+} signals conveyed by specific Ca^{2+} sensor proteins [reviewed in: 74]. The catalytic domain in each membrane guanylate cyclase resides in its intracellular region. However, topographical arrangement of this domain differs in the two sub-families. In the peptide hormone receptor, it is at the C-terminal end and in the ROS-GC and ONE-GC it is followed by a C-terminal extension [reviewed in: 75].

The present review discusses the mechanism of ANF-RGC signal transduction.

2. ANF SIGNAL RECOGNITION

2.1. Binding of Ligand-Hormone ANF

The first 28 amino acids deduced from ANF-RGC cDNA sequence represent the N-terminal hydrophobic signal peptide [7, 8] which ensures sorting of the protein to the cell membrane. These are followed by 441 amino acids which build-up the extracellular domain. Although instinctively anticipated, the first experimental evidence that the ligand, ANF, binding site in ANF-RGC is located within the extracellular domain came from site-directed and deletion mutagenesis studies. GCα, a plasma membrane guanylate cyclase, whose sequence with the exception of two amino acid residues, was identical to ANF-RGC, was cloned from rat adrenal cDNA library [8]. The two different amino acid residues were His in place of Gln^{338} and Pro in place of Leu^{364}. These amino acid substitutions resulted from single nucleotide changes, CAG→CAC and CTG→CCG, respectively. When expressed in COS cells, GCα was directed to the plasma membrane and exhibited guanylate cyclase activity; it however, did not bind and respond to ANF or other natriuretic peptides [8]. These results demonstrated that the two amino acid substitutions within the extracellular portion of ANF-RGC did not affect the expression and the basal catalytic properties of the cyclase but abolished the binding of ANF and therefore ANF signaling of the cyclase activity. By site-directed mutagenesis, the GCα cDNA was remodeled to (1) create a double mutant encoding Gln^{338} and Leu^{364} thus, to reconstruct the wild type ANF-RGC [8]; (2) create a single substitution mutant encoding Leu^{364}; and (3) delete the entire extracellular domain creating (ext^{-})ANF-RGC mutant. The double, GCαGln^{338},Leu^{364}, and the single, GCαLeu^{364}, mutants expressed both ANF binding and ANF-dependent cyclase activities identical to the wild type ANF-RGC but the deletion mutant expressed neither of these activities. Thus, it was experimentally ascertained that the ANF binding site resides within the ANF-RGC extracellular domain and Leu^{364} is necessary for proper folding of the ANF docking site.

The site of ANF binding was further analyzed independently primarily by two laboratories [76-79]. Both these groups used derivatives of ANF for cross-linking with purified full-length ANF-RGC or isolated extracellular domain of ANF-RGC. They determined that the amino terminus of ANF is in contact with the region Met^{173}-Phe^{188} of ANF-RGC, and the C-terminus, with Asp^{191}-Arg^{198} region. The fact that the identified contact regions were very close to each other was interpreted that the N- and C- termini of ANF interface distinct subunits of ANF-RGC homodimer. It also implied that one ANF-RGC dimer binds one molecule of ANF. The final identification of the ANF binding module came from crystallographic analyses of the extracellular domain of ANF-RGC co-crystallized with ANF [80, 81]. They revealed that (1) the extracellular domain of ANF-RGC exists as a dimer in the head-to-head configuration; (2) one ANF-RGC dimer binds one molecule of ANF (2:1 complex); (3) there is no preference in ANF binding to a specific monomer of the extracellular domain; bound ANF occurs in two alternate orientations of equal occurrence; (4) binding of ANF is asymmetric: the binding site in one monomer differs from that in the other; one monomer binds the N-terminal part of ANF and the other binds C-terminal part; (5) the membrane-distal portion of the extracellular domain contains chloride ion necessary for ANF binding. With the exception of enterotoxin receptor guanylate cyclase, all other membrane guanylate cyclases contain a hinge region juxtaposed to the N-terminal side of the transmembrane domain. This region contains two conserved cysteine residues separated from each other by 6-8 residues. In ANF-RGC these residues are Cys^{423} and Cys^{432} and were indicated as a critical structural motif in ANF signaling [82-84]. There was however no consensus on the mechanism of its operation [82-84]. Through the analyses Cys→Ser mutants of ANF-RGC two mechanisms were proposed: One, that the cysteins are involved in dimerization through the formation of an inter-chain S-S bridge [84, 85]; another, that they form an intra-chain disulfide bridge [83, 84]. The studies were conducted on intact cells transfected with ANF-RGC cDNA and the cyclic GMP formed in these cells was considered reflective of the membrane guanylate cyclase activity. The drawbacks with the whole-cell studies are that they are not direct measurements of the membrane guanylate cyclase activity and that other factors including soluble guanylate cyclase and its modulators, can affect the results. Most significantly, the ANF-signaling of ANF-RGC is a multi-step process. For instance, ATP is obligatory for ANF signaling and the ATP signaling events cannot be dissected out in intact cell studies. Therefore, to fully evaluate the role of the Cys^{423}-Cys^{432} motif and of the individual cysteine residues, guanylate cyclase activity of ANF-RGC mutants (1) Cys^{423}Ser, (2) Cys^{432}Ser, and (3) Cys^{423}Ser,Cys^{432}Ser was measured in isolated membranes of transfected COS cells [86]. All mutants showed significantly higher basal cyclase activity than wild type ANF-RGC; the activity of both single mutants was ~2-fold higher and the activity of the double mutant was ~4-fold higher [86]. The mutants' cyclase activities were stimulated by saturating concentrations of ANF and ATP and the achieved maximal activity was the same as the activity of ANF/ATP-stimulated wild type ANF-RGC [86]. However, because of the mutants' higher basal cyclase activities, their fold stimulation by ANF/ATP was lower than that of wild type ANF-RGC. These results were interpreted that both Cys^{423} and Cys^{432} residues independently and equally control the catalytic activity of ANF-RGC and that activation of ANF-RGC does not involve interchain-disulfide bond formation. In the basal state, the disulfide motif keeps the ANF-RGC in its repressed catalytic state and ANF/ATP signaling brings it to the fully active state [86].

2.2. Transmembrane Migration of the ANF Binding Signal

The extracellular domain of ANF-RGC is separated from the intracellular portion by a 17 amino acids long, residues 446-462, transmembrane domain followed by the Arg-Lys "transfer stop" sequence signal [8]. For receptor guanylate cyclases, the signal initiated by hormone binding to the extracellular ligand binding domain has to pass the transmembrane domain in order to be processed and transduced into cyclic GMP in the intracellular domain. A model of hormone-induced rotation has been proposed to explain how the extracellular signal is transmitted to the intracellular domain of ANF-RGC [81, 87]. According to this model, binding of ANF causes a twist motion of both ANF-RGC monomers centered on a support point close to the bound ANF. This twist motion causes that both juxtamembrane domains of the dimer are translocated with only minimal change in the distance between them and their relative orientation is altered. This movement and reorientation constitute a signal, which is transmitted through the transmembrane domain. This model was tested experimentally by introducing one to five alanine residues into the TM alpha-helix [88]. Analyses of the mutants' activities lead to the conclusion that a transmembrane domain rotation of 40 degrees leads to constitutive NPRA activation [88] suggesting that it is the rotation occurring upon ANF binding.

3. Propagation of the ANF Signal within the Intracellular Domain of ANF-RGC

3.1. ANF Signaling Requires ATP

Early studies of ANF-RGC activation by ANF performed in the crude liver and lung membranes showed that ATP significantly potentiates the cyclase activation [89-92; reviewed in: 74, 93]. Subsequently, studies from independent laboratories demonstrated that ATP is obligatory for ANF-dependent activation of ANF-RGC [94-97; reviewed in: 93]. These studies, using heterologously expressed ANF-RGC, showed that the cyclase cannot be stimulated either by ANF or ATP alone; these two together are necessary to activate ANF-RGC. The ATP effect was mimicked by its nonhydrolyzable analogs, ATPγS and AMP-PNP with the order of potency being ATPγS>ATP>AMP-PNP. Because of the effect of the ATP nonhydrolyzable analogs it was suggested that ATP is direct allosteric regulator of ANF-RGC [94-97; reviewed in: 93]. Shortly thereafter it was shown that ATP is also obligatory for the C-type natriuretic peptide dependent stimulation of CNP-RGC [14, 98]. Thus, it appears that the requirement of ATP is a common trait of natriuretic peptide receptor guanylate cyclases.

Studies on ANF-RGC deletion mutants revealed that the ATP regulated region resides in the intracellular portion of ANF-RGC between the transmembrane and the C-terminal catalytic domain [97, 99]. This ANF-RGC region was named the kinase homology domain (KHD) because of its sequence similarity with the corresponding domains of tyrosine kinases [7, 8, 99]. One laboratory proposed a model for ATP function suggesting that "binding of ANP to the extracellular domain of its receptor initiates a conformational change in the protein kinase-like domain, resulting in de-repression of guanylate cyclase activity" [99]. The central idea behind the model is that KHD in native ANF-RGC suppresses the catalytic module activity; ANF functions by relieving this suppression. Another study, however, using two KHD deletion mutants, Δ506-677 and Δ555-762, demonstrated that the mutants had significantly, 5-10-fold suppressed catalytic activity, and proposed an alternative model according through which, "(1) a signal is caused by the binding of the hormone to the receptor site..(2) there is a transmembrane migration of the signal; (3) the signal is potentiated by ATP at ARM; and (4) the amplified signal is finally transduced at the catalytic site" [97].

3.2. ATP Allosteric Effect and the ARM Domain

Within the kinase homology domain ANF-RGC contains a glycine-rich cluster sequence, G^{503}-X-G^{505}-X-X-X-G^{509} [97]. This sequence is a modified form of the protein kinases' nucleotide-binding consensus sequence G-X-G-X-X-G [84, 85]. Mutations within this glycine-rich loop result in loss of kinase activity [100, 101]. Molecular simulations show that the loss is a consequence of ATP misalignment [102]. The G^{503}-X-G^{505}-X-X-X-G^{509} motif was targeted as potentially critical for ATP function in ANF signaling. Through analyses of series of deletion and point ANF-RGC mutants involving only this motif or the entire kinase homology domain, it was determined that the G^{503}-X-G^{505}-X-X-X-G^{509} motif is critical for ATP binding to ANF-RGC and for ANF signaling of its activation [97, 98, 103; reviewed in: 74, 93]. Critical significance of the motif does not mean that it is a direct ATP binding site, but rather that it is necessary for maintaining the steric arrangements to fit the ATP molecule. For this reason, the motif has been named ATP Regulatory Module (ARM) [97] and the kinase homology domain, where the ARM resides, has been termed the ARM domain [104, 105]. The interaction of ATP with the ARM domain was analyzed by independent laboratories through photolabeling with azido-ATP [106, 107] and surface plasmon resonance (SPR) spectroscopy [107].

The photolabeling experiments were conducted with 8-azido-ATP. The isolated ARM domain was incubated with [γ^{32}P]-8-azido-ATP, UV irradiated and the radiolabeled protein was visualized by autoradiography. The results showed that the ARM domain protein was labeled with the [γ^{32}P]-8-azido-ATP probe. Specificity of the binding was determined by displacement of the [γ^{32}P]-8-azido-ATP by ATP and GTP. Unlabeled ATP displaced [γ^{32}P]-8-azido-ATP in a dose-dependent fashion: 50% bound [γ^{32}P]-8-azido-ATP was displaced by 0.4 mM ATP and in excess of 80%, was displaced by 3 mM ATP [107]. GTP did not displace [γ^{32}P]-8-azido-ATP [unpublished data]. Similar results were obtained when [γ^{32}P]-2-azido-ATP was used for the cross-linking experiments [108]. Kinetics of ATP binding to the ARM domain was determined through SPR spectroscopy. AMP-PNP, the nonhydrolyzable analog of ATP was used for the binding studies. Half-maximal binding (EC_{50}) occurred when the concentration of AMP-PNP was ~ 0.2 mM and the calculated K_D value was 0.21 mM. Similar results (EC_{50} value of 0.15 mM and K_D of 0.13 mM) were obtained when 8-azido-ATP was used for the binding analyses, indicating that the binding affinity for the isolated ARM of ANF-RGC for 8-azido-ATP and for AMP-PNP is very similar [107]. Based on these results it was concluded that that the configurations of the ARM in its isolated and native states are almost identical; and therefore, it represents the ATP signal transduction site.

To get an insight into the molecular mechanism through which ATP potentiates the ANF binding signal and transmits it to the catalytic domain, in the absence of the crystal structure of the ARM domain, its structure was simulated through computer-assisted homology based modeling [104, 105; PDB ID 1T53]. The basic structural features of the model have been experimentally validated through point mutation/expression studies [104, 105; reviewed in: 74, 93]. The essential characteristics of the model are briefly outlined below.

3.2.1. Spatial Determinants of the ARM Domain

The domain consists of two lobes: the smaller, N-terminal (91 aa residues: 496-586), and the larger, C-terminal (185 aa residues: 587-771) [104]. Four antiparallel strands β strands and one α helix form the smaller lobe; the larger lobe is predominantly helical, composed of eight α-helices and two β strands [104, 104]. The ARM motif, G^{503}-X-G^{505}-X-X-X-G^{509}, is located within the smaller N-terminal lobe.

3.2.2. The ATP-Binding Pocket

In order to identify the residues that could be potentially involved in binding of the ATP molecule components, thus in the formation of the ATP binding pocket, the model of the ARM domain in its ATP-bound form was analyzed [93]. A radius of 4 Å from the ATP molecule has been chosen as the limiting distance for the electrostatic, hydrogen or van der Waals' interactions. Two sets of residues have been identified [Figure 4 in ref. 93]: (1) those forming the floor of the ATP binding pocket; (2) those surrounding ATP molecule. The residues in the first set are: Gly^{503}, Arg^{504}, Gly^{505}, Ser^{506}, Asn^{507}, Tyr^{508}, and Gly^{509}. These residues have no direct chemical interaction with ATP; they, however, provide necessary space to accommodate the ATP molecule. The residues in the second set surround the individual components of ATP, the adenine ring, the ribose ring and the triphosphate moiety. The adenine ring is surrounded by Leu^{511}, Thr^{513}, Thr^{514}, Glu^{515}, Gln^{517}, Ala^{533}, Thr^{580}, Glu^{581}, Cys^{583}, Val^{635}, and Thr^{645}. Among these residues Leu^{511}, Thr^{513}, and Cys^{583} are within a distance shorter than 3 Å.

Gly^{503}, Leu^{511}, Thr^{513}, Thr^{514}, Gly^{580}, Ser^{587}, Asp^{590} surround the ribose ring with Leu^{511} and Thr^{514} located within 2.5 Å radius from the ribose. The phosphate groups are surrounded by Arg^{504}, Gly^{505}, Leu^{511}, Lys^{535}, Asn^{633}, Asp^{646} and Lys^{535} is the nearest residue (2.6 Å) forming a hydrogen bond with the phosphate and has been shown to be critical for ATP regulation of ANF-dependent ANF-RGC activity; Asp^{646} interacts with the triphosphate group of ATP through the formation of a coordinate bond with the metal ion Mg^{2+}.

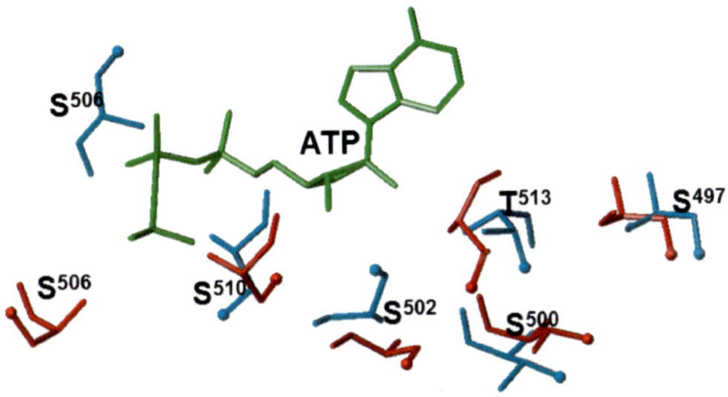

Figure 1. ATP binding to the ARM domain affects the conformation of the six phosphorylable residues. The conformation of the six phosphorylated residues is shown before (cyan) and after (red) ATP binding. The ATP molecule is shown in green. The positions of the OH groups are indicated by cyan and red balls (Reproduced with permission from ref. 108).

The model-predicted ATP binding pocket was authenticated experimentally. The 8-azido-ATP was cross-linked with the purified isolated ARM domain, the cross-linked protein was trypsin digested and the resulting peptides were sequenced. Three peptides were found to be photoaffinity modified. The longest modified peptide was identified as G^{614}MLFLHNGAICSHGNLKSSNCVVDGR^{639} and the shortest, S^{631}SNC^{634}V^{635}VDGR. In all three peptides Cys^{634} was modified indicating that this residue was the closest to the azido group. The G^{614} - R^{639} fragment contains six ATP-binding pocket-predicted residues [93]: S^{625}, K^{630}, S^{631}, S^{632}, N^{633}, and V^{635}; and the shortest S^{631} - R^{639} fragment contains three residues: S^{631}, N^{633} and V^{635} [93]. The refined ARM model [93] predicts that Val^{635} is within a 4 Å radius from the adenine ring; cross-linking of the neighboring residue to Cys^{634} validates that these residues are indeed the closest to the C-8 of the adenine ring therefore, the validity of the model was experimentally confirmed [107].

3.3.3. ATP Binding Dependent Changes in the ARM Domain

If binding of ATP to its pocket within the ARM domain serves the signaling purposes, it should induce structural changes in regions both proximal and distal to the pocket. These changes then will be propagated till they reach the catalytic domain and cause its activation. Comparative analyses of the ARM domain model in its apo and ATP-bound states identified such changes. They involve rotations of the β strands within the smaller lobe as well as movements of the β strands and α helices in the larger lobe [104, 105; reviewed in: 93]. Consequences of two such changes appear to be especially important: first, of the β1- β2 strands in the smaller lobe, and the second, of the EF and F helices in the larger lobe.

The β1 and β2 strands and the loop connecting them contains the glycine-rich cluster [104, 105; reviewed in: 93], which is meshed in and flanked by six phosphorylation sites, four serines and two threonines, S^{497}, T^{500}, S^{502}, S^{506}, S^{510}, T^{513} [109, 110]. The present consensus is that "phosphorylation of KHD is absolutely required for hormone-dependent activation of NPR-A" [110]. Without KHD phosphorylation, ANF-RGC and CNP-RGC are totally desensitized [110-112]. In this concept, a hypothetical protein kinase co-exists with ANF-RGC in the plasma membranes [113]. Another important component of the system necessary for its physiological significance is phosphatase, which dephosphorylates the six phosphorylated residues [113].

Analysis of the ATP-bound ARM domain model allows explaining how ATP binding makes it possible for the serine and threonine residues to undergo phosphorylation. After ATP binds to the ARM domain, the β1, β2 strands, and the loop between them shifts by ~3-4 Å and rotates by ~15° [104, 105; reviewed in: 93] This movement causes reorientation of the serines and threonines residues (Figure 1: red colored residues). As a consequence, the side chains and the OH groups of T^{500}, S^{502}, S^{506} and T^{513} become directed toward the protein surface (Figure 1: compare the positions of the cyan- and red-colored OH groups). The change in the positions of the side chains is most drastic for the S^{502} and S^{506} residues. Although upon ATP binding there is no toward the surface reorientation of the S^{497} and S^{510} OH groups, the entire residues are shifted toward the surface (Figure 1). This structural rearrangement permits the hypothetical protein kinase to access the side chains of the six residues and transfer the phosphate group [114]. It is noteworthy that although the concept on the key role of phosphorylation of the kinase homology domain in ANF-RGC activation was first proposed approximately two decades ago [113, 115], until today neither the kinase nor the phosphatase has been identified.

Another important outcome of ATP binding on the dynamics within the ARM domain is the translocation within the larger lobe of two helices, EF and F. Their movement was characterized by analyzing the ATP-dependent changes in the fluorescence intensity and wavelength of two tryptophan residues W^{601} and W^{669} [108]. Both residues reside in the larger lobe of the ARM domain, outside the ATP-binding pocket. The W^{601} residue is a part of the helix E structure [93, 105], its side-chain is oriented towards the protein surface, and it is flanked by several hydrophobic residues, F^{640}, L^{607}, S^{606} and S^{596} (Figure 2A). These structural features define the fluorescence λmax of W^{601} at 332 nm. W^{669} residue is located at the end of a loop connecting β8 strand and αEF helix [108]; it is a part of a conserved hydrophobic motif, ^{669}WTAPELL^{675}. The side chain of W^{669} is flanked by the polar residues: E^{699}, S^{631}, K^{667}, K^{630} and L^{696} (Figure 2B). This environment causes that the fluorescence λmax of W^{669} is at 345 nm.

Superimposition of the ATP-free and the ATP-bound forms indicates that ATP binding induces contraction of the entire ARM domain and affects the orientation and environment of W^{669} but not of W^{601}. The contraction results in shortening of the distance between W^{669}, but not W^{601}, and the ATP binding pocket (Figure 2C). ATP binding also causes reorientation of the W^{669} side chain. It turns and becomes more shielded by the surrounding amino acids. Turning of the W^{669} side chain pushes the remainder of the motif, ^{670}TAPELL^{675}, to the surface resulting in its exposure. In general, the movement of a hydrophobic motif towards the surface of the protein indicates its readiness for interaction. For the ^{669}WTAPELL^{675} motif it was proposed that it interacts with subsequent transduction motif, possibly within the catalytic domain, propagates the ANF/ATP binding signal and activates the catalytic domain [108].

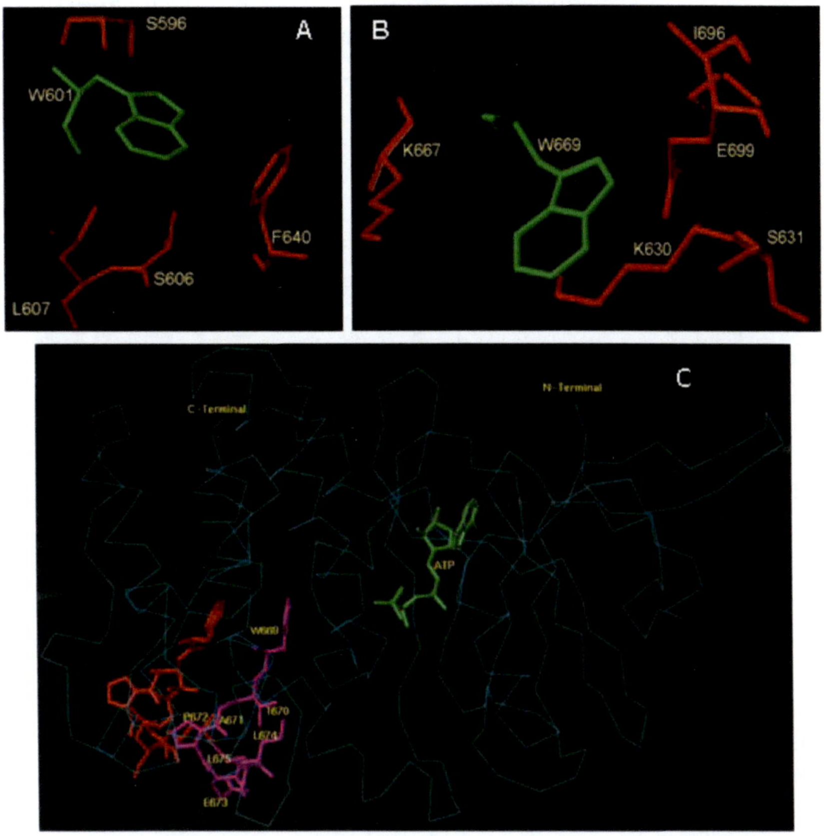

Figure 2. (A) Amino acid residues surrounding W^{601} and (B) Amino acid residues surrounding W^{669} within the ARM domain. Amino acid residues depicted in red are located within a 4 Å sphere from the respective tryptophan residue (green). (C) Conformational changes within the ^{669}WTAPELL^{675} motif induced by ATP binding to the ARM domain. The backbone structure of the ATP-bound ARM domain is shown in cyan and the ATP molecule is in green. The ^{669}WTAPELL^{675} motif is shown in magenta color. Apo structure of the ARM domain was superimposed on the ATP-bound form to assess the relative, ATP binding induced, conformational changes. For clarity, only the ^{669}WTAPELL^{675} motif (shown in red) of the apo-enzyme is visible. ATP binding results in a more compact structure of the ARM domain: the W^{669} side chain moves towards the ATP binding pocket while the side chains of T^{670}, E^{673}, L^{674} and L^{675} move toward the protein surface [compare the orientation of side chain of these amino acid residues before (in red) and after (in magenta) ATP binding]. This movement changes the surface properties of the ARM domain (*39*; reviewed in: *31*). The movement towards the surface of the protein is poised to facilitate interaction of this amino acid stretch with subsequent transduction motif, possibly within the catalytic domain, propagation of the ANF/ATP binding signal and activation of the catalytic domain (Reproduced with permission from ref.108).

Based on the studies narrated above a model for ANF/ATP signaling of ANF-RGC activity was proposed: The ANF signal originates by the binding of one molecule of ANF to the extracellular dimer domain of ANF-RGC [81, 116]. The binding modifies the juxtamembrane region where the disulfide ^{423}Cys-Cys^{432} structural motif is a key element in this modification [81, 86, 116].

The signal twists the transmembrane domain [88], induces a structural change in the ARM domain, and prepares it for the ATP activation. ARM domain binds ATP to its pocket what leads to a cascade of temporal and spatial changes [104, 105; reviewed in: 93]. They result in (1) exposure of the hydrophobic ^{669}WTAPELL^{675} motif which directly (or indirectly) interacts with the catalytic domain causing its partial activation; and (2) exposure and phosphorylation of six serine, threonine residues and full activation of ANF-RGC. Concomitantly, phosphorylation converts ATP binding site from the high to low affinity, ATP dissociates and ANF-RGC returns to its ground state 114].

Despite the overwhelming evidence in support of the direct ATP-modulated model of ANF-dependent ANF-RGC activation, there was a study claiming that ATP has no role in ANF signaling 117]. The investigators in this study used the HEK-293 cells permanently expressing ANF-RGC. And they argued that ATP in time intervals of seconds does not enhance ANF-dependent ANF-RGC activity. This conclusion was at odds with all findings in the field and its flaws have been pointed out: (1) the documented ANF-RGC activity was enormously high (nmol cyclic GMP/min/mg prot, whereas all existing data reports the order of pmol cyclic GMP/min/mg prot); (2) no nucleotide regenerating system was included in the reaction mixture thus, GTP substrate levels were not maintained, as required for enzyme kinetic studies; (3) the dose-response curve of ANF was shifted to the right and the EC_{50} value was in high micromolar (10^{-5} M) range whereas typically obtained EC_{50} values for natriuretic peptides are 50-150 pmol [106, 107]. In the follow-up studies the authors reversed their initial conclusion about ATP having no role in ANF-dependent signaling of ANF-RGC activity and showed that in ATP lowers Km for GTP, the substrate for the cyclase catalytic domain [118].

Direct support of the ATP allosteric function in the activation of ANF-RGC came from the study in which alkaloid staurosporine was used in place of ATP [119]. Staurosporine was selected because: (1) ANF-RGC ARM domain, the site of ATP binding, exhibits sequence homology with the catalytic domains of tyrosine kinases [104, 105]; (2) staurosporine binds to the same as ATP site of various protein kinases' including tyrosine kinases [120-123]; (3) staurosporine uses the same as ATP hydrogen bond interactions in its binding to protein kinases [120-123]; and 4) staurosporine exhibits higher than ATP affinity for the specific binding site [120-123]. It was, therefore, hypothesized that staurosporine, like ATP, should also bind to the ARM domain of ANF-RGC and mimic the allosteric aspect of ATP action in ANF signaling of ANF-RGC. But this process will not involve phosphorylation.

Binding of staurosporine to the ATP binding pocket was verified by functional interference [119]. It was reasoned that if this alkaloid binds to same as ATP site on the ARM domain it should inhibit ATP binding and this inhibition should be mirrored by decreased cross-linking of [γ^{32}P]-8-azido-ATP. And indeed, staurosporine prevented the ATP binding in a dose-dependent fashion with an IC_{50} of 70 μM [119].

In parallel, molecular modeling - docking of staurosporine to the ARM domain was performed. Because the majority of the residues in and around the ATP binding pocket of the ARM domain is homologous to the corresponding residues in protein kinases it was possible to use the molecular replacement logic [124] for the docking experiment. Staurosporine occupies the ATP binding site (Figure 3: the ATP molecule is shown in magenta and staurosporine in yellow). Because of its size, staurosporine induces conformational changes to fit perfectly (induced fit) in the cleft between the N- and C-terminal lobes of the ARM domain. 18 amino acid residues of the ARM domain fall within ~4 Å sphere around the staurosporine molecule; they are: L^{511}, T^{514}, Q^{517}, A^{533}, K^{535}, T^{564}, T^{580}, E^{581}, C^{583}, P^{584}, Gly^{586}, S^{632}, N^{633}, V^{635}, T^{645}, Y^{647}, D^{646} and Y^{657}. Out of these, L^{511}, T^{514}, Q^{517}, A^{533}, K^{535},

T^{580}, C^{583}, N^{33}, V^{635}, T^{645} and D^{646} are part of the original ATP binding pocket. Thus, the biochemical and modeling results were in agreement; Staurosporine binds to the ATP binding site within the ARM domain [119].

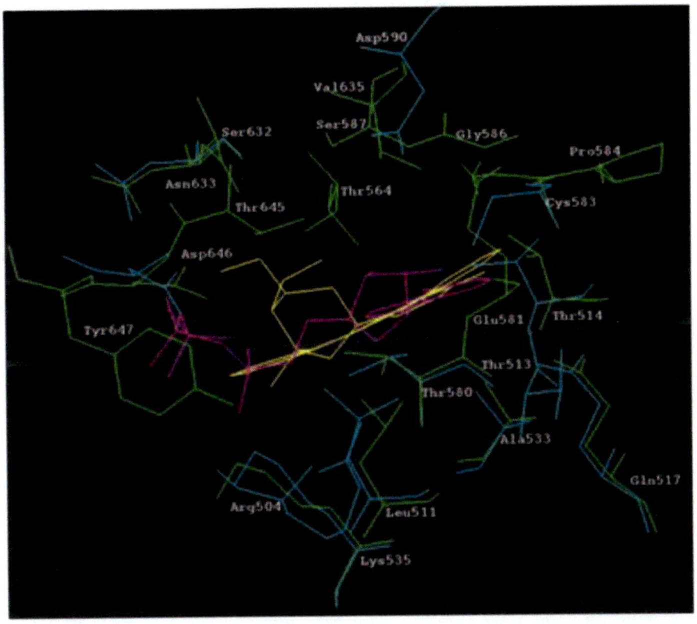

Figure 3. Docking of staurosporine into the ARM domain of ANF-RGC - Comparison with ATP. Staurosporine molecule (shown in yellow) extracted from its complex with SYK catalytic domain (PDB file 1XBC) was docked into the ARM domain. For comparison, docking of ATP (shown in magenta) is also provided. Amino acid residues constituting the staurosporine binding pocket are shown in green while those constituting the ATP binding pocket are in cyan. For clarity, only residues within 4 Å radius around the respective molecule are shown (Reproduced with permission from ref. 119).

And on functional level, can staursporine mediate the ANF-dependent ANF-RGC activation thus, mimic ATP activity? In the presence of ANF and staurosporine the ANF-RGC activity was stimulated in a staurosporine dose-dependent fashion. The half-maximal stimulation was observed at 50 nM staurosporine and the maximal stimulation at approx. 500 nM. It is noteworthy that the concentration of staurosporine resulting in the half-maximal stimulation of ANF-RGC was ~4 orders of magnitude lower than the 0.3 mM concentration of ATP necessary for half-maximal stimulation of ANF-dependent ANF-RGC activity [95, 97]. The conclusion of the study was that ATP allosteric modulation and phosphorylation of ANF-RGC are two separate steps in ANF-RGC activation. The allosteric step occurs first, it is followed by the phosphorylation step leading to full activation of the cyclase.

4. ^{669}WTAPELL^{675} Motif – the Signal Transmitter to the Catalytic Domain

The modeling analyses had shown that following ATP binding to the ARM domain there is a change in the ARM domain surface potential [93]. The most significant change occurs in the vicinity of the ^{669}WTAPELL^{675} motif and is caused by a shift of this stretch of hydrophobic residues to the surface. It was hypothesized that the motif plays a signaling role in activation of the catalytic domain [108]. If this hypothesis is correct, absence of the ^{669}WTAPELL^{675} motif from the ANF-RGC should lead to a protein which is unresponsive to ANF/ATP and this unresponsiveness should characterize the mutant cyclase both in *vitro* in the recombinant system and *in vivo* in an animal model where the motif is deleted by genetic manipulation. In the last case the loss of ANF/ATP-dependent generation of the cyclic GMP should inflict the genetically modified animal with the pathologies of physiological processes which are under the control of the ANF/ANF-RGC system, namely cardio-renal homeostasis. The hypothesis was tested and proven to be correct at the biochemical and physiological levels.

4.1. Biochemical Proof

An ANF-RGC mutant was used in which the ^{669}WTAPELL^{675} motif was deleted by mutagenesis. The mutant was expressed in COS cells and analyzed for its basal and ANF/ATP-dependent activities. Its specific basal activity was identical to the wild type cyclase [108]. When the mutant was exposed to increasing concentrations of ANF and ATP it was practically unresponsive. At 10^{-7} M ANF and ATP concentrations of 0- 1 mM the guanylate cyclase activity of the mutant increased only from 6.9 to 11 pmol cyclic GMP min^{-1}(mg prot)$^{-1}$. Under the same conditions the activity of the wild type ANF-RGC increased from 7.2 to about 90 pmol cyclic GMP min^{-1}(mg prot)$^{-1}$ [108]. Similar results were obtained when the mutant was tested in the presence of constant concentration of ATP and increasing concentrations of ANF [125]. Thus, in the recombinant system the ^{669}WTAPELL^{675} motif is indispensable for the transmission of the ANF/ATP signal to the catalytic domain but it has no control over the basal activity of the cyclase.

The modeling studies indicated that the W^{669} is the key residue within the motif. Validity of this conclusion was further assessed with a rationale that its special function is due to its bulky, aromatic ring structure. W^{699} was mutated to two aromatic ring containing amino acids: phenylalanine and tyrosine, and to an aliphatic residue, leucine. It was expected that if the W^{669} function was really due to is bulky size, aromatic side chain the W→F and W→Y mutants should respond to the ATP/ANF signal like the wild-type ANF-RGC whereas the W→L mutant should be only minimally responsive or totally unresponsive. And indeed, W^{669}F and W^{669}F mutants responded to the ANF/ATP signal like the wild-type ANF-RGC; the W^{669}L mutant was only slightly responsive to the ATP/ANF signal (29% compared to the control) and the response was comparable to that of the W^{669}A mutant [108].

4.2. Physiological Proof

A mouse model was developed in which the sequence coding for the ^{669}WTAPELL^{675} motif was deleted from the ANF-RGC gene [125]. It is noteworthy that to date it is the only ANF-RGC genetically modified animal model in which the modification is domain-specific. Viable homozygous mice, with the ^{669}WTAPELL^{675} motif deleted from both copies of ANF-RGC gene [^{669}WTAPELL$^{675(-/-)}$] prove that the deletion is not lethal and does not interfere with the reproduction. The homo- and heterozygous mice do not differ in their appearance from their wild type littermates. The main function of ANF/ANF-RGC signaling system is to regulate natriuresis, diuresis, vasodilation and prevent cardiac and renal hypertrophy. It regulates blood volume homeostasis and blood pressure by off-setting the renin-angiotensin-aldosterone system (RAAS): inhibiting renal renin secretion and aldosterone synthesis in the adrenocortical zona glomerulosa [27, 126-130]. Because biochemical analyses showed that the ANF-RGC mutant with the ^{669}WTAPELL^{675} motif deleted retains its basal activity virtually identical to that of the wild type ANF-RGC, the cyclase activity was analyzed in the membranes of the two, critical for the physiological functions of the ANF/ANF-RGC system, tissues, kidney and the adrenal gland isolated from the wild type (ANF-RGC ^{669}WTAPELL$^{675+/+}$), heterozygous (ANF-RGC ^{669}WTAPELL$^{675+/-}$) and homozygous (ANF-RGC ^{669}WTAPELL$^{675-/-}$) mice. In agreement with the *in vitro* determinations, the activities were practically indistinguishable, ~40 and ~80 pmol cyclic GMP min^{-1}(mg prot^{-1}) in the kidney and the adrenal gland, respectively of the three genotypes [125]. However, the expressed in these tissues mutated cyclase did not respond to ANF/ATP stimulation. The cyclase activity in the membranes of the homozygous (ANF-RGC ^{669}WTAPELL$^{675-/-}$) mice was the same when assayed in the presence of 10^{-7} M ANF and 0.8 mM ATP as when assayed in their absence. Interestingly, for the heterozygous (ANF-RGC ^{669}WTAPELL$^{675+/-}$) mice, the cyclase remained responsive to ANF/ATP but the achieved activity was only 47% of that achieved by the wild type cyclase in the kidney and 61% in the adrenal gland. These values, oscillating around 50% of the wild type values, clearly show that in the heterozygous mice where the product of one ANF-RGC gene copy is of the wild type and the other, of the deletion-mutated cyclase, only the wild type ANF-RGC is responsive to ANF/ATP and the mutant is not [125]. Thus, the *in vitro* activity characteristics were the mirror reflections of these *in vivo*. Almost identical values of basal guanylate cyclase activity of the ^{669}WTAPELL^{675} targeted mice genotypes implied that the expression of the mutant-cyclase does not differ from the expression of the wild type cyclase. This implication was verified by immunocytochemistry as shown in figure 4. Sections of the kidneys and adrenal glands of the wild type, heterozygous and homozygous ^{669}WTAPELL^{675} targeted mice were stained with immunopurified anti ANF-RGC antibody. Intense immunoreactivity was observed in the renal apical and basolateral aspects of the tubular elements (Figure 4A) and adrenocortical zona glomerulosa (Figure 4B) of all genotypes analyzed. Visual examination of the stained sections as well as of the respective differential interference contrast (DIC) images indicated that the targeting of the ^{669}WTAPELL^{675} motif does also not affect the integrity of these tissues. These results led to the conclusion that deletion of the sequence coding for the ^{669}WTAPELL^{675} motif from either one or two ANF-RGC gene copies does not affect the cyclase tissue-specific expression.

It is well established that elimination of ANF-RGC signaling through knocking-out either ANF-RGC or ANF gene leads to increased blood pressure, cardiac and renal hypertrophy, fibrosis, to name some of the related pathological conditions [131-135]. Since the ^{669}WTAPELL^{675} targeted mice have partially (heterozygous) and fully (homozygous) suppressed ANF-RGC response to ANF, but the basal cyclase activity is unaffected the critical question was: does this suppression lead to increased blood pressure in the genetically modified mice?

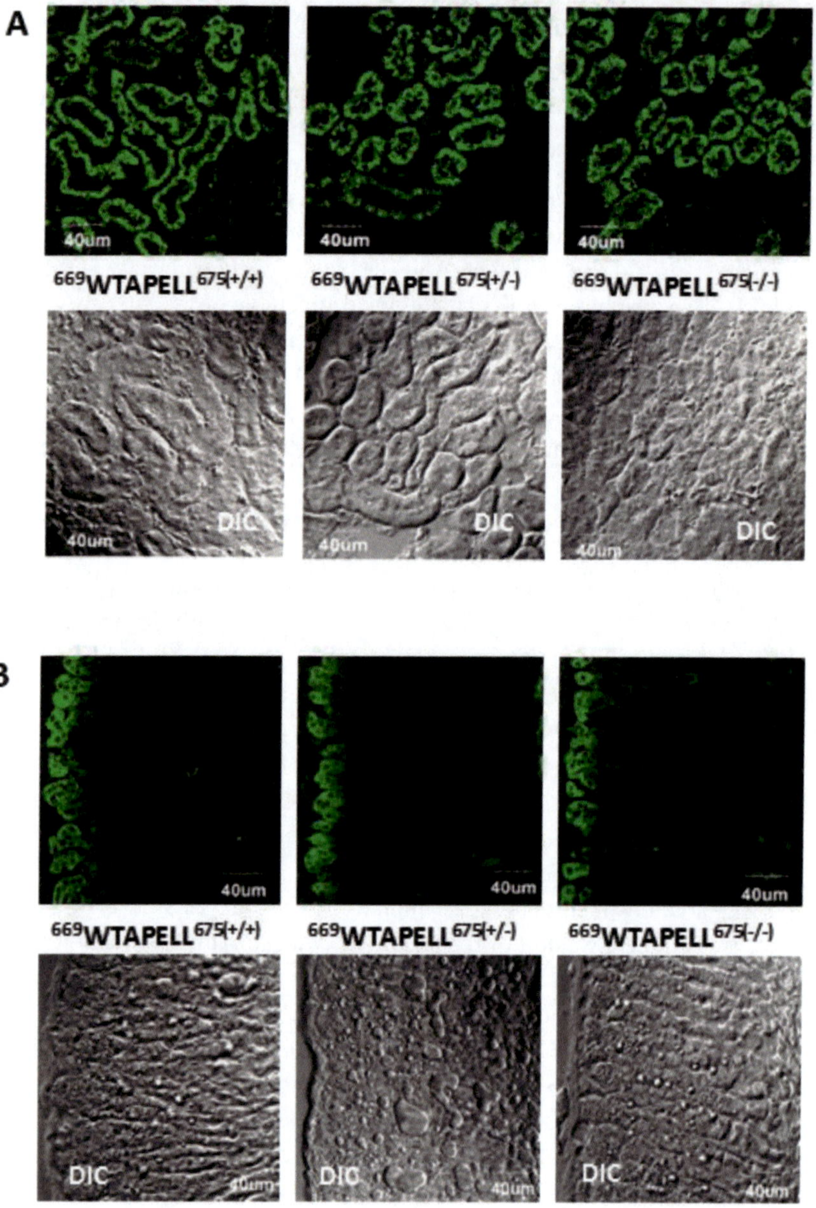

Figure 4. Deletion of the sequence coding for the ^{669}WTAPELL^{675} motif from the ANF-RGC gene does not affect expression of the mutated cyclase in the kidney (A) and adrenal gland (B). Cryosections of the kidney and adrenal gland from the wild type [^{669}WTAPELL$^{675(+/+)}$], heterozygous [^{669}WTAPELL$^{675(+/-)}$], and homozygous [^{669}WTAPELL$^{675(-/-)}$] mice were immuno-stained with ANF-RGC antibodies. The DIC image showing the integrity of the sections are also shown ("DIC"). Intense staining was observed in the renal apical and basolateral aspects of the tubular elements and adrenocortical zona glomerulosa (Reproduced with permission from ref. 125).

The answer obtained was very clear. In comparison with the wild type the blood pressure of the heterozygous mice increased by ~37% and of the homozygous mice by ~56%. The increase was statistically highly significant with P value < 0.005 [125]. It rose from 102 ± 9 mm Hg for the wild type mice to 134 ± 17 mm Hg for the heterozygous and 159 ± 11 mm Hg for the homozygous mice [125]. These values showed that the progressive elevation of blood pressure directly correlates with the number of mutated ANF-RGC gene copies.

Because volume homeostasis and blood pressure are influenced by mineralocorticoid hormone aldosterone secreted by the cells of the adrenocortical zona glomerulosa and the function of the ANF/ANF-RGC system is to inhibit this synthesis the obvious next question was whether the increased blood pressure in the ^{669}WTAPELL^{675} targeted mice is a consequence of the inability of the mutated ANF-RGC to transduce the ANF signal and to synthesize cyclic GMP sufficient quantities to inhibit aldosterone synthesis? A hint was provided by studies on ANF-RGC knockout mice showing that these mice had increased aldosterone level [136].

The plasma aldosterone concentrations were measured in both types of genetically modified mice (hetero- and homozygous) and in their isogenic controls (wild type). The results demonstrated that, in comparison with the wild type mice, the plasma aldosterone concentration increased by approximately 40% (from 147 ± 12 pg/ml to 204 ± 18 pg/ml) in the plasma of the heterozygous mice ($P < 0.005$) and by approximately 75% (up to 256 ± 22 pg/ml) in the plasma of the homozygous mice ($P < 0.005$) [125].

Chronic pressure overload leads to cardiac hypertrophy [reviewed in: 137]. The increased blood pressure in the ^{669}WTAPELL^{675} targeted mice is accompanied by cardiac hypertrophy. The ratio of the heart weight (in mg) to whole body weight (in g) was 5.6 ± 0.3 and 6.1 ± 0.5 for the ^{669}WTAPELL$^{675(+/-)}$ and ^{669}WTAPELL$^{675(-/-)}$ mice, respectively, whereas it was 5 ± 0.3 for the wild type mice [125].

Taken together the biochemical and physiological evidence prove the role of the ^{669}WTAPELL^{675} motif as the transmitter of the ANF/ATP signal from the ARM to the catalytic domain of ANF-RGC.

CONCLUSION

For over two decades the obligatory role of ATP in ANF-dependent activation of ANF-RGC was analyzed through various biochemical techniques. These techniques, however, were unable to provide an answer as to how ATP binding signals propagation of the initial ANF signal.

It was only when the model of the domain to which ATP binds, the ARM domain, was constructed, that an insight into the spatio-temporal aspects of the ATP binding was gained. The model was experimentally tested, modified and again the modifications were experimentally validated. Finally, simulations and optimizations of the favorable interactions within the ARM domain in its apo and ATP-bound states allowed proposing of a dynamic signaling cascade. The high points of the studies were (1) the establishing of the relationship between the ATP allosteric affects and phosphorylation of the consensus serine and threonine residues; and (2) the identification of the ^{669}WTAPELL^{675} motif as the possible link that signals (whether directly or indirectly remains to be determined) activation of the catalytic domain.

The signaling function of the ^{669}WTAPELL^{675} motif was tested *in vitro* and critically verified *in vivo*. The absence of the motif from the ANF-RGC molecule in the recombinant system prevents ANF-dependent activation of the cyclase. Identical prevention of the ANF signaling occurs in the ^{669}WTAPELL^{675} targeted genetically modified animals with all its pathological consequences, hyperaldosteronism, hypertension and cardiac and renal hypertrophy. Thus, the theoretically conceived model was proven correct at the physiological level in living animals.

Supported by USPHS award: HL084584 and S82701

References

[1] Paul AK. Particulate guanylate cyclase from adrenocortical carcinoma 494. Purification, biochemical and immunological characterization. *Doctoral thesis,* University of Tennessee, 1986.

[2] Paul AK, Marala RB, Jaiswal RK, Sharma RK. Coexistence of guanylate cyclase and atrial natriuretic factor receptor in a 180-kD protein. *Science* 1987; 235:1224-1226.

[3] Kuno T, Andresen JW, Kamisaki Y, Waldman SA, Chang LY, Saheki S, Leitman DC, Nakane M, Murad F: Co-purification of an atrial natriuretic factor receptor and particulate guanylate cyclase from rat lung. *J. Biol. Chem.* 1986; 261:5817-5823.

[4] Meloche S, McNicoll N, Liu B, Ong H, De Lean A. Atrial natriuretic factor R1 receptor from bovine adrenal zona glomerulosa: purification, characterization, and modulation by amiloride. *Biochemistry* 1988; 27:8151-8158.

[5] Takayanagi R, Inagami T, Snajdar RM, Imada T, Tamura M, Misono KS. Two distinct forms of receptors for atrial natriuretic factor in bovine adrenocortical cells. Purification, ligand binding, and peptide mapping. *J. Biol. Chem.* 1987; 262: 12104-12113.

[6] Sharma RK. Membrane guanylate cyclase is a beautiful signal transduction machine: overview. *Mol. Cell Biochem.* 2010; 334:3-36.

[7] Chinkers M, Garbers DL, Chang MS, Lowe DG, Chin HM, Goeddel DV, Schulz S. A membrane form of guanylate cyclase is an atrial natriuretic peptide receptor. *Nature* 1989; 338:78-83

[8] Duda T, Goraczniak RM, Sharma RK. Site-directed mutational analysis of a membrane guanylate cyclase cDNA reveals the atrial natriuretic factor signaling site. *Proc. Natl. Acad. Sci. U S A* 1991;88:7882-7886.

[9] Lowe DG, Chang MS, Hellmiss R, Chen E, Singh S, Garbers DL, Goeddel DV. Human atrial natriuretic peptide receptor defines a new paradigm for second messenger signal transduction. *EMBO J.* 1989;8:1377-13784.

[10] Marala R, Duda T, Goraczniak RM, Sharma RK. Genetically tailored atrial natriuretic factor-dependent guanylate cyclase. Immunological and functional identity with 180 kDa membrane guanylate cyclase and ATP signaling site. *FEBS Lett.* 1992; 296: 254-258.

[11] Pandey KN, Singh S. Molecular cloning and expression of murine guanylate cyclase/atrial natriuretic factor receptor cDNA. *J. Biol. Chem.* 1990; 265: 12342-12348.

[12] Chang MS, Lowe DG, Lewis M, Hellmiss R, Chen E, Goeddel DV. Differential activation by atrial and brain natriuretic peptides of two different receptor guanylate cyclases. *Nature* 1989; 341:68-72.

[13] Schulz S, Singh S, Bellet RA, Singh G, Tubb DJ, Chin H, Garbers DL. The primary structure of a plasma membrane guanylate cyclase demonstrates diversity within this new receptor family. *Cell* 1989; 1155-1162.

[14] Duda T, Goraczniak RM, Sitaramayya A, Sharma RK. Cloning and expression of an ATP-regulated human retina C-type natriuretic factor receptor guanylate cyclase. *Biochemistry* 1993; 32:1391-1395.

[15] de Sauvage FJ, Camerato TR, Goeddel DV. Primary structure and functional expression of the human receptor for Escherichia coli heat-stable enterotoxin. *J. Biol. Chem.* 1991; 266:17912-17918.

[16] Hamra FK, Forte LR, Eber SL, Pidhorodeckyj NV, Krause WJ, Freeman RH, Chin DT, Tompkins JA, Fok KF, Smith CE, Duffin KL, Siegel NR, Currie MG. Uroguanylin: structure and activity of a second endogenous peptide that stimulates intestinal guanylate cyclase. *Proc. Natl. Acad. Sci. U S A* 1993; 90:10464-10468.

[17] Khare S, Wilson D, Wali RK, Tien XY, Bissonnette M, Niedziela SM, Bolt MJ, Sitrin MD, Brasitus TA. Guanylin activates rat colonic particulate guanylate cyclase. *Biochem. Biophys. Res. Commun.* 1994; 203:1432-1437.

[18] Schulz S, Green CK, Yuen PS, Garbers DL. Guanylyl cyclase is a heat-stable enterotoxin receptor. *Cell* 1990; 63:941-948.

[19] Singh S, Singh G, Heim JM, Gerzer R. Isolation and expression of a guanylate cyclase-coupled heat stable enterotoxin receptor cDNA from a human colonic cell line. *Biochem. Biophys. Res. Commun.* 1991; 179:1455-1463.

[20] Needleman P, Blaine EH, Greenwald JE, Michener ML, Saper CB, Stockmann PT, Tolunay HE. The biochemical pharmacology of atrial peptides. *Annu. Rev. Pharmacol. Toxicol.* 1989; 29:23-54.

[21] Rosenzweig A, Seidman CE. Atrial natriuretic factor and related peptide hormones. *Annu. Rev. Biochem.* 1991; 60:229-255.

[22] de Bold AJ. Atrial natriuretic factor: a hormone produced by the heart. *Science* 1985; 230:767-770.

[23] Furuya M, Tawaragi Y, Minamitake Y, Kitajima Y, Fuchimura K, Tanaka S, Minamino N, Kangawa K, Matsuo H. Structural requirements of C-type natriuretic peptide for elevation of cyclic GMP in cultured vascular smooth muscle cells. *Biochem. Biophys. Res. Commun.* 1992; 183:964-969.

[24] Stingo AJ, Clavell AL, Heublein DM, Wei CM, Pittelkow MR, Burnett JC Jr. Presence of C-type natriuretic peptide in cultured human endothelial cells and plasma. *Am. J. Physiol.* 1992; 263:H1318-1321.

[25] Vigne P, Frelin C. C-type natriuretic peptide is a potent activator of guanylate cyclase in endothelial cells from brain microvessels. *Biochem. Biophys. Res. Commun.* 1992; 183:640-644.

[26] Atlas SA, Laragh JH. Atrial natriuretic peptide: a new factor in hormonal control of blood pressure and electrolyte homeostasis. *Annu. Rev. Med.* 1986; 37:397-414.

[27] Brenner BM, Ballermann BJ, Gunning ME, Zeidel ML. Diverse biological actions of atrial natriuretic peptide. *Physiol. Rev.* 1990; 7:665-669.

[28] Gunning ME, Brenner BM. Natriuretic peptides and the kidney: current concepts. *Kidney Int. Suppl.* 1992; 38:S127-133.

[29] John SW, Krege JH, Oliver PM, Hagaman JR, Hodgin JB, Pang SC, Flynn TG, Smithies O. Genetic decreases in atrial natriuretic peptide and salt-sensitive hypertension. Science 1995; 267:679-681 Erratum in: *Science* 1995; 267:1753.

[30] Lopez MJ, Wong SK, Kishimoto I, Dubois S, Mach V, Friesen J, Garbers DL, Beuve A. Salt-resistant hypertension in mice lacking the guanylyl cyclase-A receptor for atrial natriuretic peptide. *Nature* 1995; 378:65-68.

[31] Suda M, Ogawa Y, Tanaka K, Tamura N, Yasoda A, Takigawa T, Uehira M, Nishimoto H, Itoh H, Saito Y, Shiota K, Nakao K. Skeletal overgrowth in transgenic mice that overexpress brain natriuretic peptide. *Proc. Natl. Acad. Sci. U S A* 1998; 95:2337-2342.

[32] Centrella M, McCarthy TL. Targeted disruption of C-type natriuretic peptide: a focused assault on cartilaginous bone. *Trends Endocrinol. Metab*. 2001; 12:235-236.

[33] Goraczniak RM, Duda T, Sitaramayya A, Sharma RK. Structural and functional characterization of the rod outer segment membrane guanylate cyclase. *Biochem. J.* 1994; 302:455-461.

[34] Shyjan AW, de Sauvage FJ, Gillett NA, Goeddel DV, Lowe DG. Molecular cloning of a retina-specific membrane guanylyl cyclase. *Neuron* 1992;9: 27-737.

[35] Lowe DG. *Gene Bank accession number* M92432, 1995.

[36] Koch K-W. Purification and identification of photoreceptor guanylate cyclase. *J. Biol. Chem.* 1991; 266:8634-8637.

[37] Lowe DG, Dizhoor AM, Liu K, Gu Q, Spencer M, Laura R, Lu L, Hurley JB. Cloning and expression of a second photoreceptor-specific membrane retina guanylyl cyclase (RetGC), RetGC-2. *Proc. Natl. Acad. Sci. U S A* 1995; 92:5535-5539

[38] Goraczniak RM, Duda T, Sharma RK. Calcium modulated signaling site in type 2 rod outer segment membrane guanylate cyclase (ROS-GC2). *Biochem. Biophys. Res. Commun.* 1998; 245:447-453.

[39] Yang RB, Foster DC, Garbers DL, Fulle HJ. Two membrane forms of guanylyl cyclase found in the eye. *Proc. Natl. Acad. Sci. U S A* 1995; 92:602-606.

[40] Duda T, Jankowska A, Venkataraman V, Nagele RG, Sharma RK. A novel calcium-regulated membrane guanylate cyclase transduction system in the olfactory neuroepithelium. *Biochemistry* 2001; 40:12067-12077.

[41] Fulle HJ, Vassar R, Foster DC, Yang RB, Axel R, Garbers DL. A receptor guanylyl cyclase expressed specifically in olfactory sensory neurons. *Proc. Natl. Acad. Sci. U S A* 1995; 92:3571-3575.

[42] Schulz S, Wedel BJ, Matthews A, Garbers DL. The cloning and expression of a new guanylyl cyclase orphan receptor. *J. Biol. Chem.* 1998; 273:1032-1037.

[43] Wedel BJ, Garbers DL. New insights on the functions of the guanylyl cyclase receptors. *FEBS Lett.* 1997; 410:29-33.

[44] Pugh EN Jr, Duda T, Sitaramayya A, Sharma RK. Photoreceptor guanylate cyclases: a review. *Biosci. Rep*. 1997; 17: 429-473.

[45] Koch K-W, Duda T, Sharma RK. Ca(2+)-modulated vision-linked ROS-GC guanylate cyclase transduction machinery. *Mol. Cell Biochem*. 2010; 334;105-115.

[46] Detwiler P. Open the loop: Dissecting feedback regulation of a second messenger transduction cascade. *Neuron* 200; 36:3-4.

[47] Dizhoor AM, Olshevskaya EV, Henzel WJ, Wong SC, Stults JT, Ankoudinova I, Hurley JB. Cloning, sequencing, and expression of a 24-kDa Ca(2+)-binding protein activating photoreceptor guanylyl cyclase. *J. Biol. Chem.* 1995; 270:25200-25206.

[48] Palczewski K, Subbaraya I, Gorczyca WA, Helekar BS, Ruiz CC, Ohguro H, Huang J, Zhao X, Crabb JW, Johnson RS, Walsh KA, Gray-Keller MP, Detwiler P, Baehr W. Molecular cloning and characterization of retinal photoreceptor guanylyl cyclase-activating protein. *Neuron* 1994; 13:395-404.
[49] Duda T, Goraczniak R, Surgucheva I, Rudnicka-Nawrot M, Gorczyca WA, Palczewski K, Sitaramayya A, Baehr W, Sharma RK. Calcium modulation of bovine photoreceptor guanylate cyclase. *Biochemistry* 1996; 35:8478-8482.
[50] Pozdnyakov N, Yoshida A, Cooper NG, Margulis A, Duda T, Sharma RK, Sitaramayya A. A novel calcium-dependent activator of retinal rod outer segment membrane guanylate cyclase. *Biochemistry* 1995;34:14279-14283.
[51] Margulis A, Pozdnyakov N, Sitaramayya A. Activation of bovine photoreceptor guanylate cyclase by S100 proteins. *Biochem. Biophys. Res. Commun.* 1996; 218:243-247.
[52] Duda T, Goraczniak RM, Sharma RK. Molecular characterization of S100A1-S100B protein in retina and its activation mechanism of bovine photoreceptor guanylate cyclase. *Biochemistry* 1996; 35:6263-6266.
[53] Kumar VD, Vijay-Kumar S, Krishnan A, Duda T, Sharma RK. A second calcium regulator of rod outer segment membrane guanylate cyclase, ROS-GC1: neurocalcin. *Biochemistry* 1999; 38:12614-12620.
[54] Wen XH, Duda T, Pertzev A, Venkataraman V, Makino CL, Sharma RK. S100B serves as a Ca(2+) sensor for ROS-GC1 guanylate cyclase in cones but not in rods of the murine retina. *Cell Physiol. Biochem.* 2012; 29:417-430.
[55] Krishnan A, Venkataraman V, Fik-Rymarkiewicz E, Duda T, Sharma RK. Structural, biochemical, and functional characterization of the calcium sensor neurocalcin delta in the inner retinal neurons and its linkage with the rod outer segment membrane guanylate cyclase transduction system. *Biochemistry* 2004; 43:2708-2723.
[56] Pertzev A, Duda T, Sharma RK. Ca(2+) sensor GCAP1: A constitutive element of the ONE-GC-modulated odorant signal transduction pathway. *Biochemistry* 2010; 49:7303-7313.
[57] Duda T, Fik-Rymarkiewicz E, Venkataraman V, Krishnan A, Sharma RK. Calcium-modulated ciliary membrane guanylate cyclase transduction machinery: constitution and operational principles. *Mol. Cell Biochem.* 2004; 267:107-122.
[58] Duda T, Krishnan R, Sharma RK. GCAP1: antithetical calcium sensor of ROS-GC transduction machinery. *Calcium Binding Proteins* 2006;1:102-107.
[59] Duda T, Pertzev A, Koch KW, Sharma RK. Antithetical modes of and the Ca(2+) sensors targeting in ANF-RGC and ROS-GC1 membrane guanylate cyclases. *Front. Mol. Neurosci.* 2012; 5:44.
[60] Leinders-Zufall T, Cockerham RE, Michalakis S, Biel M, Garbers DL, Reed RR, Zufall F, Munger SD. Contribution of the receptor guanylyl cyclase GC-D to chemosensory function in the olfactory epithelium. *Proc. Natl. Acad. Sci. USA* 2007; 104: 14507-14512.
[61] Duda T, Sharma RK. ONE-GC membrane guanylate cyclase, a trimodal odorant signal transducer. *Biochem. Biophys. Res. Commun.* 2008; 367:440-445.
[62] Duda T, Sharma RK. Ca2+-modulated ONE-GC odorant signal transduction. *FEBS Lett.* 2009; 583:1327-1330.

[63] Duda T, Pertzev A, Sharma RK. Ca(2+) modulation of ANF-RGC: new signaling paradigm interlocked with blood pressure regulation. *Biochemistry* 2012; 51:9394-9405.

[64] Kuhn M, Ng CK, Su YH, Kilić A, Mitko D, Bien-Ly N, Kömüves LG, Yang RB. Identification of an orphan guanylate cyclase receptor selectively expressed in mouse testis. *Biochem. J.* 2004; 379(Pt 2):385-393.

[65] Chao Y-C, Cheng C-J, Hsieh H-T, Lin C-C, Chen C-C, Yang R-B. Guanylate cyclase-G, expressed in the Grueneberg ganglion olfactory subsystem, is activated by bicarbonate. *Biochem. J.* 2010; 432:267-273.

[66] Mamasuew K, Breer H, Fleischer J. Gruenberg ganglion neurons respond to cool ambient temperatures. *Eur. J. Neurosci.* 2008; 28:1775-1785.

[67] Duda T, Venkataraman V, Jankowska A, Lange C, Koch KW, Sharma RK. Impairment of the rod outer segment membrane guanylate cyclase dimerization in a cone-rod dystrophy results in defective calcium signaling. *Biochemistry* 2000; 39:12522-12533.

[68] Yu H, Olshevskaya E, Duda T, Seno K, Hayashi F, Sharma RK, Dizhoor AM, Yamazaki A. Activation of retinal guanylyl cyclase-1 by Ca2+-binding proteins involves its dimerization. *J. Biol. Chem.* 1999; 274:15547-15555.

[69] Wilson E M, Chinkers M. Identification of sequences mediating guanylyl cyclase dimerization. *Biochemistry* 1995; 3:4696-4701.

[70] Thorpe D S, Niu S, Morkin E. Overexpression of dimeric guanylyl cyclase cores of an atrial natriuretic peptide receptor. *Biochem. Biophys. Res. Commun.* 1991; 180:538-544.

[71] Liu Y, Ruoho A E, Rao V D, Hurley J H. Catalytic mechanism of the adenylyl and guanylyl cyclases: modeling and mutational analysis. *Proc. Natl. Acad. Sci. USA* 1997; 9:13414-13419.

[72] Yang RB, Garbers DL. Two eye guanylyl cyclases are expressed in the same photoreceptor cells and form homomers in preference to heteromers. *J. Biol. Chem.* 1997; 272:13738-13742.

[73] Qiu Y, Ogawa H, Miyagi M, Misono KS. Constitutive activation and uncoupling of the atrial natriuretic peptide receptor by mutations at the dimer interface. Role of the dimer structure in signalling. *J. Biol. Chem.* 2004; 79:6115-6123.

[74] Sharma RK. Evolution of the membrane guanylate cyclase transduction system. *Mol. Cell Biochem.* 2002; 230:3-30.

[75] Sharma RK, Duda T, Venkataraman V, Koch K-W. Calcium-modulated mammalian membrane guanylate cyclase ROS-GC transduction machinery in sensory neurons: A universal concept. *Curr. Topics Biochem. Res.* 2004; 6:111-144.

[76] McNicoll N, Escher E, Wilkes BC, Schiller PW, Ong H, De Léan A. Highly efficient photoaffinity labeling of the hormone binding domain of atrial natriuretic factor receptor. *Biochemistry* 1992; 31:4487-4493.

[77] McNicoll N, Gagnon J, Rondeau JJ, Ong H, De Léan A. Localization by photoaffinity labeling of natriuretic peptide receptor-A binding domain. *Biochemistry* 1996; 35:12950-12956.

[78] Misono KS. Natriuretic peptide receptor: structure and signaling. *Mol. Cell Biochem.* 2002; 230:49-60.

[79] He X, Nishio K, Misono KS. High-yield affinity alkylation of the atrial natriuretic factor receptor binding site. *Bioconjug Chem.* 1995; 6:541-548.

[80] Ogawa H, Zhang X, Qiu Y, Ogata CM, Misono KS. Crystallization and preliminary X-ray analysis of the atrial natriuretic peptide (ANP) receptor extracellular domain complex with ANP: use of ammonium sulfate as the cryosalt. *Acta Crystallogr. D Biol. Crystallogr.* 2003; 59:1831-1833.
[81] Ogawa H, Qiu Y, Ogata CM, Misono KS. Crystal structure of hormone-bound atrial natriuretic peptide receptor extracellular domain: rotation mechanism for transmembrane signal transduction. *J. Biol. Chem.* 2004; 279:28625-28631.
[82] Labrecque J, Mc Nicoll N, Marquis M, De Lean A. A disulfide-bridged mutant of natriuretic peptide receptor-A displays constitutive activity. Role of receptor dimerization in signal transduction. *J. Biol. Chem.* 1999; 274:9752-9759.
[83] Miyagi M, Misono KS. Disulfide bond structure of the atrial natriuretic peptide receptor extracellular domain: conserved disulfide bonds among guanylate cyclase-coupled receptors. *Biochim. Biophys. Acta* 2000; 1478:30-38.
[84] Huo X, Abe T, Misono KS. Ligand binding-dependent limited proteolysis of the atrial natriuretic peptide receptor: juxtamembrane hinge structure essential for transmembrane signal transduction. *Biochemistry* 1999; 38:16941-16951.
[85] Labrecque J, Deschênes J, McNicoll N, De Léan A. Agonistic induction of a covalent dimer in a mutant of natriuretic peptide receptor-A documents a juxtamembrane interaction that accompanies receptor activation. *J. Biol. Chem.* 2001; 276:8064-8072.
[86] Duda T, Sharma RK. Two membrane juxtaposed signaling modules in ANF-RGC are interlocked. *Biochem. Biophys. Res. Commun.* 2005; 332:149-156.
[87] Misono KS, Ogawa H, Qiu Y, Ogata CM. (2005) Structural studies of the natriuretic peptide receptor: a novel hormone-induced rotation mechanism for transmembrane signal transduction. *Peptides* 26: 957-968.
[88] Parat M, Blanchet J, De Léan A. Role of juxtamembrane and transmembrane domains in the mechanism of natriuretic peptide receptor A activation. *Biochemistry* 2010; 49:4601-4610.
[89] Chang CH, Kohse KP, Chang B, Hirata M, Jiang B, Douglas JE, Murad F. Characterization of ATP-stimulated guanylate cyclase activation in rat lung membranes. *Biochim. Biophys. Acta* 1990; 1052:159-165.
[90] Larose L, McNicoll N, Ong H, De Léan A. Allosteric modulation by ATP of the bovine adrenal natriuretic factor R1 receptor functions. *Biochemistry* 1991; 30:8990-8995.
[91] Kurose H, Inagami T, Ui M. Participation of adenosine 5'-triphosphate in the activation of membrane-bound guanylate cyclase by the atrial natriuretic factor. *FEBS Lett.* 1987; 219:375-379.
[92] Cole FE, Rondon I, Iwata T, Hardee E, Frohlich ED. Effect of ATP and amiloride on ANF binding and stimulation of cyclic GMP accumulation in rat glomerular membranes. *Life Sci.* 1989; 45:477-484.
[93] Duda T, Venkataraman V, Ravichandran S, Sharma RK. ATP-regulated module (ARM) of the atrial natriuretic factor receptor guanylate cyclase. *Peptides* 2005; 26:969-984.
[94] Chinkers M, Singh S, Garbers DL. Adenine nucleotides are required for activation of rat atrial natriuretic peptide receptor/guanylyl cyclase expressed in a baculovirus system. *J. Biol. Chem.* 1991; 266:4088-4093.
[95] Marala RB, Sitaramayya A, Sharma RK. Dual regulation of atrial natriuretic factor-dependent guanylate cyclase activity by ATP. *FEBS Lett.* 1991; 281:73-76.

[96] Wong SK, Ma CP, Foster DC, Chen AY, Garbers DL. The guanylyl cyclase-A receptor transduces an atrial natriuretic peptide/ATP activation signal in the absence of other proteins. *J. Biol. Chem.* 1995; 270:30818-30822.

[97] Goraczniak RM, Duda T, Sharma RK. A structural motif that defines the ATP-regulatory module of guanylate cyclase in atrial natriuretic factor signalling. *Biochem. J.* 1992; 282:533-537.

[98] Duda T, Goraczniak RM, Sharma RK. The glycine residue of ATP regulatory module in receptor guanylate cyclases that is essential in natriuretic factor signaling. *FEBS Lett.* 1993; 335:309-314.

[99] Chinkers M, Garbers DL. The protein kinase domain of the ANP receptor is required for signaling. *Science* 1989; 245:1392-1394.

[100] Wierenga RK, Hol WG. Predicted nucleotide-binding properties of p21 protein and its cancer-associated variant. *Nature* 1983; 302:842-844.

[101] Hanks SK, Quinn AM, Hunter T. The protein kinase family: conserved features and deduced phylogeny of the catalytic domains. *Science* 1988; 241:42-52.

[102] Bratová I, Otyepka M, Kříž Z, Koča J. The mechanism of inhibition of the cyclin-dependent kinase-2 as revealed by the molecular dynamics study on the complex CDK2 with the peptide substrate HHASPRK. *Protein Sci.* 2005; 14:445–451.

[103] Duda T, Sharma RK. ATP bimodal switch that regulates the ligand binding and signal transduction activities of the atrial natriuretic factor receptor guanylate cyclase. *Biochem. Biophys. Res. Commun.* 1995; 209:286-292.

[104] Duda T, Yadav P, Jankowska A, Venkataraman V, Sharma RK. Three dimensional atomic model and experimental validation for the ATP-Regulated Module (ARM) of the atrial natriuretic factor receptor guanylate cyclase. *Mol. Cell Biochem.* 2001; 217:165-172.

[105] Sharma RK, Yadav P, Duda T. Allosteric regulatory step and configuration of the ATP-binding pocket in atrial natriuretic factor receptor guanylate cyclase transduction mechanism. *Can. J. Physiol. Pharmacol.* 2001; 79:682-691.

[106] Joubert S, Jossart C, McNicoll N, De Léan A. Atrial natriuretic peptide-dependent photolabeling of a regulatory ATP-binding site on the natriuretic peptide receptor-A. *FEBS J.* 2005; 272:5572-5583.

[107] Burczynska B, Duda T, Sharma RK. ATP signaling site in the ARM domain of atrial natriuretic factor receptor guanylate cyclase. *Mol. Cell Biochem.* 2007; 301:193-207.

[108] Duda T, Bharill S, Wojtas I, Yadav P, Gryczynski I, Gryczynski Z, Sharma RK. Atrial natriuretic factor receptor guanylate cyclase signaling: new ATP-regulated transduction motif. *Mol. Cell Biochem.* 2009; 324:39-53.

[109] Potter LR, Hunter T. Identification and characterization of the phosphorylation sites of the guanylyl cyclase-linked natriuretic peptide receptors A and B. *Methods.* 1999; 19:506-520.

[110] Potter LR, Hunter T. Phosphorylation of the kinase homology domain is essential for activation of the A-type natriuretic peptide receptor. *Mol. Cell Biol.* 1998; 18: 2164-2172.

[111] Potter LR, Hunter T. Identification and characterization of the major phosphorylation sites of the B-type natriuretic peptide receptor. *J. Biol. Chem.* 1998; 273:15533-15539.

[112] Potter LR, Hunter T. A constitutively "phosphorylated" guanylyl cyclase-linked atrial natriuretic peptide receptor mutant is resistant to desensitization. *Mol. Biol. Cell* 1999; 10:1811-1820.

[113] Foster DC, Garbers DL. Dual role for adenine nucleotides in the regulation of the atrial natriuretic peptide receptor, guanylyl cyclase-A. *J. Biol. Chem.* 1998; 273:16311-16318.

[114] Duda T, Yadav P, Sharma RK. Allosteric modification, the primary ATP activation mechanism of atrial natriuretic factor receptor guanylate cyclase. *Biochemistry* 2011; 50:1213-1225.

[115] Potter LR, Garbers DL. Dephosphorylation of the guanylyl cyclase-A receptor causes desensitization. *J. Biol. Chem.* 1992; 267:14531-14534.

[116] Ogawa H, Qiu Y, Huang L, Tam-Chang SW, Young HS, Misono KS. Structure of the atrial natriuretic peptide receptor extracellular domain in the unbound and hormone-bound states by single-particle electron microscopy. *FEBS J.* 2009; 276:1347-1355.

[117] Antos LK, Abbey-Hosch SE, Flora DR, Potter LR.ATP-independent activation of natriuretic peptide receptors. *J. Biol. Chem.* 2005; 280:26928-26932.

[118] Antos LK, Potter LR. Adenine nucleotides decrease the apparent Km of endogenous natriuretic peptide receptors for GTP. *Am. J. Physiol. Endocrinol. Metab.* 2007; 293: E1756-1763.

[119] Duda T, Yadav P, Sharma RK.ATP allosteric activation of atrial natriuretic factor receptor guanylate cyclase. *FEBS J.* 2010; 277:2550-2553.

[120] Meggio F, Donella Deana A, Ruzzene M, Brunati AM, Cesaro L, Guerra B, Meyer T, Mett H, Fabbro D, Furet P. Different susceptibility of protein kinases to staurosporine inhibition. Kinetic studies and molecular bases for the resistance of protein kinase CK2. *Eur. J. Biochem.* 1995; 234:317-322.

[121] Prade L, Engh RA, Girod A, Kinzel V, Huber R, Bossemeyer D. () Staurosporine-induced conformational changes of cAMP-dependent protein kinase catalytic subunit explain inhibitory potential. *Structure* 1997; 5:1627-1637.

[122] Lamers MB, Antson AA, Hubbard RE, Scott RK, Williams DH. Structure of the protein tyrosine kinase domain of C-terminal Src kinase (CSK) in complex with staurosporine. *J. Mol. Biol.* 1999; 285:713-725.

[123] Kinoshita T, Matsubara M, Ishiguro, H, Okita K, Tada T. Structure of human Fyn kinase domain complexed with staurosporine. *Biochem. Biophys. Res. Commun.* 2006; 346:840-844.

[124] Rossmann MG. The molecular replacement method. *Acta Cryst. A* 1990; 46:73-82.

[125] Duda T, Pertzev A, Sharma RK. The ANF-RGC Gene Motif (669) WTAPELL(675) Is Vital for Blood Pressure Regulation: Biochemical Mechanism. *Biochemistry* 2013;52: 2337-2347.

[126] Maack T. Role of atrial natriuretic factor in volume control. *Kidney Int.* 1996; 49:1732-1737.

[127] Burnett JC Jr, Granger JP, Opgenorth TJ. Effect of synthetic atrial natriuretic factor on renal function and rennin release. *Am. J. Physiol. Renal. Fluid. Electrolyte Physiol.* 1984; 247:F863-866.

[128] Aoki H, Richmond M, Izumo S, Sadoshima J. Specific role of the extracellular signal-regulated kinase pathway in angiotensin II-induced cardiac hypertrophy in vitro. *Biochem J.* 2000; 347: 275-284.

[129] Shi S-J, Nguyen HT, Sharma GD, Navar G, Pandey KN. Genetic disruption of atrial natriuretic peptide receptor-A alters rennin and angiotensinII levels. *Am. J. Physiol. Renal. Physiol.* 2001; 281:F665-673.

[130] Olson LJ, Ho By, Cashdollar LW, Drewett JG. Functionally active catalytic domain is essential of guanylyl cyclase-linked receptor mediated inhibition of human aldosterone synthesis. *Mol. Pharmacol.* 1998; 54:761-769.

[131] Pandey KN. Guanylyl cyclase/atrial natriuretic peptide receptor-A: role in the pathophysiology of cardiovascular regulation. *Can. J. Physiol. Pharmacol.* 2011; 89: 557-573.

[132] Ellis KL, Newton-Cheh C, Wang TJ, Frampton CM, Doughty RN, Whalley GA, Ellis CJ, Skelton L, Davis N, Yandle TG, Troughton RW, Richards AM, Cameron VA. Association of genetic variation in the natriuretic peptide system with cardiovascular outcomes. *J. Mol. Cell Cardiol.* 2011; 50:695-701.

[133] Das S, Au E, Krazit ST, Pandey KN. Targeted disruption of guanylyl cyclase-A/natriuretic peptide receptor-A gene provokes renal fibrosis and remodeling in null mutant mice: role of proinflammatory cytokines. *Endocrinology* 2010; 151:5841-5850.

[134] Ellmers LJ, Scott NJ, Piuhola J, Maeda N, Smithies O, Frampton CM, Richards AM, Cameron VA. Npr1-regulated gene pathways contributing to cardiac hypertrophy and fibrosis. *J. Mol. Endocrinol.* 2007; 38:245-257.

[135] Wadei HM, Textor SC. The role of the kidney in regulating arterial blood pressure. *Nat. Rev. Nephrol.* 2012; 8:602-609.

[136] Zhao D, Vellaichamy E, Somanna NK, Pandey KN. Guanylyl cyclase/natriuretic peptide receptor-A gene disruption causes increased adrenal angiotensin II and aldosterone levels. *Am. J. Physiol. Renal. Physiol.* 2007; 293:F121-127.

[137] Katholi RE, Couri DM. Left ventricular hypertrophy: major risk factor in patients with hypertension: update and practical clinical applications. *Int. J. Hypertens.* 2011; 2011: 495349.

In: Natriuretic Peptides
Editor: Kailash N. Pandey

ISBN: 978-1-63321-080-6

Chapter 9

Natriuretic Peptides: Predictive, Diagnostic and Prognostic Implications in Cardiovascular Diseases with Particular Regard to Acute Cardiovascular Events

Speranza Rubattu* and Massimo Volpe†

[1]Department of Clinical and Molecular Medicine, School of Medicine and Psychology, Sapienza University of Rome, Sant'Andrea Hospital, Rome, Italy
[2]IRCCS, Neuromed, Pozzilli, Italy

Abstract

The involvement in the complex control of haemodynamic homeostasis and of cardiovascular remodeling underlies the pivotal role played by natriuretic peptides (NPs) in several cardiovascular diseases (CVD), with particular regard to acute cardiovascular events.

A significant predictive role for CVD, including acute vascular events and death from cardiovascular causes, in apparently healthy subjects from the general population has been described in several epidemiological studies.

NPs levels also exert a significant prognostic role towards the occurrence of major adverse cardiovascular events in patients with both stable and unstable coronary atherosclerotic disease and in patients with cerebrovascular accidents.

In heart failure elevated circulating levels of NPs exert a key diagnostic role, and they also contribute to long-term prognostic stratification of patients. NPs levels are augmented in hypertension where they behave as a useful marker of future cardiovascular events, mainly related to end-organ damage.

In addition, an altered natriuretic peptide structure, i.e., molecular variants of atrial natriuretic peptide (ANP), has been shown to contribute to the individual susceptibility to stroke and myocardial infarction in both animal models and in humans. The mechanisms

* Speranza Rubattu, M.D. Phone: 0039 06 33775979, Fax: 0039 06 33775561, E-mail: rubattu.speranza@neuromed.it.
† Massimo Volpe, M.D. Phone: 0039 06 33775654, Fax: 0039 06 33775561, E-mail: massimo.volpe@uniroma1.it.

of vascular damage directly dependent from ANP molecular variants are being actively explored in the attempt to improve our understanding of vascular diseases and to obtain key information for novel therapeutic and preventive strategies towards CVD occurrence.

Of note, a potential application of NPs as therapeutic tool for the management of several CVD, including myocardial infarction, is currently under investigation with very promising results.

1. EFFECTS OF NPs ON CONTROL OF CARDIOVASCULAR HEMODYNAMICS AND REMODELING

Natriuretic peptides (NPs) play an important role in the regulation of electrolyte and water balance through their diuretic and natriuretic effects. They also inhibit the renin–angiotensin–aldosterone system and regulate the permeability of systemic vasculature. In addition, they inhibit the activity of the sympathetic nervous system. A negative inotropic effect of ANP has been described in normal left ventricular (LV) myocytes but not in hypertrophied myocytes [1].

The primary stimulus for ANP release is atrial stretch, whereas BNP release is mainly dependent on end-diastolic wall stress. Once secreted, NPs produce important systemic effects that antagonize volume overload.

In fact, as a result of the ANP and BNP actions, heart rate and blood pressure decrease. Thereby, NPs play a major role in the regulation of cardiorenal homeostasis [1].

The different stimuli for release indicate that ANP is mainly a modulator of acute hemodynamics and blood pressure, whereas BNP modulation is associated, a part from myocardial stretch, to a broader range of disease conditions, i.e., myocardial fibrosis, myocardial and peripheral ischemia, all underlying chronic left ventricular dysfunction [2].

Precursors of natriuretic peptides, NT-proANP and NT-proBNP, have larger molecular weight, longer half-life, and are easier to measure in plasma. Of note, NT-proBNP has longer half-life than NT-proANP, thus resulting in more stable plasma levels [3].

Neutral endopeptidase 24.11 (NEP), a zinc metalloprotease, is the most important enzyme responsible for *in vivo* ANP degradation. NEP inhibition leads to increased ANP levels with consequent increases in diuresis, natriuresis and augmented cGMP urinary excretion [4].

Due to the tight control of hemodynamic functions, NT-proNPs levels have shown diagnostic and prognostic implications in several cardiovascular diseases, particularly in the setting of heart failure [5] and of hypertension [6] (Figure 1).

Of note, NPs secretion is also augmented in the presence of left ventricular hypertrophy [6-9]. The latter phenomenon has long been interpreted as a consequence of enhanced expression by an increased cardiac mass [6-9]. However, a local direct antihypertrophic role of NPs has been more recently identified [10-13] and it has been related to the interaction with specific prohypertrophic signaling pathways [14].

In addition, new evidence suggests that NPs modulate cellular growth and proliferation of both endothelial and smooth muscle cells in the vascular wall. In human coronary arterial endothelial cells and in human umbilical vein endothelial cells, a low dose of ANP (10^{-11}

mol/l) induced an increase in cell numbers and [3H]thymidine incorporation (a marker of DNA synthesis), and an increase in cell migration [15].

The effects were mediated by an augmented expression of cGK and, subsequently, Akt and ERK1/2 (p44/42 MAPK) pathways. Based on these studies, it is evident that physiological concentrations of ANP promote endothelial regeneration and could therefore be useful in the regeneration of endothelial cells after injury in atherosclerosis.

On the other hand, supraphysiological levels of ANP exert an opposite effect. In fact, they reduce endothelial cell regeneration, DNA synthesis, and cell migration [5]. Moreover, ANP also exerts anti-inflammatory properties. An inhibitory effect on atherosclerosis *in vivo* has been documented [16].

An additional important implication in the control of vascular function has been provided by a study conducted in a relatively healthy sample population with a low prevalence of cardiovascular diseases (CVD). In this study, among seven biomarkers (including BNP, renin, aldosterone, PRC, PAI-1, and urine albumin–creatinine ratio) evaluated in parallel, NTpro-ANP turned out to be the strongest correlate of brachial artery endothelial function [17]. NT pro-ANP was also inversely related to baseline mean flow velocity.

Therefore, through its potent vasodilatory action, ANP appears to be the strongest contributor to endothelial function in the general population.

As a result of their antihypertrophic and antifibrotic properties, of their ability to control cellular growth and proliferation, as well as endothelial function, NPs are currently viewed as active players in the process of cardiovascular remodeling.

2. Predictive Role of Natriuretic Peptides Levels towards Cardiovascular Risk in the General Population

Based on the above reported evidence and on the overall cardiovascular actions of NPs, the issue of a potential usefulness of NT-proANP and NT-proBNP levels for screening asymptomatic individuals at cardiovascular risk from the general population represents an interesting matter of investigation [18] (Figure 1).

Few reports have so far explored the predictive role of NPs levels in asymptomatic individuals with consistent reproducible results. In the offspring study from the Framingham population, levels of NT-proANP and NT-proBNP significantly predicted mortality from cardiovascular causes (heart failure, atrial fibrillation, stroke) in apparently healthy subjects [19]. Moreover, NT-proBNP predicted major adverse cardiovascular events (MACE) and death in a population-based study in Copenaghen [20].

Cardiovascular risk, as estimated by two validated algorithms for risk stratification, i.e., the SCORE project chart and the Framingham algorithm, increased gradually across tertiles of NT-proANP in a cohort of adult males from Southern Italy [21], thus supporting the suitability of NT-proNP levels as markers of cardiovascular risk in asymptomatic individuals from the general population.

Furthermore, in a community-based cohort from US, free from heart failure, NT-proBNP emerged as a more robust cardiac biomarker able to predict cardiovascular mortality and morbidity independently from conventional clinical risk factors and cardiac structural abnormalities [22].

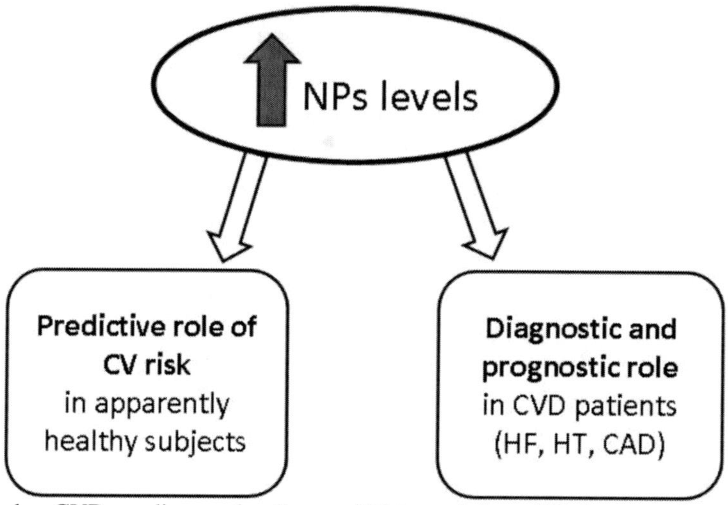

CV=cardiovascular; CVD=cardiovascular disease; HF=heart failure; HT=hypertension; CAD=coronary artery disease.

Figure 1. Schematic representation of clinical implications of elevated NPs levels in both the general population and in affected patients.

A meta-analysis conducted on 40 prospective studies by Di Angelantonio et al. demonstrated that NT-proBNP was a significant predictor of CVD (OR 2.82; CI 2.40-3.33), of stroke (OR 1.93; CI 1.58-2.37), and of coronary heart disease (OR 2.03; CI 1.54-2.66). However, assessment of either BNP or NT-proBNP levels in addition to measurement of conventional CVD risk factors (and other characteristics) yielded apparently only modest incremental improvement in risk discrimination for subsequent CVD [23].

In a study by Wannamethee et al. NT-proBNP levels turned out to be a stronger predictor than C-reactive protein in older men with and without pre-existing CVD. In this study, inclusion of NT-proBNP in a Framingham-based model yielded significant improvement in C-statistics in both men with and without CVD and resulted in a net reclassification improvement of 8.8% ($p = 0.0009$) and 8.2% ($p = 0.05$), respectively, for major CVD events [24].

In an effort to finally assess the true significance of NPs in risk prediction, a recent report from the Woscops study, a primary prevention study considering a cohort of hypercholesterolemic middle-aged men at moderate CVD risk, demonstrated that NT-proBNP, once added to conventional risk factors, almost doubled the incremental gain for prediction of CVD compared to C-reactive protein [25].

The improvement in discrimination remained significant even when excluding subjects with minor as well as with major cardiac abnormalities, thus providing a clear evidence of the valuable predictive performance of NPs among individuals free of overt CVD.

It is likely that NPs levels measurement will finally satisfy the criteria required for use as markers of CVR in apparently healthy individuals from general communities. Further large studies are still required to achieve this objective.

3. Prognostic Role of NPs towards Cardiovascular Outcome in Patients with Acute Cardiovascular Events and in Patients with Stable Atherosclerotic Disease

A release of NT-proANP is observed in patients with angiographically normal coronaries but dysfunctional endothelium [26]. NPs are significantly more expressed in human coronary explants of advanced atherosclerotic lesions in comparison with early atherosclerotic lesions [27]. Furthermore, well defined stages of atherosclerosis are characterized by progressive increases in NT-proANP and NT-proBNP levels, beginning with endothelial dysfunction and progressively more pronounced with moderate and severe coronary atherosclerosis, irrespective of the underlying myocardial disease [26]. Therefore, a tight link appears to exist between NPs and atherosclerotic vascular disease.

In a study by Squire et al. the prognostic value of NT-proNPs towards MACE occurrence was identified in subjects experiencing acute myocardial infarction (AMI) [28]. A cut off level of 509 pM and of 866 pM for NT-proANP and NT-proBNP levels, respectively, identified patients at increased risk of MACE with a higher significance for NT-proBNP. A Kaplan-Meyer curve, based on either NT-proANP or NT-proBNP or both being above the respective median value led to a significant prediction of future MACE [28]. Similar data were obtained in patients experiencing either AMI or stroke, and considering cut off levels of 850 pmol/L for NT-proANP and of 500 pmol/L for NT-proBNP [29].

Levels of NT-proNPs are able to predict cardiovascular outcome also in patients with stable atherosclerotic disease. In a 10-year follow up study performed in patients with ischemic heart disease, NT-proBNP levels belonging to the highest quartile were associated to 50% survival rate [30].

Moreover, the Claricor trial revealed that NT-proBNP levels belonging to 75-100 percentile were significantly associated with AMI, cardiovascular death and all cause death [31].

Similar prognostic value has been reported for NT-proANP levels in a study conducted in patients with stable ischemic heart disease. In particular, a cut off level of 4749 pmol/L was identified for significant prediction of death and AMI in this study [32].

Sabatine et al. recently investigated the prognostic implications of cardiovascular biomarkers (MR-proANP, MR-proADM, CT proET1, copeptin) in patients with stable ischemic heart disease [33]. In this study a multimarker model adjusted for clinical factors was created. By performing this approach, MR-proANP and MR-proADM proved to be the strongest two biomarkers, superior to NT-proBNP and cTnT measured with a highly sensitive assay, to identify patients at higher risk of cardiovascular death and heart failure.

Furthermore, elevated levels of MR-proANP, MR-proADM, Ct-proET-1 identified patients in whom, despite appearing to be at low risk clinically, therapy with an ACEI inhibitor resulted in a significant reduction in the risk of cardiovascular death or heart failure.

Thus, an important message of the latter study was that elevated levels of MR-proANP and MR-proADM in subjects that appear to be at lower risk clinically may be also useful to select patients who derive a significant benefit from ACEI therapy [33].

Thereby, clinical usefulness of NT-proNPs appears to be well established in both stable and unstable atherosclerotic diseases with particular regard to prognostic implications.

4. Diagnostic and Prognostic Implications of NPs in Heart Failure and Hypertension

A diagnostic relevance of NPs levels measurement has been well established in both acute and chronic heart failure. Both ANP and BNP levels increase in parallel with the degree of left ventricular dysfunction and hemodynamic stress. In chronic congestive heart failure (CHF), NPs levels represent useful markers to monitor the course of the disease, particularly in relation to the benefit of therapeutic strategies [34]. They also provide a reliable marker for long-term prognostic stratification of CHF. Subsequent measurements of B-type (BNP) or N-terminal-proBNP levels provide independent information regarding the risk for progression of disease across a wide spectrum of adverse outcomes: ventricular remodeling, malignant ventricular arrhythmias, hospitalization for HF, need for transplantation, and death [34, 35].

A diagnostic independent role of NPs levels has not been assessed in hypertension. On the other hand, few studies examined the significance of NPs in terms of prediction of cardiovascular risk in hypertensive patients. Higher NT-proBNP levels predicted mortality in hypertensives [36].

Midregional (MR)-NT-proANP levels were significantly associated with lower ankle-brachial index (ABI), higher urinary albumin-creatinine ratio (UACR), greater left ventricular (LV) mass index, independently of age, gender, body mass index (BMI), systolic blood pressure (SBP), glomerular filtration rate (GFR), smoking, diabetes, cholesterol levels, medication use, previous acute myocardial infarction (AMI) or stroke, particularly in African-Americans [37].

5. NPs Genetic Variants and Cardiovascular Risk

Several genetic variants of NPs are known. In particular, common genetic variants of the NPPA (ANP) and NPPB (BNP) genes, able to modulate ANP and BNP levels, such as the rs5068 SNP of NPPA gene, lead to elevated blood pressure levels and to increased risk of hypertension [38]. The role of NPs genes in favouring predisposition to hypertension, by modulating circulating levels of ANP and BNP, has been further supported by results from genome wide association studies performed in both Caucasian and Asian populations [38-40]. In fact, SNPs at the MTHFR-NPPB locus (which contains both NPPA and NPPB genes) are associated with opposite effects on ANP/BNP levels, consistent with a model in which the variants act through increased ANP/BNP production to lower blood pressure.

Therefore, cardiovascular risk can be influenced by genetic alterations of NPs contributing to hypertension. However, the effect of genetically modified NPs on cardiovascular risk can be also exerted in a blood pressure independent manner. In this regard, available evidence indicates that NPPA gene promoter variants associate directly to higher degree of left ventricular hypertrophy in hypertensive patients [41, 42]. Furthermore, the common T2238C ANP molecular variant induces endothelial, but not smooth muscle-dependent impaired vasorelaxation in young healthy human subjects [43]. Of note, the T2238C variant increased the risk of stroke and of AMI in different human cohorts [1], exerted a significant prognostic role on cardiovascular outcomes in patients with stable ischemic heart disease [44], and finally, it showed a significant gene-by-treatment interaction in a large cohort of hypertensive patients [45]. This variant, resulting from T-to-C nucleotide substitution in position 2238 of the gene, is due to a change from a stop to an arginine amino acid. The mechanisms of vascular disease promotion resulting from the structurally altered ANP molecule has been investigated in great details [43]. Thus, it was discovered that C2238/ ANP does not differently activate NPR-A, compared to wild type ANP, whereas it activates NPR-C receptor with a significantly higher affinity than wild type ANP. As a result of inappropriate NPR-C activation, it promotes endothelial cell damage *in vitro* and endothelial dysfunction *in vivo*, thus increasing the risk of cardiovascular diseases [43] (Figure 2).

The G664A variant within exon 1 has been associated to increased risk of cerebro-vascular events in the Physician's Health Study, although its mechanisms of action have never been explored [46, 47].

6. Contribution of Animal Models to Support the Role of NPs into Predisposition to Acute Cardiovascular Events

Genetically manipulated mice were used to demonstrate both the role of ANP as a determinant of salt-sensitive hypertension [48] and as a direct contributor to left ventricular hypertrophy [49, 50]. In fact, the mouse knock out for the natriuretic peptide type A receptor, the main receptor of ANP, confirmed the important contribution of ANP to left ventricular hypertrophy development, in a blood pressure independent manner [51]. Furthermore, genetic dissection of cerebrovascular accidents in hypertension was performed in the animal model of stroke-prone spontaneously hypertensive rat (SHRsp). By using a F2 hybrid population of rats derived from the crossbreeding of SHRsp and stroke resistant SHR (SHRsr) 3 quantitative trait loci (QTL) linked to stroke occurrence, independently of hypertension, were detected [52].

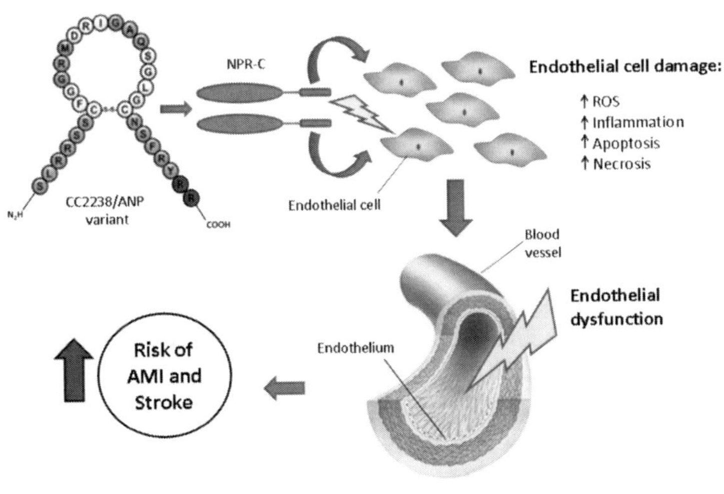

Figure 2. Schematic representation of the deleterious effects played by T2238C/ANP molecular variant in vascular endothelial cells. By activating NPR-C receptor, the CC2238/ANP mutant form promotes increased rate of inflammation, oxidative stress (ROS), apoptosis and necrosis ultimately leading to endothelial dysfunction and to increased risk of acute cardiovascular events (myocardial infarction and stroke).

The most relevant QTL mapped on rat chromosome 1, extended for about 42 cM, and exerted a 20% contribution to the stroke susceptibility phenotype. Transfer of either the whole or part of this QTL in a stroke resistant genetic configuration into the opposite genetic background led to decreased stroke susceptibility in the resulting congenic line. On the other hand, the opposite occurred in the reciprocal congenic line [53]. The gene/s responsible of stroke occurrence mapping within the chromosome 1 QTL have not been identified yet.

A second relevant QTL was identified on rat chromosome 5, and it contributed for about 10% of the overall genetic variance [52]. The gene encoding atrial natriuretic peptide (ANP) was revealed at the peak of linkage of this QTL. Both regulatory and coding mutations were described within the rat ANP gene sequence in the SHRsp parental line, with evidence of *in vitro* functional implications [54-56]. Most importantly, translation of data from the animal model to the human disease was performed. In fact, the relevance of ANP genetic variants in the occurrence of acute cardiovascular events has been reported in different human cohorts as mentioned above [42-45, 57, 58].

7. Therapeutic Approaches with the Use of NPs in Acute Myocardial Infarction

ANP can inhibit early activation of neurohormonal factors and inflammation after reperfusion therapy. It can also reverse arrhythmias, apoptosis of cardiac myocytes and endothelial cells, and limit infarct size and LV remodeling, thereby improving LV function in animal models with ischaemia/reperfusion injury [59].

The ability of ANP to prevent LV dilation and remodeling after primary coronary angioplasty in patients with AMI has been proven [60]. Patients received either ANP infusion or isosorbide dinitrate administration immediately after primary coronary angioplasty. The outcome after ANP infusion was superior to that obtained with isosorbide dinitrate. This finding is probably attributable to an increase in the myocardial uptake of norepinephrine by blockade of the renin–angiotensin–aldosterone system.

Moreover, through stimulation of ATP-sensitive potassium channel activity, ANP can favor an ischemic preconditioning effect during ischemia reperfusion and, by so doing, exert a cardioprotective effect against prolonged ischemia. Findings from a multicenter study also suggest that ANP can reduce subsequent re-hospitalization or death due to HF [61]. These cardioprotective effects of ANP were observed at a low dose of ANP without a large effect on systemic BP in patients with AMI.

Nesiritide, a recombinant form of human BNP, given soon after AMI, induced increments in plasma cyclic guanylate monophosphate (cGMP) and C-type natriuretic peptide (CNP) and decrements in other endogenous cardiac peptides with a neutral effect on renal function and a trend towards favourable ventricular remodeling [62].

Of note, few studies have shown that CNP also acts as an efficient anti-remodeling agent after AMI. In fact, serial infusions of CNP in rats after AMI were associated with reduction in LV dilatation, increase in systolic and diastolic performance, and reduction in ventricular fibrosis [1]. All mentioned studies reinforce the therapeutic potential of NPs in the treatment of AMI, and their beneficial effects on morbidity and mortality in post-MI patients.

To further pursue the achievement of NPs-derived pharmacological approaches for the treatment of CVD, including acute cardiovascular events, an alternative approach has been attempted by a strategy based on pharmacologically increased circulating ANP levels. Following the disappointing introduction of NEP inhibitors, simultaneously blocking angiotensin-converting enzyme (ACE) and responsible of frequent angioedema episodes [63, 64], a novel class of drugs (ARNi), that combine the action of both ARB and NEP inhibition, was recently introduced [65]. Among them, the LCZ696 compound provides a 1:1 ratio blockade of angiotensin type 1 receptor with a valsartan moiety and NEP inhibition with a prodrug moiety (AHU377) metabolized in the final LBQ657 molecule [65]. Its greater efficacy in controlling both sitting systolic and diastolic BP, as compared to valsartan, was reported in a randomized double-blind Phase II trial [66]. LCZ696 was well tolerated and no cases of angioedema were reported, not even in black subjects. Their application for the treatment of acute and chronic CVD and for cardiovascular risk reduction is currently under evaluation. Finally, the most recent innovation in the field of ANP-based therapy has been proposed by McKie et al. who designed a novel peptide, M-ANP, that appeared to be resistant to proteolytic degradation by NEP and possessed greater BP lowering, renal function-enhancing, and aldosterone suppressing properties than native ANP [67, 68]. In a canine model of acute hypertension, induced by continuous infusion of angiotensin II, a single intravenous infusion of M-ANP significantly lowered BP for up to 120 min. despite concomitant infusion of angiotensin II. The significant reduction in BP was associated with a significant decrease in systemic vascular resistance. In addition, there was a marked increase in diuresis and natriuresis, and aldosterone levels were significantly suppressed. Its suitability for the treatment of acute cardiovascular events deserves to be tested.

Conclusion

NPs represent important players in the occurrence of several CVD. In particular, both abnormal circulating levels and peptide structural alterations contribute to the risk of acute cardiovascular events. A part from being a simple marker of cardiac dysfunction due to their important role in cardiovascular homeostasis, they also participate directly into the processes of cardiovascular remodeling and into CVD pathogenesis. In fact, both circulating levels and peptide structural alterations serve as valid predictive, diagnostic and prognostic biomarkers for several CVD, including myocardial infarction and stroke. In addition, a potential application of NPs as therapeutic tool in the treatment of several CVD, including myocardial infarction, is being actively pursued with encouraging results.

References

[1] Rubattu, S., Sciarretta, S., Valenti, V., Stanzione, R., Volpe, M. Natriuretic peptides: an update on bioactivity, potential therapeutic use and implication in cardiovascular diseases. *Am. J. Hypertens.* 2008; 21: 733-741.

[2] Cataliotti, A., Chen, H. H., Redfield, M. M., Burnett, J. C. Jr. Natriuretic peptides as regulators of myocardial structure and function: pathophysiologic and therapeutic implications. *Heart Fail. Clin.* 2006; 2: 269-276.

[3] Potter, L. R., Yoder, R., Flora, D. R., Antos, L. K., Dickey, D. M. Natriuretic peptides: their structure, receptors, physiologic functions and therapeutic applications. *Handb Exp. Pharmacol.* 2009; 191: 341-366.

[4] Charles, C. J., Espiner, E. A., Nicholls, M. G., Richards, A. M., Yandle, T. G., Protter, A., Kosoglou, T. Clearance receptors and endopetidase 24.11: equal role in natriuretic peptide metabolism in conscious sheep. *Am. J. Physiol.* 1996; 271 (2 Pt 2): R373-R380.

[5] Schrier, R. W., Abraham, W. T. Hormones and hemodynamics in heart failure. *N. Engl. J. Med.* 1999; 341: 577–585.

[6] Hildebrandt, P., Boesen, M., Olsen, M., Wachtell, K., Groenning, B. N-terminal pro brain natriuretic peptide in arterial hypertension-a marker for left ventricular dimensions and prognosis. *Eur. J. Heart Fail.* 2004; 6: 313-317.

[7] Nishikimi, T., Yoshihara, F., Morimoto, A., Ishikawa, K., Ishimitsu, T., Saito, Y., Kangawa, K., Matsuo, H., Omae, T., Matsuoka, H. Relationship between left ventricular geometry and natriuretic peptide levels in essential hypertension. *Hypertension* 1996; 28:22–30.

[8] Cavallero, S., González, G. E., Puyó, A. M., Rosón, M. I., Pérez, S., Morales, C., Hertig, C. M., Gelpi, R. J., Fernández, B. E. Atrial natriuretic peptide behaviour and myocyte hypertrophic profile in combined pressure and volume-induced cardiac hypertrophy. *J. Hypertens.* 2007; 25:1940–1950.

[9] Qi, W., Mathisen, P., Kjekshus, J., Simonsen, S., Bjornerheim, R., Endresen, K., Hall, C. Natriuretic peptides in patients with aortic stenosis. *Am. Heart J.* 2001; 142:725–732.

[10] Kishimoto, I., Tokudome, T., Horio, T., Garbers, D. L., Nakao, K., Kangawa, K. Natriuretic peptide signaling via guanylyl cyclase (GC)-A: an endogenous protective mechanism of the heart. *Curr. Cardiol. Rev.* 2009; 5: 45–51.

[11] Kishimoto, I., Rossi, K., Garbers, D. L. A genetic model provides evidence that the receptor for atrial natriuretic peptide (guanylyl cyclase-A) inhibits cardiac ventricular myocyte hypertrophy. *Proc. Natl. Acad. Sci. US* 2001; 98: 2703–2706.

[12] Holtwick, R., van Eickels, M., Skryabin, B. V., Baba, H. A., Bubikat, A., Begrow, F., Schneider, M. D., Garbers, D. L., Kuhn, M. Pressure-independent cardiac hypertrophy in mice with cardiomyocyte-restricted inactivation of the atrial natriuretic peptide receptor guanylyl cyclase-A. *J. Clin. Invest.* 2003; 111:1399–1407.

[13] Knowles, J. W., Esposito, G., Mao, L., Hagaman, J. R., Fox, J. E., Smithies, O., Rockman, H. A., Maeda, N. Pressure-independent enhancement of cardiac hypertrophy in natriuretic peptide receptor A-deficient mice. *J. Clin. Invest.* 2001; 107: 975–984.

[14] Calvieri, C., Rubattu, S., Volpe, M. Molecular mechanisms underlying cardiac anti-hypertrophic and antifibrotic effects of natriuretic peptides. *J. Mol. Med.* 2012, 90: 5-13.

[15] Kook, H., Itoh, H., Choi, B. S., Sawada, N., Doi, K., Hwang, T. J., Kim, K. K., Arai, H., Baik, Y. H., Nakao, K. Physiological concentration of atrial natriuretic peptide induces endothelial regeneration *in vitro. Am. J. Physiol. Heart Circ. Physiol.* 2003; 284: H1388–H1397.

[16] Alexander, M. R., Knowles, J. W., Nishikimi, T., Maeda, N. Increased atherosclerosis and smooth muscle cell hypertrophy in natriuretic peptide receptor A−/− apolipoprotein E−/− mice. *Arterioscler. Thromb. Vasc. Biol.* 2003; 23:1077–1082.

[17] Kathiresan, S., Gona, P., Larson, M. G., Vita, J. A., Mitchell, G. F., Tofler, G. H., Levy, D., Newton-Cheh, C., Wang, T. J., Benjamin, E. J., Vasan, R. S. Cross-sectional relations of multiple biomarkers from distinct biological pathways to brachial artery endothelial function. *Circulation* 2006; 113: 938-945.

[18] Daniels, L. B. Natriuretic peptides and assessment of cardiovascular disease risk in asymptomatic persons. *Curr. Cardio. Risk Rep.* 2010; 4:120–127.

[19] Wang, T. J., Larson, M. G., Levy, D., Benjamin, E. J., Leip, E. P., Omland, T., Wolf, P. A., Vasan, R. S. Plasma natriuretic peptide levels and the risk of cardiovascular events and death. *N. Engl. J. Med.* 2004; 350: 655–663.

[20] Kistorp, C., Raymond, I., Pedersen, F., Gustafsson, F., Faber, J., Hildebrandt, P. N-terminal pro-brain natriuretic peptide, C-reactive protein, and urinary albumin levels as predictors of mortality and cardiovascular events in older adults. *JAMA* 2005; 293: 1609-1616.

[21] Barbato, A., Sciarretta, S., Marchitti, S., Icone, R., Di Castro, S., Stanzione, R., Cotugno, M., Ippolito, R., Palmieri, L., Calmieri, C., Battistoni, A., Volpe, M., Strazzullo, P., Rubattu, S. Aminoterminal natriuretic peptides and cardiovascular risk in an Italian male adult cohort. *Int. J. Cardiol.* 2011, 152: 245-246.

[22] McKie, P. M., Cataliotti, A., Sangaralingham, S. J., Ichiki, T., Cannone, V., Bailey, K. R., Redfiled, M. M., Rodeheffer, R. J., Burnett, J. C. Predictive utility of atrial, N-terminal pro-atrial, and N-terminal pro-B-type natriuretic peptides for mortality and cardiovascular events in the general community: a 9-year follow-up study. *Mayo Clin. Proc.* 2011; 86: 1154-1160.

[23] Di Angelantonio, E., Chowdhury, R., Sarwar, N., Ray, K. K., Gobin, R., Saleheen, D., Thompson, A., Gudnason, V., Sattar, N., Danesh, J. B-type natriuretic peptides and cardiovascular risk. Systematic review ans meat-analysis of 40 propective studies. *Circulation* 2009; 120: 2177-2187.

[24] Wannamethee, G., Welsh, P., Lowe, G. D., Gudnason, V., Di Angelantonio, E., Lennon, L., Rumley, A., Whincup, P. H., Sattar, N. N-terminal pro-brain natriuretic peptide is a more useful predictor of cardiovascular disease risk than C-reactive protein in older men with and without pre-existing cardiovascular disease. *J. Am. Coll. Cardiol.* 2011; 58: 56-64.

[25] Welsh, P., Doolin, O., Willeit, P., Packard, C., Macfarlane, P., Cobbe, S., Gudnason, V., Di Angelantonio, E., Ford, I., Sattar, N. N-terminal pro-B-type natriuretic peptide and the prediction of primary cardiovascular events: results from 15-year follow-up of WOSCOPS. *Eur. Heart J.* 2013; 34: 443-450.

[26] Casco, V. H., Veinot, J. P., Kuroski de Bold, M. L. K., Masters, R. G., Stevenson, M. M., de Bold, A. J. Natriuretic peptide system gene expression in human coronary arteries. *J. Histochem. Cytochem.* 2002; 50: 799–809.

[27] Barbato, E., Rubattu, S., Bartunek, J., Berni, A., Sarno, G., Vanderheyden, M., Delrue, L., Zardi, D., Pace, B., De Bruyne, B., Coluccia, R., Wijns, W., Volpe, M. Role of cardiac natriuretic peptides in human coronary atherosclerosis. *Atherosclerosis* 2009; 206: 258-264.

[28] Squire, I. B., O'Brien, R. J., Demme, B., Davies, J. E., Ng, L. L. N-terminal pro-atrial natriuretic peptide (N-ANP) and N-terminal pro-B-type natriuretic peptide (N-BNP) in the prediction of death and heart failure in unselected patients following acute myocardial infarction. *Cli. Sci.* 2004; 107: 309-316.

[29] Makikallio, A. M., Makikallio, T. H., Korpelainen, J. T., Vuolteenaho, O., Tapanainen, J. M., Ylitalo, K., Sotaniemi, K. A., Huikuri, H. V., Myllyla, V. V. Natriuretic peptides and mortality after stroke. *Stroke* 2005; 36: 1016–1020.

[30] Kragelund, C., Gronning, B., Kober, L., Hildebrandt, P., Steffensen, R. N-terminal pro B-type natriuretic peptide and long-term mortality in stable coronary heart disease. *New Engl. J. Med.* 2005; 352: 666-675.

[31] Harutyunyan, M. J., Mathiasen, A. B., Winkel, P., Gotze, J. P., Hansen, J. F., Hildebrandt, P., Jensen, J. B., Hilden, J., Jespersen, C. M., Kjoller, E., Kolmos, H. J., Gluud, C., Kastrup, J.; CLARICOR trial group. High-sensitivity C-reactive protein and N-terminal pro-B-type natriuretic peptide in patients with stable coronary heart disease: a prognostic study within the CLARICOR trial. *Scand. J. Cli. Lab. Invest.* 2011; 71: 52-62.

[32] Barbato, E., Bartunek, J., Marchitti, S., Mangiacapra, F., Stanzione, R., Delrue, L., Cotugno, M., Di Castro, S., De Bruyne, B., Wijns, W., Volpe, M., Rubattu, S. NT-proANP circulating level is a prognostic marker in stable ischemic heart disease. *Int. J. Cardiol.* 2012; 155: 311-312.

[33] Sabatine, M. S., Morrow, D. A., de Lemos, J. A., Omland, T., Sloan, S., Jarolim, P., Solomon, S. D., Pfeffer, M. A., Braunwald, E. Evaluation of multiple biomarkers of cardiovascular stress for risk prediction and guiding medical therapy in patients with stable coronary disease. *Circulation* 2012; 125: 233-240.

[34] Motiwala, S. R., Januzzi, J. L. The Role of Natriuretic Peptides as Biomarkers for Guiding the management of Chronic Heart Failure. *Clin. Pharm. Therap.* 2013; 93: 57-67.

[35] Volpe, M., Francia, P., Tocci, G., Rubattu, S., Cangianiello, S., Rao, M. A. E., Trimarco, B., Condorelli, M. Prediction of Long-Term Survival in Chronic Heart Failure by Multiple Biomarker Assessment: A 15-Year Prospective Follow-Up Study. *Clin. Cardiol.* 2010; 33:700-707.

[36] Paget, V., Legedz, L., Gaudebout, N., Girerd, N., Bricca, G., Milon, H., Vincent, M., Lantelme, P. N-terminal pro-brain natriuretic peptide. A powerful predictor of mortality in hypertension. *Hypertension* 2011; 57: 702-709.

[37] Kaleghi, M., Al-Omari, M. A., Kondragunta, V., Morgenthaler, N. G., Struck, J., Bergmann, A., Mosley, T. H., Kuoo, I. J. Relation of plasma midregional natriuretic peptide to target organ damage in adults with systemic hypertension. *Am. J. Cardiol.* 2009; 103: 1255-1260.

[38] The International Consortium for blood pressure Genome-Wide association studies. Genetic variants in novel pathways influence blood pressure and cardiovascular disease risk. *Nature* 2011; 478: 103-109.

[39] Newton-Cheh, C., Larson, M. G., Vasan, R. S., Levy, D., Bloch, K. D., Surti, A., Guiducci, C., Kathiresan, S., Benjamin, E. J., Struck, J., et al. Association of common variants in NPPA and NPPB with circulating natriuretic peptides and blood pressure. *Nat. Genet.* 2009; 41: 348-353.

[40] Kato, N., Takeuchi, F., Tabara, Y., Kelly, T. N., Go, M. J., Sim, X., Tay, W. T., Chen, C.-H., Zhang, Y., Yamamoto, K., et al. Meta-analysis of genome-wide association studies identifies common variants associated with blood pressure variation in east Asians. *Nat. Genet.* 2011; 43: 531-538.

[41] Rubattu, S., Bigatti, G., Evangelista, A., Lanzani, C., Stanzione, R., Zagato, L., Manunta, P., Marchitti, S., Venturelli, V., Bianchi, G., Volpe, M., Stella, P. Association of atrial natriuretic and type-A natriuretic peptide receptor gene polymorphisms with left ventricular mass in human essential hypertension. *J. Am. Coll. Cardiol.* 2006; 48; 499-505.

[42] Xue, H., Wang, S., Wang, H., Sun, K., Song, X., Zhang, W., Fu, C., Han, Y., Hui, R. Atrial natriuretic peptide gene promoter polymorphism is associated with left ventricular hypertrophy in hypertension. *Clin. Sci.* (Lond.) 2007; 114: 131–137.

[43] Sciarretta, S., Marchitti, S., Bianchi, F., Moyes, A., Barbato, E., Di Castro, S., Stanzione, R., Cotugno, M., Castello, L., Calvieri, C., Iberini, I., Sadoshima, J., Hobbs, A. J., Volpe, M., Rubattu, S. The C2238 atrial natriuretic peptide molecular variant is associated with endothelial damage and dysfunction through natriuretic peptide receptor C signaling. *Circ. Res.*, in press 2013.

[44] Barbato, E., Bartunek, J., Mangiacapra, F., Sciarretta, S., Stanzione, R., Delrue, L., Cotugno, M., Marchitti, S., Iaccarino, G., Sirico, G., Di Castro, S., Evangelista, A., Lambrechts, D., Sinnaeve, P., De Bruyne, B., Van De Werf, F., Janssens, S., Fox, K.-A.-A., Wijns, W., Volpe, M., Rubattu, S. Influence of rs5065 atrial natriuretic peptide gene variant on coronary artery disease. *J. Am. Coll. Cardiol.* 2012; 59: 1763-70.

[45] Lynch, A. I., Boerwinkle, E., Davis, B. R., Ford, C. E., Eckfeldt, J. H., Leiendecker-Foster, C., Arnett, D. K. Pharmacogenetic association of the NPPA T2238C genetic variant with cardiovascular disease outcomes in patients with hypertension. *JAMA.* 2008; 299: 296-307.

[46] Rubattu, S., Ridker, P. M., Stampfer, M., Volpe, M., Hennekens, C. H., Lindpaintner, K. The gene encoding atrial natriuretic peptide and the risk of human stroke. *Circulation* 1999; 100: 1722-1726.

[47] Lynch, A. I., Claas, S. A., Arnett, D. K. A review of the role of atrial natriuretic peptide gene polymorphisms in hypertension and its sequelae. *Curr. Hypertens. Rep.* 2009; 11: 35-42.

[48] John, S. W., Krege, J. H., Oliver, P. M., Hagaman, J. R., Hodgin, J. B., Pang, S. C., Flynn, T. G., Smithies, O. Genetic decreases in atrial natriuretic peptide and salt sensitive hypertension. *Science* 1995; 267: 679–681.

[49] Kishimoto, I., Rossi, K., Garbers, D. L. A genetic model provides evidence that the receptor for atrial natriuretic peptide (guanylyl cyclase-A) inhibits cardiac ventricular myocyte hypertrophy. *Proc. Natl. Acad. Sci. US* 2001; 98: 2703–2706.

[50] Knowles, J. W., Esposito, G., Mao, L., Hagaman, J. R., Fox, J. E., Smithies, O., Rockman, H. A., Maeda, N. Pressure-independent enhancement of cardiac hypertrophy in natriuretic peptide receptor A-deficient mice. *J. Clin. Invest.* 2001; 107: 975–984.

[51] Holtwick, R., van Eickels, M., Skryabin, B. V., Baba, H. A., Bubikat, A., Begrow, F., Schneider, M. D., Garbers, D. L., Kuhn, M. Pressure-independent cardiac hypertrophy in mice with cardiomyocyte-restricted inactivation of the atrial natriuretic peptide receptor guanylyl cyclase-A. *J. Clin. Invest.* 2003; 111: 1399–1407.

[52] Rubattu, S., Volpe, M., Kreutz, R., Ganten, U., Ganten, D., Lindpaintner, K. Chromosomal mapping of genetic loci contributing to stroke in an animal model of a complex human disease. *Nature Genetics* 1996; 13: 429-434.

[53] Rubattu, S., Hubner, N., Ganten, U., Evangelista, A., Stanzione, R., Di Angelantonio, E., Plehm, R., Langanki, R., Gianazza, E., Sironi, L., D'Amati, G., Volpe, M. Reciprocal congenic lines for a major stroke-QTL on rat chromosome 1. *Physiol. Genomics* 2006; 27: 108-113.

[54] Rubattu, S., Lee, M. A., De Paolis, P., Giliberti, R., Gigante, B., Lombardi, A., Volpe, M., Lindpaintner, K. Altered structure, regulation and function of the gene encoding atrial natriuretic peptide in the stroke-prone spontaneously hypertensive rat. *Circ. Res.* 1999; 85: 900-905.

[55] Rubattu, S., Giliberti, R., De Paolis, P., Stanzione, R., Spinsanti, P., Volpe, M. Effect of a regulatory mutation on the rat atrial natriuretic peptide gene transcription. *Peptides* 2002; 23: 555-560.

[56] De Paolis, P., Nobili, V., Lombardi, A., Tarasi, D., Barbato, D., Ganten, U., Brunetti, E., Volpe, M., Rubattu, S. Role of a molecular variant of the rat atrial natriuretic peptide gene on vascular remodeling. *Ann. Clin. Lab. Sci.* 2007; 37: 135-140.

[57] Rubattu, S., Stanzione, R., Di Angelantonio, E., Zanda, B., Evangelista, A., Tarasi, D., Gigante, B., Pirisi, A., Brunetti, E., Volpe, M. Atrial natriuretic peptide gene polymorphisms and risk of ischemic stroke in humans. *Stroke.* 2004; 35: 814-818.

[58] Gruchala, M., Ciecwierz, D., Wasag, B., Targonski, R., Dubaniewicz, W., Nowak, A., Sobiczewski, W., Ochman, K., Romanowski, P., Limon, J., Rynkiewicz, A. Association of the ScaI atrial natriuretic peptide gene polymorphism with nonfatal myocardial infarction and extent of coronary artery disease. *Am. Heart J.* 2003; 145:125-131.

[59] Kasama, S., Furuya, M., Toyama, T., Ichikawa, S., Kurabayashi, M. Effect of atrial natriuretic peptide on left ventricular remodelling in patients with acute myocardial infarction. *Eur. Heart J.* 2008; 29: 1485–1494.

[60] Kasama, S., Toyama, T., Hatori, T., Sumino, H., Kumakura, H., Takayama, Y., Ichikawa, S., Suzuki, T., Kurabayashi, M. Effects of intravenous atrial natriuretic peptide on cardiac sympathetic nerve activity and left ventricular remodeling in patients with first anterior acute myocardial infarction. *J. Am. Coll. Cardiol.* 2007; 49: 667–674.

[61] Kitakaze, M., Asakura, M., Kim, J., Shintani, Y., Asanuma, H., Hamasaki, T., Seguchi, O., Myoishi, M., Minamino, T., Ohara, T., Nagai, Y., Nanto, S., Watanabe, K., Fukuzawa, S., Hirayama, A., Nakamura, N., Kimura, K., Fujii, K., Ishihara, M., Saito, Y., Tomoike, H., Kitamura, S. J.-WIND investigators. Human atrial natriuretic peptide and nicorandil as adjuncts to reperfusion treatment for acute myocardial infarction (J-WIND): two randomised trials. *Lancet* 2007;370: 1483–1493.

[62] Chen, H. H., Martin, F. L., Gibbons, R. J., Schirger, J. A., Wright, R. S., Schears, R. M., Redfield, M. M., Simari, R. D., Lerman, A., Cataliotti, A., Burnett, J. C. Jr. Low-dose nesiritide in human anterior myocardial infarction suppresses aldosterone and preserves ventricular function and structure: a proof of concept study. *Heart* 2009; 95: 1315-1319.

[63] Burnett, J. C., Jr. Vasopeptidase inhibition: a new concept in blood pressure management. *J. Hypertens.* 1999; 17: S37-S43.

[64] Campbell, D. J. Vasopeptidase inhibition: a double-edged sword? *Hypertension* 2003; 41: 383-389.

[65] Gu, J., Noe, A., Chandra, P., Al-Fayoumi, S., Ligueros-Saylan, M., Sarangapani, R., Maahs, S., Ksander, G., Rigel, D. F., Jeng, A. Y., Kin, T. H., Zheng, W., Dole, W. P. Pharmacokinetics and pharmacodynamics of LCZ696, a novel dual-acting angiotensin receptor-neprilysin inhibitor (ARNi). *J. Clin. Pharmacol.* 2010; 50: 401-414.

[66] Ruilope, L. M., Dukat, A., Böhm, M., Lacourcière, Y., Gong, J., Lefkowitz, M. P. Blood-pressure reduction with LCZ696, a novel dual-acting inhibitor of the angiotensin II receptor and neprilysin: a randomised, double-blind, placebo-controlled, active comparator study. *Lancet* 2010; 375: 1255-1266.

[67] McKie, P. M., Cataliotti, A., Boerrigter, G., Chen, H. H., Sangaralingham, S. J., Martin, F. L., Ichiki, T., Burnett, J. C. Jr. A novel atrial natriuretic peptide based therapeutic in experimental angiotensin II mediated acute hypertension. *Hypertension* 2010; 56: 1152-1159.

[68] McKie, P. M., Ichiki, T., Burnett, J. C. Jr. M-atrial natriuretic peptide: a novel anti-hypertensive protein therapy. *Curr. Hypertens. Rep.* 2012; 14: 62-69.

In: Natriuretic Peptides
Editor: Kailash N. Pandey

ISBN: 978-1-63321-080-6

Chapter 10

NATRIURETIC PEPTIDES RECEPTORS: MOLECULAR BIOLOGY, PHISIOLOGY AND CLINICAL IMPLICATION

Daniele Masarone*[*], *Raffaella D'Alessandro*[†], *Andrea Buono*, *Rita Gravino*, *Alessandra Rea*, *Gemma Salerno*, *Ernesto Ammendola*, *Giuseppe Del Giorno*, *Lucio Santangelo*, *Antonello D'Andrea*, *Maria Giovanna Russo*, *Raffaele Calabrò*, *Giuseppe Limongelli and Giuseppe Pacileo

Department of Cardiology. Second University of Naples, Naples, Italy

ABSTRACT

Natriuretic peptides are a family of structurally related but genetically distinct hormones/paracrine/autocrine factors that regulate blood volume, blood pressure, ventricular hypertrophy, pulmonary hypertension, fat metabolism, and long bone growth.

The cellular responsiveness of natriuretic peptides is manifested through specific cell surface receptors widely spread in different target tissues.

In human three single membrane-spanning natriuretic peptide receptors have been identified. Two of this natriuretic peptide receptor type A and atriuretic peptide receptor type B, are transmembrane guanylyl cyclases, enzymes that catalyze the synthesis of cyclic guanosine monophosphate.

One natriuretic peptide receptor type C, lacks intrinsic enzymatic activity and controls the local concentrations of natriuretic peptides through constitutive receptor-mediated internalization and degradation.

Natriuretic peptide receptor type A has been studied widely. Its mode of activation by peptide ligands and mechanisms of regulation serve as prototypes for understanding the function of other particulate guanylyl cyclase.

---

[*] These two authors contributed equally for this work.

[†] These two authors contributed equally for this work.

Activation of this enzyme by its ligand is a complex process requiring oligomerization, ligand binding, kinase homology domain phosphorylation and adenosine triphosphate binding.

Finally gene knockout and genetic segregation studies have provided strong evidence for the importance of natriuretic peptide receptor type A in the regulation of blood pressure and heart and renal functions.

This chapter will focus on the physiology and molecular biology of natriuretic peptide receptor; also we discussing their role in cardiovascular disease.

1. Introduction

In 1981 De Bold e coll. showed that infusion of atrial tissue extracts in rats induced strong natriuresis [1]. Since this initial observation several groups identified peptides from atrial and ventricular tissues that possess both natriuretic and smooth muscle-relaxing activity [2-3].

These peptides are atrial natriuretic peptide (ANP), B-type natriuretic peptide (BNP) and C-type natriuretic peptide (CNP).

Natriuretic Peptides (NPs) are released in response to pressure and volume overload. They are able to induce arterial vasodilatation, natriuresis and diuresis and they reduce the activities of the renina-angiotensin-aldosterone system (RAAS) and the sympathetic nervous system. In other words their actions oppose the physiological abnormalities in heart failure (HF) which come from the neurohormonal activation. NPs are also called "myocyte stress hormones" because cardiac chambers' wall stress is the stimulus that started their synthesis.

The cellular responsiveness of natriuretic peptides (NPs) is manifested through specific cell surface receptors widely spread in different target tissues [4]. Three single membrane-spanning receptors have been characterized natriuretic peptide (NP) receptor type A (NPRA), NP receptor type B (NPRB), and NP receptor type C (NPRC).

This chapter will focus on the physiology and molecular biology of NPR; also we discussing the role of NPR in cardiovascular disease.

2. Natriuretic Peptides Receptor Structure and Molecular Biology

The discovery of the three NPs has led to the understanding that the control activity mediated by these peptides is complex and their signaling occurs via three distinct receptors: NPRA, NPRB, and NPRC.

Signaling pathways are different and receptors can be grouped into two classes:

- NPRA and NPRB have protein regions with guanylyl cyclase (GC) activity
- NPRC signaling is not dependent on GC activity.

ANP and BNP receptors share a common structure, which can be divided into three domains:

1. an extracellular domain (responsible of the binding of the natriuretic peptide)
2. a transmembrane domain
3. an intracellular domain (constituted of a kinase-like homology domain followed by the guanylyl cyclase domain) [5].

The receptors are not affine exclusively for one type of natriuretic peptide, but it is possible to observe a cross-binding among receptors and ligands:

- ANP can bind to both NPRA and NPRC
- BNP can bind to both NPRA and NPRC
- CNP can bind to both NPRB and NPRC [6].

The human NPRA gene is approximately 16 kilobase (kb), contains 22 exons and 21 introns, and is located on chromosome 1q21–22. The human NPRB gene spans about 16.5 kb, contains 22 exons, and is located on chromosome 9p21–12. Both NPRA and NPRB promoter lack a defined TATA box; but putative Sp1 binding sites have been found.

The human NPRC gene is located on chromosome 5p14-p13, spans more than 65 kb, and contains eight exons and seven introns [7].

Most of the activities of the NPs are mediated by NPRA through the conversion of guanosine triphosphate (GTP) in cyclic guanosine monophosphate (cGMP).

The synthesis of cGMP results from ANP and BNP binding NPRA and CNP binding NPRB. Such a binding in the extracellular portion of the receptors results in conformation variation of the intracellular domain which, in turn, determines an increase in the activation of the GC-coupled receptors.

The production of cGMP stimulates the activity of cGMP-dependent protein kinases (PKGs), cGMP-gated ion channels (CGN), and cGMP-regulated cyclic nucleotide phosphodiesterases [8].

Studies to investigate what is the mechanism leading to the activation of the signaling cascade have been performed on the NPRA receptor (most studies have been focused on that receptor investigation). Results about the involvement of the kinase-homology domain are discordant and do not confirm definitively the necessity of that protein sequence to activate the receptor [9].

GC activation may be caused by ANP-mediated receptor dimerization, possibly in the region comprised between the kinase homology domain (KHD) and GC catalytic domain [10].

Dimerization mechanism has been proposed also to cause a tail-to-tail or a head-to-head dimerization for NPRA and NPRC, respectively. The second one could represent the inactive state and the first one the active form [11].

The presence of the adenosine triphosphate (ATP) and its interaction with KHD is reported to increase the efficacy of the receptor and the production of cGMP; the interaction with the receptor occurs because it contains a glycine-rich nucleotide-binding motif and is suggested to provide ATP-regulatory module which promotes the signaling process. So, ATP can have a role in regulating the GC activity [12-13].

Studies on the NPRA has revealed that once the ligand has recognized the target receptor, the complex composed of ANP/NPRA is internalized in the cell and then processed in a cellular compartment (possibly in lysosomes). This process allows the down regulation of the expression of the receptors (Figure1) [14].

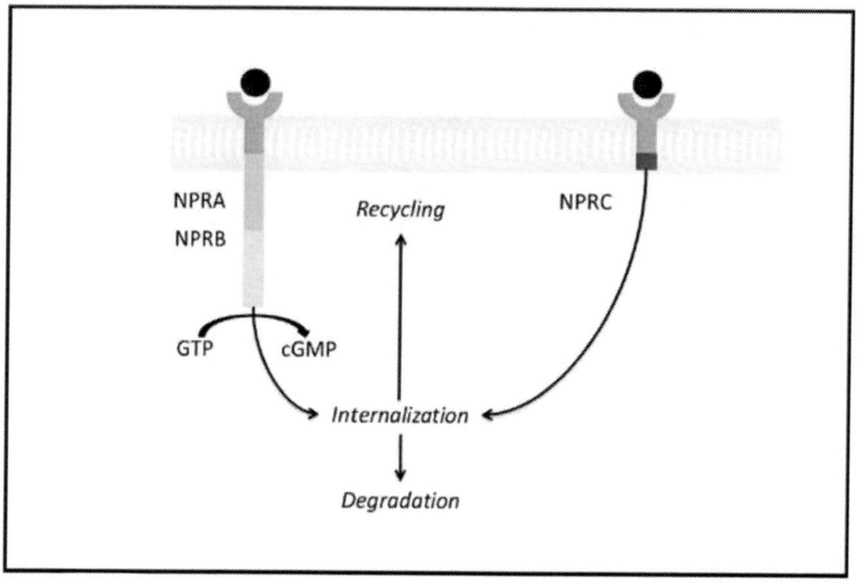

Figure 1. Processing and recycling of Natriuretic Peptides receptors. NPRA: Natriuretic Peptides Receptor-A. NPRB: Natriuretic Peptides Receptor-B. NPRC: Natriuretic Peptides Receptor-C. GTP: Guanosin Triphophate. cGMP: cyclic Guanosin Monophosphate.

The protein region involved in the internalization seems to be the KHD domain [15].

Supporting the hypothesis that internalization of the receptors is a mechanism of control of NPs signaling is the finding of the recycling process of a region of NPRC.

Similarly to NPRA, also NPRC is translocated into the cell after the binding with NPs, but the mechanism seems to be mediated by a receptor-mediated endocytosis process [16].

About the state of phosphorylation of the receptors, different studies have suggested various influences of ligand binding. Investigations about NPRA have showed a reduction of phosphate content and of the GC activity, but an induction of the phosphorylation in the GC-coupled protein [17-18]; anyway, more recent studies result in the observation of an increase in NPRA activation when the natriuretic peptide is bound to the receptor [19].

Data on molecular activation of NPRA has been collected in previous years.

In absence of natriuretic peptide, the receptor has a homodimer or homotetramer structure, is phosphorylated, its GC activity is not present. Ligand binding is not the sufficient signal to receptor activation: receptor oligomerization following the structural variation induced by NPs does not lead to receptor activation [20]. ANP receptor identification was carried out through the specific binding of ^{125}I-ANP to plasma membranes of several tissues and cells. Different ANP receptors have been identified with molecular weight ranging from 60–180 kiloDalton (kDa) (Table1).

Subsequent researches revealed the presence of high affinity binding sites for ANP in 120–180 kDa, presenting also GC activity. Conversely, 70 kDa ANP receptor did not present GC activity. Full length ANP allowed the identification of a specific amino acid sequence (Asn-Phe-Arg-Tyr) located at the carboxyl-terminus of ANP molecule. Smaller residues lacking that specific sequence, did not showed the same catalytic activity. Molecular studies through cloning and expression analysis of complementary DNA (cDNA) led to the identification and characterization of the peptide sequence of three different subtypes of natriuretic peptide receptors. Based on their molecular properties, a separated description can be reported for NPRA/NPRB and for NPRC receptors. NPRA and NPRB structural features revealed the presence of four distinct domains: a single transmembrane spanning region, an extracellular ligand-binding domain and the intracellular KHD and GC catalytic domain.

Table 1. Different type of Natriuretic peptide receptor type A

Molecular weight	Tissue	Reference
125 kDa	bovine adrenal cortex	Misono KS, Grammer RT, Rigby JW, et al. Photoaffinity labeling of atrial natriuretic factor receptor in bovine and rat adrenal cortical membranes. *Biochem. Biophys. Res. Commun.* 1985: 130:994–1001.
135–180 kDa	kidney cortex, Leydig tumor and neuroblastoma cells	Yip CC, Laing LP, Flynn TG. Photoaffinity labeling of atrial natriuretic factor receptors of rat kidney cortex plasma membranes. *J. Biol. Chem.* 1985:260:8229–32. Pandey KN, Inagami T, Misono KS. Atrial natriuretic factor receptor on cultured Leydig tumor cells: ligand binding and photoaffinity labeling. *Biochemistry* 1986:25:8467–8472. Pandey KN, Inagami T, Misono KS. Three distinct forms of atrial natriuretic factor receptors: kidney tubular epithelium cells and vascular smooth muscle cells contain different types of receptors. *Biochem. Biophys. Res. Commun.* 1987:147:1146–52.
60–70 kDa	various plasma membrane preparations and intact cultured cells	Hirose S, Akiyama F, Shinjo M, et al. Solubilization and molecular weight estimation of atrial natriuretic factor receptor from bovine adrenal cortex. *Biochem. Biophys. Res. Commun.* 1985:130:574–579. Meloche S, Ong H, Cantin M, et al. Affinity cross-linking of atrial natriuretic factor to its receptor in bovine adrenal zona glomerulosa. *J. Biol. Chem.* 1986:261:1525–1528. Vandlen RL, Arcuri KE, Napier MA. Identification of a receptor for atrial natriuretic factor in rabbit aorta membranes by affinity cross-linking. *J. Biol. Chem.* 1985:260:10889–10892. Schenk DB, Phelps MN, Porter JG, et al. Identification of the receptor for atrial natriuretic factor on cultured vascular cells. *J. Biol. Chem.* 1985:260:14887– 47890.

The comparisons of the amino acid sequences show a 62% identity among NPRA and NPRB proteins. Intracellular regions are more similar (78% identity residues) than extracellular domains of these two receptors (43% identity). NPRA and NPRB are also known as GC-A and GC-B, respectively. Extracellular ligand-binding domains of NPRA and NPRB share 44% of sequence identity. The GC catalytic domain is composed of 250-amino acids located at the carboxyl-terminus region. The KHD is constituted of about 280-amino acids, structurally following the transmembrane region.

The two receptors share a higher sequence identity in the GC catalytic region, reaching a percent value of 88%, KHD regions have a 60% of sequence identity. Some differences in receptor sequences have also been found.

NPRA contains putative glycosylation sites and while these have not been recognized in NPRB intracellular domain.

Cysteine residues are present in the extracellular domain of both NPRA and NPRB and are located in the similar positions; in this domain N-liked glycosylation sites are partially conserved.

The kinase domain comparison revealed a lower identity percent. NPRC presents a large extracellular domain of 496-amino acids, a single transmembrane domain, and a very short 37-amino acid cytoplasmic tail that bears no homology to any other known receptor protein.

NPRC lacks the GC catalytic domain.

The extracellular domain of the third receptor shares a 30% of identity to the other ones.

Four cysteine residues, domain potential signals for N-linked glycosylation and several serines and threonines are present in the extracellular for O-glycosylation sites.

A function for clearance has been recognized for NPRC receptor [21-22].

Molecular mechanisms of action have been recognized for NPRA.

Two ECD monomers can bind one ligand molecule, leading to the formation of a 2:1 complex.

The ligation of ANP seems to be mediated by the presence of a chloride atom occurs, possibly through a reversible bound mechanism. The dependence of ANP binding on chloride presence has been confirmed by *in vitro* studies. The ANP-receptor complex induces changes in the quaternary structure of the receptor, reorienting the intracellular domain, leading GC domains to the active configuration [23]. Different protein structures correspond to different molecular properties.

NPRA is activated by the three natriuretic peptides in the following order: ANP, BNP and CNP.

In NPRA, studies have revealed that residues Asn-2, Asn-13, Asn-180, Asn-306, Asn-347, and Asn-395 of NPRA are glycosylated; although the role of protein glycosylation is not completely fulfilled, some indication for an effect on ligand binding has also been suggested.

Phosphorylation on NPRA residues occurs on four serines (Ser-497, Ser-502, Ser-506, and Ser-510) and two threonines (Thr-500 and Thr-513) located at the amino-terminal portion of its kinase homology domain.

Other than natriuretic peptides, NPR-A is reported to bind heat shock proteins 70 and 90 (HSP70 and HSP90), molecular chaperones required for proper protein folding.

Mutagenesis studies suggest that five of the seven extracellular asparagines.

In NPRB are glycosylated leading to a 90% loss in CNP binding (mutation of Asn-24 to Asp), probably due to improper folding or cellular targeting of the receptor.

NPR-B is phosphorylated on Thr-513, Thr-516, Ser-518, Ser- 523, and Ser-526.

NPR-B is activated by CNP, ANP and BNP in the reported order.

The extracellular domain of human NPR-C is glycosylated on Asn-41, Asn-248, and Asn-349; the affinity order of NPR-C for natriuretic peptides is ANP, CNP and BNP [24].

In fine a different distribution of NPR is recognized (Table 2).

NPR-A and/or its mRNA is expressed in kidney glomerulus, adrenal zona glomerulosa, pituitary, cerebellum, and endocardial endothelial cells of the right and left atrium and right ventricle [25].

Table 2. Distribution of natriuretic peptide receptor. NPRA: Natriuretic Peptides Receptor-A. NPRB: Natriuretic Peptides Receptor-B. NPRC: Natriuretic Peptides Receptor-C

Type of NPR	Tissue	Reference
NPRA	kidney glomerulus, adrenal zona glomerulosa, pituitary, cerebellum, endocardial endothelial cells of the right and left atrium and right ventricle	Goy MF, Oliver PM, Purdy KE, et al. Evidence for a novel natriuretic peptide receptor that prefers brain natriuretic peptide over atrial natriuretic peptide. *Biochem. J.* 2001: 358*:379–387.* Mukhopadhyay AK, Speth RC, Guidone G et al. Spatiotemporal regulation of the two atrial natriuretic peptide receptors in testis. *Endocrinology* 2004 :145:1392–1401. Takayanagi R, Snajdar RM, Imada T, et al. Purification and characterization of two types of atrial natriuretic factor receptors from bovine adrenal cortex: guanylate cyclase-linked *Int. J. Bioch.* 1997:22:8-17.
NPRB	bone, brain, fibroblast, heart, kidney, lung, ovarian, uterine, and vascular smooth muscle tissue	Wilcox JN, Augustine A, Goeddel DV, et al. 1991 Differential regional expression of three natriuretic peptide receptor genes within primate tissues. *Mol. Cell Biol.* 1991: 11:3454–3462. Chrisman TD, Garbers DL. Reciprocal antagonism coordinates C-type natriuretic peptide and mitogen-signaling pathways in fibroblasts. *J. Biol. Chem.* 199:274:4293–4299.
NPRC	adrenal, brain, heart, kidney, lung, mesentery, placenta, and vascular smooth muscle tissue	Nagase M, Katafuchi T, Hirose S, et al. Tissue distribution and localization of natriuretic peptide receptor subtypes in stroke-prone spontaneously hypertensive rats. *J. Hypertens.* 1997:15:1235–1243. Porter JG, Arfsten A, Fuller F, et al. Isolation and functional expression of the human atrial natriuretic peptide clearance receptor cDNA. *Biochem. Biophys. Res. Commun.* 1990:171:796–803.

NPR-B and/or its mRNA is expressed in bone, brain, fibroblast, heart, kidney, lung, ovarian, uterine, and vascular smooth muscle tissue [28].

NPR-C and/or its mRNA is expressed in adrenal, brain, heart, kidney, lung, mesentery, placenta, and vascular smooth muscle tissue [29].

3. Natriuretic Peptides Receptor Signaling

cGMP is the intracellular second messenger that mediates the action of NPs. They activate membrane-bound guanylyl cyclase (GC), to produce cGMP [30]. Then cGMP exerts its physiological action activating three types of effector molecules: cGMP-dependent protein kinase (PKG), cGMP-regulated phosphodiesterases (PDEs) and cGMP-gated cation channels, among which PKG and PDEs might be the primary mediator, predominating in the cardiovascular system. The third type of effector molecule, cGMP-gated cation channels, exists in retinal and olfactory neuroepithelium and nephrons, but neither protein expression nor physiological function of these channels have been established in the cardiovascular system. cGMP-PKG regulates a broad array of physiologic processes in the cardiovascular system, acting on vascular smooth muscle cells, vascular endothelium and directly on cardiac myocytes, and modulating cardiac function, hypertrophy/remodeling and cell survival response [31]. Next to NPs pathway, cGMP's synthesis is regulated by nitritic oxide (NO), that express its function through a similar pathway, consisting of different isoenzymes of the same family of kinases and phosphodiesterases involved in NPs signaling way [32]. In fact the biosynthesis of cGMP from GTP is catalyzed by two different isoforms of guanylyl cyclase, one which functions as the biosensor for nitric oxide and the other, as the plasma membrane receptor for natriuretic peptides.

Once generated, cGMP interacts with PKGs and PDEs, amplifying the signal due to the involvement of other molecules downstream cGMP-dependent protein kinases. So far, three isotypes of cGMP-dependent protein kinases have been identified, two of which are splice variants of a single gene [33]. The prominent isotype in the cardiovascular system is PKG type I (PKG-I), which consists of an α and a β isoform. Its concentration is very high (>0.1μM) in vascular smooth muscle cells (both PKG-Iα and PKG-Iβ) and endothelial cells (PKG-Iβ) and lower in cardiac myocytes (PKG-Iα). Instead PKG type II (PKG-II) is mainly expressed in the kidney, brain, and intestine. Both PKG-I and PKG-II exist as homodimers with identical structures. Each subunit consists of three functional domains: an N-terminal domain, a regulatory domain, and a kinase domain [34]. PKG homodimerization is mediated by the N-terminal domain, which also suppresses kinase domain activity in the absence of cGMP, and interacts with target substrate proteins. Amino-terminal modifications, acetylation in the case of PKG-I and myristoylation in the case of PKG-II, appear to control intracellular localization and hence function of PKG. One difference between PKG-Iα and PKG-Iβ isoforms regards their N-terminus domain, which also regulates cooperativity between the cGMP-binding sites of PKG-Iα [35]. Consequently, PKG-Iβ requires ten-fold higher concentrations of cGMP for kinase activation than does PKG-Iα, witnessing the main role of this isoform The first step, after cGMP binding to specific sites in the regulatory domain, is a conformational change of PKG, resulting in the release of the N-terminus inhibition of the kinase domain. Once activated, the kinase domain is then ready to express its action: it catalyzes the phosphorylation of a serine/threonine side chain of the target substrate proteins. Several protein targets undergo the phosphorylation by PKG-I, including phospholamban (PLB), vasodilatory-stimulated phosphoprotein (VASP), regulator of G protein signaling 2 (RGS2), and the L-type calcium channel, representing the last mediators of the transductional cascade, being directly responsible of the effects generated by the binding hormone-receptor, in different tissues.

PDEs comprise a 21-gene super family categorized into 11 isoenzymes (PDE1-PDE11) with a total of 48 isoforms. The main action is to break the phosphodiester bond in cGMP (and/or cAMP) resulting in the linear GMP (or AMP), playing an important role in cGMP catabolism. The various isoenzymes are differentially expressed in tissues, are selective or non-selective for the two cyclic nucleotides, and can be activated or inhibited by them as well. In particular, cGMP regulates the activity of PDE-2, -3, -5, and -9. Of these, PDE-2, -3, and -5 are known to be expressed in cardiac myocytes, PDE-3 and -5 are expressed in vascular smooth muscle cells, and PDE-2, -3, and -5 are expressed in vascular endothelial cells [36-37]. While expressed in the heart, the cardiovascular role of PDE9 remains unknown at present [38]. The action of cGMP is expressed thanks to a simultaneously activation of some PDEs isoforms, such as PDE2 and PDE5, and inhibition of other ones, like PDE3 isoform. The balance of that system is the basis of a proper transmission of the intracellular signal. So, cGMP selectively activates PDE2 and PDE5 by binding to regulatory GAF domains in the N-terminus site. Three different N-terminal splice variants (PDE2A1-3) have been identified. PDE2A3 is the human variant and thought to be membrane associated. PDE2 hydrolyzes both cAMP and cGMP. Upon cGMP-GAF binding, PDE2 undergoes a conformational change and increases its enzymatic activity for cAMP [39].

Its dual substrate specificity allows it to mediate negative cross-talk between the cGMP and cAMP signaling pathways. In fact, through PDE2 activation, cGMP is able to reduce the cAMP signal and affect cardiac function. Moreover, recent studies suggest PDE2 could be involved into regulation of the L-type calcium channel in cardiac myocytes [40].

Unlike PDE2, PDE5 selectively hydrolyzes cGMP, thanks to the high affinity for cGMP.

Localization of PDE5 is mainly cytosolic, with particular localization in the cardiac myocyte Z bands, suggesting an association with other Z band-localized proteins such as PKG. In fact, PKG is able to enhance catalytic activity of PDE5 by phosphorylation, at a S92, which stabilizes cGMP binding. PKA can also phosphorylate PDE5, particularly when cGMP levels are high and cGMP is already bound to PDE5. This provides a positive feedback mechanism that can be initially triggered by cGMP synthesis [41-42].

Known as regulator of vascular smooth muscle contraction through regulation of cGMP, with high levels of expression in the lung and corpus cavernosum, more recently, several studies have demonstrated PDE5 was implicated also in cardiovascular system as an important regulator of cGMP in cardiac myocytes, including the hypertrophic response to pressure-overload stress, cell survival signaling and apoptosis associated with ischemia/reperfusion [43], and doxorubicin toxicity [44].

One study performed by Castro et al. [45] has established the specificity of PDE2 and PDE5 for the distinct NP-cGMP and NO-cGMP pools: NP-induced cGMP pool is subplasmalemmal and specifically hydrolyzed by the PDE2 isoform. In contrast, NO-induced cGMP pool appears to be more cytosolic, not localized to the subplasmalemmal region, and is specifically hydrolyzed by PDE5 isoform.

Like PDE2, PDE3 is also a dual substrate enzyme and can hydrolyze both cAMP and cGMP with relatively high affinities. But, unlike PDE2, the Vmax of PDE3 for cAMP is 10-fold greater than for cGMP, so the enzyme largely targets cAMP and can be competitively inhibited by cGMP. Otherwise, at low cGMP levels, PDE3 may have a larger role in controlling cGMP levels.

In the cardiovascular system, PDE3 is involved in regulating cardiac myocyte and vascular smooth muscle contractility as well as vascular smooth muscle phenotype switch and stress response. PDE3 regulates cardiac myocyte contractility and relaxation respectively via L-type calcium channels [46-47] and the cardiac sarcoplasmic reticulum calcium pump (SERCA2) [48]. In addition to the above mentioned PDE isoforms (PDE-2, -3, and -5), PDE1 and PDE9, two other PDE isoforms, are also involved into the hydrolyzazion of cGMP, expressing a relevant role in the cardiovascular system. PDE1 has dual specificity for cAMP and cGMP catabolism.

PDE1 isoforms are cytosolic, calcium- and calmodulin-dependent phosphodiesterases.

It is also expressed in cardiac myocytes, and recent studies have found it plays an important role in both cAMP and cGMP hydrolysis in vitro in human myocardium, though its physiologic role remains unknown [49].

Instead, PDE9 is expressed in brain and heart, exists in both cytosol and cell nucleus, and has the highest affinity for cGMP of all the cGMP-hydrolyzing PDEs. Its function remains unknown and is an area of active investigation.

3.1. cGMP Regulation of Cardiovascular System

The effects of cGMP signal pathway are expressed both in vascular (action on vascular smooth muscle cells and vascular endothelium) and cardiac (cardiomyocytes) system (Figure 2). cGMP signaling is able to direct vascular tone and smooth muscle cell proliferation and differentiation [50-51]. Vascular tone is regulated by changes in intracellular free calcium concentrations within smooth muscle cells (SMC), so the reduction in intracellular calcium concentration results in vasorelaxation. cGMP regulates the cytosolic free calcium level within vascular SMC through several mechanisms: inhibition of IP3-mediated calcium release from intracellular stores; removal and sequestration of intracellular calcium through calcium pump mechanisms; and both direct and indirect inhibition of the influx of extracellular calcium through voltage-gated calcium channels. The inhibition of IP3-mediated calcium release is obtained thanks to a phosphorylation mediated by PKG-Iβ of both the IP3 receptor (IP3R) and the IP3R-associated PKG-I substrate (IRAG). This modify is responsible in the decreased calcium release from the sarcoplasmic and endoplasmic reticulum [52-53].

PKG-Iα guarantees the modulation of the amount of intracellular calcium also by phosphorylating phospholamban, an important modulator of calcium transport of sarcoplasmic reticulum, in vascular smooth muscle cells and cardiac myocytes [54]. Phosphorylated phospholamban is unable to inhibit SERCA2 and the result is that calcium is taken up into the endoplasmic reticulum, reducing intracellular free calcium levels. Similarly, the decrease of cytoplasmic calcium concentration derived from cGMP-PKG mediated phosphorylation of the plasmalemmal calcium ATPase pump, that stimulates the extrusion of calcium [55].

Ultimately, PKG also phosphorylates the large conductance, calcium-activated potassium channel, thereby opening the channel, hyperpolarizing the cell membrane, and closing a number of other channels, including the L-type calcium channel.

To complete the role on vascular tone regulation, recent evidences seems to suggest that cGMP-PKG signaling modulates vascular tone also by altering calcium sensitization and thin filament interaction. In fact, PKG-Iα phosphorylates the myosin-binding subunit (MBS) of MLC phosphatase, thus activating the catalytic subunit of the phosphatase [56-57].

Dephosphorylation of MLC reduces the calcium sensitivity of the smooth muscle contractile machinery to inhibit smooth muscle contraction. Vasodilatory-stimulated phosphoprotein (VASP) and the 20-kDa heat shock-related protein (HSP20) are other two target substrates of PKG. These two thin filament-actin binding proteins are phosphorylated by PKG, with a resultant decreased ability in binding to actin filaments [58], and associated with vasorelaxation [59].

Another important role of cGMP-PKG signaling in vascular smooth muscle cells is its regulation of proliferation and differentiation. cGMP-PKG activation directs an anti-proliferative signal, down regulating mRNA levels of cyclins-A, D1, and E, and vascular endothelial growth factor (VEGF, under hypoxic conditions) and upregulating mRNA levels of MAPK phosphatase-1 (MKP-1) and p16 (gene encoding Cdk inhibitor 2A).

In vascular endothelial cells, cGMP-PKG signaling regulates cell motility, migration, and proliferation, all of which are vital to angiogenesis, and vascular permeability. As in vascular smooth muscle cells, the physiological actions of cGMP in vascular endothelial cells are determined by the downstream phosphorylation target substrates of PKG.

As demonstrated by Chen et al. RNA-mediated knockdown of NPR-A, PKG, or VASP all prevent ANP-induced endothelial tube formation, demonstrating the importance of ANP-mediated cGMP-signaling in endothelial function [60].

In the end, cGMP signaling has been shown to improve endothelial barrier function and protect against vascular injury in the systemic [61] and pulmonary vasculature [62].

But the effect of cGMP signal acts also directly on cardiac myocyte, negatively modulating contractility and hypertrophy and mediates apoptosis.

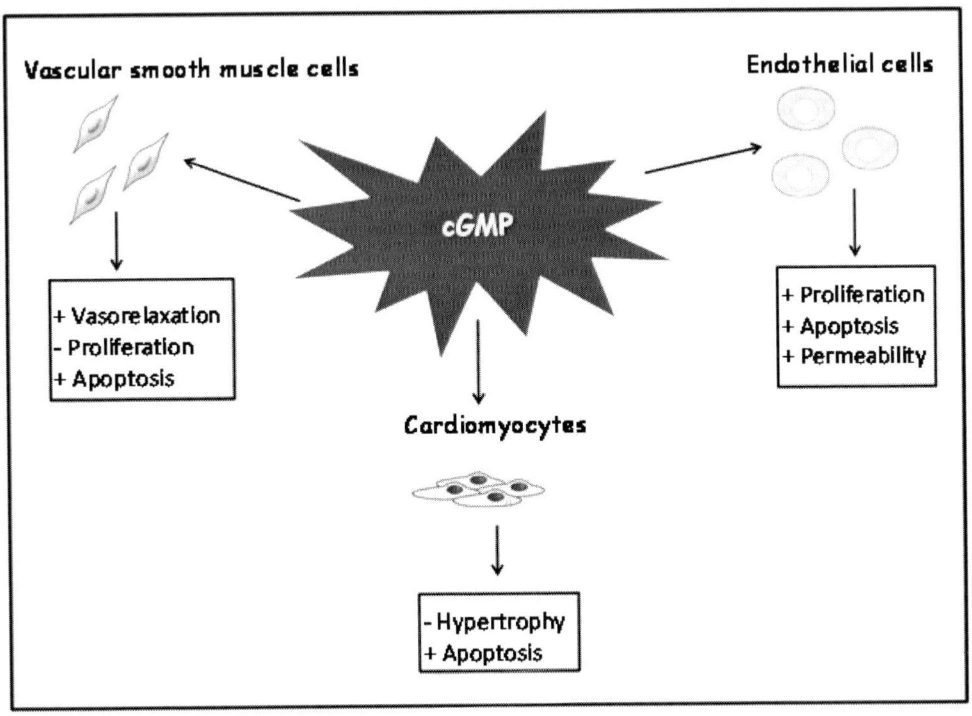

Figure 2. Effects of cGMP on cardiovascular system.

The role of cGMP-PKG signaling in cardiac contractility was determined by studies of isolated myocytes from conventional and cardiac-specific PKG-I knockout mice [63]. Whereas cGMP analogues reduced the force of contraction in electrically stimulated myocardium from wild-type control mice, they had no effect on the force of contraction in myocardium from conventional or cardiac-specific PKG-I knockout mice.

The exact role of cGMP-PKG activation in protecting against hypertrophic signaling remains to be confirmed. That cGMP has an antihypertrophic cardiac effect was initially suggested by the exacerbation of cardiac hypertrophy in various knockout mice lacking elements of the cGMP signaling pathway.

The cardiac hypertrophic response involves a complex web of signaling pathways. The best studied downstream mechanism by which cGMP-PKG signaling exerts its antihypertrophic action is the inhibition, at multiple levels, of the calcium-calcineurin-NFAT pathway [64]. Calcium activation of calcineurin results in the dephosphorylation and nuclear translocation of cytoplasmic NFAT (nuclear factor of activated T-cells), resulting in expression of a hypertrophic gene program. cGMP-PKG signaling may reduce intracellular calcium concentrations to inhibit calcineurin activation, though other mechanisms appear to apply as well.

cGMP signal seems to be also at the basis of the antifibrogenic effects of natriuretic peptides in the heart. Recently, ANP-cGMP-PKG activation was shown to inhibit transforming growth factor (TGF)-β1–induced extracellular matrix expression in cardiac fibroblasts via phosphorylation of Smad3 [65].

Ultimately, it has been reported a pro-apoptotic effects of cGMP in vascular smooth muscle cells, endothelial cells, and cardiac myocytes and appear to be mediated by PKG [66]. The downstream mechanisms have not been fully determined, but in vascular cells they may involve activation of c-Jun N-terminal kinase (JNK) and/or the phosphorylation and inactivation of β-catenin. Instead, in cardiac myocytes, cGMP decreases the mRNA levels of the anti-apoptotic Bcl-2 homologue Mcl-1 [67].

4. Clinical Implication of NPs Receptors

4.1. NPs Receptor in Hypertension

Since many physiological systems are involved in blood pressure control, it is plausible to think that a mutation in any of the genes involved in these systems is potentially responsible for the development of hypertension. However, hypertension is a complex and heterogeneous pathology with variable impact of genetic components and strong influence of the environment [68].

As the natriuretic peptides have profound cardiovascular and renal effects, clinical studies of their regulation and effects in hypertension have received early interest [69].

It has been shown that ANP infusion for short time periods in patients with mild hypertension produces enhanced increases of plasma cGMP but normal natriuresis and diuresis while in genetically hypertensive rats (SHR), normal cGMP elevation is accompanied by exaggerated diuresis and natriuresis [70]. On the other hand, prolonged infusion of low ANP doses in mildly hypertensive patients leads to exaggerated diuretic and natriuretic responses but with normal cGMP elevations [71]. Genetic mouse models with disruption of both ANP and NPRA genes have provided strong support for the role of NPR system in the regulation of arterial pressure and other physiological functions. The involvement of NPRA in hypertension is further supported by studies showing that knock-out mice lacking GC-A develop hypertension, which is not altered by salt intake [72], Therefore, the genetic defects that reduce the activity of ANP and its receptor system can be considered as candidate contributors to essential hypertension. Interestingly, complete absence of NPR-A causes hypertension in mice and leads to altered renin and ANG II levels, cardiac hypertrophy, and lethal vascular events similar to those seen in untreated human hypertensive patients [73]. In contrast, increased expression of NPRA (corresponding to the increasing number of *Npr1* gene copies) reduces blood pressure and increases the second messenger cGMP [74].

The mechanistic role of ANP/NPRA system in counter-acting the pathophysiology of hypertension is still not well understood. However the studies in both SHR and GC-A gene-knockout hypertensive mouse models suggest that in hypertension, both kidney and circulatory renin concentrations are decreased, and as a compensatory event, the adrenal renin is increased.

4.2. NPR in Congestive Heart Failure

HF is the most widely recognized pathology with alteration of the NPs system. Vasoconstriction, fluid and salt retention are the hallmarks of HF syndromes, and are associated with high plasma NPs levels. In fact increased levels of circulating ANP and BNP are hallmarks of HF in the clinic and are routinely used to diagnose heart failure. Despite the increased levels of natriuretic peptides in the failing circulation, the physiological responses to these agents are blunted. Several theories have been proposed to explain this resistance in the kidney. One hypothesis is that there is increased local renal ANP degradation by neutral endopeptidases (NEPs). However, NEPs inhibitor treatment of severe HF in dogs was unable to maintain natriuresis [75]. Another explanation for the blunted response to ANP and BNP is increased NPRC expression. However, blocking ANP binding to NPR-C with a specific ligand did not alter renal cGMP production or sodium excretion in dogs with severe CHF.

The hypothesis currently favour is that NPR is inactivated through down regulation in target organ of NPs such renal and pulmonary tissue.

In a mice model of HF, Brian e co-worker demonstrate that renal ANP-dependent guanylyl cyclase activity is reduced; also since mRNA levels are elevated in tissues where NPR-A activity is decreased, the authors proposed that increased degradation, not reduced synthesis, of NPR-A likely explains the down regulation.

In addition GC activity in type I and type II pneumocytes has been shown to be reduced in early hamster CHF, suggesting that lung NPR are implicated in heart failure [76].

A study of 200–230-day-old hamsters (advanced stage of the disease) revealed that lungs NPR are down regulated [77].

The significance of pulmonary NPR down regulation in pulmonary tissue is not clear. Physiological concentrations of ANP may protect the lungs from chemically induced pulmonary edema [78-79], also this anti-edematous action is thought to be a receptor-mediated mechanism involving cGMP and positive correlation has also been found between cGMP levels and the anti-edematous action of ANP [80]. Whether the down regulation of pulmonary NPR makes the lungs more vulnerable to edema formation remains to be determined.

Otherwise the down regulation of pulmonary receptors in HF could also contribute to the decreased pulmonary clearance of NPs observed in dogs with chronic HF [81].

However at day the role of NPR in the development of HF is not well understood and better understanding of cross-talk with other signalling systems is required.

While other avenues of exploration are to be contemplated, we are now at the beginning of its pharmacotherapeutic implications.

Conclusion

Over the past 25 years, natriuretic peptides and their cognate receptors were discovered and purified, and the genes encoding these proteins were cloned and "knocked out." Additional studies investigated the structure and regulation of the individual participants in these signal transduction pathways; ultimately yielding a tremendous body of literature on natriuretic peptide-dependent regulation of physiological and pathophysiological processes. During the next decade, we anticipate that further discoveries will be made regarding the molecular nature of these pathways as well as their clinical applications. Regarding the former, structural information on the guanylyl cyclase and kinase homology domains of NPR-A and NPR-B would be informative, as would be the identity of molecules such as kinases and phosphatases that regulate these receptors. On the clinical side, we expect the discovery of small molecule activators or inhibitors (antagonist) of these pathways that may be used to treat diseases like systemic hypertension, and heart failure, as well as diseases yet to be associated with these pleiotropic signaling systems.

References

[1] De Bold AJ, Borenstein HB, Veress AT, Sonnenberg H. A rapid and potent natriuretic response to intravenous injection of atrial myocardial extract in rats. *Life Sci.* 1981; 28:89-94.

[2] Misono KS, Grammer RT, Fukumi H, Inagami T. Rat atrial natriuretic factor: isolation, structure and biological activities of four major peptides. *Biochem. Biophys. Res. Commun.* 1984; 123:444-51.

[3] Kangawa K, Tawaragi Y, Oikawa S, et al. Identification of rat gamma atrial natriuretic polypeptide and characterization of the cDNA encoding its precursor. *Nature* 1984; 312:152-5.

[4] Pandey KN. Biology of natriuretic peptides and their receptors. *Peptides* 2005; 26:901-32.

[5] Garbers DL. Guanylyl Cyclase Receptors and Their Endocrine, Paracrine, and Autocrine Ligands. *Cell* 1992; 71:1-4.

[6] Drewett JG, Garbers DL. The Family of Guanylyl Cyclase Receptors and Their Ligands. *Endocr. Rev.* 1994; 15:135-62.

[7] Potter LR, Abbey-Hosch S, Dickey DM. Natriuretic peptides, their receptors, and cyclic guanosine monophosphate-dependent signaling functions. *Endocr. Rev.* 2006; 27:47-72.
[8] Lincoln TM, Cornwell TL. Intracellular Cyclic-Gmp Receptor Proteins. *Faseb. J.* 1993; 7:328-38.
[9] Duda T, Goraczniak RM, Sharma RK. The Glycine Residue of Atp Regulatory Module in Receptor Guanylate Cyclases That Is Essential in Natriuretic Factor Signaling. *Febs. Lett.* 1993; 335:309-14.
[10] Garbers DL, Lowe DG. Guanylyl Cyclase Receptors. *J. Biol. Chem.* 1994;269:30741-4.
[11] van den Akker F. Structural insights into the ligand binding domains of membrane bound guanylyl cyclases and natriuretic peptide receptors. *J. Mol. Biol.* 2001; 311: 923-37.
[12] Chinkers M, Singh S, Garbers DL. Adenine-Nucleotides Are Required for Activation of Rat Atrial-Natriuretic-Peptide Receptor Guanylyl Cyclase Expressed in a Baculovirus System. *J. Biol. Chem.* 1991; 266:4088-93.
[13] Duda T, Goraczniak RM, Sharma RK. Core Sequence of Atp Regulatory Module in Receptor Guanylate Cyclases. *Febs. Lett.* 1993; 315:143-8.
[14] Rathinavelu A, Isom GE. Differential Internalization and Processing of Atrial-Natriuretic-Factor B-Receptors and C-Receptors in Pc12 Cells. *Biochem. J.* 1991; 276: 493-7.
[15] Pandey KN, Nguyen HT, Li M, Boyle JW. Natriuretic peptide receptor-A negatively regulates mitogen-activated protein kinase and proliferation of mesangial cells: Role of cGMP-dependent protein kinase. *Biochem. Bioph. Res. Co.* 2000; 271:374-9.
[16] Pandey KN. Kinetic-Analysis of Internalization, Recycling and Redistribution of Atrial-Natriuretic-Factor Receptor Complex in Cultured Vascular Smooth-Muscle Cells - Ligand-Dependent Receptor down-Regulation. *Biochem. J.* 1992; 288:55-61.
[17] Potter LR, Garbers DL. Dephosphorylation of the Guanylyl Cyclase-a Receptor Causes Desensitization. *J. Biol. Chem.* 1992; 267:14531-4.
[18] Ballermann BJ, Marala RB, Sharma RK. Characterization and Regulation by Protein Kinase-C of Renal Glomerular Atrial Natriuretic Peptide Receptor-Coupled Guanylate-Cyclase. *Biochem. Bioph. Res. Co.* 1988; 157:755-61.
[19] Foster DC, Garbers DL. Dual role for adenine nucleotides in the regulation of the atrial natriuretic peptide receptor, guanylyl cyclase-A. *J. Biol. Chem.* 1998; 273:16311-8.
[20] Potter LR, Hunter T. Guanylyl cyclase-linked natriuretic peptide receptors: Structure and regulation. *J. Biol. Chem.* 2001; 276:6057-60.
[21] Pandey KN. Biology of natriuretic peptides and their receptors. *Peptides* 2005; 26:901-32.
[22] Pandey KN. The functional genomics of guanylyl cyclase/natriuretic peptide receptor-A: Perspectives and paradigms. *Febs. J.* 2011; 278:1792-807.
[23] Misono KS, Philo JS, Arakawa T, et al. Structure, signaling mechanism and regulation of the natriuretic peptide receptor guanylate cyclase. *Febs. J.* 2011; 278:1818-29.
[24] Potter LR, Abbey-Hosch S, Dickey DM. Natriuretic peptides, their receptors, and cyclic guanosine monophosphate-dependent signaling functions. *Endocr. Rev.* 2006; 27:47-72.
[25] Kalra PR, Anker SD, Coats AJS. Water and sodium regulation in chronic heart failure: the role of natriuretic peptides and vasopressin. *Cardiovasc. Res.* 2001; 51:495-509.
[26] Wong SKF, Garbers DL. Receptor Guanylyl Cyclases. *J. Clin. Invest.* 1992; 90:299-305.

[27] Drewett JG, Garbers DL. The Family of Guanylyl Cyclase Receptors and Their Ligands. *Endocr. Rev.* 1994; 15:135-62.

[28] van den Akker F. Structural insights into the ligand binding domains of membrane bound guanylyl cyclases and natriuretic peptide receptors. *J. Mol. Biol.* 2001; 311: 923-37.

[29] Gerzer R, Witzgall H, Tremblay J, Gutkowska J, Hamet P. Rapid Increase in Plasma and Urinary Cyclic-Gmp after Bolus Injection of Atrial Natriuretic Factor in Man. *J. Clin. Endocr. Metab.* 1985; 61:1217-9.

[30] Tremblay J, Desjardins R, Hum D, Gutkowska J, Hamet P. Biochemistry and physiology of the natriuretic peptide receptor guanylyl cyclases. *Mol. Cell Biochem.* 2002; 230:31-47.

[31] Tsai EJ, Kass DA. Cyclic GMP signaling in cardiovascular pathophysiology and therapeutics. *Pharmacol. Therapeut.* 2009; 122:216-38.

[32] Lugnier C, Keravis T, Eckly-Michel A. Cross talk between NO and cyclic-nucleotide phosphodiesterases in the modulation of signal transduction in blood vessel. *J. Physiol. Pharmacol.* 1999; 50:639-52.

[33] Conti M, Beavo J. Biochemistry and physiology of cyclic nucleotide Phosphocliesterases: Essential components in cyclic nucleotide signaling. *Annu. Rev. Biochem.* 2007; 76:481-511.

[34] Hofmann F, Feil R, Kleppisch T, Schlossmann J. Function of cGMP-dependent protein kinases as revealed by gene deletion. *Physiol. Rev.* 2006; 86:1-23.

[35] Das A, Smolenski A, Lohmann SM, Kukreja RC. Cyclic GMP-dependent protein kinase I alpha attenuates necrosis and apoptosis following ischemia/reoxygenation in adult cardiomyocyte. *J. Biol. Chem.* 2006; 281:38644-52.

[36] Maurice DH. Cyclic nucleotide phosphodiesterase-mediated integration of cGMP and camp signaling in cells of the cardiovascular system. *Front. Biosci.* 2005; 10:1221-8.

[37] Netherton SJ, Maurice DH. Vascular endothelial cell cyclic nucleotide phosphodiesterases and regulated cell migration: Implications in angiogenesis. *Mol. Pharmacol.* 2005; 67:263-72.

[38] Osadchii OE. Myocardial phosphodiesterases and regulation of cardiac contractility in health and cardiac disease. *Cardiovasc. Drug Ther.* 2007; 21:171-94.

[39] Martins TJ, Mumby MC, Beavo JA. Purification and characterization of a cyclic GMP-stimulated cyclic nucleotide phosphodiesterase from bovine tissues. *J. Biol. Chem.* 1982; 257:1973-9.

[40] Nagayama T, Zhang M, Hsu S, Takimoto E, Kass DA. Sustained soluble guanylate cyclase stimulation offsets nitric-oxide synthase inhibition to restore acute cardiac modulation by sildenafil. *J. Pharmacol. Exp. Ther.* 2008; 326:380-7.

[41] Zhang M, Koitabashi N, Nagayama T, et al. Expression, activity, and pro-hypertrophic effects of PDE5A in cardiac myocytes. *Cell Signal* 2008; 20:2231-6.

[42] Borlaug BA, Melenovsky V, Marhin T, Fitzgerald P, Kass DA. Sildenafil inhibits beta-adrenergic-stimulated cardiac contractility in humans. *Circulation* 2005; 112:2642-9.

[43] Salloum FN, Abbate A, Das A, et al. Sildenafil (Viagra) attenuates ischemic cardiomyopathy and improves left ventricular function in mice. *Am. J. Physiol. Heart Circ. Physiol.* 2008; 294:H1398-406.

[44] Fisher PW, Salloum F, Das A, Hyder H, Kukreja RC. Phosphodiesterase-5 inhibition with sildenafil attenuates cardiomyocyte apoptosis and left ventricular dysfunction in a chronic model of doxorubicin cardiotoxicity. *Circulation* 2005; 111:1601-10.

[45] Castro LR, Verde I, Cooper DM, Fischmeister R. Cyclic guanosine monophosphate compartmentation in rat cardiac myocytes. *Circulation* 2006; 113:2221-8.

[46] Jurevicius J, Skeberdis VA, Fischmeister R. Role of cyclic nucleotide phosphodiesterase isoforms in cAMP compartmentation following beta2-adrenergic stimulation of ICa,L in frog ventricular myocytes. *J. Physiol.* 2003; 551:239-52.

[47] Malecot CO, Bers DM, Katzung BG. Biphasic Contractions Induced by Milrinone at Low-Temperature in Ferret Ventricular Muscle - Role of the Sarcoplasmic-Reticulum and Transmembrane Calcium Influx. *Circ. Res.* 1986; 59:151-62.

[48] Gaide MS, Fitterman WS, Wiggins JR, Myerburg RJ, Cameron JS, Bassett AL. Amrinone relaxes potassium-induced contracture of failing right ventricular muscle of cats. *J. Cardiovasc. Pharmacol.* 1983; 5:335-40.

[49] Vandeput F, Wolda SL, Krall J, et al. Cyclic nucleotide phosphodiesterase PDE1C1 in human cardiac myocytes. *J. Biol. Chem.* 2007; 282:32749-57.

[50] Munzel T, Feil R, Mulsch A, Lohmann SM, Hofmann F, Walter U. Physiology and pathophysiology of vascular signaling controlled by guanosine 3',5'-cyclic monophosphate-dependent protein kinase [corrected]. *Circulation* 2003; 108:2172-83.

[51] Murad F, Rapoport RM, Fiscus R. Role of cyclic-GMP in relaxations of vascular smooth muscle. *J. Cardiovasc. Pharmacol.* 1985; 7 Suppl 3:S111-8.

[52] Ammendola A, Geiselhoringer A, Hofmann F, Schlossmann J. Molecular determinants of the interaction between the inositol 1,4,5-trisphosphate receptor-associated cGMP kinase substrate (IRAG) and cGMP kinase I beta. *J. Biol. Chem.* 2001; 276:24153-9.

[53] Schlossmann J, Ammendola A, Ashman K, et al. Regulation of intracellular calcium by a signalling complex of IRAG, IP3 receptor and cGMP kinase I beta. *Nature* 2000; 404:197-201.

[54] Raeymaekers L, Hofmann F, Casteels R. Cyclic Gmp-Dependent Protein-Kinase Phosphorylates Phospholamban in Isolated Sarcoplasmic-Reticulum from Cardiac and Smooth-Muscle. *Biochem. J.* 1988; 252:269-73.

[55] Eggermont JA, Vrolix M, Wuytack F, Raeymaekers L, Casteels R. The (Ca-2+-Mg-2+)-Atpases of the Plasma-Membrane and of the Endoplasmic-Reticulum in Smooth-Muscle Cells and Their Regulation. *J. Cardiovasc. Pharm.* 1988; 12:S51-S5.

[56] Nakamura M, Ichikawa K, Ito M, et al. Effects of the phosphorylation of myosin phosphatase by cyclic GMP-dependent protein kinase. *Cellular Signalling* 1999; 11:671-6.

[57] Surks HK, Mendelsohn ME. Dimerization of cGMP-dependent protein kinase 1 alpha and the myosin-binding subunit of myosin phosphatase: role of leucine zipper domains. *Cellular Signalling* 2003; 15:937-44.

[58] Reinhard M, Giehl K, Abel K, et al. The Proline-Rich Focal Adhesion and Microfilament Protein Vasp Is a Ligand for Profilins. *Embo J.* 1995; 14:1583-9.

[59] Beall AC, Kato K, Goldenring JR, Rasmussen H, Brophy CM. Cyclic nucleotide-dependent vasorelaxation is associated with the phosphorylation of a small heat shock-related protein. *J. Biol. Chem.* 1997; 272:11283-7.

[60] Chen HJ, Levine YC, Golan DE, Michel T, Lin AJ. Atrial natriuretic peptide-initiated cGMP pathways regulate vasodilator-stimulated phosphoprotein phosphorylation and angiogenesis in vascular endothelium. *J. Biol. Chem.* 2008; 283:4439-47.

[61] Furst R, Bubik MF, Bihari P, et al. Atrial natriuretic peptide protects against histamine-induced endothelial barrier dysfunction *in vivo*. *Mol. Pharmacol.* 2008; 74:1-8.

[62] Brovkovych V, Gao XP, Ong E, et al. Augmented inducible nitric oxide synthase expression and increased NO production reduce sepsis-induced lung injury and mortality in myeloperoxidase-null mice. *Am. J. Physiol.-Lung C* 2008; 295:L96-L103.

[63] Wegener JW, Nawrath H, Wolfsgruber W, et al. cGMP-dependent protein kinase I mediates the negative inotropic effect of cGMP in the murine myocardium. *Circ. Res.* 2002; 90:18-20.

[64] Fiedler B, Lohmann SM, Smolenski A, et al. Inhibition of calcineurin-NFAT hypertrophy signaling by cGMP-dependent protein kinase type I in cardiac myocytes. *P. Natl. Acad. Sci. USA* 2002; 99:11363-8.

[65] Li P, Wang DJ, Lucas J, et al. Atrial natriuretic peptide inhibits transforming growth factor beta-induced Smad signaling and myofibroblast transformation in mouse cardiac fibroblasts. *Circ. Res.* 2008; 102:185-92.

[66] Chiche JD, Schlutsmeyer SM, Bloch DB, et al. Adenovirus-mediated gene transfer of cGMP-dependent protein kinase increases the sensitivity of cultured vascular smooth muscle cells to the antiproliferative and pro-apoptotic effects of nitric oxide cGMP. *J. Biol. Chem.* 1998; 273:34263-71.

[67] Wu CF, Bishopric NH, Pratt RE. Atrial natriuretic peptide induces apoptosis in neonatal rat cardiac myocytes. *J. Biol. Chem.* 1997; 272:14860-6.

[68] Hamet P, Pausova Z, Adarichev V, Adaricheva K, Tremblay J. Hypertension: Genes and environment. *J. Hypertens.* 1998; 16:397-418.

[69] Cusson JR, Tremblay J, Larochelle P, Schiffrin EL, Gutkowska J, Hamet P. Clinical relationships of cyclic GMP. *Adv. Pharmacol.* 1994; 26:305-19.

[70] Pang SC, Hoang MC, Tremblay J, et al. Effect of Natural and Synthetic Atrial Natriuretic Factor on Arterial Blood-Pressure, Natriuresis and Cyclic-Gmp Excretion in Spontaneously Hypertensive Rats. *Clin. Sci.* 1985; 69:721-6.

[71] Cusson JR, Thibault G, Kuchel O, Hamet P, Cantin M, Larochelle P. Cardiovascular, Renal and Endocrine Responses to Low-Doses of Atrial Natriuretic Factor in Mild Essential-Hypertension. *J. Hum. Hypertens.* 1989; 3:89-96.

[72] Lopez MJ, Wong SKF, Kishimoto I, et al. Salt-Resistant Hypertension in Mice Lacking the Guanylyl Cyclase-a Receptor for Atrial-Natriuretic-Peptide. *Nature* 1995; 378:65-8.

[73] Swithers SE, Stewart RE, Mccarty R. Binding-Sites for Atrial-Natriuretic-Factor (Anf) in Kidneys and Adrenal-Glands of Spontaneously Hypertensive (Shr) Rats. *Life Sciences* 1987; 40:1673-81.

[74] Takayanagi R, Imada T, Grammer RT, Misono KS, Naruse M, Inagami T. Atrial-Natriuretic-Factor in Spontaneously Hypertensive Rats - Concentration Changes with the Progression of Hypertension and Elevated Formation of Cyclic-Gmp. *J. Hypertens.* 1986; 4:S303-S7.

[75] Chen HH, Schirger JA, Chau WL, et al. Renal response to acute neutral endopeptidase inhibition in mild and severe experimental heart failure. Circulation 1999;100:2443-8.

[76] Vagnetti D, Secca T, Roscani C, Santarella B, P DN. Atrial natriuretic factor stimulation of particulate guanylate cyclase in the alveolar zone of hamster lung: A cytochemical study. *Med. Sci. Res.* 1991; 19:47-8.

[77] Mukaddam-Daher S, Tremblay J, Fujio N, et al. Alteration of lung atrial natriuretic peptide receptors in genetic cardiomyopathy. *Am. J. Physiol.* 1996; 271:L38-45.

[78] Imamura T, Ohnuma N, Iwasa F, et al. Protective effect of alpha-human atrial natriuretic polypeptide (alpha-hANP) on chemical-induced pulmonary edema. *Life Sci.* 1988; 42:403-14.

[79] Lofton CE, Newman WH, Currie MG. Atrial-Natriuretic-Peptide Regulation of Endothelial Permeability Is Mediated by Cgmp. *Biochem. Bioph. Res. Co* 1990; 172:793-9.

[80] Perreault T, Gutkowska J. Role of Atrial-Natriuretic-Factor in Lung Physiology and Pathology. *Am. J. Resp. Crit. Care* 1995; 151:226-42.

[81] Brandt RR, Redfield MM, Aarhus LL, Lewicki JA, Burnett JC. Clearance Receptor-Mediated Control of Atrial-Natriuretic-Factor in Experimental Congestive-Heart-Failure. *American Journal of Physiology* 1994; 266:R936-R43.

In: Natriuretic Peptides
Editor: Kailash N. Pandey

ISBN: 978-1-63321-080-6

Chapter 11

Natriuretic Peptides Modulate Inner Ear Function

George J. Trachte [*] ***and Janet L. Fitzakerley***
Department of Biomedical Sciences, University of Minnesota Medical School,
Duluth, MN, US

Abstract

Natriuretic peptides dramatically increase urine volume and sodium content by influencing renal transport mechanisms for sodium, potassium and chloride. Many of the identical transporters provide vital contributions to fluid and electrolyte balance in the cochlea of the inner ear. Recent results from experiments using knockout mouse models support the hypothesis that the natriuretic peptide pathway has a critical role in hearing. In this review, the components of the natriuretic peptide system known to be present in the inner ear and their ability to regulate the function of the cochlea will be detailed.

Components of the Natriuretic Peptide Pathway

Natriuretic peptides: As detailed in other chapters of this book, natriuretic peptides were initially identified as cardiac hormones causing a rapid and marked diuresis and natriuresis (de Bold et al., 1981). Atrial natriuretic peptide (ANP) was subsequently identified as a 28 amino acid cyclic peptide (Flynn et al., 1983). Brain natriuretic peptide (BNP) (Sudoh et al., 1988) and C-type natriuretic peptide (CNP) (Sudoh et al., 1990) were identified as additional natriuretic peptides with a conserved cyclic 17 amino acid ring structure. Both ANP and BNP activate the same particulate guanylyl cyclase, termed natriuretic peptide receptor A (NPR-A), whereas CNP activates a distinct particulate guanylyl cyclase (NPR-B) (Chinkers et al., 1989; Chang et al., 1989; Koller et al., 1991). All three natriuretic peptides are capable of

[*] Corresponding author: George Trachte, Professor, Department of Biomedical Sciences, University of MN Medical School, Duluth Campus, 1035 University Drive, Duluth, MN 55812. Email :gtracht1@d.umn.edu.

influencing sodium, potassium and fluid transport and, therefore, are of interest for their effects on inner ear fluid and electrolyte balance and, ultimately, on sound processing.

Natriuretic peptide synthesizing enzymes: Natriuretic peptides are synthesized in a variety of tissues as pro-hormones and undergo a proteolytic cleavage to liberate a biologically active peptide (Maki et al., 1984; Oikawa et al., 1984; Yamanaka et al., 1984; Glembotski et al., 1988; Takahashi et al., 1992). The heart is the major site for ANP production, with ANP being synthesized in the atria (Glembotski et al., 1988) and BNP in the ventricles (Takahashi et al., 1992). The third major member of the natriuretic peptide family, CNP, is primarily synthesized in brain and vascular endothelium, as well as other sites (Takahashi et al., 1992). The proteolytic enzyme mediating the activation of the pro-hormone form of ANP and BNP is a transmembrane serine protease named corin (Semenov et al., 2010; Szabo and Bugge, 2011; Yan et al., 2000). pro-CNP is processed to its active peptide by furin, a cytosolic protease (Wu et al., 2003) that is found in many tissues (Van de Ven, 1990), including the brain (Day et al., 1993).

Natriuretic peptide receptors: In the original experiments conducted in heart tissue, natriuretic peptides were found to produce elevations in cyclic guanosine monophosphate (cGMP) concentrations (Hamet et al., 1984). Chinkers et al. (1989) eventually cloned a membrane associated guanylyl cyclase that functions as the receptor for ANP, which is now universally referred to as natriuretic peptide receptor-A (NPR-A). NPR-A also responds to BNP with elevated guanylyl cyclase activity (Koller et al., 1991; Chang et al., 1989). A second particulate guanylyl cyclase was subsequently cloned and identified as NPR-B (Koller et al., 1991; Chang et al., 1989). The NPR-B is a relatively selective receptor for CNP (Koller et al., 1991; Chang et al., 1989). Maack et al. (1987) identified a third natriuretic peptide receptor (NPR-C) that binds all three natriuretic peptides and was initially shown to function as a clearance receptor. Subsequently, NPR-C has been shown to have transducing activity as it alters the activity of adenylyl cyclase, phospholipase C and calcium channels (described by Anand-Srivastava in this book).

Downstream effectors: The NP signal transduction cascade has not been completely characterized in any system. However, diffusible cGMP is thought to mediate its actions primarily through cyclic nucleotide dependent protein kinase I (abbreviated PKG, cGK1 or Pkgi; reviewed by Silberach and Roberts, 2001). Downstream actions of natriuretic peptides that are mediated through phosphorylation include actions on the $Na^+/K^+/2Cl^-$ cotransporter, Na^+-K^+-ATPase and several types of ion channels (including potassium channels) (reviewed by Kourie and Rive, 1999). Forms of all three of these types of effector proteins are critical for inner ear function.

The reported actions of ANP on the $Na^+/K^+/2Cl^-$ cotransporter are tissue dependent. In vascular smooth muscle cells, elevations of intracellular cGMP stimulate the $Na^+/K^+/2Cl^-$ cotransporter via a PKG-dependent mechanism (reviewed by Adragna et al., 2004), resulting in the movement of fluid and ions across the cell membrane. It has been shown that this increase in cGMP concentration can be initiated by ANP, and the ANP action can be inhibited by furosemide (O'Donnell and Owen, 1986; Owen et al., 1987; Fujita et al., 1989). Similarly, in neuroblastoma cells, 1 μM ANP increased $Na^+/K^+/2Cl^-$ cotransport by 82%, with a subsequent indirect stimulation of Na^+-K^+-ATPase (Delporte et al., 1993).

However, in several other tissues, ANP and/or cGMP have been shown to decrease $Na^{+}/K^{+}/2Cl^{-}$ cotransporter activity or to have no effect. In atrial and ventricular myocytes, ANP-induced reductions in cell volume could be blocked by inhibiting the cotransporter with bumetanide, suggesting that the decrease in cell volume may act as a negative feedback mechanism to inhibit the stretch-induced release of ANP from atrial and ventricular cells (Clemo and Baumgarten, 1991). ANP and cGMP have been shown to reduce Na^{+} and Cl^{-} influx in the flounder intestine (Rao and Nash, 1988; O'Grady, 1989), and an ANP-initiated decrease in cotransporter activity has been shown to occur in HeLa cells (Kort and Koch, 1990a,b) . In the skin, BNP inhibits the $Na^{+}/K^{+}/2Cl^{-}$ cotransporter, while ANP has no effect (Bourgoigne, 1987). ANP also has no effect on cotransporter activity in brain capillary endothelial cells (Vigne et al., 1994).

Natriuretic peptides have been shown to interact with several types of potassium channels, including inward- and delayed-rectifier K^{+} channels (Kourie and Rive, 1999). More specifically, it has been reported that ANP mimics the moderately prolonged suppression of an inward rectifier current typically produced by stimulation of metabotropic glutamate receptors (Dixon and Copenhagen, 1997). ANP has also been shown to potentiate the slow component of the delayed rectifier current in sino-atrial node cells in a PKG dependent manner (Shimizu et al., 2002). This region of the heart has been shown to express KCNQ1 (Brahmajothi et al., 1997). Increased cGMP is also thought to open an ATP-sensitive potassium channel, causing hyperpolarization and relaxation of muscle cells (White et al., 1993).

These downstream effects are extremely relevant to auditory function because of the key roles played by the $Na^{+}/K^{+}/2Cl^{-}$ cotransporter, Na^{+}-K^{+}-ATPase and potassium channels in sound processing.

Phosphodiesterases: Phosphodiesterases degrade cyclic nucleotides, such as cGMP; therefore, they represent the terminal step in the natriuretic peptide pathway in that the phosphodiesterases eliminate the natriuretic peptide signal involving NPR-A or NPR-B. Phosphodiesterases (PDEs) represent a large family of enzymes that inactivate cyclic nucleotides such as cGMP. There are 11 members of the PDE superfamily, resulting in 22 PDE isoforms that have been identified in mammals (Bender and Beavo, 2006; reviewed by Lugnier et al., 2006). Types 5, 6 and 9 are selective for degradation of cGMP and types 1, 2, 3, 10 and 11 degrade both cAMP and cGMP. Inhibitors of PDE5 have been used clinically to treat erectile dysfunction for over a decade. Hearing loss has been identified as a significant side effect of these agents (see below), clearly highlighting cGMP as a potential modulator of cochlear function.

The Natriuretic Peptide Pathway in the Cochlea

Cochlear morphology and function: The cochlea is a spiraling, bone-covered sensory organ possessing three extracellular fluid compartments of dramatically differing ionic compositions. Two compartments, the scalae vestibuli and tympani, contain perilymph, which has a high Na^{+}/low K^{+} composition that is similar to other extracellular fluids. The third cochlear compartment (scala media) contains endolymph, which differs from virtually all other mammalian extracellular fluids by virtue of its high potassium (145 mM), and very low sodium content (Bosher and Warren, 1968). The endolymph bathes the sensory hair cells in the organ of Corti that are responsible for the generation of the electrical signals that ultimately carry acoustic information to the brain via the auditory nerve. As a result of the process that produces the high potassium ion concentrations, the scala media has a high positive electrical potential relative to the surrounding tissue, called the endocochlear potential (EP; Davis, 1965). Cells found in the lateral wall of the cochlea in a region termed the stria vascularis contain the ion transporters and channels required for the generation of the unique extracellular environment of the scala media (reviewed by Wangemann, 2002, 2006). Maintenance of the cochlear fluid balance is essential for normal inner ear function, and disruptions in these processes may play a role in hearing loss and in Ménière's disease.

We will emphasize the localization of natriuretic peptide system components to 1) the stria vascularis (i.e., endolymph forming region), 2) the organ of Corti (i.e., sound sensing region), or 3) the modiolius (i.e., the region where the cell bodies and axons of the auditory nerve are found) (Figure 1). Somewhat predictably, most elements of the natriuretic peptide system are found in the fluid-forming lateral wall of the cochlea, and not in the sound-transducing organ of Corti.

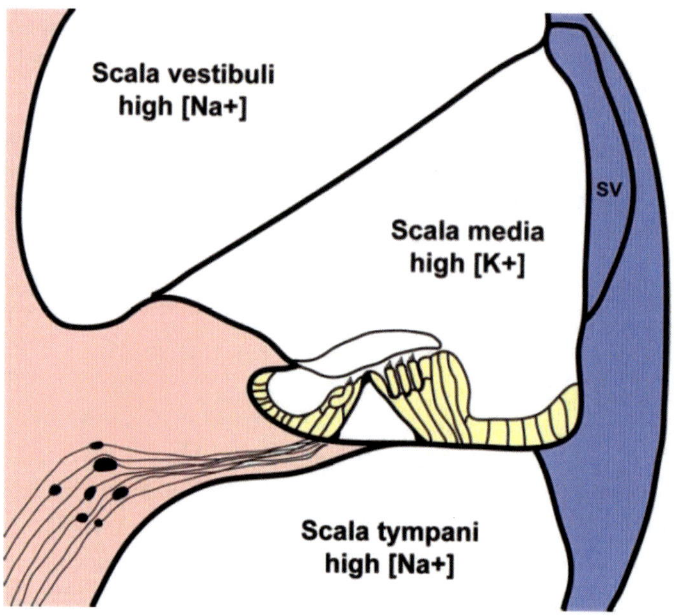

Figure 1. Natriuretic peptides are positioned to modulate the composition of the unique extracellular fluid of the inner ear. A representative cross-section of the cochlea is shown, indicating the three fluid compartments (scala tympani, scala media and scala vestibuli). The scala media contains the endolymph which has an exceptionally high K+ concentration (145 mM). The lateral wall of the cochlea (blue), contains the stria vascularis (SV) and is responsible for the formation and circulation of the endolymph. The organ of Corti (yellow) is the sound transducing region of the cochlea. It contains the sensory hair cells, which are in contact with the endolymph on their apical surface. The center of the cochlea is formed by the modiolus (pink) which contains the cell bodies and axons of the auditory nerve. NPs and NPRs are absent from the organ of Corti, suggesting that the natriuretic peptide system influences endolymph formation and afferent neurotransmission rather than sound transduction.

Natriuretic peptide synthesis in the cochlea: It is likely that ANP is synthesized in the inner ear because Guipponi et al. (2008) demonstrated that corin mRNA transcripts are found solely in the cochlea, heart and testes of mice. Cogently, corin mRNA was not expressed anywhere else in the brain or other sensory organs, nor was it found in any of 9 other organs that were tested in that study. This very selective localization of corin to the ear, heart and testes provides strong indirect evidence in support of a functional role for ANP in the auditory system. Within the cochlea, corin mRNA was found in the stria vascularis, organ of Corti and modiolus.

Pro-CNP is converted to CNP by furin (Wu et al., 2003). The distribution of furin is relatively ubiquitous, but there are no specific reports of its localization in the inner ear. The presence of CNP in the cochlea (see below) suggests that either furin is present to account for its generation, or that circulating CNP enters the inner ear in an endocrine manner.

Natriuretic peptides in the cochlea: Both ANP and CNP mRNA transcripts and immunoreactive peptides have been identified in the inner ear, although there are differences among reports of their specific localization within the cochlea. ANP mRNA has been found in the cochlea using RT-PCR (Suzuki et al., 1998) and reported in cultured marginal cells of the stria vascularis (Qiao et al., 2011). However, most reports have localized ANP and/or pro-ANP protein to the spiral ligament of the lateral wall, but not to the stria vascularis or the organ of Corti (Meyer zum Gottesberge et al., 1995; Yoon and Hellstrom; 1992; Yoon et al., 2012). Meyer zum Gottesberge et al. (1991) localized ANP immunoreactive products to supporting cells of the guinea pig organ of Corti but not to the hair cells or the stria vascularis. A recent study has localized ANP mRNA and protein to the spiral ganglion neurons in the modiolus (Sun et al., 2013). Although these studies are not in agreement on the details, the bulk of the evidence supports the hypothesis that ANP is synthesized in the inner ear and its tissue distribution is most consistent with an effect on fluid and ion transport and on primary neurotransmission rather than direct actions on cochlear hair cells. In addition, the concentration of ANP in endolymph is 2 orders of magnitude higher than that typically found in plasma (i.e., 10,000 pg/ml vs. 100 pg/ml) (Yoon et al., 2012). The elevated ANP concentrations in endolymph also are consistent with ANP exerting a major influence on auditory function.

Immunoreactive CNP products have a similar distribution to that of ANP (Suzuki et al., 2000), being concentrated in the spiral ligament but not the stria vascularis or organ of Corti. Suzuki et al. (1997) confirmed the presence of mRNA transcripts for CNP in rat inner ear via RT-PCR. The cochlear transcript levels of CNP were as high as in brain and much greater than in heart. These reports suggest that CNP is synthesized in cochlea and that it is likely to modify fluid and/or electrolyte transport in the formation of endocochlear fluid rather than affect hair cell transduction of mechanical signals. Some caveats to this conclusion include the dearth of data on pro-CNP or furin, its activating enzyme. Furthermore, no data regarding the functional role of CNP and/or NPR-B in sound processing have been reported.

In contrast to the data regarding ANP and CNP, reports of BNP immunoreactivity in the inner ear are conspicuously absent. Suzuki et al. (1998) failed to find BNP mRNA transcripts in the rat inner ear, suggesting an absence of synthesis. Studies to date have failed to localize BNP transcripts or BNP immunoreactive proteins in the cochlea or vestibular apparatus; thus, no role for BNP in maintenance of hearing is anticipated at this time.

Natriuretic peptide receptors in the cochlea: All three known natriuretic peptide receptors are present in the inner ear. Both NPR-A and NPR-B transcripts are present in the inner ear at levels comparable to or greater than those in heart and brain and NPR-C transcripts are present at levels comparable to areas of the brain but less than levels in the heart (Suzuki et al., 1998; Dornhoffer et al., 2002; Long et al., 2008; Long et al., 2010; Qiao et al., 2011). Furata et al. (1995; 1998) localized NPR-A and NPR-B transcripts to the spiral ligament and not the stria vascularis by in situ hybridization. Krause et al. (1997) also found both NPR-A and NPR-B transcripts in guinea pig cochlea but the NPR-A transcripts were more abundant. NPR-C transcripts also are present in rat inner ear at levels comparable to those in brain (Suzuki et al., 1998; Dornhoffer et al., 2002) although Krause et al. (1997) failed to find NPR-C transcripts in guinea pig cochlea. There is evidence that NPR expression is upregulated by injection of ANP through the round window (Dornhoffer et al., 2002; Boone et al., 2005), suggesting that some of the variability in the data may be due to the dynamics of the system.

The actual presence of NPR-A protein in the stria vascularis of guinea pigs has been demonstrated by ANP binding (Koch et al., 1992; Meyer zum Gottesberge et al., 1995), and shown in the modiolus (Sun et al., 2013). Koch et al. (1992) also demonstrated an increase in cGMP production in response to ANP in an in vitro stria vascularis preparation. The combination of cochlear RT-PCR, in situ hybridization in the spiral ligament, ANP binding and functional activation of NPR-A in the stria vascularis identify the lateral wall of the cochlea as one of the few locations of a complete natriuretic peptide system in the body.

Downstream effectors in the cochlea: The cGMP generated in the inner ear is assumed to alter auditory function by affecting PKG activity, although only one published study has tested this assertion directly. Jaumann et al. (2012) showed the presence of PKG1 mRNA transcripts and immunoreactive protein in the inner ear of mice. Furthermore, they investigated the effect of PKG1 knockouts and found the mice to be more susceptible to noise-induced hearing loss. These data are consistent with the hypothesized model for effects of natriuretic peptides on hearing but there is a clear need for additional studies utilizing either transgenic mice or PKG inhibitors to ascertain the precise role of PKG in mediating effects of natriuretic peptides on hearing.

Information from studies investigating the actions of soluble guanylyl cyclases (sGCs) and their agonists (particularly nitric oxide) may shed some light on the effects of cGMP in the cochlea. sGCs have been found in many cochlear regions, including the auditory nerve, the organ of Corti, the stria vascularis and cochlear blood vessels.

As reviewed by Takumida and Anniko (2002), NO has been found to have a role in the regulation of cochlear blood flow (discussed further in the section describing cochlear phosphodiesterases), afferent neurotransmission and outer hair cell function. sGCs are found in afferent terminals and supporting cells in the organ of Corti (Heinrich et al., 2000; Takumida et al., 2000). In the normal inner ear, NO has been shown to modulate the discharge of afferent fibers (Flores et al., 2001; Heiniger-Burki et al., 1995) and/or modify the calcium currents and gap junctions in supporting cells (Blasits, 2000; Matsunobu and Schacht, 2000). NO has also been implicated in cochlear pathology. More specifically, NO may potentiate glutamate release, exacerbating the neurotoxicity produced by noise exposure, and may directly cause more widespread tissue damage in a cGMP-independent manner that involves formation of peroxynitrite (reviewed by Takumida et al., 2001). NO has also been shown to induce apoptosis in outer hair cells (Yamane et al., 2004).

sGCs are also found in the lateral wall of the cochlea, although their normal role in stria vascularis physiology is unknown. Inducible NO synthase (Shi and Nuttall, 2003) and NO concentrations (Chen et al., 2008) increase in the lateral wall following noise exposure, and inducible NO synthase activity is increased in the stria vascularis following cisplatin administration (Lee et al., 2004), suggesting a role for NO in stria vascularis pathology in a manner similar to the contribution to noise-induced organ of Corti damage outlined above.

Inner ear sensory transduction is absolutely dependent on the cycling of K^+ from the stria vascularis into the scala media, through the hair cells into the supporting cells, and back through the spiral ligament into the stria (reviewed by Wangemann 2002, 2006). As part of this process, K^+ release from strial cells generates the endocochlear potential. Ion transport mediated by $Na^+/K^+/2Cl^-$ cotransporter and Na^+/K^+-ATPase is a fundamental component of this process, as is flow through the specific K^+ channels KCNQ1 and KCNQ4. The importance of K^+ cycling is underscored by the fact that mutations of KCNQ1 (Lee et al., 2000), KCNQ4 (Kharkovets et al., 2006) and the $Na^+/K^+/2Cl^-$ cotransporter NKCC1 (Delpire et al., 1999) cause deafness. In addition, the $Na^+/K^+/2Cl^-$ cotransporter is blocked by loop diuretics such as furosemide, which have been shown to significantly reduce the endocochlear potential (Green et al., 1981; Rybak and Whitworth, 1986), and hearing loss is known to be an undesirable side effect of clinical administration of these commonly used drugs (Vargish et al., 1970). Many of these transporters and channels have been shown to be downstream targets of natriuretic peptides in other tissues, although the exact interactions among these proteins remains to be demonstrated in the inner ear. It is anticipated that some, or all, of these transporters and channels will be affected by ANP to alter K^+ concentrations in the endolymph.

Phosphodiesterases in the cochlea: 6 PDE isoforms are expressed in the guinea pig cochlea: PDE3a, PDE4d, PDE8a, PDE8b, PDE9a and PDE11a (Huang et al., 2010). Notably absent from this list is PDE5, the target of the PDE inhibitors that are in clinical use. Of the 6 isoforms found in the cochlea, 3 are known to hydrolyze cGMP: PDE3 and 11, which hydrolyze both cAMP and cGMP (Degerman et al., 1997; Liu and Maurice, 1998), and PDE9, which hydrolyzes only cGMP (Fisher et al., 1998).

In contrast to these results in the guinea pig, Jaumann et al. (2012) demonstrated the presence of mRNA transcripts for phosphodiesterase 5 in the inner ear of mice and rats, as well as the presence of phosphodiesterase 5 immunoreactive products. These authors also found that the selective phosphodiesterase 5 inhibitor, vardenafil, preserved auditory function that had been compromised by noise exposure. Vardenafil did not alter hearing in control mice without exposure to noise.

Clinical treatments indicate that PDE inhibitors can have beneficial effects on hearing. Pentoxyifylline, a nonspecific PDE inhibitor, is one of several vasodilators used in the treatment of sudden hearing loss, particularly idiopathic sudden hearing loss (de Barros et al., 2003; Racic et al., 2003). In addition, the results of a small experimental trial indicated that 60% of patients reported substantial reduction or elimination of Ménière's disease symptoms following administration of sildenafil, a specific PDE5 inhibitor (Black, 2000).

The fact that PDE inhibitors cause vasodilation is the rationale behind the use of pentoxyifylline (a nonspecific PDE inhibitor) to treat idiopathic sudden hearing loss (de

Barros et al., 2003; Racic et al., 2003). Pharmacological studies, involving both NO donors and inhibitors, have established that NO increases blood flow in both the cochlear and vestibular arteries (Arenberg et al., 1997; Brechtelbauer et al., 1994; Fessenden et al., 1994), as well as spiral modiolar artery (Wangemann, 2002). NO also plays a key role in maintaining cochlear blood flow under pathological conditions, such as focal cochlear microcirculation disorder (Nagura et al., 2001). Similarly, ANP infusion results in a regional increase in blood flow in the vestibular artery, despite systemic decreases in blood pressure (Rachel et al., 1995). ANP administration can also protect the cochlea from ischemia and reperfusion injury (Xiong et al., 2006).

In contrast to these beneficial actions, the initial case reports regarding the effects of PDE5 inhibitors documented sudden sensorineural hearing loss and vestibular system dysfunction following self-administration of extremely large doses of sildenafil (Hamzavi et al., 2002; Mukherjee and Shivakumar, 2006). Systematic examination of information in the FDA adverse effects database indicates that there is a potential link between PDE5 inhibitor use and sudden sensorineural hearing loss, although the incidence of hearing loss is only slightly higher in the patients taking the PDE5 inhibitors than in the general population (Maddox et al., 2009). Long-term administration of high daily doses of sildenafil has been shown to result in significant hearing loss in mice (Hong et al., 2008). This may correlate with a recent report of an increased incidence of self-reported hearing impairment in human patients taking sildenafil, but not vardenafil or tadalafil (McGwin, 2010). An objective study evaluating the effects of PDE inhibitors on hearing in humans (Okuyucu et al., 2009) demonstrated a significant increase in thresholds 1 hour following vardenafil administration in 4 of 18 subjects, but not 5 and 72 hours later. Overall, this evidence indicates that high doses of specific inhibitors of PDE5 may damage the cochlea, although some reports of beneficial effects of lower doses exist.

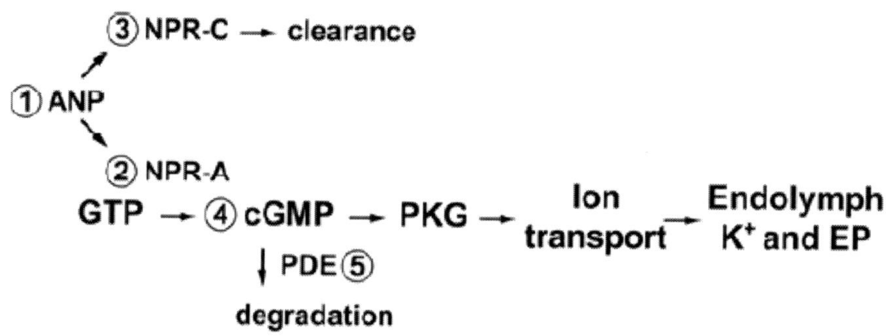

Figure 2. Multiple stages of the natriuretic peptide system have been tested functionally in the inner ear. A hypothetical scheme showing how the major components of the natriuretic peptide pathway could influence hearing is shown. The actions of ANP (labelled #1) are proposed to be mediated by the natriuretic peptide A receptor (NPR-A, labelled #2), resulting in an elevation of cochlear cyclic guanosine monophosphate (cGMP, labelled #4) concentrations. cGMP then activates a cGMP-dependent protein kinase (PKG) that could subsequently alter ion transport and endolymph composition and the endocochlear potential (EP), which would alter auditory transduction. The actions of the proposed pathway can be attenuated either by ANP binding to the natriuretic peptide C receptor (NPR-C, labelled #3) to enhance clearance or by degradation of cGMP by phosphodiesterases (labelled #5).

Functional Demonstration of the Role of Natriuretic Peptides in Hearing

A hypothetical scenario that could be drawn from the data presented thus far involves ANP activating NPR-A in the cochlea to generate cGMP (Figure 2). If the cochlea is similar to other tissues, the cGMP likely activates protein kinase G to alter ion transport via the $Na^+/K^+/2Cl^-$ cotransporter, resulting in an increase in endocochlear potassium ion concentrations and an improvement in hearing. In the last section of this review, we will outline recent work detailing the auditory effects of: 1) direct infusion of ANP; 2) elimination of NPR-A; 3) elimination of NPR-C; 4) cGMP in aging animals and 5) phosphodiesterase inhibitors. These studies can be summarized as indicating a beneficial effect of ANP and phosphodiesterase inhibitors on hearing thresholds.

1) *Positive effect of ANP infusion on hearing:* Continuous infusion (0.25 ng/g/hr) of ANP into the carotid artery of CBA/J mice resulted in a temporary improvement in hearing of approximately 10 dB (Figure 3). The duration of the effect was ~20 min, with hearing thresholds returning to normal levels despite continued infusion. ANP infusion elevated plasma ANP concentrations up to 10 fold and the impact on hearing was proportional to the effect on plasma ANP levels (Fitzakerley et al., 2010). Preliminary data have indicated that ANP infusion has a mild, sustained depressor effect of 10 mmHg on blood pressure, thus dissociating the vasoactive effects of ANP from its effects on hearing. In addition, infusion of ANP did not produce any threshold improvement in homozygote NPR-A knockout mice despite an increase in plasma concentrations of ANP (Clarke et al., 2012). These data provide compelling evidence that ANP can influence hearing via effects on NPR-A, as well as raising the possibility that there is a potential endocrine, as well as paracrine, role for NPR-A in the cochlea.
2) *Hearing loss in NPR-A knockout mice:* A pivotal step in the proposed pathway for ANP effects on hearing involves activation of NPR-A to generate cGMP. The involvement of this receptor was elucidated using transgenic mice lacking functional NPR-A (Lopez et al., 1995). Homozygote NPR-A knockout mice were significantly more likely to have elevated thresholds than control mice (Figure 4A), and a significantly higher percentage of them had high frequency hearing loss at a young age (Figure 4B) (Fitzakerley and Trachte, 2006). Responses of heterozygous NPR-A were intermediate between those of homozygote and control (wild-type and CBA/J) mice. These data support the hypothesized scenario for ANP having a beneficial role in hearing by virtue of its effect on NPR-A. However, a critical caveat is that this knockout of NPR-A increased the susceptibility of mice to hearing loss, but did not prevent the cochlea from developing normally or universally cause deafness. Thus, NPR-A appears to be a factor in maintaining normal auditory function but is not essential for hearing.
3) *NPR-C ablation protects against hearing loss:* A common strategy used to increase ANP effects on the cardiovascular or renal systems involves eliminating the actions of NPR-C, reulting in decreased ANP clearance (Maack et al., 1987). We hypothesized that ablation of NPR-C should have the opposite effect of NPR-A knockout (i.e., that elimination of NPR-C could protect against hearing loss). We used *lgj* nice, that have a spontaneous 36 bp deletion in the NPR-C gene (Jaubert et

al., 1999) to test this hypothesis (Fitzakerley et al., 2003). This mutant strain was backcrossed by the Jackson Laboratory (Bar Harbor, ME) to a C57Bl/6 line for embryonic storage. Wild-type mice of the Balb/C X C57Bl/6 strain had a significant hearing loss at 12 kHz compared to CBA/J controls and to heterozygous and homozygous mutants (Figure 5A), implying that a mutation of one copy of the NPR-C gene is sufficient to protect against the early onset hearing loss observed in wild-type animals. This protection against hearing loss in the NPR-C +/- and -/- mice also was evident when the assessment of normal hearing vs. hearing loss was made, as less than 50% of wild-type mice had normal hearing compared to more than 95% of heterozygote and homozygote animals (Figure 5B).

4) *cGMP concentrations are correlated with degree of age-related hearing loss:* We found that there is a correlation between the amount of age-releated hearing loss and cochlear cGMP concentrations (Figure 6). CBA/J mice exhibited a gradual, statistically significant increase in 16 kHz pure tone thresholds, from approximately 20 dB SPL at 25 postnatal days of age (P25) to 60 dB SPL at P455 (Figure 6A). The time course of this threshold increase paralleled a decrease in cochlear cGMP concentrations, from 60 fmoles/mg tissue at P25 to <20 fmoles/mg tissue at P455 (Figure 6A). When measured in the same animal, there was a strong correlation between ABR threshold and cochlear [cGMP] (Figure 6B). This correlation describes a simple exponential relationship of lower thresholds in CBA/J mice with higher levels of cochlear cGMP. These data are consistent with the hypothesized pathway including cGMP as a critical mediator of ANP effects. One complication in interpreting these results involves the redundant origins of cGMP in the cochlea. The exact contributions of different guanylyl cyclase enzymes has not been established in the cochlea under normal conditions or during aging at this time. However, the general downward trend in cochlear cGMP concentrations as hearing sensitivity declines with age is consistent with our hypothesis.
5) *Inhibition of phosphodiesterase has complicated effects on hearing:* We and others have investigated the impact of phosphodiesterase inhibitors on hearing (Holz et al., 2010). In our study, administration of the cGMP-specific PDE inhibitor zaprinast (Manallack et al., 2005) had complex effects on cochlear function (Figure 7). In animals with a mild to moderate hearing loss, a low dose of zaprinast improved thresholds by approximately 10 dB, while a higher dose produced a significant hearing loss of >15 dB. The hearing improvement was dependent upon control hearing thresholds, with the largest threshold improvements being seen in mice with a preexisting mild to moderate hearing loss. These data provide further support for the hypothesis that cGMP plays a significant role in hearing, and may facilitate the development of phosphodiesterase inhibitors that can be used to treat moderate hearing loss.

There are at least three potential explanations for the complicated nature of the cochlear response to zaprinast: 1) non-selective inhibition of PDE isozymes, 2) differential effects of cGMP in the organ of Corti versus the lateral wall, and 3) vasodilatory effects on cochlear blood flow.

Zaprinast is a relatively non-selective PDE inhibitor, blocking PDE1, PDE5, PDE6, PDE9, PDE10 and PDE11 with differing potency (reviewed by Manallack et al., 2005). Therefore, these results would appear to indicate that the improvement in thresholds and increases in cochlear cGMP observed in response to low doses of zaprinast may be mediated by inhibition of PDE9 and/or PDE11, while the hearing loss observed in response to higher doses of zaprinast (and of other more selective PDE5 inhibitors) may be due to non-selective effects on other PDEs, or on other enzymes. The data obtained with the higher zaprinast dose are consistent with known deleterious effects of high doses of PDE inhibitors on hearing in human populations. The beneficial effects of lower doses of zaprinast support our hypothesis predicting the improvement of hearing with higher cGMP concentrations (Figure 2).

In isolated pulmonary arteries, zaprinast is known to increase the vasodilation caused by NO, resulting in increased blood flow (Thusu et al., 1995), suggesting that effects on cochlear blood flow must be considered as an explanation for the effect of this PDE inhibitor on hearing. However, it should be noted that the doses of zaprinast used in our experiments were chosen to be below the levels that produce significant systemic vasodilation. In addition, the evidence from both the NO and ANP studies indicates that cGMP is involved in the maintenance of cochlear blood flow in the face of perturbations of systemic blood pressure. It is, therefore, unlikely that vasodilation (whether systemic or local) can explain both the threshold improvements *and* the hearing loss observed in these experiments.

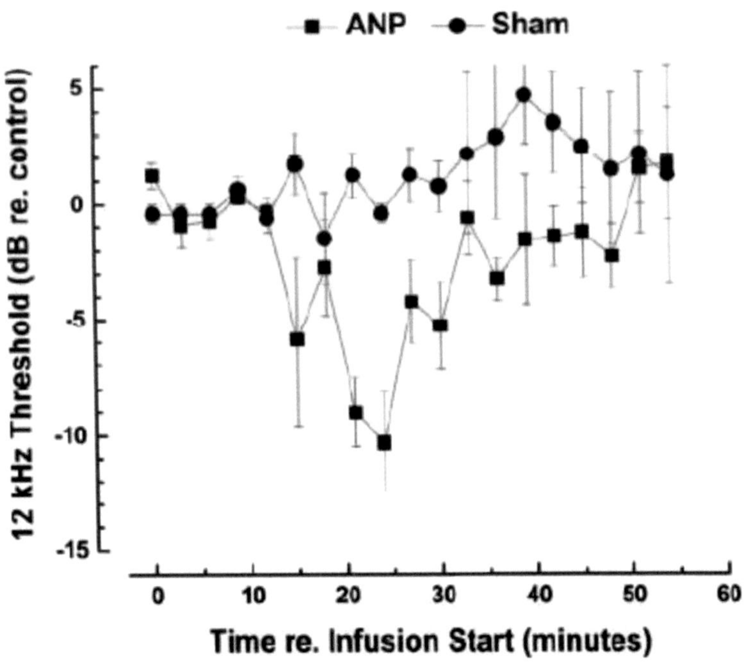

Figure 3. ANP infusion temporarily improves hearing. The effect of continuous infusion of ANP (0.25 ng/g/hr) on auditory brainstem response (ABR) thresholds is shown. ANP was infused into the carotid artery of anesthetized mice, and ABRs were recorded in response to 12 kHz tone pips. Intraarterial ANP significantly decreased sound thresholds (i.e., improved hearing) compared to control values and sham infusion ($p = 0.016$). There were 16 mice in the ANP group and 7 sham infused mice. Data shown are means ± SEM.

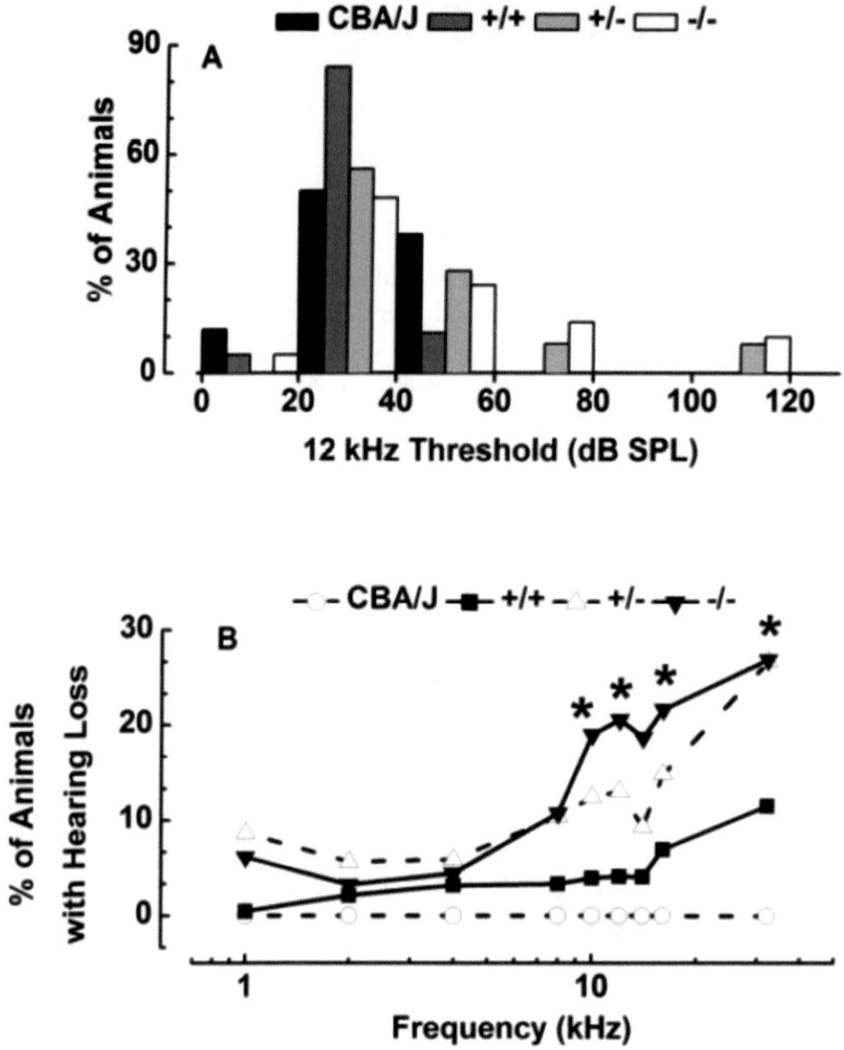

Figure 4. Knocking out the NPR-A can cause an early onset, high frequency hearing loss. A. The impact of NPR-A knockout on hearing thresholds is shown by plotting the minimum intensity (threshold) required to produce a response to a 12 kHz tone burst in 25-35 day old CBA/J (control) and NPR-A knockout mice. Only heterozygous and homozygous knockout mice were profoundly deaf (i.e., had thresholds above 60 dB SPL). n=14, 18, 24 and 14 for CBA/J, wild-type (+/+), heterozygote (+/-) and homozygote (-/-) animals, respectively. B. The incidence of profound hearing loss is shown as a function of stimulus frequency for CBA/J and NPR-A knockout mice. Greater than 15% of heterozygous and homozygous NPR-A knockout mice had a profound hearing loss at all frequencies above 4 kHz, a significant difference from control animals (asterisks indicating $p<0.05$). By definition no CBA/J mice had a profound hearing loss. n=14, 20-33, 38-52 and 15-22 for CBA/J, +/+, +/- and -/- animals, respectively.

Summary

This chapter has detailed compelling evidence supporting a beneficial effect of natriuretic peptides, specifically ANP, on auditory function. The pathway for this effect appears to be consistent with the actions of natriuretic peptides in other tissues, involving ANP acting on

NPR-A to generate cGMP, presumably affecting PKG, ion transport and endocochlear fluid composition to impact hearing. The pathway includes diverging points in that ANP also can interact with the NPR-C to lessen its effect on hearing and cGMP can be degraded by phosphodiesterase to eliminate the beneficial effect of natriuretic peptides. The proposed model is shown in its entirety in figure 2. Each divergent point represents a potential site of action for therapeutics that could replicate the actions of ANP. These therapeutics could be of benefit in Ménière's disease, as well as idiopathic sensorineural hearing loss.

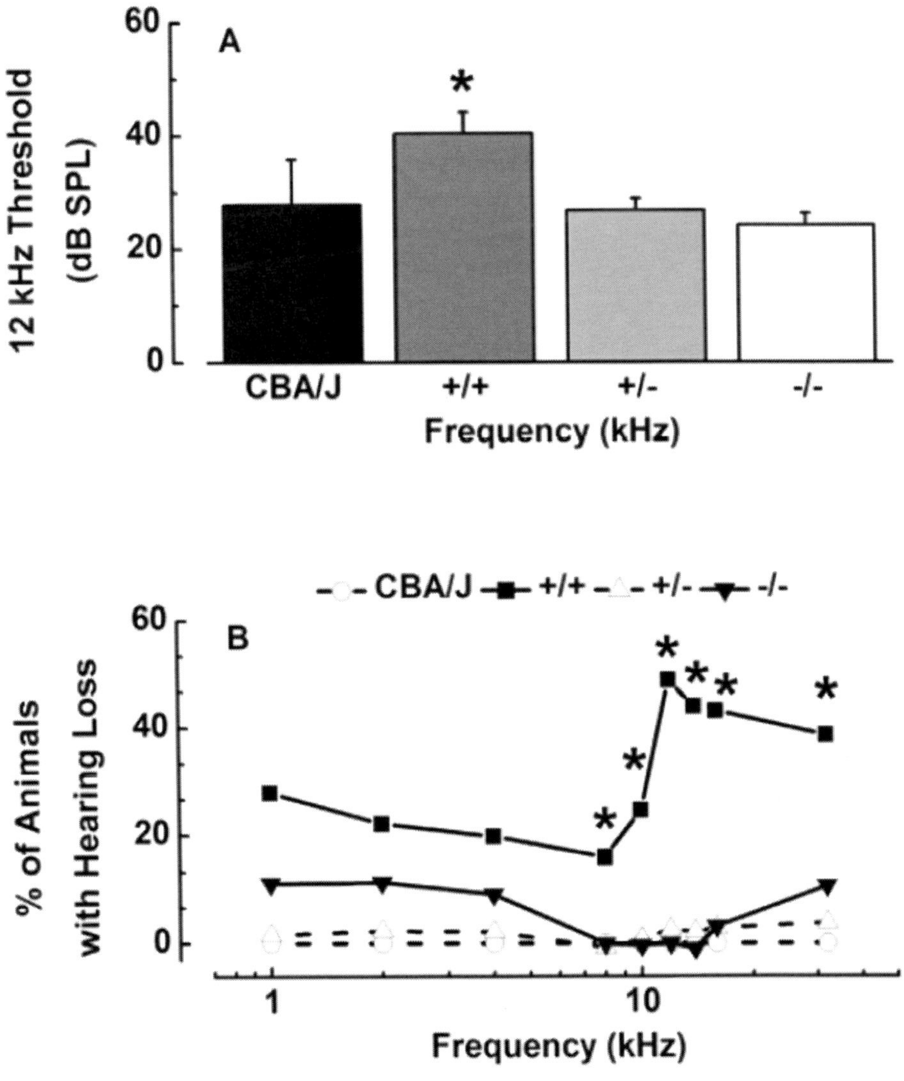

Figure 5. Mutation of the NPR-C can protect against early onset high frequency hearing loss. A. Mean 12 kHz thresholds are shown for 25-60 day old CBA/J and NPR-C mutant mice. Wild-type mice exhibited significantly higher hearing thresholds than CBA/J mice (asterisk indicating $p<0.05$). Error bars indicate ± SEM. n=14, 7, 18 and 9 for CBA/J, +/+, +/- and -/- animals, respectively. B. Incidence of profound hearing loss is shown. Greater than 20% of all wild-type mice exhibited a significant hearing loss at all frequencies above 8 kHz, whereas there was no difference between +/- and -/- NPR-C mutant mice and CBA/J mice at any frequency.

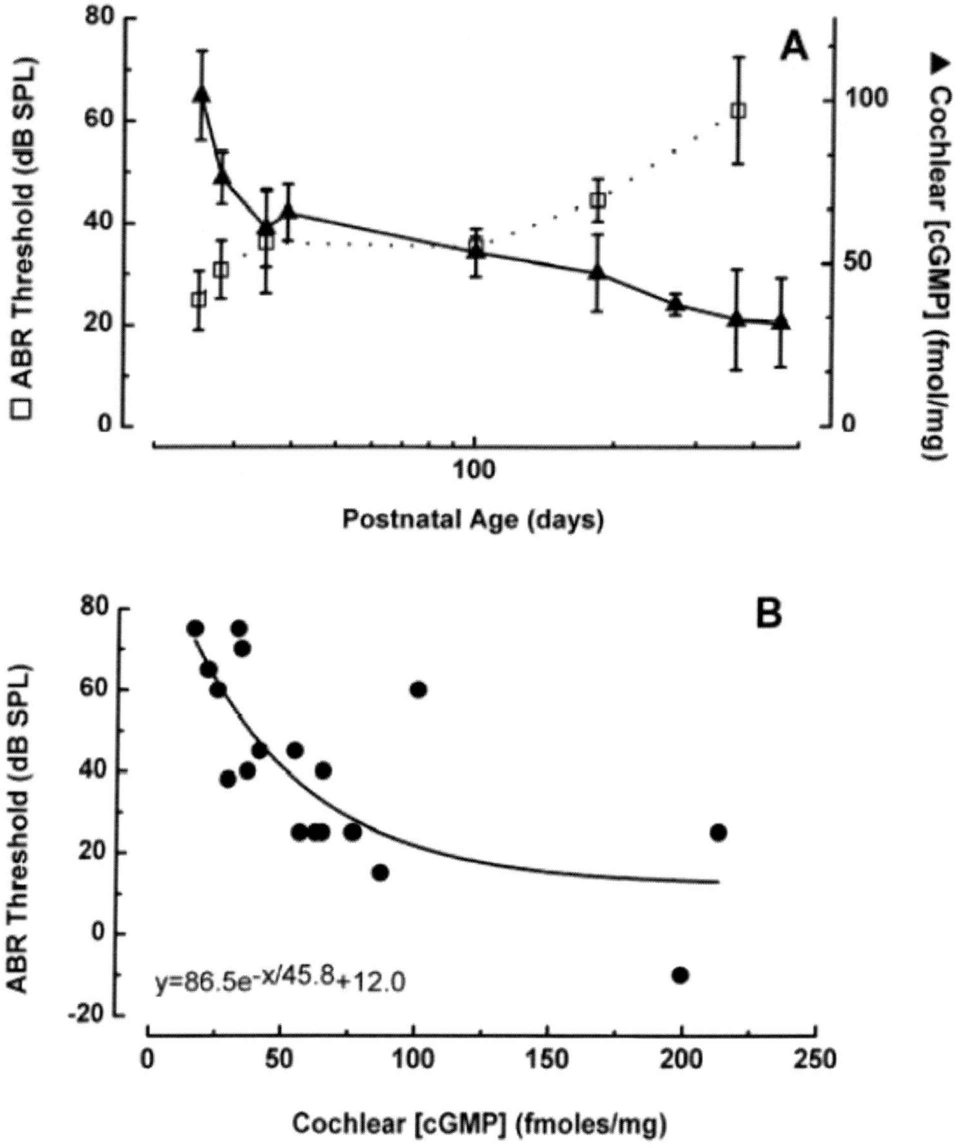

Figure 6. Cochlear cGMP concentrations and ABR thresholds are correlated in aging CBA/J mice. A. 16 kHz thresholds and cochlear [cGMP] are plotted as a function of postnatal age. The time course of increases in acoustic thresholds parallelled decreases in cochlear [cGMP], from 60 fmoles/mg tissue at postnatal day (P)25 to <20 fmoles/mg tissue at P455 (differences between P25 and P455 were significant; $p<0.001$). Values shown are means ± SEM with the open squares representing ABR thresholds and the closed traingles representing cochlear cGMP concentrations. n=3-6 for thresholds; n=3-21 for [cGMP]. B. When measured in the same animals, CBA/J thresholds were strongly correlated with cochlear [cGMP]. An exponential fit to the data is shown ($r=0.76$, $p<0.001$).

Specific deficiencies remain in our understanding of natriuretic peptide effects on hearing, including the role of other cochlear natriuretic peptides, such as CNP. Although the CNP-specific guanylyl cyclase, NPR-B, is present in the cochlea, its role in hearing has been largely unexplored. The precise effect of cGMP on PKG and ion transport in the cochlea is another major area that requires study. Other future opportunities for this area of research include confirmation of the source of cochlear natriuretic peptides as endogenous to the cochlea (paracrine) or blood borne (endocrine). Although the presence of synthesizing enzymes and pro-hormones in the cochlea suggests that natriuretic peptides are synthesized there, circulating natriuretic peptides could represent a mechanism whereby somatic processes modify cochlear function. The site of action of all components of the natriuretic peptide system also is incompletely resolved but appears to involve structures in the lateral cochlear wall and modiolus, implicating this system in endolymph production and afferent neurotransmission.

Hearing loss is currently the most common sensory deficit in human populations, and there are currently no successful pharmacological therapies for inner ear diseases. The improvements in hearing thresholds mediated by ANP and by low doses of some phosphodiesterase inhibitors are an encouraging step in the development of medical treatments for inner ear disorders. However, the complicated response of the cochlea to phosphodiesterase inhibition implies that moderate activation of the natriuretic peptide system may be beneficial but that overactivation could have detrimental effects. Although the investigations described in this chapter have added significantly to our understanding of sound processing and have the potential to lead to novel therapies, a more complete understanding of the function of all of the components of the cochlear natriuretic peptide system will be essential for the development of therapeutic agents.

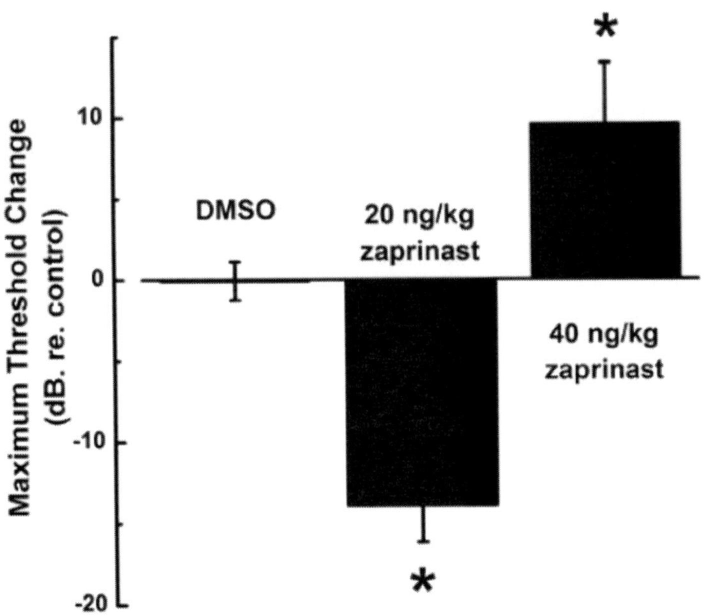

Figure 7. The PDE inhibitor zaprinast has complex effects on hearing. Zaprinast affected ABR thresholds in a dose-dependent manner in animals that had a mild to moderate hearing loss. Zaprinast or its diluent (dimethylsulfoxide; DMSO) was injected i.p. in anesthetized mice and the effect on hearing thresholds was measured using ABR techniques. DMSO did not significantly alter thresholds, but a low dose (20 µg/g) of zaprinast significantly lowered thresholds (i.e., improved hearing) by approximately 10 dB, while a high dose (40 µg/g) raised hearing thresholds (i.e., worsened hearing). The data represent means SEM; asterisks indicate $p<0.05$.

Acknowledgments

The authors wish to thank Jeremy Anderson, Brittni Balogh, Aaron Clarke, Melissa Hanson, Nina Holz, Crystal Lambert, Sam McDonald, Donna Parkes, Matt Ruona, Shelby Rhyberg, Mary Kay Smith, and Matt Stellar for assistance with the experiments described. We also thank Drs. Steven Juhn and Eric Javel (Department of Otolaryngology, University of Minnesota Medical School) for assistance and support in establishing the mouse auditory brainstem monitoring procedure described in this chapter. We are also grateful to Dr. David Garbers (Southwestern Medical School) for supplying the NPR-A deficient mice. Experiments described in this chapter were supported by the National Institutes of Health (R25 GM 086669 and 5R21DC6063), the American Otological Society and the University of Minnesota Academic Health Center.

References

Adragna, NC; Chen, Y; Delpire, E; Lauf, PK; Morris, M. Hypertension in K-Cl cotransporter-3 knockout mice. *Adv Exp Med Biol*, 559, 379-85, 2004.

Arenberg, JG; Komjathy, DA; Seidman, MD; Quirk, WS. Local effects of nitric oxide on vestibular blood flow in the Mongolian gerbil. *Eur Arch Otorhinolaryngol*, 254, 367-71, 1997.

Bender, AT; Beavo; JA. Cyclic nucleotide phosphodiesterases: Molecular regulation to clinical use. *Pharm Rev*, 58, 488-520, 2006.

Black, B. Sildenafil (Viagra) and endolymphatic hydrops. *Am J Otol*, 21, 771, 2000.

Blasits, S; Maune, S; Santos-Sacchi, J. Nitric oxide uncouples gap junctions of supporting Deiters cells from Corti's organ. *Pflugers Arch Eur J Physiol*, 440, 710-212, 2000.

Boone, RT; Zuo, C; Fan, CY; Dornhoffer, J. Modification of atrial natriuretic peptide receptor expression in the rat inner ear. *Otol Neurotol*, 26, 534-537, 2005.

Boscher, SK; Warren, RL. Observations on the electrochemistry of the cochlear endolymph of the rat: a quantitative study of its electrical potential and ionic composition as determined by means of flame spectrophotometry. *Proc R Soc Lond B Biol Sci*, 171, 227-249, 1968.

Bourgoignie, JJ. Natriuretic hormones: Comparison of renal effects. *Klin-Wochenschr*, 65(Suppl. 8), 14-20, 1987.

Brahmajothi, MV; Morales, MJ; Rasmusson, RL; Campbell, DL; Strauss, HC. Heterogeneity in K^+ channel transcript expression detected in isolated ferret cardiac myocytes. *Pacing Clin Electrophysiol*, 20, 388-396, 1997.

Brechtelsbauer, PB; Nuttall, AL; Miller, JM. Basal nitric oxide production in regulation of cochlear blood flow. *Hear Res*, 77, 38-42, 1994.

Chang, M-S; Lowe, DG; Lewis, M; Hellmiss, R; Chen, E; Goeddel, DV. Differential activation by atrial and brain natriuretic peptides of two different receptor guanylate cyclases. *Nature*, 341, 68-72, 1989.

Chen, YS; Tseng, FY; Lin, KN; Yang, TH; Lin-Shiau, SY; Hsu, CJ. Chronologic changes of nitric oxide concentration in the cochlear lateral wall and its role in noise-induced permanent threshold shift. *Laryngoscope*, 118, 832-836, 2008.

Chinkers, M; Garbers, DL; Chang, M-S; Lowe, DG; Chin, H; Goeddel, DV; Schulz, S. A membrane form of guanylate cyclases is an atrial natriuretic peptide receptor. *Nature*, 338, 78-83, 1989.

Clemo, HF; Baumgarten, CM. Atrial natriuretic factor decreases cell volume of rabbit atrial and ventricular myocytes. *Am J Physiol*, 260, C681-C690, 1991.

Crouch, JJ; Sakaguchi, N; Lytle, C; Schulte, BA. Immunohistochemical localization of the Na-K-Cl co-transporter (NKCC1) in the gerbil inner ear. *J Histochem Cytochem*, 45, 773-778, 1997.

Davis, H. A model for transducer action in the cochlea. *Cold Spring Harbor Symp Quant Bio*, 30, 181-190, 1965.

Day, R; Schafer, MK; Cullinan, WE; Watson, SJ; Chrétien, M; Seidah, NG. Region specific expression of furin mRNA in the rat brain. *Neurosci Lett*, 149, 27-30, 1993.

de Barros, FA; Penido, NO; Ramos, HV; Sanchez, ML; Fukuda, Y. Audiological evaluation of twenty patients receiving pentoxifylline and prednisone after sudden deafness: prospective study. *Int Tinnitus J*, 9, 17-22, 2003.

deBold, AJ; Borenstein, HB; Veress, AT; Sonnenberg, H. A rapid and potent natriuretic response to intravenous injection of atrial myocardial extract in rat. *Life Sci*, 28, 89-94, 1981.

Degerman, E; Belfrage, P; Manganiello, VC. Structure, localization, and regulation of cGMP-inhibited phosphodiesterase (PDE3). *J Biol Chem*, 272, 6823-6826, 1997.

Delpire, E; Lu, J; England, R; Dull, C; Thorne, T. Deafness and imbalance associated with inactivation of the secretory Na-K-2Cl co-transporter. *Nature Genetics*, 22, 192-195, 1999.

Delporte, C; Winand, J; Poloczek, P; Christophe, J. Regulation of Na-K-Cl cotransport, Na,K-adenosine triphosphatase, and Na/H exchanger in human neuroblastoma NB-OK-1 cells by atrial natriuretic peptide. *Endocrinology*, 133, 77-82, 1993.

Dixon, DB; Copenhagen, DR. Metabotropic glutamate receptor-mediated suppression of an inward rectifier current is linked via a cGMP cascade. *J Neurosci*, 17, 8945-8954, 1997.

Dornhoffer, JL; Danner, C; Zhou, L; Li, S. Atrial natriuretic peptide receptor upregulation in the rat inner ear. *Ann Otol Rhinol Laryngol*, 111, 1040-1044, 2002.

Fessenden, JD; Coling, DE; Schacht, J. Detection and characterization of nitric oxide synthase in the mammalian cochlea. Brain Res, 668, 9-15, 1994. Erratum in: *Brain Res*, 675, 394, 1995.

Fisher, DA; Smith, JF; Pillar, JS St.; Denis, SH; Cheng, JB. Isolation and characterization of PDE9A, a novel human cGMP-specific phosphodiesterase. *J Biol Chem*, 273, 15559-15564, 1998.

Fitzakerley, JL; Trachte, GT. The influence of natriuretic peptide A receptor knockout on hearing. *Abstr Soc Neurosci*, 36, 2006.

Fitzakerley, JL; Trachte, GJ; Wu, T; Marcus, DC; Javel, E. Functional consequences of natriuretic peptide C-receptor mutation. *ARO Midwinter Meeting*, 2003.

Fitzakerley, JL; Gilbert, JS; Trachte, GJ. Administration of atrial natriuretic peptide improves hearing thresholds: Implications for Ménière's disease treatment. *6th International Meeting on Ménière's disease and Inner Ear Disorders*, 2010.

Flores, A; Soto, E; Vega, R. Nitric oxide in the afferent synaptic transmission of the axolotl vestibular system. *Neurosci*, 103, 457-464, 2001.

Flynn, TG; de Bold, ML; de Bold, AJ. The amino acid sequence of an atrial natriuretic peptide with potent diuretic and natriuretic properties. *Biochem Biophys Res Commun*, 117, 859-865, 1983.

Fujita, T; Hagiwara, H; Ohuchi, S; Kozuka, M; Ishido, M; Hirose, S. Stimulation of Na-K-Cl cotransport in cultured vascular endothelial cells by atrial natriuretic peptide. *Biochem Biophys Res Commun*, 159, 734-40, 1989.

Furata, H; Luo, L; Ryan, AF; Mori, N. Expression of mRNA encoding vasopressin V1a, vasopressin V2 and ANP-B receptors in the rat cochlea. *Hear Res*, 117, 140-148, 1998.

Furata, H; Mori, N; Luo, L; Ryan, AF. Detection of mRNA encoding guanylate cyclase A/atrial natriuretic peptide receptor in the rat cochlea by competitive polymerase chain reaction and in situ hybridization. *Hear Res*, 92, 78-84, 1995.

Glembotski, CC; Dixon, JE; Gibson, TR. Secretion of atrial natriuretic factor-(1-98) by primary cardiac myocytes. *J Biol Chem*, 263, 16073-16081, 1988.

Green, TP; Rybak, LP; Mirkin, BL; Juhn, SK; Morizono, T. Pharmacologic determinants of ototoxicity of furosemide in the chinchilla. *J Pharmacol Exp Ther*, 216, 537-542, 1981.

Guipponi, M; Toh, M-Y; Tan, J; Park, D; Hanson, K; Ballana, E; Kwong, D; Cannon, PZF; Wu, Q; Gout, A; Delornzi, M; Speed, TP; Smith, RJH; Dahl, HH; Petersen, M; Teasdale, RD; Estivill, X; Park, WJ; Scott, HS. An integrated genetic and functional analysis of the role of type II transmembrane serine proteases (TMPRSSs) in hearing loss. *Human Mutation*, 29, 130-141, 2008.

Hamet, P; Tremblay, J; Pang, SC; Carcia, R; Thibault, G; Gutkowska, J; Cantin, M. Genest Effect of native and synthetic atrial natriuretic factor on cyclic GMP. *Biochem Biophys Res Comm*, 123, 515-527, 1984.

Hamzavi, J; Schmetterer, L; Formanek, M. Vestibular symptoms as a complication of sildenafil: a case report. *Wien Klin Wochenschr*, 114, 54-55, 2002.

Heiniger-Bürki, CB; Ehrenberger, K; Felix, D. Is ntiric oxide involved in inner ear neurotransmission? *Prim Sens Neuron*, 1, 41-46, 1995.

Heinrich, U; Maurer, J; Koesling, D; Mann, W; Förstermann, U. Immuno-electron microscopic localization of the alpha(1) and beta(1)-subunits of soluble guanylyl cyclase in the guinea pig organ of Corti. *Brain Res*, 885, 6-13, 2000.

Holz, N; Trachte, GJ; Fitzakerley, JL. Improvement in acoustic thresholds following administration of a phosphodiesterase (PDE) inhibitor in mice. *ARO Midwinter Meeting*, 2010.

Hong, BN; Yi, TH; Kim, SY; Kang, TH. High dosage sildenafil induces hearing impairment in mice. *Bio Pharm Bull*, 31, 1981-1984.

Huang, MZ; Zhang, JG; Long, YS. Identification of novel cyclic nucleotide phosphodiesterase gene cDNAs in the cochlea of guinea pig (Cavia porcellus) through conserved homologous sequences. *Mol Biotechnol*, 44, 66-70, 2010.

Jaubert, J; Jaubert, F; Martin, N; Washburn, LL; Lee, BK; Eicher, EM; Guénet, JL. Three new allelic mouse mutations that cause skeletal overgrowth involve the natriuretic peptide receptor C gene (NPR3). *Proc Natl Acad Sci USA*, 96, 10278-83, 1999.

Jaumann, M; Dettling, J; Gubelt, M; Zimmermann, U; Gerling, A; Paquet-Durand, F; Feil, S; Wolpert, S; Franz, C; Varakina, K; Xiong, H; Brandt, N; Kuhn, S; Geisler, H-S; Rohbock, K; Ruth, P; Schlossmann, J; Hutter, J; Sandner, P; Feil, R; Engel, J; Knipper, M; Ruttiger, L. cGMP-PrkG1 signaling and Pde5 inhibition shelter cochlear hair cells and hearing function. *Nature Medicine*, 18, 252-259, 2012.

Kharkovets, T; Dedek, K; Maier, H; Schweizer, M; Khimich, D; Nouvian, R; Vardanyan, V; Leuwer, R; Moser, T; Jentsch, TJ. Mice with altered KCNQ4 K+ channels implicate sensory outer hair cells in human progressive deafness. *EMBO J*, 25, 642-652, 2006.

Koch, T; Gloddek, B; Gutzke, S. Binding sites of atrial natriuretic peptide (ANP) in the mammalian cochlea and stimulation of cyclic GMP synthesis. *Hear Res*, 63, 197-202, 1992.

Kort, JJ; Koch, G. The Na+,K+,2Cl-cotransport system in HeLa cells: aspects of its physiological regulation. *J Cell Physiol*, 145, 253-261, 1990.

Koller, KJ; Lowe, DG; Bennett, GL; Minamino, N; Kangawa, K; Matsuo, H; Goeddel, DV. Selective activation of the B natriuretic peptide receptor by C-type natriuretic peptide. Science, 252, 120-123, 1991.

Kourie, JI; Rive, MJ. Role of natriuretic peptides in ion transport mechanisms. *Med Res Rev*, 19, 75-94, 1999.

Krause, G; Meyer zum Gotteberge, AM; Wolfram, G; Gerzer, R. Transcripts encoding three types of guanylyl cyclase coupled transmembrane receptors in inner ear tissues of guinea pigs. *Hear Res*, 11, 95-106, 1997.

Lee, JE; Nakagawa, T; Kita, T; Kim, TS; Iguchi, F; Endo, T; Shiga, A; Lee, SH; Ito, J. Mechanisms of apoptosis induced by cisplatin in marginal cells in mouse stria vascularis. ORL *J Otorhinolaryngol Relat Spec*, 66, 111-118, 2004.

Lee, MP; Ravenel, JD; Hu, RJ; Lustig, LR; Tomaselli, G; Berger, RD; Brandenburg, SA; Litzi, TJ; Bunton, TE; Limb, C; Francis, H; Gorelikow, M; Gu, H; Washington, K. targeted disruption of the Kvlqt1 gene causes deafness and gastric hyperplasia in mice. *J Clin Invest*, 106, 1447-1455, 2000.

Liu, H; Maurice, DH. Expression of cyclic GMP-inhibited phosphodiesterases 3A and 3B (PDE3A and PDE3B) in rat tissues: differential subcellular localization and regulated expression by cyclic AMP. *Br J Pharmacol*, 125, 1501-1510, 1998.

Long, L; Tang, Y; Xia, Q; Xi, Z; Liu, J. Detection of atrial natriuretic peptide receptor in the labyrinth of the mouse inner ear. *Neuro Endocrinol Lett*, 29, 577-580, 2008.

Long, L; Tang, Y; Xia, Q; Xia, Z; Liu, J. The expression of atrial natriuretic peptide receptor in the mouse inner ear labyrinth. *Neuro Endocrinol Lett*, 31, 126-130, 2010.

Lopez, MJ; Wong, SK; Kishimoto, I; Dubois, S; Mach, V; Friesen, J; Garbers, DL; Beuve, A. Salt-resistant hypertension in mice lacking the guanylyl cyclase-A receptor for atrial natriuretic peptide, *Nature* 378, 65-68, 1995.

Lugnier, C. Cyclic nucleotide phosphodiesterase (PDE) superfamily: A new target for the development of specific therapeutic agents. *Pharmacol Therapeutics*, 109, 366-398, 2006.

Maack, T; Suzuki, M; Almeida, FA; Nussenzveig, D; Scarborough, RM; McEnroe, GA; Lewicki; JA. Physiological role of silent receptors of atrial natriuretic factor._*Science*, 238, 675-678, 1987.

Maddox, PT; Saunders, J; Chandrasekhar, SS. Sudden hearing loss from PDE5 inhibitors: A possible cellular stress etiology. *Laryngoscope*, 119, 1586-1589, 2009.

Maki, M; Tkayanagi, R; Misono, KS; Pandey, KN; Tibbets, C; Inagami, T. Structure of rat atrial natriuretic factor precursor deduced from cDNA sequence. *Nature*, 309, 722-724, 1984.

Manallack, DT; Hughes, RA; Thompson, PE. The next generation of phosphodiesterase inhibitors: Structural clues to ligand and substrate selectivity of phosphodiesterases. *J. Med Chem*, 48, 3449-3462, 2005.

Matsunobu, T; Schacht, J. Nitric oxide/cyclic GMP pathway attenuates ATP-evoked intracellular calcium increase in supporting cells of the guinea pig cochlea. *J Comp Neurol*, 423, 452-61, 2000.

McGwin, G, Jr.; Phosphodiesterase type 5 inhibitor use and hearing impairment. Arch. Otolaryngol. *Head Neck Surg*, 136, 488-492, 2010.

Meyer zum Gottesberge, AM; Gagelmann, M; Forssmann, WG. Atrial natriuretic peptide-like immunoreactive cells in the guinea pig inner ear. *Hear Res*, 56, 86-92, 1991.

Meyer zum Gottesberge, AM; Schleicher, A; Drummer, C; Gerzer, R. The volume protective natriuretic peptide system in the inner ear. *Acta Otolaryngol*, (Suppl) 520, 170-173, 1995.

Mukherjee, B; Shivakumar, T. A case of sensorineural deafness following ingestion of sildenafil. *J Laryngol Otol*, 121, 395-7, 2007 Epub 2006 Dec 14.

Nagura, M; Iwasaki, S; Mizuta, K; Mineta, H; Umemura, K; Hoshino, T. Role of nitric oxide in focal microcirculation disorder of guinea pig cochlea. *Hear Res*, 153, 7-13, 2001.

O'Donnell, ME; Owen, NE. Atrial natriuretic factor stimulates Na/K/Cl cotransport in vascular smooth muscle cells. *Proc Natl Acad Sci USA*, Aug, 83(16), 6132-6136, 1986.

O'Grady, SM. Cyclic nucleotide-mediated effects of ANF and VIP on flounder intestinal ion transport. *Am J Physiol*, 256, C142-C146, 1989.

Oikawa, S; Imai, M; Ueno, A; Tanaka, S; Noguchi, T; Nakazato, H; Kangawa, K; Fukuda, A; Matsuo, H. Cloning and sequence analysis of cDNA encoding a precursor for human atrial natriuretic polypeptide. *Nature*, 309, 724-726, 1984.

Okuyucu, S; Guven, OE; Aroglu, E; Ucar, E; Dagli, S. Effect of phosphodiesterase-5 inhibitor on hearing. *J. Laryngol. Otol.*, 123, 218-222, 2009.

Owen, NE; Bush, EN; Holleman,W; O'Donnell, ME. Effect of atrial natriuretic factor on Na+-K+-Cl- cotransport of vascular smooth muscle cells. *Hypertension*, 10, I128-130, 1987.

Quiao, L; Han, Y; Zhang, P; Cao, Z; Qiu, J. Detection of atrial natriuretic peptide and its receptor in marginal cells and cochlea tissues from developing rats. *Neuro Endocrinol Lett*, 32, 187-192, 2011.

Rachel, JD; Dziadziola, JK; Quirk, WS. Atrial natriuretic peptide participates in the regulation of vestibular blood flow. *Hear Res*, 89, 181-186, 1995.

Racic , G; Maslovara, S; Roje, Z; Dogas, Z; Tafra, R. Hyperbaric oxygen in the treatment of sudden hearing loss. *ORL J Otorhinolaryngol Relat Spec*, 65, 317-20, 2003.

Rao, MC; Nash, NT. 8-BrcAMP does not affect Na-K-2Cl cotransport in winter flounder intestine. *Am J Physiol*, 255, C246-C251, 1988.

Rybak, LP; Whitworth, C. Changes in endolymph chloride concentration following furosemide injection. *Hear Res.*, 24, 133-136, 1986.

Semenov, AG; Tamm, NN; Seferian, KR; Postnikov, AB; Karpova, NS; Serebryanaya, DV; Koshkina, EV; Krasnoselsky, MI; Katukha, AG. Processing of Pro-B-type natriuretic peptide: furin and corin as candidate convertases. *Clin Chem*, 56, 1166-1176, 2010

Shi, X; Nuttall, AL. Upregulated iNOS and oxidative damage to the cochlear stria vascularis due to noise stress. *Brain Res*, 967, 1-10, 2003.

Shimizu, K; Shintani, Y; Ding, WG; Matsuura, H; Bamba, T. Potentiation of slow component of delayed rectifier K(+) current by cGMP via two distinct mechanisms: inhibition of phosphodiesterase 3 and activation of protein kinase G. *Br J Pharmacol*, 137, 127-137, 2002.

Silberbach, M; Roberts, CT Jr. Natriuretic peptide signaling: molecular and cellular pathways to growth regulation. *Cell Signal*, 13, 221-31, 2001.

Sudoh, T; Kangawa, K; Minamino, N; Matsuo, H. A new natriuretic peptide in porcine brain. *Nature*, 332, 78-81, 1988.

Sudoh, T; Minamino, N; Kangawa, K; Matsuo, H. C-type natriuretic peptide (CNP): a new member of natriuretic peptide family identified in porcine brain. *Biochem Biophys Res Commun*, 168, 863-870, 1990.

Sun, F; Zhou, K; Want, SJ; Liang, PF; Wu, YX; Zhu, GX; Qiu, JH; Zhu, MZ. Expression and localization of atrial natriuretic peptide and its receptors in rat spiral ganglion neurons. *Brain Res Bull* 95: 28-32, 2013.

Suzuki, M; Kitanishi, T; Kitano, H; Yazawa, Y; Kitajima, K; Takeda, T; Tokunaga, Y; Maeda, T; Kimura, H; Tooyama, I. C-type natriuretic peptide-like immunoreactivity in the rat inner ear. *Hear Res*, 139, 51058, 2000.

Suzuki, M; Kitano, H; Kitanishi, T; Yazawa, Y; Kitajima, K; Takeda, T; Kimura, H; Tooyama, I. Detection of C-type natriuretic peptide (CNP) and atrial natriuretic peptide (ANP-B) receptor mRNA in rat inner ear. *NeuroReport*, 8, 439-443, 1997.

Suzuki, M; Kitano, H; Kitanishi, T; Yazawa, Y; Kitajima, K; Takeda, T; Kimura, H; Tooyama, I. RT-PCR analysis of mRNA expression of natriuretic peptide family and their receptors in rat inner ear. *Mol Brain Res*, 55, 165-168, 1998.

Szabo, R; Bugge, TH. Membrane-anchored serine proteases in vertebrate cell and developmental biology. *Annu Rev Cell Dev Biol*, 27, 213-35, 2011.

Takumida, M; Anniko, M. Nitric oxide in the inner ear. *Curr Opin Neurol*, 15, 11-15, 2002.

Takumida, M; Anniko, M; Popa, R; Zhang, DM. Localization of soluble guanylate cyclase activity in the guinea pig inner ear. *Acta Otolaryngol*, 120, 28-33, 2000.

Takumida, M; Anniko, M; Popa, R; Zhang, DM. Pharmacological models for inner ear therapy with emphasis on nitric oxide. *Acta Otolaryngol* (Stockh), 121, 16-20, 2001.

Takahashi, T; Allen, PD; Izumo, S. Expression of A-, B- and C-type natriuretic peptide genes in failing and developing human ventricles. *Circ. Res.*, 71, 9-17, 1992.

Thusu, KG; Morin, FC 3rd; Russell, JA; Steinhorn, RH. The cGMP phosphodiesterase inhibitor zaprinast enhances the effect of nitric oxide. *Am. J. Respir. Crit. Care Med.*, 152, 1605-1610, 1995.

Van de Ven, WJM; Voorberg, J; Fontijn, R; Pannekoek, H; Van den Ouweland, AMW; Van Duijnhoven, HLP; Roebroek, AJM; Siezen, RJ. Furin is a subtilisin-like proprotein processing enzyme in higher eukaryotes, *Mol. Biol. Rep.*, 14, 265-275, 1990.

Vargish, T; Benjamin, R; Shenkman, L. Deafness from furosemide. *Ann Intern Med*, 72, 761, 1970.

Vigne, P; Lund, L; Frelin, C. Cross talk among cyclic AMP, cyclic GMP, and Ca(2+)-dependent intracellular signalling mechanisms in brain capillary endothelial cells. *J Neurochem*, 62, 2269-2274, 1994.

Wangemann, P. K+ cycling and the endocochlear potential. *Hear Res*, 165, 1-9, 2002.

Wangemann, P. Supporting sensory transduction: cochlear fluid homeostasis and the endocochlear potential. *J Physiol*, 576, 11-21, 2006.

White, RE; Lee, AB; Shcherbatko, AD; Lincoln, TM; Schonbrunn, A; Armstrong, DL. Potassium channel stimulation by natriuretic peptides through cGMP-dependent dephosphorylation. *Nature*, 361, 263-266, 1993.

Wu, C; Wu, F; Pan, J; Morser, J; Wu, Q. Furin-mediated processing of Pro-C-type natriuretic peptide. *J. Biol. Chem.*, 278, 25847-25852, 2003

Xiong, W; Qiu, JH; Lin, Y; Qiao, L; Liu, SL. [Atrial natriuretic factor's effects on the reperfusion process after cochlea ischemia]. *Zhonghua Er Bi Yan Hou Tou Jing Wai Ke Za Zhi*, 41, 293-296, 2006.

Yamanaka, M; Greenberg, B; Johnson, L; Seilhamer, J; Brewer, M; Friedemann, T; Miller, J; Atlas, S; Laragh, J; Lewicki, J; Fiddes, J. Cloning and sequence analysis of the cDNA for the rat atrial natriuretic factor precursor. *Nature*, 309, 719-722, 1984.

Yamane, H; Takayama, M; Sunami, K; Iguchi, H; Kanazawa, A; Tokuhara, Y; Nishiura, H; Nishimura, K; Tao, SD. Nitric oxide induces apoptosis of the hair cells of cochlea. *Acta Otolaryngol Suppl*, 554, 6-11, 2004.

Yan, W; Wu, F; Morser, J; Wu, Q. Corin, a transmembrane cardiac serine protease, acts as a pro-atrial natriuretic peptide-converting enzyme. *Proc Natl Acad Sci (USA)*, 97, 8525-8529, 2000.

Yoon, Y-J; Hellstrom, S. Immunohistochemical localization of alpha-atrial natriuretic polypeptide in the rat cochlea. *Acta Otolaryngol*, 112, 604-610, 1992.

Yoon, YJ; Lee, EJ; Kim, SH. Synthesis of atrial natriuretic peptide in the rabbit inner ear. *Laryngoscope*, 122, 1605-1608, 2012.

In: Natriuretic Peptides
Editor: Kailash N. Pandey

ISBN: 978-1-63321-080-6

Chapter 12

MECHANISMS OF ACTION OF GUANYLIN PEPTIDES

Aleksandra Sindic*

Associate Professor, Department of Physiology, School of Medicine, University of Zagreb, Zagreb, Croatia

ABSTRACT

Guanylin peptides (GPs) are proposed to be intestinal natriuretic peptides. Ingestion of a salty meal induces secretion of guanylin (GN) and uroguanylin (UGN) into the intestinal lumen, where they inhibit sodium absorption and induce anion and water secretion. At the same time, those hormones stimulate renal electrolyte excretion by inducing natriuresis, kaliuresis, and diuresis and therefore participate in the prevention of hypernatremia and hypervolemia after salty meals.

In the gut, GPs activate guanylate cyclase C (GC-C) and increase intracellular cGMP which inhibits apical sodium/proton exchange (NHE) and increases the secretion of chloride and bicarbonate via activation of the Cystic Fibrosis Transmembrane Conductance Regulator (CFTR). However, GC-C deficient mice, renal effects of GPs are unaltered, which could be by explained by recently described, new cGMP-independent signaling pathways (G-protein coupled receptor). This is not unusual while ANP also activates two different receptor types: guanylate cylcase A and clearance receptor which is also G-protein coupled receptor.

It is known that GPs and their signaling pathway exist in other organs like liver, pancreas, lung, sweet glands, and male reproductive system and they are involve in different pathological conditions like cystic fibrosis, asthma, intestinal tumors, kidney and heart failure. To determine those physiological and pathophysiological roles of GPs recent research is done.

* Corresponding author: Aleksandra Sindic, MD, PhD, Assistant Professor, Department of Physiology, School of Medicine, University of Zagreb, Salata 3, 10000 Zagreb, Croatia. Email: asindic@mef.hr Phone: ++ 385 1 4566 766 or ++ 385 1 4596 875.

1. INTRODUCTION

The existence of intestinal hormones that regulates kidney function was first proposed by Lennane et al. in 1975 [1]. After a salty meal, intestinal natriuretic peptides also known as guanylin peptides (GPs), to prevent postprandial hypernatremia, inhibit sodium absorption from intestinal lumen [2, 3] and increase renal sodium secretion [4-6]. GPs appeared early in evolution as they are found in all examined animals (mammals, birds, fishes) which indicates the importance of those hormones in the maintains of the homeostasis of water and electrolyte homeostasis in parallel to other natriuretic peptides and renin–angiotensin–aldosterone system and arginine vasopressin.

The GPs family has, as known to date, four members: guanylin (GN) was isolated first from rat intestine [7]. One year later uroguanylin (UGN) was discovered in opossum urine [8]. We have less knowledge about last two members of the GPs family: lymphoguanylin isolated from opossum lymphatic tissues [9] and renoguanylin founded in eels [10].

2. STRUCTURE AND METABOLISM OF GUANYLIN PEPTIDES

Genes for GPs are located at the first human chromosome (p33 – p36) and the fourth mouse chromosome. They are synthesized as any other peptide (preprohormone, prohormone, hormone) and have 15 – 19 amino acids with two disulfide bonds which are essential for their activity (Figure 1.) [11-13]. Receptor for GPs was first described as receptor for the heat-stable enterotoxin of Escherichia coli (STa) [14] which has three disulfide bonds and it is presumed that this is the reason for its stronger and unregulated activation on the receptor which produces secretory diarrhea [15, 16]. After the identification of a receptor for the exogenous enterotoxin in the intestine and in other tissues we started to look for endogenous ligands and first discovered GN and soon after UGN. Human GN (known as Guanylat cyclase activating peptide – 1, GCAP–1).consists of 15 amino acids and possesses two disulfide bonds between two cysteins (between 4 - 12 and 7 - 15 positions) (Figure 1.) [11, 13]. Human UGN (known as Guanylat cyclase activating peptide – 2, GCAP-2) has 16 amino acids and has two disulfide bonds (between 4 - 12 and 7 -15 positions) (Figure 1.) [12]. Structure differences for GN and UGN in different species are given in Table 1. A physiologically important structural difference between GN and UGN lies in their sensitivity towards proteases. GN has a tyrosine as its ninth amino acid which makes it sensitive to proteases, mostly chymotrypsin, while UGN has an asparagine at the same position [17, 18]. Therefore GN is degraded by chymotrypsin present in different parts of the kidney nephrons [4, 19] and is consequently not present in urine.

The intestine is producing GPs in the gut lumen (effects on the water and electrolyte absorption) and it is the main source of plasma GPs which induce natriuresis in the kidney [20-22]. Recently, circulating plasma prouroguanylin was suggested to mediate entero-renal signaling by conversion to the UGN in the kidney [23, 24]. The both types of the intestinal secretion of GPs are stimulated by parasympathetic stimulation (n.vagus) [25, 26]. Interestingly, from still unknown reason, Zn^{2+} deficiency leads to increase in UGN expression in rat enterocytes [27-30] which could lead to secretory diarrhea which is present in patients with Zn^{2+} deficiency.

Table 1. Sequences of guanylin and uroguanylin in different species

Guanylin	
Eel	Y D E **C E I C** M F **A A C** T **G C**
Opossum	S H T **C E I C** A F **A A C** A **G C**
Human	P G T **C E I C** A Y **A A C** T **G C**
Rat/mouse	P N T **C E I C** A Y **A A C** T **G C**
Pig	P S T **C E I C** A Y **A A C** A **G C**
Guinea pig	P S T **C E I C** A Y **A A C** A **G C**
Uroguanylin	
Eel	P D P **C E** I **C** A **N** A **A C T G C** L
Opossum	Q E D **C E** L **C** I **N** V **A C T G C**
Human	N D D **C E** L **C** V **N** V **A C T G C** L
Rat/Mouse	T D E **C E** L **C** I **N** V **A C T G C**
Pig	G D D **C E** L **C** V **N** V **A C T G C** S
Guinea pig	N D E **C E** L **C** V **N** I **A C T G C**

Identical amino acids in these proteins of different species are indicated in bold [7, 8, 17, 38, 123].

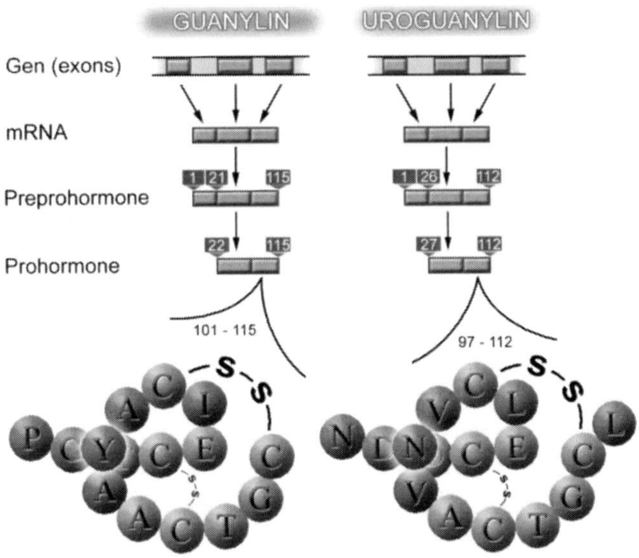

Figure 1. Structure and synthesis of human guanylin peptides. Human preproguanylin has 115, proguanylin 94 and guanylin 15 amino acids. Human preprouroguanylin 112 amino acids. Human preprouroguanylin has 112, prouroguanylin 86 and uroguanylin 16 amino acids. Structures of human guanylin and uroguanylin are shown at the bottom of the Figure. green colored circles show identical amino acids in both peptides. Light circled amino acids are responsible for peptide sensitivity to chymotrypsin [11, 119-122].

GPs mRNA could be found in other organs as well, like adrenal glands, reproductive system, lung, pancreas, and kidneys where the importance of GPs secretion is, as far as we know, local regulation of membrane transport systems and it is still not very well understood [31-33].

3. Signaling Pathways of Guanylin Peptides in the Intestine

Salt ingestion leads to the secretion of GN and UGN from enterochromaffin cells into the intestinal lumen [2, 3, 34, 35]. mRNA for GN is located from duodenum to distal colon from lower to higher expression along the intestine [36, 37]. UGN is present along the gastrointestinal tract with the highest expression in the duodenum [33, 34, 36, 38-40] where could play a role in regulation of the duodenal pH. At low duodenal pH after emptying stomach content (pH 5) UGN could more potently activate it's receptor compared to the activity at pH 8. As you can see at Figure 2 UGN increase bicarbonate secretions and inhibits H^+ secretion and therefore decrease H^+ concentration in the duodenum [41].

When secreted in the gut lumen, GPs stimulate enterocytes via the membrane bound guanylate cyclase C (GC-C) and increase the intracellular concentration of cGMP [7, 8]. The increase in cGMP activates the protein kinase G II [42, 43] and inhibits the phosphodiesterase III which leads to increase in cellular concentration of cAMP and activation of protein kinase A [42, 44, 45]. Furthermore, cGMP slowing down the Na^+ absorption by inhibiting apical Na^+ /H^+ exchange (NHE2) and therefore preventing fast increase in blood sodium concentrations after the salty meal [46, 47]. Protein kinase G II and protein kinase A increase the secretion of Cl^-, and HCO_3^- via the Cystic Fibrosis Transmembrane Conductance Regulator (CFTR) and preventing to fast water absorption and increase in blood volume which could not be quickly excreted from the organism by the kidneys (Figure 2) [44, 48-50].

Main GPs receptor, GC-C, belongs to the family of guanylate cyclases which consists of a cytoplasmatic guanylate cyclase (receptor for NO and CO) and six types of membrane bound guanylate cyclases. GC-A and GC-B are receptors for the natriuretic peptides atrial natriuretic peptide (ANP), brain natriuretic peptide (BNP), and C type natriuretic peptide (CNP). Guanylate cyclases D, E and F (GC-D, GC-E, GC-F) are located in sensory organs and recently GC-D was proposed to be a GPs receptor in olfactory system [51]. Guanylate cyclase G (GC-G or GC-1) is an orphan receptor present in the skeletal muscle, lung, intestine, testis, and kidney [52-54]. GC-C is localized in adrenal glands, brain, the embryonic or regenerating but not adult liver, placenta, testis, airways, spleen, thymus, and lymphatic nodes [39, 55, 56]. GC-C expression is increased in infants and prematures which are therefore more sanative to presence of STa in intestine [57]. Like other GCs, GC-C has extracellular, juxtamembrane domain, kinase homology and catalytic domain (Figure 3).

Juxtamembrane domain is located intracellular close to cell membrane and has similar structure to parts of IGF (insulin-like growth factor) and EGF (epidermal growth factor) receptors which are responsible for binding the G-proteins [58, 59]. Kinase homology domain is similar to catalytic domain of PDGF (platelet-derived growth factor) receptors. For pronouns catalytic activity this domain should be phosphorylated. C-terminal tail of the GC-C is unique for this member of guanylate cyclase family and it is involved in binding the GC-C to cytoskeleton and important for endocytoses of ligand-receptor complexes [60] and phosphorylation of this part of GC-C receptor by protein kinase C (PKC) leads to 70% increase in intracellular concentration of cGMP after STa stimulation [61-65].

The existence of additional receptors for GPs became evident when intestinal and renal effects were observed in GC-C – deficient mice. These mice are resistant to intestinal secretion produced by STa [66-68], however, 10% of the STa intestinal binding sites are still present possibly involving Ca^{2+} signaling pathway with activation of PKC [69, 70].

Furthermore, binding sites for GPs that are distinct from GC-C are located on the basolateral membrane of colonocytes [71] and those receptors and signaling pathways are still not identified. Furthermore, GPs effects are still present in GC-C–deficient mice and this additional signaling pathway is investigated in more details in mouse and human cortical collecting ducts [6, 72, 73]. Existence of GC-C independent signaling pathway is shown as well by knowing that UGN deficient mice are developing hypertension compared to GC-C knock out animals which blood pressure is normal [74, 75]. Further research should be done to investigate importance of GC-C independent signaling pathway in extra renal tissues.

4. Signaling Pathways of Guanylin Peptides in the Kidney

The action of the GPs in the kidney can be secreted by the intestine and paracrine produced locally by the kidney itself. Similar to the difference in the axial expression along the intestine, GN and UGN are not expressed equally along the different nephron segments. mRNA for UGN is present mostly in the proximal tubule, again comparable to the proximal intestine, and GN is more pronounced in the collecting duct, in analogy to GN expression in the colon [76].

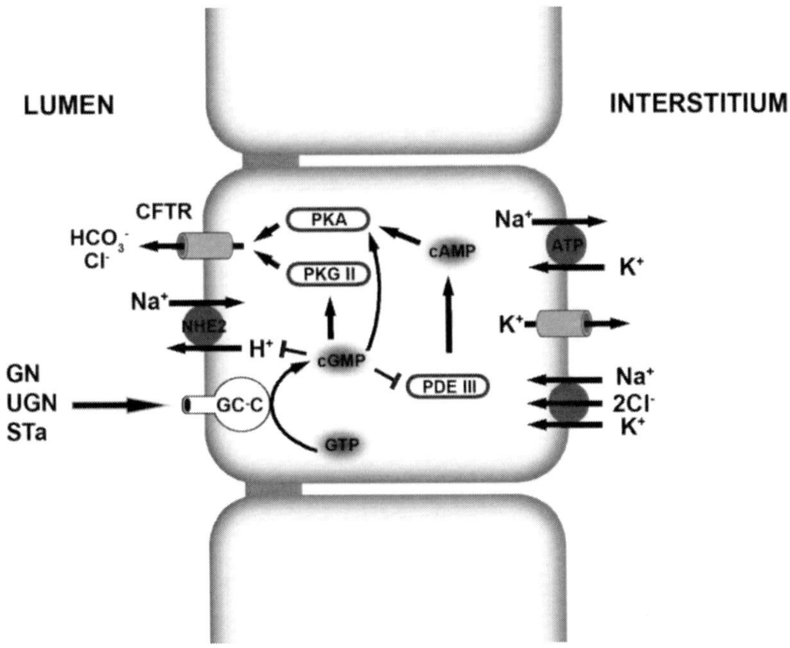

Figure 2. Signaling pathway of guanylin peptides in the intestine. Guanylin (GN), uroguanylin (UGN), and heat-stable enterotoxin of E. coli (STa) activate GC-C, and increases of the intracellular concentration of cGMP follows. Functions of the cGMP in the enterocytes are several: inhibits the Na^+/H^+ exchanger type 2 (NHE2) and activates protein kinase G type II (PKG II). Furthermore, cGMP activates protein kinase A directly or indirectly by inhibition of phosphodiesterase III (PDE III) followed by increase in intracellular cAMP concentration. Both protein kinases activate Cl^- and HCO_3^- secretion via activation of cystic fibrosis transmembrane regulator (CFTR).

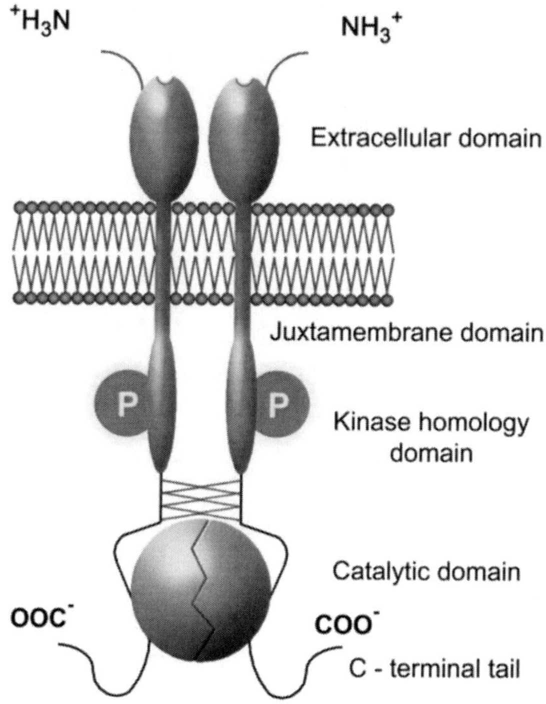

Figure 3. Guanylat cyclase C.

Effects of the GN and UGN on the kidney nephron segments are slightly different. UGN is proposed to be mostly natriuretic hormone for numerous reasons [77]. After high salt diet animals have higher UGN and cGMP concentrations in the urine as well as increased UGN expression in the neprhons [76, 78]. Finally in UGN deficient mice natriuresis produced by oral salt load is decreased [74]. In contrast to predominant natriuretic effects of UGN, GN expression in the kidney does not change on high Na^+ diet and GN mainly caused kaliuresis and diuresis with less pronounced natriuretic effect [5, 76].

The first discovered localization of GPs action is proximal tubule. GPs increase intracellular cGMP concentration in a proximal tubule cell line of opossum kidney (OK cells), in kangaroo proximal tubule cells (PtK-2 cells) and in a human proximal tubule cell line (IHKE1) where inhibits Na^+/H^+ exchanger and sodium reapsorption [39, 79-81]. GC-C is expressed in kidney proximal tubules however its specific localization along the nephron varies from species to species [76]. Involvement of cGMP and GC-C in the signaling pathway of GPs in proximal tubule cells is obviously not present in the effects of GPs described in GC-C deficient mice and interestingly another GC-C independent signaling pathway in the kidneys were discovered.

Final hormonal regulation of renal electrolyte and water homeostasis takes place in the collecting duct (CD). They have at least 2 types of the cells: principal cells which are responsible for K^+ secretion (ROMK channels), Na^+ and water reabsorption (ENaC and aquaporins 2, 3, and 4, respectively) and intercalated cells responsible for K^+ reapsorption. GPs induce changes in membrane voltages of principal cells of wild-type and in GC-C–deficient mice and human cortical CD is in the line with the observation that GC-C–deficient mice still exert natriuresis, kaliuresis, and diuresis upon GPs infusion [6, 72, 73]. The source of UGN reaching the lumen of the collecting duct is coming from the blood and/or from secretion by proximal tubule cells [76]. It looks like the GN present in lumen of CDs could be secreted only by local cells since GN is degradated with chymotripsin. Since UGN, in contrast to GN, is resistant to degradation by luminal proteases, UGN will be concentrated along the nephron due to volume reabsorption and will be present in the finall urine (Figure 4.) [17].

In principal cells of the cortical CD GPs activate a G-protein coupled receptor which leads to an increase in phospholipase A2 (PLA_2) and inhibition of ROMK channels via arachidonic acid [72, 73, 82]. A reduction in the potassium conductance of the principal cells and the consequent depolarizations of the cell membrane will reduced the driving force for Na^+ and consecutive water reapsorption which could at least in the part explain the natriuresis and diuresis seen *in vivo*. Remaining question is if GPs could inhibit potassium reapsoription in intercalated cells and therefore induce kaliuresis as well. Further more ROMK deficient mice show natriuresis, kaliuresis, and diuresis, a pattern of responses so similar to mice stimulated with GPs which is still not fully understood [83]. These observations propose ROMK channels as possible final targets for GN and UGN actions in the cortical CD.

It is less known about function of GPs in last part of the nephron, medullary CD. It was shown that GPs decrease the cell volume and increase the luminal space which suggests secretion of water and consequently diuresis [84], however GPs do not change water permeability in inner medullary CD cells [85].

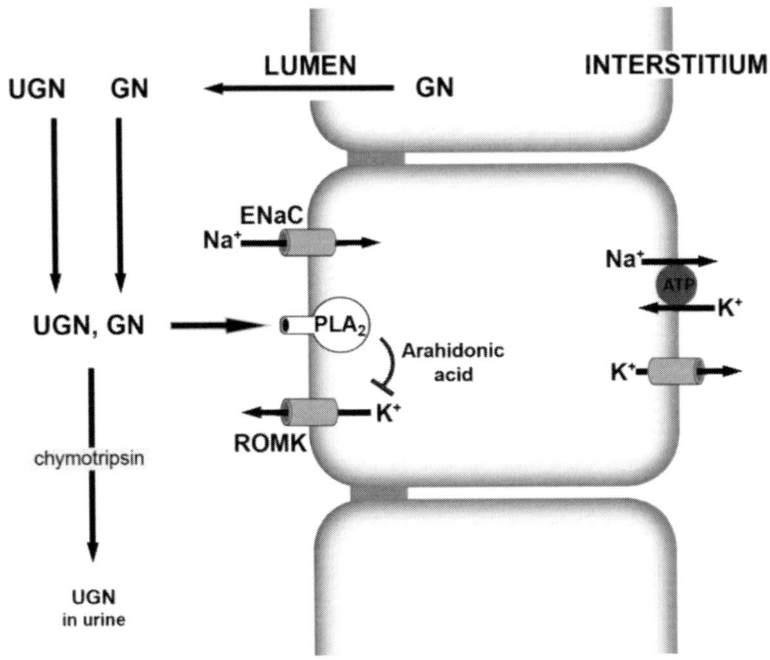

Figure 4. Signaling pathways of GN and UGN in a principal cell of the cortical collecting ducts. Guanylin peptides activate phospholipase A2 (PLA_2), which leads to an increase in the intracellular concentration of arachidonic acid and an inhibition of luminal ROMK K^+ channels. ENaC, epithelial sodium channels.

5. Cellular Effects of Guanylin Peptides in Other Organs

Patients with cystic fibrosis have pathological changes in water and electrolyte transport in the pancreas, the liver, the lung, and the intestine. It might be speculated that GPs are also involved in the physiology of those organs. GPs and parts of their signaling pathway are present in the pancreas. mRNA for GC-C and UGN are detected in the cells of the exocrine part of the pancreas and GN increases intracellular cGMP *via* GC-C in a human pancreatic cell line [86-88]. In the liver, mRNA for GC-C is present in fetuses and newborn rats, after partial hepatectomies, and during liver regeneration and it looks like that GPs have no physiological role in normal adult liver [56, 67, 89, 90]. Lungs are one of the most damaged organs in cystic fibrosis. In healthy individuals GN is secreted by airways where activates well known GC-C signaling pathway and leads to CFTR activation [55, 56, 91-93]. Furthermore, GPs relax tracheal smooth muscle cells and mucus production in large and small airways which suggests UGN for a new treatment of asthma [94, 95].

GPs and parts of the GC-C–dependent signaling pathway are also found in sweat glands, the adrenal medulla, and the male reproductive system [96-98]. Recently was discovered that UGN relaxes human corpus cavernosus and could be used in treatment of erectile dysfunction [99]. Even the appetite is proposed to be regulated by circulatory proUGN secreted from the intestine and binding to GC-C in hypothalamus [100]. According to this limited information on effects of GPs in various organs, the GC-C–dependent signaling pathway seems to be the predominant signaling pathway in most organs and only in the kidney does the signaling involve different receptors and signaling mechanisms.

6. Guanylin Peptides in Pathophysiology

The most investigated pathological conditions which are involving GPs are intestinal tumors. GPs are less expressed in colon adenocarcinoma, adenoma, and intestine polyps in human and mouse compared to healthy individuals suggesting involvement of GMs in development of intestinal tumors [101-103]. Indeed, GPs regulate proliferation and differentiation, prolonging the cell cycle, and induce apoptosis of T84, CaCo2 cells, and mouse intestinal cells and oral application of UGN leads to an decrease in number and size of polyps in mice that develop intestinal polyposis [104-107]. Like UGN, non-steroidal anti-inflammatory drug present their anti-cancer effects via increase in cellular levels of cGMP [103, 108, 109].

Furthermore, ectopic expression of GC-C could be used as marker for development of metastasis in lymphatic system of esophageal, stomach and colon cancer [110-113].

GPs are also involved in different kidney diseases. Concentrations in plasma GPs increase in patients with chronic renal failure, glomerulonephritis and in patients on hemodialysis [20, 21, 114-116]. In patients with nephritic syndrome, UGN plasma concentration was higher and urinary concentration was lower compared with values in healthy volunteers possibly because kidney is damaged and it is reduced capability to metabolize and excrete GPs [20]. Recently in experimental nephrotic syndrome, changes in UGN concentrations in urine and plasma corresponded to changes in Na^+ excretion therefore UGN could be used as natriuretic peptide in nephritic syndrome [117, 118].

CONCLUSION

GPs are well known as intestinal natriuretic peptides with well established function in the intestine and the kidneys. The main signaling pathway for GPs includes GC-C and increase intracellular concentration of cGMP. This signaling pathway is not important in GPs function in the kidneys. We known that GPs and their signaling pathway exist in other organs like liver, pancreas, lung, sweet glands, and male reproductive system and they are involve in different pathological conditions like cystic fibrosis, asthma, intestinal tumors, kidney and heart failure. Future research should investigate importance of GC-C independent signaling pathways in extra-renal tissues and determining new treatments of pathological conditions.

REFERENCES

[1] Lennane, RJ; Peart, WS; Carey, RM; Shaw, J. A comparison on natriuresis after oral and intravenous sodium loading in sodium-depleted rabbits: evidence for a gastrointestinal or portal monitor of sodium intake. *Clin Sci Mol Med*, 1975, 49, 433-6

[2] Kita, T; Kitamura, K; Sakata, J; Eto, T. Marked increase of guanylin secretion in response to salt loading in the rat small intestine. *Am J Physiol*, 1999, 277, G960-6

[3] Li, Z; Knowles, JW; Goyeau, D; Prabhakar, S; Short, DB; Perkins, AG; Goy, MF. Low salt intake down-regulates the guanylin signaling pathway in rat distal colon. *Gastroenterology*, 1996, 111, 1714-21

[4] Greenberg, RN; Hill, M; Crytzer, J; Krause, WJ; Eber, SL; Hamra, FK; Forte, LR. Comparison of effects of uroguanylin, guanylin, and Escherichia coli heat-stable enterotoxin STa in mouse intestine and kidney: evidence that uroguanylin is an intestinal natriuretic hormone. *J Investig Med*, 1997, 45, 276-82

[5] Fonteles, MC; Greenberg, RN; Monteiro, HS; Currie, MG; Forte, LR. Natriuretic and kaliuretic activities of guanylin and uroguanylin in the isolated perfused rat kidney. *Am J Physiol*, 1998, 275, F191-7

[6] Carrithers, SL; Ott, CE; Hill, MJ; Johnson, BR; Cai, W; Chang, JJ; Shah, RG; Sun, C; Mann, EA; Fonteles, MC; Forte, LR; Jackson, BA; Giannella, RA; Greenberg, RN. Guanylin and uroguanylin induce natriuresis in mice lacking guanylyl cyclase-C receptor. *Kidney Int*, 2004, 65, 40-53

[7] Currie, MG; Fok, KF; Kato, J; Moore, RJ; Hamra, FK; Duffin, KL; Smith, CE. Guanylin: an endogenous activator of intestinal guanylate cyclase. *Proc Natl Acad Sci U S A*, 1992, 89, 947-51

[8] Hamra, FK; Forte, LR; Eber, SL; Pidhorodeckyj, NV; Krause, WJ; Freeman, RH; Chin, DT; Tompkins, JA; Fok, KF; Smith, CE; et al. Uroguanylin: structure and activity of a second endogenous peptide that stimulates intestinal guanylate cyclase. *Proc Natl Acad Sci U S A*, 1993, 90, 10464-8

[9] Forte, LR; Eber, SL; Fan, X; London, RM; Wang, Y; Rowland, LM; Chin, DT; Freeman, RH; Krause, WJ. Lymphoguanylin: cloning and characterization of a unique member of the guanylin peptide family. *Endocrinology*, 1999, 140, 1800-6

[10] Yuge, S; Inoue, K; Hyodo, S; Takei, Y. A novel guanylin family (guanylin, uroguanylin, and renoguanylin) in eels: possible osmoregulatory hormones in intestine and kidney. *J Biol Chem*, 2003, 278, 22726-33

[11] de, Sauvage, FJ; Keshav, S; Kuang, WJ; Gillett, N; Henzel, W; Goeddel, DV. Precursor structure, expression, and tissue distribution of human guanylin. *Proc Natl Acad Sci U S A*, 1992, 89, 9089-93

[12] Kita, T; Smith, CE; Fok, KF; Duffin, KL; Moore, WM; Karabatsos, PJ; Kachur, JF; Hamra, FK; Pidhorodeckyj, NV; Forte, LR; et al. Characterization of human uroguanylin: a member of the guanylin peptide family. *Am J Physiol*, 1994, 266, F342-8

[13] Nokihara, K; Wray, V; Ando, E; Naruse, S; Hayakawa, T. Synthesis, solution structure, binding activity, and cGMP activation of human guanylin and its disulfide isomer. *Regul Pept*, 1997, 70, 111-20

[14] Schulz, S; Green, CK; Yuen, PS; Garbers, DL. Guanylyl cyclase is a heat-stable enterotoxin receptor. *Cell*, 1990, 63, 941-8

[15] Field, M; Graf, LH; Laird, WJ; Smith, PL. Heat-stable enterotoxin of Escherichia coli: in vitro effects on guanylate cyclase activity, cyclic GMP concentration, and ion transport in small intestine. *Proc Natl Acad Sci U S A*, 1978, 75, 2800-4

[16] Hughes, JM; Murad, F; Chang, B; Guerrant, RL. Role of cyclic GMP in the action of heat-stable enterotoxin of Escherichia coli. *Nature*, 1978, 271, 755-6

[17] Hamra, FK; Fan, X; Krause, WJ; Freeman, RH; Chin, DT; Smith, CE; Currie, MG; Forte, LR. Prouroguanylin and proguanylin: purification from colon, structure, and modulation of bioactivity by proteases. *Endocrinology*, 1996, 137, 257-65

[18] Hamra, FK; Krause, WJ; Eber, SL; Freeman, RH; Smith, CE; Currie, MG; Forte, LR. Opossum colonic mucosa contains uroguanylin and guanylin peptides. *Am J Physiol*, 1996, 270, G708-16

[19] Arao, M; Yamaguchi, T; Sugimoto, T; Fukase, M; Chihara, K. Characterization of a chymotrypsin-like hydrolytic activity in the opossum kidney cell. *Biochem Cell Biol*, 1994, 72, 157-62

[20] Kinoshita, H; Fujimoto, S; Fukae, H; Yokota, N; Hisanaga, S; Nakazato, M; Eto, T. Plasma and urine levels of uroguanylin, a new natriuretic peptide, in nephrotic syndrome. *Nephron*, 1999, 81, 160-4

[21] Kinoshita, H; Fujimoto, S; Nakazato, M; Yokota, N; Date, Y; Yamaguchi, H; Hisanaga, S; Eto, T. Urine and plasma levels of uroguanylin and its molecular forms in renal diseases. *Kidney Int*, 1997, 52, 1028-34

[22] Kinoshita, H; Nakazato, M; Yamaguchi, H; Matsukura, S; Fujimoto, S; Eto, T. Increased plasma guanylin levels in patients with impaired renal function. *Clin Nephrol*, 1997, 47, 28-32

[23] Moss, NG; Fellner, RC; Qian, X; Yu, SJ; Li, Z; Nakazato, M; Goy, MF. Uroguanylin, an intestinal natriuretic peptide, is delivered to the kidney as an unprocessed propeptide. *Endocrinology*, 2008, 149, 4486-98

[24] Qian, X; Moss, NG; Fellner, RC; Goy, MF. Circulating prouroguanylin is processed to its active natriuretic form exclusively within the renal tubules. *Endocrinology*, 2008, 149, 4499-509

[25] Martin, S; Adermann, K; Forssmann, WG; Kuhn, M. Regulated, side-directed secretion of proguanylin from isolated rat colonic mucosa. *Endocrinology*, 1999, 140, 5022-9

[26] Moro, F; Levenez, F; Nemoz-Gaillard, E; Pellissier, S; Plaisancie, P; Cuber, JC. Release of guanylin immunoreactivity from the isolated vascularly perfused rat colon. *Endocrinology*, 2000, 141, 2594-9

[27] Blanchard, RK; Cousins, RJ. Differential display of intestinal mRNAs regulated by dietary zinc. *Proc Natl Acad Sci U S A*, 1996, 93, 6863-8

[28] Blanchard, RK; Cousins, RJ. Upregulation of rat intestinal uroguanylin mRNA by dietary zinc restriction. *Am J Physiol*, 1997, 272, G972-8

[29] Cui, L; Blanchard, RK; Coy, LM; Cousins, RJ. Prouroguanylin overproduction and localization in the intestine of zinc-deficient rats. *J Nutr*, 2000, 130, 2726-32

[30] Blanchard, RK; Cousins, RJ. Regulation of intestinal gene expression by dietary zinc: induction of uroguanylin mRNA by zinc deficiency. *J Nutr*, 2000, 130, 1393S-8S

[31] Li, Z; Perkins, AG; Peters, MF; Campa, MJ; Goy, MF. Purification, cDNA sequence, and tissue distribution of rat uroguanylin. *Regul Pept*, 1997, 68, 45-56

[32] Miyazato, M; Nakazato, M; Matsukura, S; Kangawa, K; Matsuo, H. Uroguanylin gene expression in the alimentary tract and extra-gastrointestinal tissues. *FEBS Lett*, 1996, 398, 170-4

[33] Nakazato, M; Yamaguchi, H; Date, Y; Miyazato, M; Kangawa, K; Goy, MF; Chino, N; Matsukura, S. Tissue distribution, cellular source, and structural analysis of rat immunoreactive uroguanylin. *Endocrinology*, 1998, 139, 5247-54

[34] Perkins, A; Goy, MF; Li, Z. Uroguanylin is expressed by enterochromaffin cells in the rat gastrointestinal tract. *Gastroenterology*, 1997, 113, 1007-14

[35] Cetin, Y; Kuhn, M; Kulaksiz, H; Adermann, K; Bargsten, G; Grube, D; Forssmann, WG. Enterochromaffin cells of the digestive system: cellular source of guanylin, a guanylate cyclase-activating peptide. *Proc Natl Acad Sci U S A*, 1994, 91, 2935-9

[36] London, RM; Krause, WJ; Fan, X; Eber, SL; Forte, LR. Signal transduction pathways via guanylin and uroguanylin in stomach and intestine. *Am J Physiol*, 1997, 273, G93-105

[37] Wiegand, RC; Kato, J; Currie, MG. Rat guanylin cDNA: characterization of the precursor of an endogenous activator of intestinal guanylate cyclase. *Biochem Biophys Res Commun*, 1992, 185, 812-7

[38] Fan, X; Hamra, FK; London, RM; Eber, SL; Krause, WJ; Freeman, RH; Smith, CE; Currie, MG; Forte, LR. Structure and activity of uroguanylin and guanylin from the intestine and urine of rats. *Am J Physiol*, 1997, 273, E957-64

[39] Fan, X; Wang, Y; London, RM; Eber, SL; Krause, WJ; Freeman, RH; Forte, LR. Signaling pathways for guanylin and uroguanylin in the digestive, renal, central nervous, reproductive, and lymphoid systems. *Endocrinology*, 1997, 138, 4636-48

[40] Whitaker, TL; Witte, DP; Scott, MC; Cohen, MB. Uroguanylin and guanylin: distinct but overlapping patterns of messenger RNA expression in mouse intestine. *Gastroenterology*, 1997, 113, 1000-6

[41] Hamra, FK; Eber, SL; Chin, DT; Currie, MG; Forte, LR. Regulation of intestinal uroguanylin/guanylin receptor-mediated responses by mucosal acidity. *Proc Natl Acad Sci U S A*, 1997, 94, 2705-10

[42] Vaandrager, AB; Bot, AG; Ruth, P; Pfeifer, A; Hofmann, F; De, Jonge, HR. Differential role of cyclic GMP-dependent protein kinase II in ion transport in murine small intestine and colon. *Gastroenterology*, 2000, 118, 108-14

[43] Vaandrager, AB; Bot, AG; De, Jonge, HR. Guanosine 3',5'-cyclic monophosphate-dependent protein kinase II mediates heat-stable enterotoxin-provoked chloride secretion in rat intestine. *Gastroenterology*, 1997, 112, 437-43

[44] Chao, AC; de, Sauvage, FJ; Dong, YJ; Wagner, JA; Goeddel, DV; Gardner, P. Activation of intestinal CFTR Cl- channel by heat-stable enterotoxin and guanylin via cAMP-dependent protein kinase. *Embo J*, 1994, 13, 1065-72

[45] Dousa, TP. Cyclic-3',5'-nucleotide phosphodiesterase isozymes in cell biology and pathophysiology of the kidney. *Kidney Int*, 1999, 55, 29-62

[46] Fawcus, K; Gorton, VJ; Lucas, ML; McEwan, GT. Stimulation of three distinct guanylate cyclases induces mucosal surface alkalinisation in rat small intestine in vitro. *Comp Biochem Physiol A Physiol*, 1997, 118, 291-5

[47] Toriano, R; Ozu, M; Politi, MT; Dorr, RA; Curto, MA; Capurro, C. Uroguanylin regulates net fluid secretion via the NHE2 isoform of the Na/H+ exchanger in an intestinal cellular model. *Cell Physiol Biochem*, 2011, 28, 733-42

[48] Cuthbert, AW; Hickman, ME; MacVinish, LJ; Evans, MJ; Colledge, WH; Ratcliff, R; Seale, PW; Humphrey, PP. Chloride secretion in response to guanylin in colonic epithelial from normal and transgenic cystic fibrosis mice. *Br J Pharmacol*, 1994, 112, 31-6

[49] Goldstein, JL; Sahi, J; Bhuva, M; Layden, TJ; Rao, MC. Escherichia coli heat-stable enterotoxin-mediated colonic Cl- secretion is absent in cystic fibrosis. *Gastroenterology*, 1994, 107, 950-6

[50] Tien, XY; Brasitus, TA; Kaetzel, MA; Dedman, JR; Nelson, DJ. Activation of the cystic fibrosis transmembrane conductance regulator by cGMP in the human colonic cancer cell line, Caco-2. *J Biol Chem*, 1994, 269, 51-4

[51] Basu, N; Visweswariah, SS. Defying the stereotype: non-canonical roles of the Peptide hormones guanylin and uroguanylin. *Front Endocrinol (Lausanne),* 2011, 2, 14

[52] Hirsch, JR; Kruhoffer, M; Adermann, K; Heitland, A; Maronde, E; Meyer, M; Forssmann, WG; Herter, P; Plenz, G; Schlatter, E. Cellular localization, membrane distribution, and possible function of guanylyl cyclases A and 1 in collecting ducts of rat. *Cardiovasc Res*, 2001, 51, 553-61

[53] Kuhn, M; Ng, CK; Su, YH; Kilic, A; Mitko, D; Bien-Ly, N; Komuves, LG; Yang, RB. Identification of an orphan guanylate cyclase receptor selectively expressed in mouse testis. *Biochem J*, 2004, 379, 385-93

[54] Schulz, S; Wedel, BJ; Matthews, A; Garbers, DL. The cloning and expression of a new guanylyl cyclase orphan receptor. *J Biol Chem*, 1998, 273, 1032-7

[55] Range, SP; Holland, ED; Basten, GP; Knox, AJ. Regulation of guanosine 3':5'-cyclic monophosphate in ovine tracheal epithelial cells. *Br J Pharmacol*, 1997, 120, 1249-54

[56] Schulz, S; Chrisman, TD; Garbers, DL. Cloning and expression of guanylin. Its existence in various mammalian tissues. *J Biol Chem*, 1992, 267, 16019-21

[57] Laney, DW; Mann, EA; Dellon, SC; Perkins, DR; Giannella, RA; Cohen, MB. Novel sites for expression of an Escherichia coli heat-stable enterotoxin receptor in the developing rat. *Am J Physiol*, 1992, 263, G816-21

[58] Gudermann, T; Nurnberg, B; Schultz, G. Receptors and G proteins as primary components of transmembrane signal transduction. Part 1. G-protein-coupled receptors: structure and function. *J Mol Med (Berl),* 1995, 73, 51-63

[59] Okamoto, T; Katada, T; Murayama, Y; Ui, M; Ogata, E; Nishimoto, I. A simple structure encodes G protein-activating function of the IGF-II/mannose 6-phosphate receptor. *Cell*, 1990, 62, 709-17
[60] Lucas, KA; Pitari, GM; Kazerounian, S; Ruiz-Stewart, I; Park, J; Schulz, S; Chepenik, KP; Waldman, SA. Guanylyl cyclases and signaling by cyclic GMP. *Pharmacol Rev*, 2000, 52, 375-414
[61] Crane, JK; Wehner, MS; Bolen, EJ; Sando, JJ; Linden, J; Guerrant, RL; Sears, CL. Regulation of intestinal guanylate cyclase by the heat-stable enterotoxin of Escherichia coli (STa) and protein kinase C. *Infect Immun*, 1992, 60, 5004-12
[62] Wada, A; Hasegawa, M; Matsumoto, K; Niidome, T; Kawano, Y; Hidaka, Y; Padilla, PI; Kurazono, H; Shimonishi, Y; Hirayama, T. The significance of Ser1029 of the heat-stable enterotoxin receptor (STaR): relation of STa-mediated guanylyl cyclase activation and signaling by phorbol myristate acetate. *FEBS Lett*, 1996, 384, 75-7
[63] Roy, N; Guruprasad, MR; Kondaiah, P; Mann, EA; Giannella, RA; Visweswariah, SS. Protein kinase C regulates transcription of the human guanylate cyclase C gene. *Eur J Biochem*, 2001, 268, 2160-71
[64] Crane, JK; Shanks, KL. Phosphorylation and activation of the intestinal guanylyl cyclase receptor for Escherichia coli heat-stable toxin by protein kinase C. *Mol Cell Biochem*, 1996, 165, 111-20
[65] Vaandrager, AB; van, der, Wiel, E; De, Jonge, HR. Heat-stable enterotoxin activation of immunopurified guanylyl cyclase C. Modulation by adenine nucleotides. *J Biol Chem*, 1993, 268, 19598-603
[66] Mann, EA; Jump, ML; Wu, J; Yee, E; Giannella, RA. Mice lacking the guanylyl cyclase C receptor are resistant to STa-induced intestinal secretion. *Biochem Biophys Res Commun*, 1997, 239, 463-6
[67] Schulz, S; Lopez, MJ; Kuhn, M; Garbers, DL. Disruption of the guanylyl cyclase-C gene leads to a paradoxical phenotype of viable but heat-stable enterotoxin-resistant mice. *J Clin Invest*, 1997, 100, 1590-5
[68] Charney, AN; Egnor, RW; Alexander-Chacko, JT; Zaharia, V; Mann, EA; Giannella, RA. Effect of E. coli heat-stable enterotoxin on colonic transport in guanylyl cyclase C receptor-deficient mice. *Am J Physiol Gastrointest Liver Physiol*, 2001, 280, G216-21
[69] Ganguly, U; Chaudhury, AG; Basu, A; Sen, PC. STa-induced translocation of protein kinase C from cytosol to membrane in rat enterocytes. *FEMS Microbiol Lett*, 2001, 204, 65-9
[70] Weikel, CS; Spann, CL; Chambers, CP; Crane, JK; Linden, J; Hewlett, EL. Phorbol esters enhance the cyclic GMP response of T84 cells to the heat-stable enterotoxin of Escherichia coli (STa). *Infect Immun*, 1990, 58, 1402-7
[71] Albano, F; Brasitus, T; Mann, EA; Guarino, A; Giannella, RA. Colonocyte basolateral membranes contain Escherichia coli heat-stable enterotoxin receptors. *Biochem Biophys Res Commun*, 2001, 284, 331-4
[72] Sindic, A; Velic, A; Basoglu, C; Hirsch, JR; Edemir, B; Kuhn, M; Schlatter, E. Uroguanylin and guanylin regulate transport of mouse cortical collecting duct independent of guanylate cyclase C. *Kidney Int*, 2005, 68, 1008-17
[73] Sindic, A; Hirsch, JR; Velic, A; Piechota, H; Schlatter, E. Guanylin and uroguanylin regulate electrolyte transport in isolated human cortical collecting ducts. *Kidney Int*, 2005, 67, 1420-7

[74] Lorenz, JN; Nieman, M; Sabo, J; Sanford, LP; Hawkins, JA; Elitsur, N; Gawenis, LR; Clarke, LL; Cohen, MB. Uroguanylin knockout mice have increased blood pressure and impaired natriuretic response to enteral NaCl load. *J Clin Invest*, 2003, 112, 1244-54

[75] Kuhn, M. Function and dysfunction of mammalian membrane guanylyl cyclase receptors: lessons from genetic mouse models and implications for human diseases. *Handb Exp Pharmacol*, 2009, 47-69

[76] Potthast, R; Ehler, E; Scheving, LA; Sindic, A; Schlatter, E; Kuhn, M. High salt intake increases uroguanylin expression in mouse kidney. *Endocrinology*, 2001, 142, 3087-97

[77] Moss, NG; Riguera, DA; Fellner, RC; Cazzolla, C; Goy, MF. Natriuretic and antikaliuretic effects of uroguanylin and prouroguanylin in the rat. *Am J Physiol Renal Physiol*, 2010, 299, F1433-42

[78] Fukae, H; Kinoshita, H; Fujimoto, S; Kita, T; Nakazato, M; Eto, T. Changes in urinary levels and renal expression of uroguanylin on low or high salt diets in rats. *Nephron*, 2002, 92, 373-8

[79] Forte, LR; Krause, WJ; Freeman, RH. Receptors and cGMP signalling mechanism for E. coli enterotoxin in opossum kidney. *Am J Physiol*, 1988, 255, F1040-6

[80] London, RM; Eber, SL; Visweswariah, SS; Krause, WJ; Forte, LR. Structure and activity of OK-GC: a kidney receptor guanylate cyclase activated by guanylin peptides. *Am J Physiol*, 1999, 276, F882-91

[81] Sindice, A; Basoglu, C; Cerci, A; Hirsch, JR; Potthast, R; Kuhn, M; Ghanekar, Y; Visweswariah, SS; Schlatter, E. Guanylin, uroguanylin, and heat-stable euterotoxin activate guanylate cyclase C and/or a pertussis toxin-sensitive G protein in human proximal tubule cells. *J Biol Chem*, 2002, 277, 17758-64

[82] Macica, CM; Yang, Y; Hebert, SC; Wang, WH. Arachidonic acid inhibits activity of cloned renal K+ channel, ROMK1. *Am J Physiol*, 1996, 271, F588-94

[83] Lu, M; Wang, T; Yan, Q; Yang, X; Dong, K; Knepper, MA; Wang, W; Giebisch, G; Shull, GE; Hebert, SC. Absence of small conductance K+ channel (SK) activity in apical membranes of thick ascending limb and cortical collecting duct in ROMK (Bartter's) knockout mice. *J Biol Chem*, 2002, 277, 37881-7

[84] Furuya, S; Naruse, S; Ando, E; Nokihara, K; Hayakawa, T. Effect and distribution of intravenously injected 125I-guanylin in rat kidney examined by high-resolution light microscopic radioautography. *Anat Embryol (Berl),* 1997, 196, 185-93

[85] Klokkers, J; Langehanenberg, P; Kemper, B; Kosmeier, S; von, Bally, G; Riethmuller, C; Wunder, F; Sindic, A; Pavenstadt, H; Schlatter, E; Edemir, B. Atrial natriuretic peptide and nitric oxide signaling antagonizes vasopressin-mediated water permeability in inner medullary collecting duct cells. *Am J Physiol Renal Physiol*, 2009, 297, F693-703

[86] John, M; Wiedenmann, B; Kruhoffer, M; Adermann, K; Ankorina-Stark, I; Schlatter, E; Ahnert-Hilger, G; Forssmann, WG; Kuhn, M. Guanylin stimulates regulated secretion from human neuroendocrine pancreatic cells. *Gastroenterology*, 1998, 114, 791-7

[87] Kulaksiz, H; Schmid, A; Honscheid, M; Eissele, R; Klempnauer, J; Cetin, Y. Guanylin in the human pancreas: a novel luminocrine regulatory pathway of electrolyte secretion via cGMP and CFTR in the ductal system. *Histochem Cell Biol*, 2001, 115, 131-45

[88] Kulaksiz, H; Cetin, Y. Uroguanylin and guanylate cyclase C in the human pancreas: expression and mutuality of ligand/receptor localization as indicators of intercellular paracrine signaling pathways. *J Endocrinol*, 2001, 170, 267-75

[89] Scheving, LA; Russell, WE. Guanylyl cyclase C is up-regulated by nonparenchymal cells and hepatocytes in regenerating rat liver. *Cancer Res*, 1996, 56, 5186-91

[90] Mann, EA; Shanmukhappa, K; Cohen, MB. Lack of guanylate cyclase C results in increased mortality in mice following liver injury. *BMC Gastroenterol*, 2010, 10, 86

[91] Cetin, Y; Kulaksiz, H; Redecker, P; Bargsten, G; Adermann, K; Grube, D; Forssmann, WG. Bronchiolar nonciliated secretory (Clara) cells: source of guanylin in the mammalian lung. *Proc Natl Acad Sci U S A*, 1995, 92, 5925-9

[92] Krause, WJ; Freeman, RH; Fort, LR. Autoradiographic demonstration of specific binding sites for E. coli enterotoxin in various epithelia of the North American opossum. *Cell Tissue Res*, 1990, 260, 387-94

[93] Zhang, ZH; Jow, F; Numann, R; Hinson, J. The airway-epithelium: a novel site of action by guanylin. *Biochem Biophys Res Commun*, 1998, 244, 50-6

[94] Ohbayashi, H; Yamaki, K; Suzuki, R; Takagi, K. Effects of uroguanylin and guanylin against antigen-induced bronchoconstriction and airway microvascular leakage in sensitized guinea-pigs. *Life Sci*, 1998, 62, 1833-44

[95] Ohbayashi, H; Yamaki, KI. Both inhalant and intravenous uroguanylin inhibit leukotriene C4-induced airway changes. *Peptides*, 2000, 21, 1467-72

[96] Reinecke, M; David, I; Loffing-Cueni, D; Ablinger, P; Cetin, Y; Kuhn, M; Forssmann, WG. Localization, expression, and characterization of guanylin in the rat adrenal medulla. *Histochem Cell Biol*, 1996, 106, 367-74

[97] Spreca, A; Simonetti, S; Rambotti, MG. Atrial natriuretic peptide and guanylin-activated guanylate cyclase isoforms in human sweat glands. *Histochem J*, 2000, 32, 725-31

[98] Jaleel, M; London, RM; Eber, SL; Forte, LR; Visweswariah, SS. Expression of the receptor guanylyl cyclase C and its ligands in reproductive tissues of the rat: a potential role for a novel signaling pathway in the epididymis. *Biol Reprod*, 2002, 67, 1975-80

[99] Sousa, CM; Havt, A; Santos, CF; Arnaud-Batista, FJ; Cunha, KM; Cerqueira, JB; Fonteles, MC; Nascimento, NR. The relaxation induced by uroguanylin and the expression of natriuretic peptide receptors in human corpora cavernosa. *J Sex Med*, 2010, 7, 3610-9

[100] Valentino, MA; Lin, JE; Snook, AE; Li, P; Kim, GW; Marszalowicz, G; Magee, MS; Hyslop, T; Schulz, S; Waldman, SA. A uroguanylin-GUCY2C endocrine axis regulates feeding in mice. *J Clin Invest*, 2011, 121, 3578-88

[101] Steinbrecher, KA; Tuohy, TM; Heppner, Goss, K; Scott, MC; Witte, DP; Groden, J; Cohen, MB. Expression of guanylin is downregulated in mouse and human intestinal adenomas. *Biochem Biophys Res Commun*, 2000, 273, 225-30

[102] Cohen, MB; Hawkins, JA; Witte, DP. Guanylin mRNA expression in human intestine and colorectal adenocarcinoma. *Lab Invest*, 1998, 78, 101-8

[103] Shailubhai, K; Yu, HH; Karunanandaa, K; Wang, JY; Eber, SL; Wang, Y; Joo, NS; Kim, HD; Miedema, BW; Abbas, SZ; Boddupalli, SS; Currie, MG; Forte, LR. Uroguanylin treatment suppresses polyp formation in thc Apc(Min/+) mouse and induces apoptosis in human colon adenocarcinoma cells via cyclic GMP. *Cancer Res*, 2000, 60, 5151-7

[104] Pitari, GM; Di, Guglielmo, MD; Park, J; Schulz, S; Waldman, SA. Guanylyl cyclase C agonists regulate progression through the cell cycle of human colon carcinoma cells. *Proc Natl Acad Sci U S A*, 2001, 98, 7846-51

[105] Liu, L; Li, H; Underwood, T; Lloyd, M; David, M; Sperl, G; Pamukcu, R; Thompson, WJ. Cyclic GMP-dependent protein kinase activation and induction by exisulind and CP461 in colon tumor cells. *J Pharmacol Exp Ther*, 2001, 299, 583-92

[106] Suhasini, M; Li, H; Lohmann, SM; Boss, GR; Pilz, RB. Cyclic-GMP-dependent protein kinase inhibits the Ras/Mitogen-activated protein kinase pathway. *Mol Cell Biol*, 1998, 18, 6983-94

[107] Steinbrecher, KA; Wowk, SA; Rudolph, JA; Witte, DP; Cohen, MB. Targeted inactivation of the mouse guanylin gene results in altered dynamics of colonic epithelial proliferation. *Am J Pathol*, 2002, 161, 2169-78

[108] Camici, M. Guanylin peptides and colorectal cancer (CRC). *Biomed Pharmacother*, 2008, 62, 70-6

[109] Li, P; Lin, JE; Schulz, S; Pitari, GM; Waldman, SA. Can colorectal cancer be prevented or treated by oral hormone replacement therapy? *Curr Mol Pharmacol*, 2009, 2, 285-92

[110] Wolfe, HR; Mendizabal, M; Lleong, E; Cuthbertson, A; Desai, V; Pullan, S; Fujii, DK; Morrison, M; Pither, R; Waldman, SA. *In vivo* imaging of human colon cancer xenografts in immunodeficient mice using a guanylyl cyclase C--specific ligand. *J Nucl Med*, 2002, 43, 392-9

[111] Park, J; Schulz, S; Haaf, J; Kairys, JC; Waldman, SA. Ectopic expression of guanylyl cyclase C in adenocarcinomas of the esophagus and stomach. *Cancer Epidemiol Biomarkers Prev*, 2002, 11, 739-44

[112] Cagir, B; Gelmann, A; Park, J; Fava, T; Tankelevitch, A; Bittner, EW; Weaver, EJ; Palazzo, JP; Weinberg, D; Fry, RD; Waldman, SA. Guanylyl cyclase C messenger RNA is a biomarker for recurrent stage II colorectal cancer. *Ann Intern Med*, 1999, 131, 805-12

[113] Waldman, SA; Cagir, B; Rakinic, J; Fry, RD; Goldstein, SD; Isenberg, G; Barber, M; Biswas, S; Minimo, C; Palazzo, J; Park, PK; Weinberg, D. Use of guanylyl cyclase C for detecting micrometastases in lymph nodes of patients with colon cancer. *Dis Colon Rectum*, 1998, 41, 310-5

[114] Nakazato, M; Yamaguchi, H; Shiomi, K; Date, Y; Fujimoto, S; Kangawa, K; Matsuo, H; Matsukura, S. Identification of 10-kDa proguanylin as a major guanylin molecule in human intestine and plasma and its increase in renal insufficiency. *Biochem Biophys Res Commun*, 1994, 205, 1966-75

[115] Nakazato, M; Yamaguchi, H; Kinoshita, H; Kangawa, K; Matsuo, H; Chino, N; Matsukura, S. Identification of biologically active and inactive human uroguanylins in plasma and urine and their increases in renal insufficiency. *Biochem Biophys Res Commun*, 1996, 220, 586-93

[116] Fukae, H; Kinoshita, H; Fujimoto, S; Nakazato, M; Eto, T. Plasma concentration of uroguanylin in patients on maintenance dialysis therapy. *Nephron*, 2000, 84, 206-10

[117] Kikuchi, M; Fujimoto, S; Fukae, H; Kinoshita, H; Kita, T; Nakazato, M; Eto, T. Role of uroguanylin, a Peptide with natriuretic activity, in rats with experimental nephrotic syndrome. *J Am Soc Nephrol*, 2005, 16, 392-7

[118] Baba, A; Fujimoto, S; Kikuchi, M; Kita, T; Kitamura, K. Effects of uroguanylin on natriuresis in experimental nephrotic rats. *Nephrology (Carlton)*, 2009, 14, 80-5

[119] Hill, O; Kuhn, M; Zucht, HD; Cetin, Y; Kulaksiz, H; Adermann, K; Klock, G; Rechkemmer, G; Forssmann, WG; Magert, HJ. Analysis of the human guanylin gene

and the processing and cellular localization of the peptide. *Proc Natl Acad Sci U S A*, 1995, 92, 2046-50

[120] Magert, HJ; Reinecke, M; David, I; Raab, HR; Adermann, K; Zucht, HD; Hill, O; Hess, R; Forssmann, WG. Uroguanylin: gene structure, expression, processing as a peptide hormone, and co-storage with somatostatin in gastrointestinal D-cells. *Regul Pept*, 1998, 73, 165-76

[121] Wiegand, RC; Kato, J; Huang, MD; Fok, KF; Kachur, JF; Currie, MG. Human guanylin: cDNA isolation, structure, and activity. *FEBS Lett*, 1992, 311, 150-4

[122] Miyazato, M; Nakazato, M; Yamaguchi, H; Date, Y; Kojima, M; Kangawa, K; Matsuo, H; Matsukura, S. Cloning and characterization of a cDNA encoding a precursor for human uroguanylin. *Biochem Biophys Res Commun*, 1996, 219, 644-8

[123] Magert, HJ; Hill, O; Zucht, HD; Martin, S; Meyer, M; Forssmann, WG; Adermann, K. Porcine guanylin and uroguanylin: cDNA sequences, deduced amino acid sequences, and biological activity of the chemically synthesized peptides. *Biochem Biophys Res Commun*, 1999, 259, 141-8

In: Natriuretic Peptides
Editor: Kailash N. Pandey
ISBN: 978-1-63321-080-6

Chapter 13

N-Terminal Pro-B-Type Natriuretic Peptide: A Biomarker of Cardiac Dysfunction during Human Fetal Life

Waltraut M. Merz and Ulrich Gembruch
University Bonn Medical School
Department of Obstetrics and Prenatal Medicine, Bonn, Germany

Abstract

Cardiovascular dysfunction during human intrauterine life is a rare but serious condition. Structural cardiac defects, malformations resulting in increased cardiac output or intrathoracic volume rearrangement, and uteroplacental dysfunction have the potential to provoke cardiac failure. When prematurity precludes delivery and postnatal treatment, intrauterine therapy is the only option to prevent fetal demise. At present assessment of the fetal cardiovascular status is limited to ultrasound-based methods. N-terminal pro-B-type natriuretic peptide (nt-proBNP) serves as biomarker of cardiovascular dysfunction in adult and pediatric cardiology. We investigated its suitability during human fetal life.

Normal range in fetal venous blood was 1998 (242 - 3754) ng/L (mean ± 2SD). Increased concentrations were present in conditions associated with myocardial pressure or volume load. In structural cardiac defects nt-proBNP reflected the functional status of the ventricles, with highest values (median / IQR 17,165 / 31,031 ng/L) in left-ventricular outflow tract obstructions with intact ventricular septum and high pressure gradients, whereas in outflow tract obstructions with loss of ventricular function, shunt and complex defects only mild to moderately elevated levels were detected. In fetal anemia circulating levels of nt-proBNP correlated with the degree of anemia. Intrauterine treatment was associated with nt-proBNP normalization. A threshold of 10,000 ng/L nt-proBNP had a negative predictive value of 100% for the development of hydrops.

Our results show that nt-proBNP is a suitable biomarker of fetal cardiovascular dysfunction. It may be a useful diagnostic and prognostic adjunct in the monitoring of intrauterine therapy of outflow tract obstructions and treatment of fetal anemia. Nt-proBNP action during fetal life and its involvement in cardiac remodeling point to a role in fetal programming of adult cardiovascular disease. Long-term studies are required to further elucidate this association.

1. INTRODUCTION

1.1. Intrauterine Cardiovascular Dysfunction: Relevance and Management Considerations

Technical progress in the development of high-resolution ultrasound machines and new methods in molecular genetics allowed for major advances in the diagnosis of fetal conditions associated with cardiovascular dysfunction. In contrast to diagnostic achievements, options for prenatal therapy have been lagging behind, transplacental drug treatment of arrhythmias and intrauterine transfusion in anemia being the only established intrauterine treatments. However, a number of other fetal conditions are associated with cardiovascular dysfunction. These include malformations with direct effect on the circulation like cardiac defects and cardiomyopathies. Disorders with secondary development of cardiovascular dysfunction comprise malformations associated with increased cardiac output (e.g., arterio-venous malformations, sacrococcygeal teratoma) or intrathoracic volume rearrangements (e.g., congenital diaphragmatic hernia, cystic-adenomatoid pulmonary malformation, pulmonary sequestration). Furthermore, changes in pre- and afterload are underlying cardiovascular dysfunction in intrauterine growth restriction, twin-twin transfusion syndrome (TTTS) and some causes of fetal hydrops. A number of these malformations are amenable to surgical intervention. Open fetal surgery however is characterized by significant maternal morbidity and high preterm delivery rates. Progress in the development of fetoscopic instruments and innovative therapeutic concepts involving minimally-invasive intrauterine interventions resulted in a rapid increase in the number of procedures being performed, e.g., laser photocoagulation of placental anastomoses in TTTS, thoracoamniotic shunting in hydrothoraces and cystic adenomatoid malformation of the lung, balloon occlusion of the trachea in severe diaphragmatic hernia, valvuloplasty and stenting in some cardiac malformations [1,2]. Certain aspects have to be taken into consideration when preparing for an intervention. For cardiac defects e.g., these include:

(a) Incidence and timing of diagnosis

Although structural cardiac defects are the commonest congenital malformations (1% of life births) [3] they comprise a heterogeneous group with low frequencies of particular defects. Atrial septum defects or persistent foramen ovale, frequently diagnosed after delivery, are both not amenable to prenatal diagnosis. Establishment of diagnosis at an early gestational age (GA) is difficult, detection rates vary between 23 - 80% [4,5].

(b) Case selection and timing of intervention

The development of the fetal cardiovascular system is usually undisturbed due to the parallel circulation with its intra- and postcardiac connections. In the majority of cases structural cardiac defects result in a redistribution of the blood flow between the left and the right chamber. However, lesions that alter loading patterns, e.g., outflow tract obstructions, have the potential for rapid progression. Substantial myocardial damage or irreversible loss of ventricular function may occur at a GA where prematurity precludes delivery and postnatal treatment [4,6]. Fetal percutaneous balloon valvuloplasty of the affected valve is aiming to prevent irreversible loss of function. The plasticity of the fetal heart and the dynamics of intrauterine development complicate the choice of the ideal intervention timing.

(c) Methods of investigation

Diagnosis and prognosis, case selection and treatment evaluation are based exclusively on ultrasound, and include morphologic and functional criteria [7-9].

(d) Treatment evaluation

A systematic evaluation of minimally-invasive fetal cardiac interventions is lacking; the same applies for other minimally-invasive interventions with the exception of laser photocoagulation in twin-twin-transfusion syndrome [10,11]. Comparisons between treatment algorithms and data on long-term outcome are lacking.

These issues illustrate the difficulties encountered in the design of evidence-based protocols for intrauterine treatment of cardiovascular dysfunction. In adult cardiology various classes of biomarkers are available for diagnosis, differential diagnosis, case selection, intervention, prognosis and follow-up of cardiovascular diseases [12]. Major features of a biomarker include: the availability of an analytical method with reliable measurement; consistent association between biomarker and disease progress and prognosis; relevance for diagnosis of subclinical disease, risk stratification, treatment choice and surveillance. A detailed understanding of the pathobiology of the biomarker is not essential [13,14]. N-terminal pro-B-type natriuretic peptide (nt-proBNP), the inactive cleavage product of brain natriuretic peptide (BNP) satisfies these criteria and is established biomarker in pediatric and adult cardiology [15,16].

1.2. Natriuretic Peptides: Pre- and Postnatal Physiology and Metabolism

In 1981, de Bold discovered atrial natriuretic peptide (ANP), thereby confirming the endocrine function of the heart [17]. ANP, BNP and C-type natriuretic peptide (CNP) are members of the natriuretic peptide (NP) family. Main production sites of ANP are the atria, whereas BNP is synthesized predominantly in the ventricles; brain and skeleton are the sites of CNP production. NPs counteract plasma-volume expansion by shifting intravascular fluid into the extravasal space, increasing venous capacity and natriuresis, and reducing the sympathetic tone in peripheral blood vessels; additionally, NPs are anti-mitotic [18].

NPs share a common molecular structure: They are synthesized as pre-pro-hormones, activation is achieved by two-fold cleavage at the n-terminal end. The ring structure consists of a conserved sequence of 11 amino acids and contains the active site. Metabolization occurs by ring cleavage. Plasma half life of ANP, BNP and nt-proBNP are 2, 20, and 120 minutes, respectively. Pre-proBNP consists of 134 amino acids and is coded by natriuretic precursor protein B (NPPB), located on chromosome one (1p36.22). Molecular analyses of promoters and other regulatory domains demonstrate spacial and temporal regulation of NP transcription [19]. Cleavage of the 26 amino acids-containing signal sequence results in proBNP-108. This cleavage happens before completion of the synthesis of the C-terminal part of the pro-hormone in the sarcoplasmatic reticulum [20]. The second cleavage results in the formation of the active hormone BNP-32 and the inactive n-terminal fragment nt-proBNP-76. Furin, a membrane-associated serine endopeptidase localized in the Golgi network and expressed in various tissues as well as corin, a membrane-bound serine peptidase exclusively expressed in the heart catalyze the cleavage. A higher expression of both proteases has been demonstrated in hypertrophied cardiomyocytes [20-24]. ProBNP-108 contains several glycosylation binding sites. The relevance of these sites is subject of intense research, since glycosylation impairs activation of the pro-hormone [20].

NP release is stimulated by cardiomyocyte distension, myocardial hypoxia or ischemia, and in response to increasing levels of circulating angiotensin. Receptors include natriuretic peptide-receptor A (NPR-A) and B (NPR-B), both membrane-bound guanyl-cyclase receptors, with cGMP-dependent kinase as second messenger. Receptor affinity varies; compared to BNP, ANP has a higher affinity for NPR-A, and substrate affinity for NPR-B is CNP >> ANP >> BNP. Receptors are located in blood vessels, lung, kidney, adrenal, fat tissue, brain, and heart [21]. Mode of action is different for ANP and BNP; ANP is involved in blood pressure regulation by antagonizing the renin-angiotensin-aldosterone system, inhibiting endothelin secretion, and reducing sympathetic tone in the peripheral vasculature [23]. BNP acts predominantly in a paracrine manner. Knock-out mice develop myocardial fibrosis, particularly after pressure load, which mainly affects the ventricles. BNP therefore is attributed a role in the regulation of cardiac remodeling [21].

NPs are metabolized by receptor-binding: natriuretic peptide-receptor C (NPR-C) is localized in vascular endothelium; it lacks the intracellular part, after internalization NPs are therefore degraded. Affinity for NPR-C decreases in the order ANP > CNP >> BNP. A second clearance mechanism includes hydrolysis by meprin A and neutral endopeptidase, enzymes that are abundant on the cell surface of various organs, particularly within the kidney and blood vessels [21]. Passive renal excretion is a third way of inactivation.

During human development, NPs are involved in the morphogenesis of various organ systems [25,26]. In-situ hybridization revealed coincidence of high ANP and BNP mRNA concentrations during crucial stages of cardiac morphogenesis, and knock-out studies confirmed involvement of NPs in heart development. In the second half of pregnancy the NP-system appears to be functional, reacting to stimuli in analogy to postnatal life.

1.3. Nt-proBNP in Fetal and Neonatal Umbilical Blood

Nt-proBNP measurement is preferred over BNP due to its longer half life and higher stability at room temperature. Additionally BNP standardization is hampered by utilization of various antibodies [27]. Cord blood BNP / nt-proBNP concentration in full-term neonates is 6 - 11 times higher compared to maternal plasma levels; maternal and umbilical vein concentrations do not correlate [28,29]. The difference is due to a higher production rate in the fetus, transplacental transfer does not seem to occur. Expression studies of human placentae revealed ANP, but no BNP synthesis in the cytotrophoblast layer [26]. Investigations in circulating BNP / nt-proBNP during fetal life are restricted to two studies comprising 31 samples [30,31]. Other analyses were performed on umbilical cord blood samples in newborns (see Table 1). Reference values for fetal nt-proBNP values have not yet been established.

We studied the suitability of nt-proBNP as biomarker of fetal cardiac dysfunction. Because the NP system is functional from 20 weeks of gestation onwards, and fetuses with suspected cardiovascular dysfunction beyond 34 weeks are delivered and treated postnatally, investigations were limited to cases with GA of 20 - 34 weeks. After construction of reference values, nt-proBNP was analyzed in conditions associated with cardiac dysfunction and during the course of intrauterine therapy.

Table 1. Nt-proBNP concentration in umbilical cord blood after delivery

Author	Details (GA, if mentioned, as mean/SD or median /IQR or range)	Assay or Antibody	ng/L Mean with range/SD or median with IQR/range	pmol/L Mean with range/SD or median with IQR/range
TERM DELIVERIES				
[32]	63 newborns; n=29 NVD, n=34 C/S (elective and emergency); mean (SD) GA NVD 39.0 (1.0), C/S 38.3 (1.2)	Modular Analytics, Roche	Mean (SD) *766 (537)*	Mean (SD) 90 (63)
[33]	49 newborns; n=46 NVD, n=3 C/S; mean (SD) GA 40.9 (1.0)	Spec. RIA (AB against nt-proBNP AA 5-24)	Mean (range) 2634 (1321-15,839)	Mean (range) *311 (156–1870)*
[30]	87 newborns; n=46 NVD, n=32 elective C/S; n=9 emergency C/S; GA 36-42	Elecsys Roche	Mean (range) *623 (222-1839)*	Mean (range) 74 (26-217)
[28]	122 newborns; n=110 NVD, n=12 C/S; GA not mentioned	Elecsys Roche	Mean (SD) *579 (351)*	Mean (SD) 68 (41)
[29]	42 newborns; n=42 NVD; median (IQR) GA 39 (39-40)	Elecsys Roche	Median (IQR) *553 (414-833)*	Median (IQR) 65 (49-98)
[34]	71 newborns; n=67 NVD, n=4 C/S; GA 32-42	Elecsys Roche	Mean (SD) 678 (381)	Mean (SD) *80 (45)*
[35]	155 newborns; n=107 NVD, n=48 C/S; median (range) GA 39 (37-42)	Elecsys Roche	Median (range) *626 (153-2518)*	Median (range) 74 (18-297)
[36]	229 newborns; n=229 NVD; median (range) GA 39+3 (37-42)	Elecsys Roche	Mean (SD) *831 (1022)*	Mean (SD) 98 (121)
[36]	51 newborns; n=51 emergency C/S; median (range) GA 39+3 (37-42)	Elecsys Roche	Mean (SD) *1201 (1223)*	Mean (SD) 142 (144)
[37]	51 newborns; n=51 NVD (37 umbilical cord samples); term	Biomedica (AB against nt-proBNP AA 8-29)	Mean (SD) 1880 (491-6598)	Mean (SD) *222 (58-779)*
[38]	62 newborns; mode of delivery not mentioned; term	Elecsys Roche	Median (range) *668 (281-2595)*	Median (range) 79 (33-306)
[39]	16 newborns; n=16 NVD; GA median (range) 39 (37-43)	Biomedica	Mean (range) 1.9 (0.7-4.1)	Mean (range) *0.223 (0.088-0.480)*[a]
PRETERM DELIVERIES				
[40]	21 newborns; mode of delivery not mentioned; GA mean (SD) 28.2 (2.9)	Elecsys Roche	Median (IQR) 6344 (1728-14,272)	Median (IQR) *749 (204-1685)*
[32]	21 newborns; n=7 NVD, n=14 C/S; GA mean (SD) SPP 30.2 (3.5), C/S 34.4 (1.4)	Modular Analytics, Roche	*2299 (SD 1400)*	271 (SD 165)

ng/L x 0.118 = pmol/L; pmol/L x 8.47 = ng/L. Published figures appear in Italic; [a] figures published in fmol/L.

AA, amino acid; AB, antibody; C/S, cesarean section; GA, gestational age (in weeks); IQR, interquartile range; NVD, normal vaginal delivery; RIA, radio-immuno-assay; SD, standard deviation.

2. Patients and Methods

2.1. Patients

Pregnant women between 20 - 34 weeks of gestation who attended our Center for Prenatal Medicine between May 2006 and October 2011 and for whom percutaneous puncture of the umbilical vein was indicated were recruited. Complication rates of fetal blood sampling are 1.2 - 2.3%, depending on indication and GA; they consist of fetal bleeding, bradycardia, and preterm premature rupture of membranes with consecutive chorioamnionitis or premature delivery [41,42]. In our series no complications occurred. No puncture was performed exclusively for study purposes. Indications for fetal blood sampling (FBS) were as follows:

(a) Rapid karyotyping
These samples included fetuses with advanced GA and morphologic abnormalities suggestive of chromosomal aberrations. Direct preparation from circulating lymphocytes yields a result within 48 - 72 hours, whereas conventional cytogenetics (cultivation of fetal cells from amniotic fluid) takes 10 - 14 days.

(b) Serologic investigations
In suspected fetal anemia, thrombocytopenia or infection serologic investigations are indicated.

(c) Fetocide
Termination of pregnancy at advanced GA entails fetocide as part of the procedure.

GA was confirmed by first-trimester ultrasonographic examination in all cases. Written consent was obtained from all women. Exclusion criteriae were applied as follows: For the establishment of reference values fetuses with structural malformations of the cardiovascular and urogenital system and fetal conditions with potential impact on nt-proBNP concentration like tumors, neuromuscular disorders, infections, anemia, and hydrops were excluded. Additionally, monochorionic twin pregnancies were excluded. All fetuses were appropriate for GA, and Doppler indices of umbilical artery and ductus venosus were within normal range at the time of blood sampling, confirming the absence of any strain on the fetal cardiovascular system. For the investigation of cardiac malformations, cases with urogenital and thoracic malformations, tumors, neuromuscular disorders, infections, anemia, effusions / hydrops other than cardiac, monochorionic twin pregnancies, and also cases with abnormal middle cerebral artery (MCA) blood velocity waveforms were excluded. For the analysis in anemia multiple pregnancies, malformations except tumors, and cases with fetal growth restriction were excluded.

2.2. Methods

2.2.1. Nt-proBNP Analysis

Nt-proBNP was analyzed by chemiluminescence assay with a commercially available kit (Siemens Dimension Vista System, Flex Reagent Cartridge PBNP, Siemens Healthcare

Diagnostics Ltd., Frimley, Camberley, UK, produced under licence by Roche Diagnostics) on a Dimension Vista® 1500 laboratory system (Siemens Healthcare Diagnostics, Eschborn, Germany). This system is widely used, thereby facilitating comparison with other reports. The antibody is directed against the central part of the molecule, epitopes include amino acids 22 - 28 and 42 - 46. All samples were analyzed within 2 hours, the inter- and intra-assay coefficient of variation was 3.5% and 2.3%, respectively.

Normal values are assay-dependent. The mean difference between two assays is cited as 1649.7 ng/L [43]. Compared to BNP, nt-proBNP levels are 3.5 times higher [44]. For healthy adults various normal values are quoted. According to the manufacturer's specification the 95th percentile is 124 ng/L for women below 55 years of age. Umbilical cord arterial and venous concentrations correlate closely (R^2=0.9587) [34]. Studies of nt-proBNP concentrations in umbilical cord blood after delivery are listed in Table 1.

2.2.2. Ultrasound and Doppler Examination, Echocardiography

High-resolution ultrasound equipment (Philips iu22, Philips Hamburg, Germany; GE logiq 9, Voluson E8, Voluson 730 expert, GE Munich, Germany) was used in all cases, with 5 to 9 and 2 to 6, 7, and 8 MHz respectively, convex transducers. The high-pass filter was set at 100 Hz, and the spatial peak temporal average power output was kept at <100 W/cm^2. All Doppler recordings were obtained at an insonation angle <10° to flow. Angle correction was not performed. At least five consecutive uniform Doppler velocity waveforms with the highest velocities and a narrow band of frequencies were recorded, and one cycle was analyzed.

A detailed assessment of the fetal anatomy and cardiovascular status including echocardiography and Doppler examination was performed in subjects and controls. Estimated fetal weight (EFW) was calculated [45]; appropriate for gestational age (AGA) was defined as EFW 10th - 90th percentile, fetal growth restriction (FGR) as EFW <10th percentile. Hydrops was defined as fluid accumulation in at least two of the following fetal compartments: skin, serous cavities (pleural or pericardial effusion), ascites, or polyhydramnios. Velocity waveforms of the umbilical artery (UA), middle cerebral artery (MCA), ductus venosus (DV), and uterine artery (utA) were obtained using standard positioning of the respective sample volumes, and indices were calculated as follows: arterial pulsatility index (PI = [systolic peak velocity – end-diastolic peak velocity] / time-averaged velocity), venous pulsatility index (PIV = [systolic peak velocity – peak velocity during atrial systole] / time averaged maximum velocity), and arterial resistance index (RI = [systolic peak velocity – end-diastolic peak velocity] / systolic peak velocity). Abnormal values were defined as UA PI >90th percentile for GA or absent or reversed end-diastolic flow [46]; peak systolic velocity of the middle cerebral artery (PSV-MCA) >1.28 multiples of the median (MoM) or PI >90th percentile for GA [47]; DV PIV >90th percentile for GA or negative a-wave [48]; and utA RI >90th percentile for GA or bilateral notching [49].

Echocardiographic examination followed a standardized protocol [50]. Left-ventricular systolic performance was assessed by measurement of the peak systolic velocity (V) in the ascending aorta, with the sample volume positioned directly behind the valve leaflets. The pressure gradient ΔP was calculated according to the modified Bernoulli equation ($\Delta P = 4 \times V^2$). The aortic root inner diameter was measured in frozen real-time images during systole; lateral long axis view was used for calculation.

2.2.2. Invasive Procedures

Invasive procedures were performed under ultrasound control without fetal relaxation or sedation. Diagnostic and therapeutic interventions were completed in one procedure if feasible.

For fetal blood sampling the umbilical vein near its insertion at the placenta was punctured and one mL of venous blood aspirated for nt-proBNP analysis. The indicated procedure was performed thereafter. Intrauterine transfusion (IUT) was carried out using packed red cells with a hemoglobin concentration of 23 - 24 g/dL and a hematocrit of 72%. Transfusion volume was calculated from the EFW (30 - 40 mL/kg body weight), aiming for a maximum blood volume expansion of 30 - 40 mL/kg body weight/IUT. If the degree of anemia necessitated transfusion of a higher volume, a second IUT was arranged 48 hours later.

2.2.3. Statistical Analysis

Statistical analysis was performed with SPSS 17.0 for Windows (SPSS, Chicago, Ill.). Evaluation of fetal data needs to account for changes that may occur during the course of pregnancy. Recommendations for the construction of fetal normal values have been published; however, they are not applicable to the cross-sectional design of our study [51]. Additionally, ethical considerations preclude sampling from healthy fetuses and multiple specimens from affected cases. Modifications in the construction of normal values were therefore undertaken as follows: Double analyses were not performed, and reference values were established despite small cohort size. Healthy fetuses did not serve as reference population; instead, samples were taken from cases with disorders which do not impact on nt-proBNP concentration. According to our knowledge these include neural tube defects, brain malformations, skeletal dysplasias and numerical or structural chromosomal aberrations without associated malformations.

Hemoglobin values were transformed into MoMs. For this, expected hemoglobin values were calculated with the following formula: $e^{(2.84-8.55/GA)}$. MoMs were then calculated by dividing the measured values by the expected values. Anemia was classified as mild (0.84 - 0.65 MoM), moderate (0.64 - 0.55 MoM) and severe (<0.55 MoM) [47]. Values for PSV-MCA were MoM-transformed accordingly, using the following formula for expected PSV-MCA: $e^{(2.31+0.046\ GA)}$. Anemia according to PSV-MCA values was classified as mild (1.29 - 1.49 MoM), moderate (1.50 - 1.54 MoM) and severe (≥1.55 MoM) [47].

For normally distributed values, between group comparisons of continuous variables were performed by independent samples t-test or one-way ANOVA with post-hoc analysis by Tamhane-T2 test. Otherwise, Mann-Whitney-U or Kruksal-Wallis test was used. Within group comparisons were performed by paired samples t-test and Wilcoxon signed-rank test was applied to the post-ante differences. Categorical variables were compared by χ^2 test. For correlation analysis Spearman's or Pearsons's coefficient was calculated.

3. RESULTS

3.1. Nt-proBNP Reference Values

56 cases were recruited; fetal diagnoses are listed in Table 2. Mean gestational age (±2 SD) was 26.1 weeks (18.2 - 33.9). Kolmogorov-Smirnov test revealed normal distribution for nt-proBNP with a mean concentration (±2SD) of 1998 (242 - 3754) ng/L; median concentration (IQR) was 1874 (1181) ng/L, KS-Z 0.755; asymptotic significance, 2-sided 0.619. At a confidence niveau of 95% ($p<0.05$) a significant negative correlation between nt-proBNP and GA could be demonstrated ($p=0.012$; Pearson's correlation, 2-sided -0.333; $r^2=0.111$), see Figure 1.

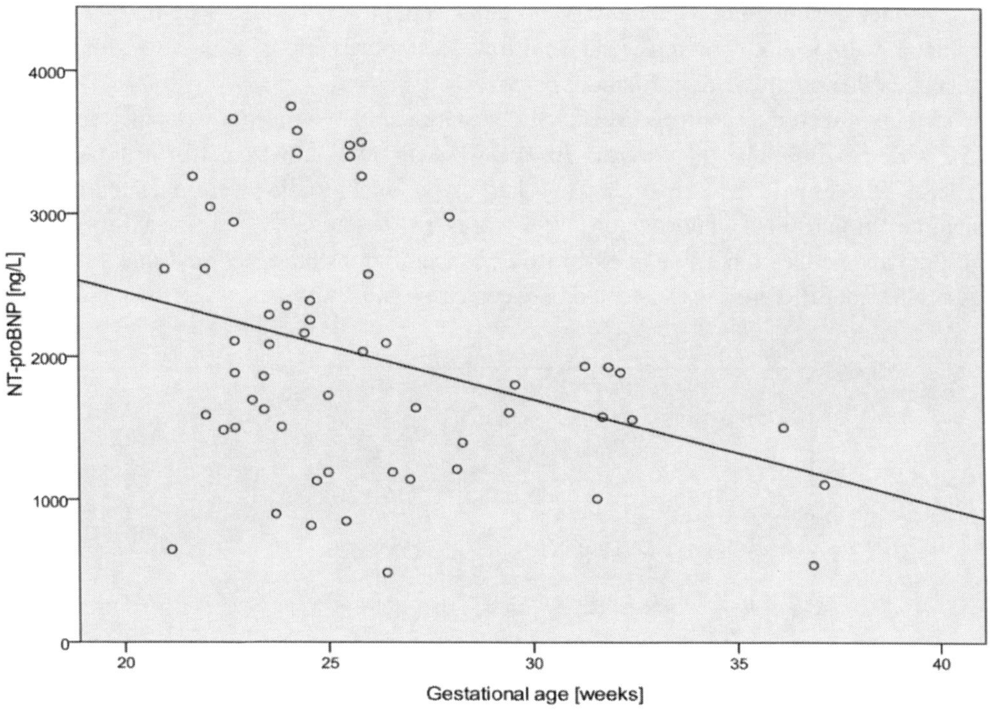

Figure 1. Nt-proBNP concentration in fetal plasma. Equation for the line: $y = -74.8x + 3946$ ng/L.

Table 2. Fetal diagnoses in samples used for analysis of reference values

Diagnosis	n (%)
Neural tube defect	19 (33.9)
Brain malformation	18 (32.1)
Trisomy 21	7 (12.5)
Structural chromosomal aberration	4 (7.1)
Skeletal dysplasia	4 (7.1)
Platelet alloimmunization	2 (3.6)
Preterm premature rupture of membranes	2 (3.6)

Using ANOVA the following variables were investigated for their effect on nt-proBNP concentrations: maternal age, maternal disease, gestational age, dichorionic-diamniotic twin pregnancy, aneuploidy, and fetal malformations that had not been excluded. None of these reached statistical significance. 22 cases were recruited after the first analysis. Reference values changed non-significantly: mean (±2SD) 1923 (149 - 3698) ng/L; median (IQR) 1874 (1092) ng/L. Validity of the initial analysis was therefore confirmed.

3.2. Nt-proBNP in Fetal Cardiac Malformations

45 cases were recruited. They were classified according to defect type and effect on cardiac function into: volume load (shunt defects), pressure load (left- or right-ventricular outflow tract obstructions with intact ventricular septum) (LVOT/IVS or RVOT/IVS), and combined volume and pressure load (outflow tract obstructions with ventricular septum defect and other complex malformations).

75 fetuses served as controls. Mean (SD) gestational age in the study population was 26.9 (3.5), with no difference to controls (p=0.137). Karyotyping was performed in 42 cases (93.3%). 19 fetuses (42.2%) had a normal karyotype, aneuploidies were detected in 14, and structural chromosomal abnormalities in 9 cases (31.9 and 20.0%, respectively). Cardiac defect details are listed in Table 3. Nt-proBNP concentrations are listed in Table 4 and Figure 2. A significant difference was detected between cases and controls.

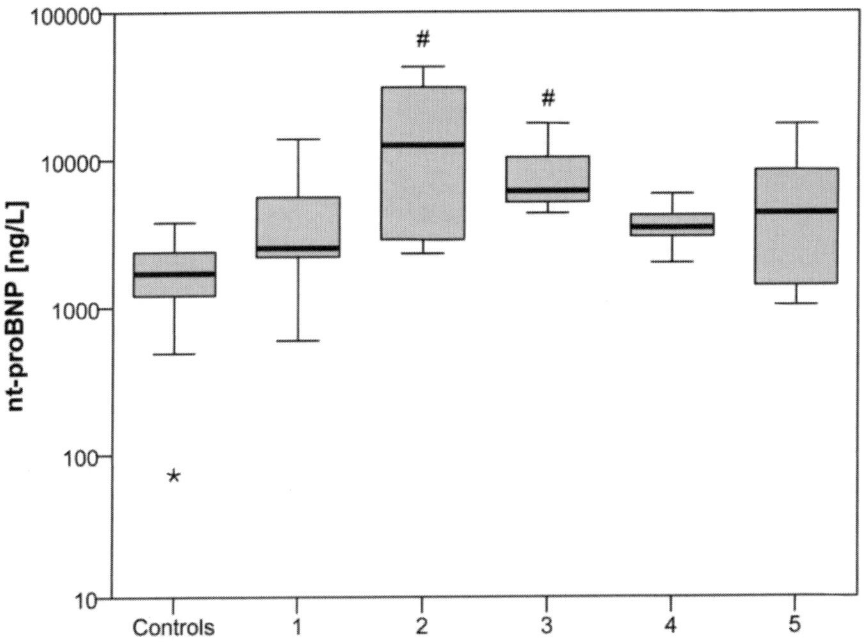

Figure 2. Nt-proBNP concentration in cardiac malformations according to defect type, and controls. Logarithmic presentation of nt-proBNP values. For group details see text and Table 3. #p<0.02 (group 2+3 vs. group 1).

Table 3. Cardiac malformation details

Group 1, Shunt defects (n=20)
ASD
AVSD
DILV
TAC
Group 2, LVOT obstruction with IVS (n=6)
AA
AS
HLHS in AA or AS or TGA with PA
Group 3, RVOT obstruction with IVS (n=3)
PA and severe PS with /without TR and VCC
Group 4, RVOT obstruction with VSD (n=8)
ToF
DORV
DORV with PA and unbalanced AVSD
PA with VSD
Group 5, other complex malformations (n=6)
right isomerism with unbalanced AVSD, DORV, PS and agenesis of DV or with common atrium, VSD, AA and agenesis of DV
CoA
unbalanced AVSD with CoA
right ventricular and pulmonary hypoplasia

additionally, the study group included one case of each, HCMP and isolated LPSVC.

AA, aortic atresia; AS, aortic stenosis; AVSD, atrioventricular septum defect; CoA, coarctation of the aorta; DILV, double inlet left ventricle; DORV, double outlet right ventricle; DV, ductus venosus; HCMP, hypertrophic cardiomyopathy; HLHS, hypoplastic left heart syndrome; IVS, intact ventricular septum; LPSVC, left-persistent superior vena cava; LVOT, left ventricular outflow tract; PA, pulmonary atresia; PS, pulmonary stenosis; RVOT, right ventricular outflow tract; TAC, truncus arteriosus communis; TGA, transposition of the great arteries; ToF, Tetralogy of Fallot; TR, tricuspid regurgitation; VCC, ventriculo-coronary communications; VSD, ventricular septum defect.

Table 4. Nt-proBNP concentration in fetuses with structural cardiac defects and controls

	n	malformation	nt-proBNP concentration (ng/L)[a]
controls	75	-	1696 (1189)
cases[b]	45	-	3658 (3705)[c]
group 1	20	shunt	2507 (3442)
group 2	6	LVOT / IVS	17,165 (31,031)[d]
group 3	3	RVOT / IVS	6086[d]
group 4	8	RVOT / VSD	3446 (1418)
group 5	6	others, complex	4355 (9324)

[a]median (IQR); [b]including one case of hypertrophic cardiomyopathy and left-persistent superior vena cava (22,768 ng/L and. 1804 ng/L, respectively); [c]$p<0.001$ vs. controls; [d]$p=0.013$ (group 2+3 vs. group 1).

LVOT/IVS, left-ventricular outflow tract obstruction with intact ventricular septum.

RVOT/IVS, right-ventricular outflow tract obstruction with intact ventricular septum.

RVOT/VSD, right-ventricular outflow tract obstruction with ventricular septum defect.

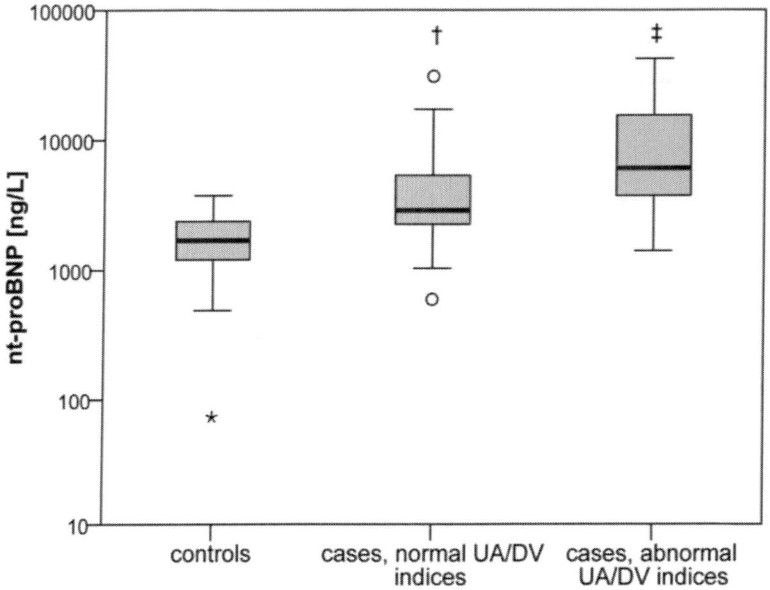

Figure 3. Nt-proBNP concentration and Doppler indices in cardiac malformations. Logarithmic presentation of nt-proBNP values. †p<0.001 vs. controls, ‡p<0.05 vs. cases with normal UA/DV Doppler indices. UA/DV, umbilical artery/ductus venosus.

Table 5. Cardiac morphology and functional status of the left ventricle and nt-proBNP concentration in LVOT/IVS

Echocardiographic finding	Vmax (cm/s)	pressure gradient aortic valve (mm Hg)	DV (PIV)	FGR	nt-proBNP (ng/L)
High LV pressure					
AS, PS, HLH, EFE, restrictive FO, MV dysplasia	228	21	abnormal	no	42,479
valvular AS, LV dilatation, holosystolic MR, restrictive FO	355	50	normal	no	30,848
valvular AS, HLH, LV dilatation, EFE, MV dysplasia, MR, TR, heart failure, hydrops	190	14	abnormal	no	28,957
Chronic hypoplastic LV with low intraventricular pressure					
AA, HLH, EFE, VCC	-	-	normal	no	5372
AA, HLH	-	-	normal	no	2866
HLH, TGA, PA	-	-	abnormal	no	2301

AA, aortic atresia; AS, aortic stenosis; DV, ductus venosus; EFE, endocardial fibroelastosis; FO, foramen ovale; FGR, fetal growth restriction; HLH, hypoplastic left heart; LV, left ventricle; MR, mitral regurgitation; MV, mitral valve; PA, pulmonary atresia; PIV, pulsatility index for veins; PS, pulmonary stenosis; TGA, transposition of the great arteries; TR, tricuspid regurgitation; VCC, ventriculo-coronary communication.

Abnormal blood flow velocities in the UA or DV were present in 15 fetuses (33.3%). Nt-pro-BNP concentrations were significantly higher in these cases, see Figure 3. No difference was detected between FGR and AGA fetuses. Karyotype had no effect on nt-proBNP concentration. Soubgroup analysis showed significantly higher levels of circulating nt-proBNP in LVOT/IVS or RVOT/IVS (p<0.02), i.e., malformations associated with pressure load, in contrast to shunt defects which are associated with volume load. In cases with outflow tract obstructions different functional states of the ventricle were present, see Table 5. Nt-proBNP concentrations reflected the actual ventricular state accurately, with high values in cases with high left-ventricular pressure and preserved cardiac function and normal to somewhat increased values in the others.

3.3. Nt-proBNP in Fetal Anemia before and after Intrauterine Treatment

27 cases were studied and compared with 78 controls; 7 fetuses were hydropic. Anemia was caused by the following disorders: Rhesus alloimmunization (n=10); Parvovirus-B19 infection (n=8); tumors with associated hemorrhage (n=5); chronic fetomaternal hemorrhage (n=2); elliptocytosis and unknown (one case each). All fetuses survived, complications did not occur. Mean time interval between first and second IUT was 6.8 days (range 1 - 21 days), mean number of IUTs per fetus 3.2 (range 1 - 7). On initial diagnosis the degree of anemia was mild in 6, moderate in 4, and severe in 12 cases. Mean GA (±SD) at the first presentation was 24.1 weeks (19.2 - 29.1), GA was comparable in cases and controls. Nt-proBNP levels correlated with the degree of anemia, see Table 6; hydropic fetuses had significantly higher nt-proBNP concentrations, see Figure 4.

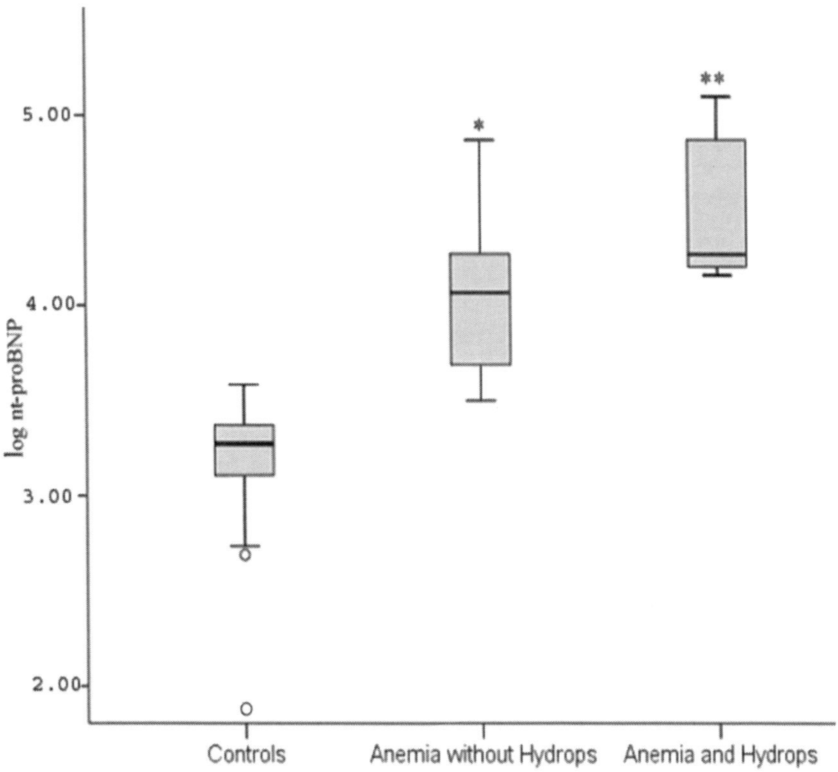

Figure 4. Nt-proBNP concentration in anemic fetuses with and without hydrops, and controls. Nt-proBNP levels are presented after logarithmic transformation.

Table 6. Nt-proBNP in fetal anemia at initial presentation

	n	nt-proBNP (ng/L)[a]	
controls	78	1874 (1092)	
anemia	22	15,167 (27,758)	$p<0.01^b$
mild	6	4876 (6735)	$p<0.01^b$
moderate	4	17,302 (48,662)	$p<0.02^c$
severe	12	18,152 (40,755)	

[a]median (IQR); [b]vs. controls; [c]vs. mild anemia. In 5 cases samples could not be retrieved before the first IUT.

At a threshold of 10,000 ng/L nt-proBNP had a negative predictive value for the development of hydrops of 100%, with a positive predictive value of 44%. Neither hemoglobin nor PSV-MCA discriminated between hydropic and non-hydropic fetuses. Nt-proBNP and hemoglobin, and PSV-MCA and hemoglobin correlated well (r=-0.637 and r=-0.626, respectively). No correlation was present between nt-proBNP and PSV-MCA.

During the course of treatment the following changes occurred (see Table 7): nt-proBNP concentrations decreased significantly after the first IUT whereas hemoglobin and PSV-MCA changed non-significantly. Additionally, circulating nt-proBNP concentrations normalized after the third IUT (nt-proBNP before the fourth IUT vs. controls: $p>0.1$) whereas hemoglobin and PSV-MCA values were still beyond their respective reference ranges.

Table 7. Nt-proBNP, hemoglobin and PSV-MCA during the course of intrauterine treatment for fetal anemia

	n	nt-proBNP[a] (ng/L)	Hb[a] (g/dL)	Hb MoM	PSV-MCA (cm/s)	PSV-MCA MoM
before 1. IUT	27	15,167 (27,758)	5.9 (3.9)	0.49 (0.26)	55 (22)	1.75 (0.42)
before 2. IUT	16	12,305 (15,993)[b]	8.9 (3.5)	0.69 (0.31)	65 (28)	1.54 (0.77)
before 3. IUT	14	5830 (7400)	8.4 (4.0)	0.64 (0.37)	67 (37)	1.67 (0.51)
before 4. IUT	8	2528 (1200)	8.9 (2.5)	0.70 (0.19)[c]	60 (25)	1.76 (0.71)[d]

[a]median (IQR); [b]$p<0.05$ (vs. concentration before treatment); [c]$p<0.02$ (vs. MoM for GA); [d]$p<0.05$ (vs. MoM for GA).

GA, gestational age; Hb, hemoglobin; IUT, intrauterine transfusion; MoM, multiples of the median; PSV-MCA peak systolic velocity of the middle cerebral artery.

4. Discussion

4.1. Nt-proBNP Reference Values

Compared to umbilical cord concentrations in term neonates levels of circulating nt-proBNP are elevated in fetuses during the second half of gestation; the plasma concentrations correlate inversely with gestational age. Since increased nt-proBNP concentrations during human intrauterine life are not due to maternal transfer or differences in metabolic pathways our finding may be explained by the physiologic changes occurring during the course of pregnancy which are characterized by improvement in diastolic cardiac function (compliance and relaxation time), decrease in afterload, and increase in systolic myocardial function.

Two studies investigated fetal nt-proBNP concentrations: Fortunato et al. 2006 [30] examined 22 fetuses between 20 - 25 weeks of gestation and found similar nt-proBNP concentrations. Walther et al. 2004 [31] examined nine fetal blood samples at GA 21 weeks. Application of a different assay precludes direct comparison, fetal plasma concentrations were only two-fold increased over corresponding maternal blood samples.

A correlation between GA and nt-proBNP concentration in umbilical cord blood samples after delivery was described by Kocylowsky et al. 2009 [36] (38 - 42 weeks of gestation, 280 cases) and Seong et al. 2010 [32] (30 - 34 weeks, 84 cases).

Elevated levels of BNP have been found in umbilical cord blood samples of newborns with FGR [33,36,52]. The etiology of FGR is heterogeneous and includes genetic and infectious causes; the majority of cases however develop as a result of restricted uteroplacental perfusion. In the presence of uteroplacental dysfunction the fetal cardiovascular system adapts to the increased vascular resistance of the fetoplacental circulation and the reduced substrate supply. These changes can be monitored by Doppler ultrasound examination of the fetoplacental and fetal blood vessels, they proceed in a predictable sequence: The increased resistance in the placental vasculature causes a rise in the umbilical artery resistance; this is followed by a relative rise in the left- and a corresponding fall in the right-ventricular cardiac output. Further deterioration results in 'brain sparing', characterized by preferential perfusion of the essential organs brain, heart, and adrenals [53]. Subsequently diastolic dysfunction ensues, resulting in abnormal Doppler flow velocity indices in the precordial veins. Elevated levels of nt-proBNP in fetuses with FGR may therefore be taken as indirect evidence of an operating NP system.

The following pre- and post-analytic features have to be taken into consideration:

(a) Molecular heterogeneity of circulating BNP [20,24,54,55,56]

Release of BNP into the blood circulation usually includes cleavage of proBNP-108 into BNP-32 and nt-proBNP-76. However, proBNP-108 has been detected in the circulation, particularly in patients with cardiac dysfunction [23,24]. Compared to BNP-32 proBNP-108 has a 6 - 8-fold lower mode of action in vitro [55]. The increase of plasma proBNP-108 in cases with cardiac dysfunction therefore results in relative BNP-32-deficiency. The observation of a lack of BNP action despite increased plasma concentrations of the immunoreactive molecule may be owed to this mechanism [55].

(b) Molecular heterogeneity of circulating nt-proBNP [54]

In addition to the 76 amino acid molecule smaller fragments have been detected, with truncations on both sides. Using an antibody against the central portion of the nt-proBNP molecule enhances test precision.

(c) BNP glycosylation

Seven specific glycosylation sites for the pro-hormone are known [55]. Glycosylation of proBNP-108 precludes cleavage into the 32 and 76 amino acid molecules [22]. The clinical relevance of glycosylation remains to be established [20].

(d) Cross-reactivity of various assays with respect to proBNP, glycosylated BNP, and additional molecular forms [55,56,57,58]

Various types of cross-reactivity have been demonstrated. These features result in difficulties converting nt-proBNP into BNP and comparing results performed with different assays [57].

(e) Mode of delivery in analyses of umbilical cord blood concentrations

Results are conflicting: Bakker et al. 2004 [34], Bar-Oz et al. 2005 [28], and Seong et al. 2009 [32] did not find an association between mode of delivery and umblilical cord blood concentration. In contrast, Fortunato et al. 2006 [30] and Kocylowski et al. 2009 [36] reported differences. However, whereas the former authors measured highest values in elective cesarean section and lower concentrations after vaginal delivery and secondary cesarean section, the latter found higher values in secondary cesarean section compared to vaginal delivery. Other studies do not mention mode of delivery [29,33,35,37,38,39,40]. Further investigations are necessary to clarify this issue.

4.2. Nt-proBNP Concentration in Fetal Cardiac Malformations

Our investigation revealed markedly elevated levels of circulating nt-proBNP in fetuses with structural cardiac defects. The increase was most pronounced in cases with pressure load.

The distinctive features of the fetal cardiovascular system preclude a comparison between hemodynamic sequelae of cardiac defects in pre- and post-natal life. Pure pressure or volume load conditions are rarely present as a result of the parallel arrangement of the fetal circulation and the additional communications (foramen ovale and arterial duct). The fetal circulation is characterized by right ventricular (RV) dominance and venous recirculation of approximately one third of the combined ventricular output. Some 45% of the combined ventricular output is ejected into the placenta, which acts as a large arterio-venous shunt. RV afterload consists of placental and lower body impedance whereas the left ventricle (LV) works against the high impedance of the brain and upper body. LV preload differs in particular because it depends on the patency of the foramen ovale.

We classified the malformations according to the hemodynamic effect evoked by the underlying cardiovascular morphology. Samples from fetuses with shunt lesions, RVOT obstructions / VSD and complex lesions had lower nt-proBNP levels compared to cases with right or LVOT obstruction / IVS. This may be attributed to the parallel arrangement of the fetal circulation. The presence of the shunt equalizes chamber pressures and reduces myocardial wall stress of the respective ventricle, even if an outflow tract obstruction is present as in group four. Additionally, the reservoir function of the placenta may attenuate the impact of volume load. Accordingly, none of these cases had signs of increased volume load. Outflow tract obstructions / IVS were associated with markedly increased nt-proBNP concentrations. Elevated levels were related to impaired function. We assume this to be a consequence of the reduced compliance and contractility of the fetal myocardium, confirming the vulnerability towards afterload. The intrauterine hemodynamics of outflow tract obstructions / IVS results in exceedingly high pressure gradients in early stages followed by irreversible loss of ventricular function. For LVOT / IVS this is followed by a reduction in the LV preload resulting from interatrial left to right shunting and minimal forward flow through the mitral valve. The presence of mitral insufficiency in left and of tricuspid insufficiency in right ventricular outflow tract obstructions may distinctly reduce the corresponding intraventricular pressure load and preserve ventricular function. Nt-proBNP concentrations reflected the prevailing status of the affected chamber.

Abnormal Doppler indices in the precordial veins are indicative of reduced filling capacity and cardiac compromise. Nt-proBNP concentrations were significantly higher in this group of patients compared to cases with cardiac defects and normal blood flow velocity waveforms. Owing to small numbers a subgroup analysis of nt-proBNP levels and DV indices was not performed.

The effect of FGR on our study population was evaluated. A significant correlation of nt-proBNP and early markers of cardiac dysfunction in FGR has been demonstrated [33,36,52]. We attribute the difference to our results to the low GA of our study group. Mean GA was considerably higher in the aforementioned reports. In contrast to our cases the stage of hypoxemia and deterioration was substantially more advanced in these series of FGR fetuses with signs of cardiac dysfunction.

Two studies investigated nt-proBNP concentrations in umbilical cord blood of neonates with congenital heart defects. Kocylowski et al. 2009 [36] found elevated nt-proBNP levels in seven of 10 newborns with cardiac defects and assumed a correlation with volume overload. Values are not mentioned. Lechner et al. 2009 [35] detected significantly higher nt-proBNP concentrations in cord blood of newborns with structural cardiac defects compared to controls (median 1589 ng/L, range 378 - 7626 ng/L, n=46). Further analysis revealed a difference between uni- and biventricular defects (median 2001 ng/L, range 378 – 762 6ng/L vs. 963 ng/L, 443 - 2585 ng/L). Compared to our results, nt-proBNP levels at birth were considerably lower.

The invasiveness of FBS with its procedure-related risks precludes the application of nt-proBNP for monitoring cardiovascular function in fetuses with cardiac malformations. Nt-proBNP does however provide evidence of changes in myocardial characteristics and cardiac remodeling in a variety of cardiac defects as already described for FGR due to uteroplacental dysfunction. In fetuses undergoing cardiac interventions, particularly aortic valvuloplasty, nt-proBNP measurement may provide pathophysiologic insight in addition and correlation to existing tests of myocardial function [4,6]. Non-invasive tests like myocardial performance indices (Doppler, tissue Doppler) and myocardial velocity, strain and strain rates (tissue Doppler, speckle tracing) can give clues of cardiac dysfunction, but their clinical relevance for fetal monitoring and surveillance requires further evaluation [9,59,60].

4.3. Nt-proBNP in the Diagnosis and Management of Fetal Anemia

We found elevated levels of circulating nt-proBNP in fetuses with high cardiac output secondary to anemia. Concentrations correlated well with the degree of anemia, highest levels were present in cases with hydrops. Treatment resulted in a significant decrease and eventually normalization of nt-proBNP values despite persistently abnormal hemoglobin and PSV-MCA measurements.

Anemia exerts work load on the fetal heart; in order to maintain tissue oxygenation the combined cardiac output is increased. The hyperdynamic circulation is characterized by a rise in myocardial stretching and filling pressures, reduced afterload, and cardiomegaly. Myocardial perfusion in particular is dependent on oxygen supply. In acute hypoxemia, adaptive mechanisms allow a 4-5-fold increase in coronary perfusion and include increased coronary perfusion pressure, decreased coronary resistance and autoregulation. If these adaptations fail to meet oxygen demand myocardial ischemic changes occur, eventually resulting in decreased myocardial function and dilated cardiomyopathy. Although the precise mechanisms are debated, development of hydrops in fetal anemia heralds deterioration of the fetal status and impending death.

IUT, initially published in 1963 by Liley, has been the first invasive fetal therapy [61]. Blood group incompatibility, particularly of the Rhesus system, is the most frequent cause. Untreated fetal anemia results in high cardiac output failure, end-stage disease erythroblastosis fetalis is characterized by fetal hydrops and intrauterine death. Previously diagnosis was established by spectrophotometric measurement of the bilirubin concentration in amniotic fluid or hemoglobin determination in fetal blood; this has been superseded by non-invasive diagnosis of the measurement of the PSV-MCA. Management has advanced from intraperitoneal injection of red blood cells after radiographic fetal localization to percutaneous intravenous transfusion under ultrasound control. Serial IUTs until fetal maturity are the treatment of choice. Since then, survival rates have been increasing from 50 to >90%, even in already hydropic fetuses with hemoglobin concentrations <4 g/dL. The implementation of ante- and postnatal RhD prophylaxis further contributed to the reduction in perinatal mortality from 120 to 1.3 deliveries / 100,000. Aspects of intrauterine therapy still under investigation include prognostic parameters of hydrops formation and criteria for scheduling reassessment and subsequent transfusion.

Only one study reported nt-proBNP levels in anemic fetuses (n=9) at 30 weeks of gestation before and after a single IUT [31]. The application of a different assay precludes direct comparison. Nt-proBNP values in anemic cases before IUT in the aforementioned study were 6-fold higher compared to controls, and increased by 13% after IUT.

Our study is the first longitudinal analysis of nt-proBNP levels in volume load secondary to anemia during intrauterine life. The close correlation between hemoglobin and nt-proBNP values allows the conclusion that nt-proNP is secreted in response to the increased cardiac output required to compensate for the reduced oxygen-carrying capacity of the blood.

The development of hydrops is associated with worse fetal outcome. Current concepts on the pathophysiology of hydrops in anemia include genuine myocardial failure, high cardiac output failure, reduced plasmatic colloid-oncotic pressure, increased capillary permeability, and obstruction of venous and consecutively lymphatic flow. We found significantly higher nt-proBNP concentrations in hydropic fetuses; these results support a myocardial cause of hydrops development in anemia. The increased total feto-placental blood volume present only in hydropic fetuses with anemia may exceed the myocardial capacity resulting in high cardiac output failure [62]; on the other hand increased end-diastolic ventricular pressure and reduced coronary perfusion pressure may induce hypoxic myocardial dysfunction. Our data do not allow a differentiation between these mechanisms. Investigating parameters of myocardial injury may aid to clarify this issue.

Our study confirms previous results of ductus venosus blood flow indices in fetal anemia. A rise in precordial vein pressure is likely to be a late finding, confirming the adaptability of the fetal myocardium to the hyperdynamic circulation. Parvovirus-B19-associated myocarditis may impair myocardial function. We could not detect an impact of infection on nt-proBNP concentrations.

The normalization of circulating nt-proBNP after three IUTs despite persistently low hemoglobin levels is remarkable given that the oxygen dissociation curve is more unfavorable for adult red blood cells (RBC), and consequently a higher cardiac output is required for tissue oxygenation. This finding may be an indicator of the myocardial adjustment to increased work load, illustrating the plasticity of the fetal heart. Adaptive cardiac mechanisms to chronic anemia have been investigated in animals. Chronically anemic sheep fetuses show increases in myocardial mass and vascularization as well as changes in the expression of various angiogenic, hypoxia-related and glycolytic genes [63,64]. Furthermore, cardiac remodeling induced by chronic intrauterine anemia results in increased contractile response and coronary conductance in adult animals [65,66]. Rheological changes secondary to the presence of adult RBCs in the fetal circulation may be an alternative explanation for the normalization of nt-proBNP despite persistent anemia. Compared to fetal RBCs, adult RBCs are smaller with less rigid cell membranes; additionally the proportion of nucleated RBCs is lower. These factors may alleviate cardiac work load even in the presence of increased cardiac output.

In severely anemic fetuses hemoglobin and PSV-MCA measurements are not able to predict hydrops. Likewise, the decreasing sensitivity of PSV-MCA to detect fetal anemia after a previous IUT precludes its application in the management of cases with multiple IUTs. In these situations nt-proBNP may provide valuable information given its high negative predictive value.

CONCLUSION

Our investigations demonstrate that nt-proBNP is suitable as a biomarker of cardiovascular dysfunction during fetal life. Clinical applications may be feasible in structural cardiac malformations and fetal anemia. In fetuses undergoing cardiac interventions, particularly aortic valvuloplasty, nt-proBNP measurement may provide pathophysiologic insight in addition and correlation to existing tests of myocardial function. In severely anemic fetuses hemoglobin and PSV-MCA measurements are not able to predict hydrops; nt-proBNP may provide valuable information given its high negative predictive value.

The association between myocardial changes, cardiac remodeling and nt-proBNP concentrations alludes to fetal programming, defined as a change in the pattern of gene expression which results in morphologic, metabolic and hormonal alterations that imply vulnerability for development of adult disease. Changes in pre- and afterload are capable to trigger cardiovascular re-programming. The paracrine action of nt-proBNP results in changes of the cardiac gene expression program. Nt-proBNP may therefore be involved in fetal programming, and may serve as prenatal biomarker of adult cardiac disease. This may be of particular interest in fetuses without malformations who develop growth restriction secondary to uteroplacental dysfunction. Long-term follow-up studies are required to further investigate this association.

References

[1] Deprest JA, Flake AW, Gratacos E, Ville Y, Hecher K, Nicolaides K, Johnson MP, Luks FI, Adzick NS, Harrison MR. The making of fetal surgery. *Prenat Diagn* 2010;30:653-67.

[2] Van Mieghem T, Baud D, Devlieger R, Lewi L, Ryan G, de Catte L, Deprest J. Minimally invasive fetal therapy. *Best Pract Res Clin Obstet and Gynaecol* 2012; 26:711-25.

[3] Nyberg DA, McGahan JP, Pretorius DH, Pilu G (eds). Diagnostic imaging of fetal anomalies. 421-450, 2003, *Lippincott Williams & Wilkins, Philadelphia.*

[4] Gardiner HM. The case for fetal cardiac intervention. *Heart* 2009;95:1648-52.

[5] Allan L. Fetal cardiac scanning today. *Prenat Diagn* 2010;30:639-43.

[6] McElhinney DB, Tworetzky W, Lock JE. Current status of fetal cardiac intervention. *Circulation* 2010;121:1256-63.

[7] McElhinney DB, Marshall AC, Wilkins-Haug LE, Brown DW, Benson CB, Silva V, Marx GR, Mizrahi-Arnaud A, Lock JE, Tworetzky W. Predictors of technical success and postnatal biventricular outcome after in utero aortic valvuloplasty for aortic stenosis with evolving hypoplastic left heart syndrome. *Circulation* 2009;120:1482-90.

[8] McElhinney DB, Vogel M, Benson CB, Marshall AC, Wilkins-Haug LE, Silva V, Tworetzky W. Assessment of left ventricular endocardial fibroelastosis in fetuses with aortic stenosis and evolving hypoplastic left heart syndrome. *Am J Cardiol* 2010;106:1792-7.

[9] Van Mieghem T, DeKoninck P, Steenhaut P, Deprest J. Methods for prenatal assessment of fetal cardiac function. *Prenat Diagn* 2009;29:1193-203.

[10] Roberts D., Neilson JP., Kilby M., Gates S. Interventions for the treatment of twin-twin transfusion syndrome. *Cochrane Database Syst Rev*. 2008; Jan 23;(1):CD 002073.

[11] van Klink JMM, Koopman HM, van Zwet, EW, Oepkes D, Walther FJ, Lopriore E. Cerebral injury and neurodevelopmental impairment after amnioreduction versus laser surgery in twin-twin transfusion syndrome: a systematic review and meta-analysis. *Fetal Diagn Ther* 2013;33:81-89.

[12] Braunwald E. Biomarkers in heart failure. *N Engl J Med* 2008;358:2148-59.

[13] Morrow DA, de Lemos JA. Benchmarks for the assessment of novel cardiovascular biomarkers. *Circulation* 2007;115:949-52.

[14] Böhm M, Voors AA, Ketelslegers J-M, Schirmer SH, Turgonyi E, Bramlage P, Zannad F. Biomarkers: optimizing treatment guidance in heart failure. *Clin Res Cardiol* 2011;11:973-81.

[15] Hunt SH, Abraham WT, Chin MH, Feldman AM, Francis GS, Ganiats TG, Jessup M et al. ACC/AHA 2005 Guideline update for the diagnosis and management of chronic heart failure in the adult – summary article: A report of the American College of Cardiology/American Heart Association Task Force on Practice Guidelines. *Circulation* 2005;112:1825-52.

[16] Dickstein K, Cohen-Solal A, Filippatos G, McMurray JJV, Ponikowski P, Poole-Wilson PA, et al. ESC Guidelines for the diagnosis and treatment of acute and chronic heart failure 2008. *Eur Heart J* 2008;29:2388-442.

[17] De Bold AJ, Borenstein HB, Veress AT, Sonnenberg H. A rapid and potent natriuretic response to intravenous injection of atrial myocardial extracts in rats. *Life Sci* 1981;28:89-94.

[18] Levin ER, Gardner DG, Samson WK. Natriuretic peptides. *N Eng J Med* 1998;339:321-8.

[19] Hayek S, Nemer M. Cardiac natriuretic peptides: from basic discovery to clinical practice. *Cardiovasc Ther* 2011;29:362-76.

[20] Mair J. Clinical significance of pro-B-type natriuretic peptide glycosylation and processing. *Clin Chem* 2009;55:394-7.

[21] Potter LR, Yoder AR, Flora DR, Antos LK, Dickey DM. Natriuretic peptides: Their structures, receptors, physiologic functions and therapeutic applications. *Handb Exp Pharmacol* 2009;191:341-66.

[22] Semenov AG, Postnikov AB, Tamm NN, Seferian KR, Karpova NS, Bloshchitsyna MN, Koshkina EV et al. Processing of pro-brain natriuretic peptide is suppressed by O-glycosylation in the region close to the cleavage site. *Clin Chem* 2009;55:489-98.

[23] Xu-Cai YO, Wu Q. Molecular forms of natriuretic peptides in heart failure and their implications. *Heart* 2010;96:419-24.

[24] Nishikimi T, Kuwahara K, Nakao K. Current biochemistry, molecular biology, and clinical relevance of natriuretic peptides. *J Cardiol* 2011;57:131-40.

[25] Waschek JA. Developmental actions of natriuretic peptides in the brain and skeleton. *Cell Mol Life Sci* 2004;61:2332-42.

[26] Cameron VA, Ellmers LJ. Minireview: Natriuretic peptides during development of the fetal heart and circulation. *Endocrinology* 2003;144:2191-4.

[27] Ordonez-Llanos J, Collinson PO, Christenson RH Amino-terminal pro-B-type natriuretic peptide: Analytic considerations. *Am J Cardiol* 2008;101[suppl]:9A-15A.

[28] Bar-Oz B, Lev-Sagie A, Arad I, Salpeter L, Nir A. N-terminal pro-B-type natriuretic peptide concentrations in mothers just before delivery, in cord blood, and in newborns. *Clin Chem* 2005;51:926-7.

[29] Hammerer-Lercher A, Mair J, Tews G, Puschendorf B, Sommer R. N-terminal pro-B-type natriuretic peptide concentrations are markedly higher in the umbilical cord blood of newborns than in their mothers. *Clin Chem* 2005;51:913-915.

[30] Fortunato G, Giarrusso PC, Martinelli P, Sglavo G, Vassallo M, Tomeo L, Rea M, Paladini D. Cardiac troponin T and amino-terminal pro-natriuretic peptide concentrations in fetuses in the second trimester and in healthy neonates. *Clin Chem Lab Med* 2006;44:834-836.

[31] Walther T, Stepan H, Pankow K, Gembardt F, Faber R, Schultheiss H-P, Siems WE. Relation of ANP and BNP to their N-terminal fragments in fetal circulation: evidence for enhanced neutral endopeptidase activity and resistance of BNP to neutral endopeptidase in the fetus. *BJOG* 2004;111:452-5.

[32] Seong WJ, Yoon DH, Chong GO, Hong DG, Koo TB, Lee TH, Chun SS. Umbilical cord blood amino-terminal pro-brain natriuretic peptide levels according to the mode of delivery. *Arch Gynecol Obstet* 2010;281:907-12.

[33] Girsen A, Ala-Kopsala M, Mäkikallio K, Vuolteenaho O, Räsänen J. Cardiovascular hemodynamics and umbilical artery N-terminal peptide of proB-type natriuretic peptide in human fetuses with growth restriction. *Ultrasound Obstet Gynecol* 2007;29:296-303.

[34] Bakker J, Gies I, Slavenburg B, Bekers O, Delhaas T, van Dieijen-Visser M. Reference values for N-terminal pro-B-type natriuretic peptide in umbilical cord blood. *Clin Chem* 2004;50:2465-67.

[35] Lechner E, Wiesinger-Eidenberger G, Wagner O, Weissensteiner M, Schreier-Lechner E, Leibetseder D, Arzt W, Tulzer G. Amino terminal pro B-type natriuretic peptide levels are elevated in the cord blood of neonates with congenital heart defect. *Pediatr Res* 2009;66:466-9.

[36] Kocylowski RD, Dubiel M, Gudmudsson S, Sieg I, Fritzner E, Alkasi Ö, Breborowicz GH, von Kaisenberg CS. Biochemical tissue-specific injury markers of the heart and brain in postpartum cord blood. *Am J Obstet Gynecol* 2009;200:273.e1-273.e25.

[37] Mir TS, Laux R, Hellwege HH, Liedke B, Heinze C, von Buelow H, Läer S, Weil J. Plasma concentrations of aminoterminal pro atrial natriuretic peptide and aminoterminal pro brain natriuretic peptide in healthy neonates: marked and rapid increase after birth. *Pediatrics* 2003;112:896-9.

[38] Schwachtgen L, Herrmann M, Georg T, Schwarz P, Marx N, Lindinger A. Reference values of NT-proBNP serum concentrations in the umbilical cord blood and in healthy neonates and children. *Z Kardiol* 2005;94:399-404.

[39] Fleming SM, O'Gorman T, O'Byrne L, Grimes H, Daly KM, Morrison JJ. Cardiac troponin I and N-terminal pro-brain natriuretic peptide in umbilical artery blood in relation to fetal heart rate abnormalities during labor. *Pediatr Cardiol* 2001;22:393-6.

[40] Ramakrishnan S, Heung YM, Round J, Morris TP, Collinson P, Williams AF. Early N-terminal pro-brain natriuretic peptide measurements predict clinically significant ductus arteriosus in preterm infants. *Acta Paediatrica* 2009;98:1254-9.

[41] Holzgreve W, Tercanli S, Surbek D, Miny P. Invasive diagnostic methods. In: Rodeck CH, Whittle MJ (eds). Fetal medicine. *Basic science and clinical practice.* 417-35, 2001, Churchill Livingstone.

[42] Nanal T, Kyle P, Soothill PW. A classification of pregnancy losses after invasive prenatal diagnostic procedures: an approach to allow comparison of units with a different case mix. *Prenat Diagn* 2003;23:488-92.

[43] Nir A, Lindinger A, Rauh M, Bar-Oz B, Laer S, Schwachtgen L, Koch A, Falkenberg J, Mir TS. NT-pro-B-type natriuretic peptide in infants and children: reference values based on combined data from four studies. *Pediatr Cardiol* 2009;30:3-8.

[44] Nir A. Bar-Oz B, Perles Z, Brooks R, Korach A, Rein AJJT. N-terminal pro-B-type natriuretic peptide: reference plasma levels from birth to adolescence. Elevated levels at birth and in infants and children with heart diseases. *Acta Paediatr* 2004;93:603-607.

[45] Hadlock FP, Harrist RB, Martinez-Poyer J (1991) In utero analysis of fetal growth: a sonographic weight standard. *Radiology* 181:129-133.

[46] Kurmanavicius J, Florio I, Wisser J, Hebisch G, Zimmermann R, Müller R et al. (1997) Reference resistance indices of the umbilical, fetal middle cerebral and uterine arteries at 24-42 weeks of gestation. *Ultrasound Obstet Gynecol* 10:112-120.

[47] Mari G, Deter RL, Carpenter RL, Rahman F, Zimmerman R, Moise KJ Jr et al. (2000) Noninvasive diagnosis by Doppler ultrasonography of fetal anemia due to maternal red-cell alloimmunization. Collaborative Group for Doppler Assessment of the Blood Velocity in Anemic Fetuses. *N Engl J Med* 342:9-14.

[48] Hecher K, Campbell S, Snijders R, Nicolaides K (1994) Reference ranges for fetal venous and atrioventricular blood flow parameters. *Ultrasound Obstet Gynecol* 4:381-390.

[49] Gómez O, Figueras F, Fernández S, Bennasar M, Martínez JM, Puerto B et al. (2008) Reference ranges for uterine artery mean pulsatility index at 11-41 weeks of gestation. *Ultrasound Obstet Gynecol* 32:128-132.

[50] Chaoui R, Heling K, Mielke G, Hofbeck M, Gembruch U. Quality standards of the DEGUM for performance of fetal echocardiography. *Ultraschall Med* 2008;29:197-200.

[51] Royston P, Wright EM. How to construct 'normal ranges' for fetal variables. *Ultrasound Obstet Gynecol* 1998;11:30-8.

[52] Crispi F, Hernandez-Andrade E, Pelsers M, Plasencia W, Benavides-Serralde JA, Eixarch E, Le Noble F, Ahmed A et al. Cardiac dysfunction and cell damage across clinical stages of severity in growth-restricted fetuses. *Am J Obstet Gynecol* 2008;199:254.e1-254.e8.

[53] Baschat AA, Gembruch U, Harman CR. The sequence of changes in Doppler and biophysical parameters as severe fetal growth restriction worsens. Ultrasound Obstet Gynecol 2001;18:571-7.

[54] Ala-Kopsala M, Maga J, Peuhkurinen K, Leipälä J, Ruskoaho H, Leppäluoto J, Vuolteenaho O. Molecular heterogeneity has a major impact on the measurement of circulating n-terminal fragments of A- and B-type natriuretic peptides. *Clin Chem* 2004;50:1576-88.

[55] Liang F, O'Rear J, Schellenberger U, Tai L, Lasecki M, Schreiner GF, Apple FS, Maisel AS, Pollitt NS, Protter AA. Evidence for functional heterogeneity of circulating B-type natriuretic peptide. *J Am Coll Cardiol* 2007;49:1071-80.

[56] Luckenbill KN, Christenson RH, Jaffe AS, Mair J, Ordonez-Llanos J, Pagani F, Tate J, Wu AHB, Ler R, Apple FS. Cross-reactivity of BNP, NT-proBNP, and proBNP in commercial BNP and NT-proBNP assays: preliminary observations from the IFCC committee.

[57] Clerico A, Emdin M. Diagnostic accuracy and prognostic relevance of the measurement of cardiac natriuretic peptides: A review. *Clin Chem* 2004;50:33-50.

[58] Apple FS, Panteghini M, Ravkilde J, Mair J, Wu AHB, Tate J, Pagani F, Christenson RH, Jaffe AS. Quality specifications for B-type natriuretic peptide assays. *Clin Chem* 2005;51:486-93.

[59] Comas M, Crispi F, Cruz-Martinez R, Martinez JM, Figueras F, Gratacos E (2010) Usefulness of myocardial tissue Doppler vs conventional echocardiography in the evaluation of cardiac dysfunction in early-onset intrauterine growth restriction. *Am J Obstet Gynecol* 203:45.e1-7.

[60] Willruth AM, Geipel A, Berg C, Fimmers R, Gembruch U. Assessment of fetal global and regional ventricular function in congenital heart disease using a novel feature tracking technique. *Ultraschall in Med* 2012;33:251-257.

[61] Liley AW. Intrauterine transfusion of the foetus in haemolytic disease. *BMJ* 1963;2:1107-9.

[62] Pasman SA, Bil-van den Brink CP, Kamping MA, Adama van Scheltema PN. Total blood volume is maintained in nonhydropic fetuses with severe hemolytic anemia. *Fetal Diagn Ther* 2009;26:10-5.

[63] Jonker SS, Giraud MK, Giraud GD, Chattergoon NN, Louey S, Davis LE, Faber JJ, Thornburg KL. Cardiomyocyte enlargement, proliferation and maturation during chronic fetal anaemia in sheep. *Exp Physiol* 2009;95:131-9.

[64] Mascio CE, Olison AK, Ralphe JC, Tomanek, RJ, Scholz TD, Segar JL. Myocardial vascular and metabolic adaptations in chronically anemic fetal sheep. *Am J Physiol Regul Integr Comp Physiol* 2005;289:R1736-R45.

[65] Broberg CS, Giraud GD, Schultz JM, Thornburg KL, Hohimer AR, Davis LE. Fetal anemia leads to augmented contractile response to hypoxic stress in adulthood. *Am J Physiol Regul Integr Comp Physiol* 2003;285:R649-R55.

[66] Davis L, Thornburg KL, Giraud GD. The effects of anaemia as a programming agent in the fetal heart. *J Physiol* 2005;565:35-41.

In: Natriuretic Peptides
Editor: Kailash N. Pandey
ISBN: 978-1-63321-080-6

Chapter 14

TRANSCRIPTIONAL CONTROL MECHANISMS OF GUANYLYL CYCLASE/NATRIURETIC PEPTIDE RECEPTOR A GENE

Prerna Kumar and Kailash N. Pandey*
Department of Physiology, Tulane University Health Sciences Center and School of Medicine, New Orleans, LA, US

ABSTRACT

Guanylyl cyclase/natriuretic peptide receptor-A (GC-A/NPRA) is the principal loci involved in the regulatory actions of cardiac hormones namely, atrial and brain natriuretic peptide (ANP and BNP). Interaction of ANP and BNP with GC-A/NPRA produces the intracellular second messenger cGMP, which plays a central role in the pathophysiology of hypertension and cardiovascular disorders. Various *in vitro* and *in vivo* studies have shown that NPRA expression is regulated by a number of factors including hormones, growth factor stimulation, changes in extracellular osmolality, and certain physiological and patho-physiological conditions. The molecular basis for these changes in NPRA expression is still poorly understood, and they are likely to be mediated by a number of processes, including changes in gene transcription, mRNA stability, and protein synthesis or degradation. Emerging studies have begun to identify mechanisms responsible for changes in *Npr1* (coding for GC-A/NPRA) gene expression and have defined *trans*-acting regulatory factors, which are essential for basal *Npr1* gene transcription. The proximal promoter region of the murine *Npr1* gene contains functional *cis*-elements for Sp1, Ets-1, p300, and an inverted CCAAT box, but lacks TATA box. The 5' flanking region of the *Npr1* gene contains several putative binding sites for transcription factors but their role is yet to be elucidated. In this review we focus mainly on recent advances in understanding the transcriptional regulation of *Npr1* gene with an emphasis on various components of gene regulatory pathways (e.g., the basal transcription machinery, DNA-

* Address for Correspondence to: Dr. Kailash N. Pandey, Department of Physiology, SL-39, Tulane University Health Sciences Center and School of Medicine, 1430 Tulane Ave, New Orleans, LA 70112 Tel: (504) 988-1628; Fax: (504) 988-2675; Email: kpandey@tulane.edu.

binding transcription factors, coregulators, histone modifications, and DNA methylation), polymorphism, and various external and internal stimuli.

Keywords: Natriuretic peptide receptor A, guanylyl cyclase A, gene transcription, transcription factors, polymorphism

ABBREVIATIONS

ANP, Atrial natriuretic peptide; GC-A/NPRA, guanylyl cyclase/natriuretic peptide receptor-A; TF, transcription factor; TSS, transcription start site; RT-PCR, reverse transcription polymerase chain reaction; IP, immunoprecipitation; EMSA, electrophoretic mobility shift assay; siRNA, small interference RNA; ChIP, chromatin immunoprecipitation assay; MMCs, mouse mesangial cells; RASM, rat aortic smooth muscle cells; IMCD, inner medullary collecting duct; Ang II, angiotensin II; RE, response element; ATRA, all-trans retinoic acid; RAR, retinoic acid receptor; RXR, retinoid X receptor; RARE, retinoic acid response element; VDRE, vitamin D response element.

1. INTRODUCTION

Natriuretic peptides (NP) are a family of structurally related but genetically distinct hormones that play important roles in the control of renal, cardiovascular, endocrine, and skeletal homeostasis. Atrial natriuretic peptide (ANP) is the first identified member of the NP family which consists of ANP, B-type natriuretic peptide (BNP), C-type natriuretic peptide (CNP), and urodilatin [38, 89, 121]. ANP and BNP are cardiac hormones that are produced mainly in the atrium and ventricle, respectively, and circulate in the plasma, whereas CNP is largely present in the vasculature [38, 39, 54, 89, 125]. Three subtypes of NP receptors (NPRA, NPRB, and NPRC) belong to the family of seven mammalian plasma membrane bound receptors; however, these receptors are variable in their ligand specificity and signal transduction activity [54, 79]. Both ANP and BNP binds to its receptor natriuretic peptide receptor-A (NPRA) and produce the second messenger cGMP and elicit natriuretic, diuretic, and vasorelaxant effects, whereas CNP primarily activates NPRB, but all 3 NPs bind to NPRC [76, 78, 107, 124, 150]. The receptors guanylyl cyclase A (GC-A) also known as GC-A/NPRA and GC-B/NPRB are 135-kDa transmembrane proteins; ligand binding to these receptors generates the second messenger cGMP [76, 78, 128, 141]. Both NPRA and NPRB contain at least 4 distinct regions: an extracellular ligand binding domain, a single transmembrane spanning region, an intracellular protein kinase-like homology domain, and a GC catalytic domain [78, 140]. NPRC lacks the intracellular catalytic domain and has been termed as NP clearance receptor [52]. The genes encoding these receptor subtypes are designated as *Npr1*, *Npr2*, and *Npr3* corresponding to NPRA, NPRB, and NPRC, respectively, which have been shown to be located on separate chromosomes in genome [41, 68, 101, 126, 152]. The chromosomal loci of human *Npr1* gene at 1q21-22, *Npr2* at 9p12-21, and *Npr3* at 5p14-p13 has been mapped [101].

NPRA is one of the principal receptor of ANP and BNP as most of the physiological effects of these hormones are mediated by generation of intracellular second messenger cGMP or its cell permeable analogs, which plays a central role in the pathophysiology of hypertension and cardiovascular disorders [1, 102, 121, 123, 153]. Recent studies with mice lacking *Npr1* gene, demonstrated that genetic disruption of NPRA increases the blood pressure and causes hypertension in these animals [98, 118, 145]. A number of studies have shown that disruption of the *Npr1* gene activates proinflammatory cytokines leading to fibrosis, extracellular matrix production, hypertrophic growth, and remodeling of heart and kidneys in *Npr1*-null mutant mice and all of these pathophysiological changes have an impact on clinically relevant diseases in humans [36, 91, 122, 159]. GC-A/NPRA and /or its mRNA is expressed in kidney, adrenal, adipose, heart, lung, brain, lung, liver, vascular beds, and testis tissues. The gene encoding NPRA designated as *Npr1* has been cloned and characterized in human [100, 151], mouse [57, 128], rat [32, 169], bull frog [142] and medaka fish [168]. The *Npr1* gene in different species is structurally very similar and consists of 22 exons interrupted by 21 introns. A number of repeat elements are present in the non-coding regions of the gene which include 5'- and 3'-flanking regions and the introns [57, 169]. The 5'-regulatory region of the *Npr1* gene has been cloned and characterized in different species, human [114], rat [169], mouse [57], and medaka fish [168]. This region contains several putative cis-acting motifs of known transcription factors (TFs), suggesting that these multiple regulatory elements may exert a cooperative influence in order to closely regulate *Npr1* gene expression. The proximal promoter region shows length polymorphism a $(CT)_n$ dinucleotide repeat which may have a significant regulatory role in *Npr1* expression [77]. Although *Npr1* promoter regulation is not very well studied but reports show that the promoter is regulated by number of factors including, auto-regulation by ANP and cGMP via cGMP response element (cGMP-RE) [18, 66, 105], other hormones such as glucocorticoids [96, 97, 116], vitamin D [28, 29], all-trans retinoic acid [82], and angiotensin (Ang II) [5, 58], extracellular osmolarity [24-26], transcription factors [80, 81, 83, 92, 93], and polymorphism [113, 134, 154, 157]. The regulation of *Npr1* gene expression and activity is complex and involves various signaling network at both transcriptional and posttranscriptional levels which needs further extensive characterization of the promoter region and its regulation under normal and pathophysiological conditions.

2. *NPR1* PROMOTER SEQUENCE AND CIS-REGULATORY ELEMENTS

The 5' flanking region of the *Npr1* gene containing the regulatory elements has been isolated and analyzed in different species including human [114], rat [169], mouse [57], and medaka fish [168]. The 5' flanking region of the *Npr1* gene lacks a typical TATA box but contains binding sites for Sp1 and an inverted CCAAT box which is common in all the species. The CCAAT is a crucial component of several promoters as it binds to different TFs and can operate in reverse orientation [104, 108]. It has been shown that a CCAAT sequence positioned between -141 and -137 of the rat *Npr1* promoter, when mutated, reduces promoter activity by 90% in rat aortic smooth muscle (RASM) cells [92]. The transcription of genes lacking TATA box often begins by an element called initiator (Inr) positioned between -3 and +5 relative to the initiation site [148]. The transcription start site (TSS) of mouse and rat *Npr1* genes are located at 362 and 370 nucleotide upstream of the respective start codons and contain Inr elements CATACTCC and CACACTCC, respectively, which shows 75-90% sequence homology to the consensus Inr sequence CTCANTCT as defined by Smale et al. [148]. The *Npr1* gene homologs of medaka fish named as OLGC2 and OLGC7 lacks classical TATA box and have TSS at 567 bp (G) and 631 bp (C) upstream of the initiation codon, respectively, as showed by primer extension analysis [168]. The 2.1-kb 5' flanking region of the human *Npr1* gene was isolated by thermal asymmetric interlaced polymerase chain reaction (TAIL-PCR) method and contains TSS at 88 nucleotide upstream of the start codon and lacks an Inr element at this position, suggesting species specific differential regulation of the gene [114]. Moreover, human *Npr1* gene contains several putative TF binding sites, including an inverted CCAAT box, Sp1, AP2, and zeste and there is a novel (TC)n sequence repeats approximately 300 bp upstream of the start codon [114]. The murine *Npr1* promoter contains binding sites for several TFs including, SRY, Nkx-2.5, AP-4, HFH-3, GATA1/2, Sp1, CREB, CRE-BP, and c/EBPα/β [57].

The comparative analysis of 5' flanking region of *Npr1* gene among different species showed that mouse and rat genes share almost 80% sequence homology in this region, while only 58% homology is observed in the 5' flanking region of rodent genes compared with that of human *Npr1* gene [57]. There are some common putative TF binding sites for *cis*-acting elements (LYF1, MZF-1, c-ETS, p300, and δEF1) in both rodents and human *Npr1* gene that are present within 500 bp region upstream of the start codon, suggesting they might play a significant role in regulating *Npr1* gene transcription. Functional analysis of *Npr1* promoter was studied by transient transfection of *Npr1* promoter and the 5'-untranslated region (UTR) deletion constructs of the mouse *Npr1* gene [58]. The construct containing -496 bp was found to contain all the basal promoter elements to allow basal expression of the gene and the construct containing only the 5' UTR showed some basal promoter activity. The 5' UTR significantly repressed the promoter activity when included with the -1841 bp construct suggesting the presence of negative regulatory elements (NRE) in 5' UTR region and contains a putative binding site for δEF1 present in *Npr1* gene of different species. Another strong repressor element was found in the region -1841 to -916 bp as revealed by promoter activity assay. Some positive regulatory elements were also found in the promoter region -916 to -496 bp containing binding sites for Ap-1, Ap-4, GATA-1/2, and C/EBP α/ β and the region -1982 to -1841 bp containing tandem repeat element of $(CA)_{18}$ relative to the TSS [58]. It was shown that deletion of 150 bp in the *Npr1* promoter construct -496/+55 (-356/+55) at the 5' end significantly increased the basal promoter activity [80]. Further analysis of the promoter region -356 to +55 showed the presence of several TF binding sites such as GATA1/2, Ets-1, Sp1, Lyf-1, SRY, and p300 (Figure 1). Deletion of 90 bp from the 3' end of the construct -356/+55 (-46 to +55) significantly decreased the promoter activity strongly suggesting that the region -46 to +55 having Ets-1 and p300 binding sites is essential for *Npr1* gene transcription. Initial experiments with overexpression of Lyf-1 and GATA-1 along with *Npr1* promoter considerably decreased promoter activity and Ets-1 significantly increased promoter activity suggesting they might be having important regulatory impact on the *Npr1* gene expression [80].

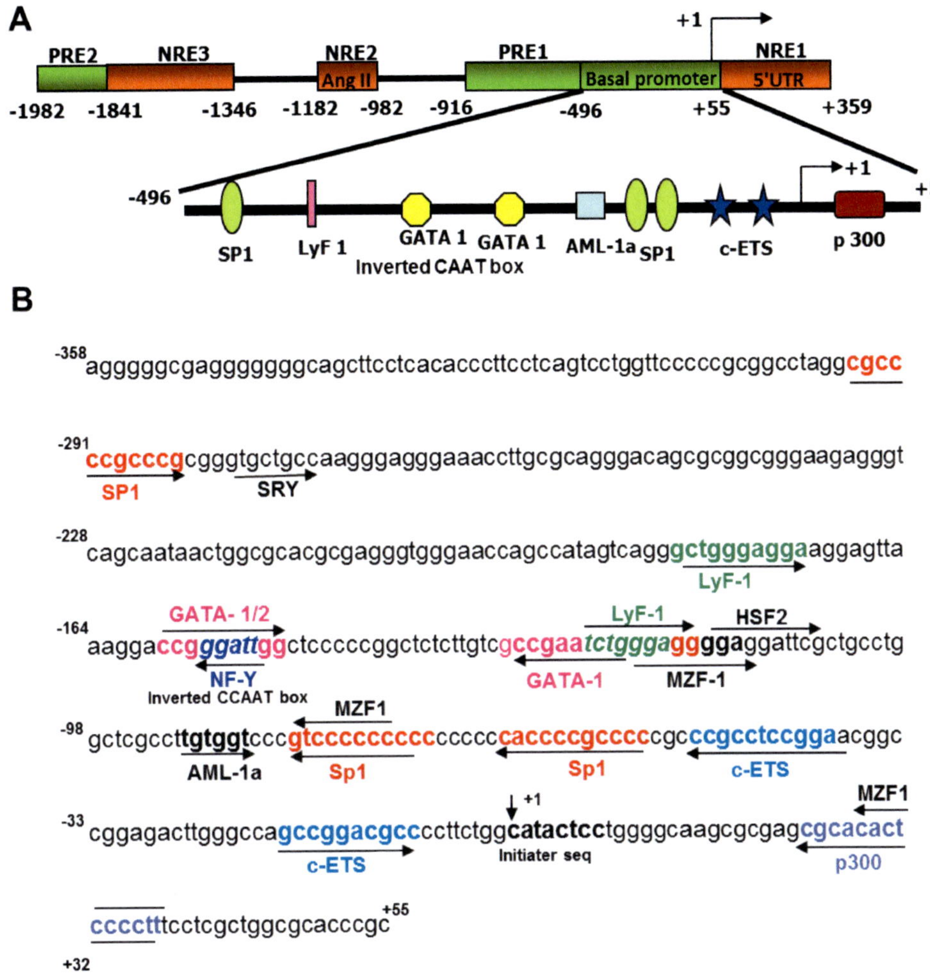

Figure 1. Schematic representation of murine *Npr1* gene promoter and nucleotide sequence of the proximal promoter region. A) Schematic map of *Npr1* promoter region (-1982 to +359) showing various positive and negative regulatory elements. B) Nucleotide positions are numbered with respect to transcription start site +1. Putative binding sites are shown by arrows above the appropriate sequences and the nucleotide sequence are bolded. Transcription factors and their *cis*-regulatory elements that have been demonstrated to regulate *Npr1* promoter activity in experiments are in color. Transcription factors having putative binding sites which are not yet elucidated are shown in black. GenBank accession no. AJ307712.

3. Transcription Factors

3.1. E Twenty-Six Transformation Specific Sequence (Ets-1)

The 5' flanking region of rodent and human *Npr1* gene shows presence of highly conserved two Ets binding sites near the TSS [57, 83]. Ets factors interact with the 5'-GGA (T/A)-3' motif and share the winged helix-turn-helix DNA-binding domain and Ets-1(a family member) binds to the consensus sequence PuCC/a-GGAA/T-GCPy [45, 47]. It has been shown that overexpression of Ets-1 in mouse mesangial cells (MMCs) enhanced murine *Npr1* proximal promoter activity by 12-fold [80]. The individual contribution of each Ets-1 motif was determined by introducing mutation in either element which reduced *Npr1* promoter activity by almost 50% and simultaneous mutation of both the sites showed an additive effect in loss of *Npr1* transcriptional activity (~ 90%) as compared with wild type construct. Mutation of both the sites together abolished the Ets-1-dependent *Npr1* promoter activity by almost 90%. UV cross-linking assay confirmed the binding of endogenous Ets-1 protein to its consensus sequence in the MMCs nuclear extract [80]. Further investigation showed a significant increase in *Npr1* mRNA levels in Ets-1-transfected MMCs and mouse leydig tumor (MA-10) cells [81]. Depletion of endogenous Ets-1 by small interfering RNA (siRNA) dramatically decreased promoter activity by 80%. Furthermore, chromatin immunoprecipitation (ChIP), gel shift, and super shift assays confirmed the *in vivo* and *in vitro* binding of Ets-1 to the two putative Ets-1 binding sites in MMCs and MA-10 cells. Overexpression of Ets-1 significantly enhanced GC activity and ANP-stimulated intracellular accumulation of cGMP in transfected cells. [81]. Ets family members have been suggested to play a primary function in formation of the initiation complex on core promoters lacking the TATA sequence and are frequently located in the vicinity of the TSS in various TATA-deficient gene promoters [6, 135]. The TATA-less *Npr1* gene can be added to this group of genes because it exhibits features typical of TATA-less GC-rich promoters and the two Ets motifs present in the murine *Npr1* promoter region -46 to +55 shows high sequence homology with the Ets-1 consensus sequence, suggesting that this arrangement might serve as a core for recruiting and/or stabilizing transcriptional complexes. Earlier studies have shown that Ets-1 protein is essential for normal coronary, myocardial, and kidney development and regulate transcription of several gene involved in angiogenesis, remodeling of the extracellular matrix, human stress response gene GADD153, and forkhead-related transcription factor detected during nephrogenesis [20, 65, 94, 109, 143, 144].

3.2. p300

Coactivators p300 and cAMP RE-binding (CREB)-binding protein (CBP) play a central role in coordinating various signal-specific events with the transcription machinery, allowing an appropriate level of gene activity to occur in response to diverse physiological stimuli [160]. The rodents and human *Npr1* proximal promoter contains a highly conserved inverted p300 binding site which is 100% homologous to p300 consensus DNA binding sequence GGGAGTG [57, 83, 133]. The *Npr1* proximal promoter construct -356/+55 containing Sp1, Ets-1, and p300 binding sites exhibited 60- to 70-fold increase in luciferase activity in rat thoracic aortic smooth muscle cells and MMCs, as compared with the pGL3-basic vector [83]. Deletion of the region +29 to +55 bp containing p300 binding site from the -356/+55 construct significantly decreased *Npr1* promoter activity in both cell lines. Knockdown of endogenous p300 protein reduced the *Npr1* promoter activity and ChIP assay showed recruitment of p300 to the *Npr1* promoter. Moreover, overexpression of adenovirus E1A significantly decreased *Npr1* promoter activity by 40%, whereas the mutant E1A lacking the p300-binding domain showed no significant effect on *Npr1* gene transcription. Furthermore, it was seen that Ets-1 and p300 cooperatively activates *Npr1* promoter activity (Figure 2). Overexpression of Ets-1 and p300 synergistically increased *Npr1* expression by almost 12-

fold, *Npr1* mRNA levels by 5-fold, and ANP-stimulated intracellular accumulation of cGMP levels by 26-fold in transfected MMCs. Knockdown of endogenous Ets-1 and p300 protein expression by siRNA inhibited 90% of the *Npr1* promoter activity. Sequential ChIP assay demonstrated a physical association between p300 and Ets-1 when bound to *Npr1* promoter, suggesting that physical interaction between Ets-1 and p300 is important to stimulate *Npr1* gene transcription and expression. The histone acetyltransferase (HAT) activity of p300 was required for their cooperative interaction as mutant p300 lacking HAT activity showed no functional synergism with Ets-1. Moreover, immunoprecipitation (IP) assay showed enhanced Ets-1 acetylation in p300-transfected cells [83]. p300 has been shown to catalyze acetylation of many non-histone proteins including transcription factors, p53, Ets-1, Ets-2, GATA1, Sp1, Smad2/3 and Smad7 [59, 69, 147, 149]. Recent evidences suggest that p300 is critical for the development of embryonic heart and plays a critical role in quantitative control of physiologically adaptive cardiac hypertrophy that is dependent on its HAT activity [37, 162]. Studies have shown that ablation of *Npr1* gene promotes cardiac hypertrophy and fibrosis in *Npr1* null mutant mice [122, 145, 159]. *Npr1* gene is under a tight control of Ets-1 and p300 transcription factors, which plays an important role in hypertension and cardiovascular regulation.

3.3. Specificity Protein 1 (Sp1)

Sp1 is ubiquitously expressed, DNA binding TF that can activate or repress gene transcription in response to physiologic and pathological stimuli. Sp1 binds to GC boxes with the consensus sequence 5'-G/T-GGGCGG-G/A-G/A-C/T-3' or 5'-G/T-G/A-GGCG-G/T-G/A-G/A-C/T-3' [17, 88]. The proximal promoter region of *Npr1* gene contains two Sp1 binding sites in the near vicinity of initiator element [57]. It has been shown that in the absence of classical TATA box, generally the Inr element positioned at the TSS determines the beginning of transcription but needs stimulation by Sp1-dependent RE placed upstream of Inr element [148]. The TATA-less rat *Npr1* gene promoter contains three putative Sp1-binding sites (positioned between -341 and -51 bp) and shows almost 90-95% sequence homology to Sp1 consensus site [93]. It has been shown that mutation of any of these binding sites reduced the promoter activity by 50-75% and simultaneous mutation led to a 90% reduction in the promoter activity. Gel shift assay confirmed the binding of Sp1 and Sp3 to the *Npr1* promoter and overexpression of Sp1 or Sp3 induced wild-type, but not the mutant *Npr1* promoter [93]. The three putative Sp1 binding sites in the murine *Npr1* promoter is positioned between -295 and -53 shows almost 85-90% sequence homology with the predicted Sp1 consensus sequence [57]. It has been shown that overexpression of Sp1 protein induced luciferase activity of the *Npr1* proximal promoter -356/+55 in MMCs [82]. Quantitative ChIP assay demonstrated association of Sp1 protein to the *Npr1* promoter region -138 to +90 bp. Sp1 expression has been shown to be modulated during pathologic cardiac hypertrophy [7, 8]. Sp1 and Sp3 are required for trans-activation of the human SERCA2 promoter and Na,K-ATPase beta 1 subunit gene transcription in cardiomyocytes [16, 172]. Ischemia/reperfusion has been shown to reduce Sp1-mediated cystathionine-synthase expression in the kidney and Sp1 transactivates the murine H (+)-K (+)-ATPase alpha [2]-subunit gene which plays a central role in potassium and acid-base homeostasis [165, 170]. These studies indicate an important role of Sp1 in controlling pathophysiology of renal and cardiac tissues and regulation of *Npr1* gene transcription could be one of the protective mechanisms involved.

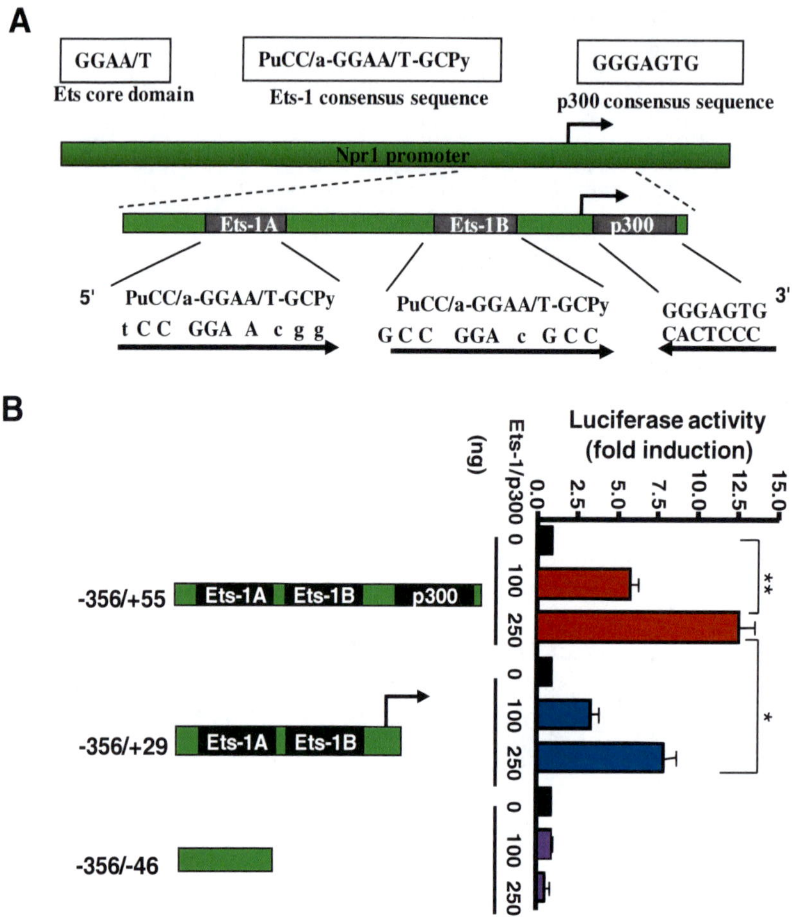

Figure 2. Effect of overexpression of Ets-1 and p300 on *Npr1* promoter activity in mesangial cells. A) Schematic map of Ets-1 and p300 consensus binding site on Npr1 promoter. B) Luciferase activity of *Npr1* promoter deletion constructs in cells overexpressing Ets-1 and p300. * $p < 0.05$; ** $p < 0.01$.

3.4. Nuclear Transcription Factor (NF-Y)

The CCAAT box promoter element and the TF NF-Y that binds to it are among the first *cis*-elements and *trans*-acting factors identified and their interplay is required for transcriptional activation of number of CCAAT promoters [46, 104]. NF-Y has been shown to regulate fibroblast procollagen alpha 1(I) transcription involved in fibroblast activation and renal 25-hydroxyvitamin D-24-hydroxylase gene which is important for controlling the level of circulating 1,25-dihydroxyvitamin D3 (1,25(OH)2D3) implicating its role in growth regulation [53, 95]. It has been shown that NF-Y binds specifically to wild type CCAAT sequence present in the rat *Npr1* promoter by electrophoretic mobility shift assay (EMSA) and supershift experiments [92]. Cotransfection of dominant negative mutant of NF-YA subunit resulted in a decrease in the *Npr1* promoter activity confirming that endogenous NF-Y is an activator of the promoter. Simultaneous mutation of CCAAT element in conjunction with Sp1 sites, repressed almost 99% of the *Npr1* promoter activity in RASM cells. Overexpression of NF-Y subunits together with Sp1 synergistically activated *Npr1* promoter activity in Drosophila cells deficient in these TFs. A direct physical association between NF-Y and Sp1 was shown both *in vitro* and *in vivo* by glutathione S-transferase pull down and coimmunoprecipitation assays suggesting *Npr1* promoter activity is regulated through functional and physical association between NF-Y and Sp1 [92]. Cooperative interaction between NF-Y and Sp1 gene is required for transcriptional activation for many genes including, pro-alpha3(V) collagen, peptidylarginine deiminase type III, and human neuronal alpha3 Na+,K+-ATPase gene [11, 48, 166].

4. Nuclear Receptors

4.1. Vitamin D Receptor

Vitamin D, or its polar metabolite 1,25 dihydroxyvitamin D3 (VD), through association with its nuclear receptor, has been shown to have a significant role in regulation of growth and function in the cardiovascular system and kidney [2, 35, 174]. The liganded vitamin D receptor (VDR) forms a complex with its heterodimeric partner, the retinoid X receptor, and associates with VDRE present on target gene promoters to stimulate or inhibit gene expression [63, 103, 136]. It has been shown that liganded VDR inhibits human ANP promoter activity [23, 27]. On the contrary, it has been shown that VD induces NPRA expression and activity in target cells [28]. Pretreatment with VD significantly induced ANP-stimulated cGMP levels and *Npr1* mRNA levels in cultured RASM cells. A VD and VDR-dependent significant increase in *Npr1* promoter activity was observed by luciferase assay and 5' deletion mutant analysis of *Npr1* promoter suggested a VD response region located between -706 and -387 harboring 5 putative VDRE which showed similarity to the consensus VDRE (two AGGTCA motifs as direct repeats separated by three spacer nucleotides) [156]. Gel shift assay and mutation analysis of the 5 putative VDRE confirmed a functional VDRE located 495 bp upstream of TSS which was critical for VD-mediated increase in *Npr1* promoter activity and gene expression [28]. It was demonstrated that VDR increased NPRA activity and gene expression through a transcriptional mechanism in a ligand-dependent manner in inner medullary collecting duct (IMCD) cells [29]. Treatment with physiological concentration range for VD led to 2.5-fold increase in *Npr1* promoter activity and 3-fold increase in *Npr1* mRNA levels and protein expression. VDR-dependent stimulatory activity of NPRA was further confirmed by using siRNA directed against endogenous VDR in the cells, which showed reduction in VD-dependent increase in *Npr1* promoter activity. It has been suggested that low levels of VD or VDR-dependent activity may contribute to sodium retention seen in congestive heart failure and salt sensitive hypertension through a reduction in NPRA activity [29].

4.2. Retinoic Acid Receptor

All-trans retinoic acid (ATRA), a biologically active vitamin A metabolite, plays a protective role in various pathophysiological conditions of numerous organ systems [85-87, 120, 139, 173]. The physiological effects of retinoic acid (RA) is mediated by binding to RA receptors (RAR) and retinoid X receptors (RXR) via RARE present on the target gene promoters and regulate gene transcription [10, 40, 74]. Recently it has been shown that ATRA activates *Npr1* gene transcription and GC activity of the receptor via interactive role of its receptors, Ets-1, and Sp1 [82]. Deletion analysis of the murine *Npr1* promoter and luciferase assays showed that ATRA enhanced a 16-fold *Npr1* promoter activity of the construct -356/+55, *Npr1* mRNA levels by 7-fold, and NPRA protein expression by 4-fold. ATRA significantly induced intracellular accumulation of cGMP levels and GC activity of the receptor protein in both ANP-dependent and -independent manner. Cells treated with a RAR-specific agonist, 4-[E-2- (5,6,7,8-tetrahydro-5,5,8,8-tetra-methyl-2-naphthalenyl)-1-propenyl]benzoic acid (TTNPB), mimicked the effect of ATRA in inducing *Npr1* promoter activity. Treatment with ATRA in RXRα, RARα, or both receptors-transfected cells showed an increase in luciferase activity of the *Npr1* promoter by 15-, 24-, and 45-fold, respectively, compared with untreated control cells. Knockdown of RARα or/and RXRα by siRNA reduced the ATRA-mediated activation of *Npr1* promoter activity by 61–90%. Cells treated with Ro-41-5253, a RARα-specific antagonist, attenuated ATRA-induced *Npr1* promoter activity by 80%. Retinoic acid induced Ets-1 protein expression and its binding to the *Npr1* promoter. ChIP assays showed that ATRA enhanced binding of Sp1 to the *Npr1* promoter. In the absence of RARE the RARα was recruited by Ets-1 and Sp1 to form a transcriptional activator complex with their binding sites in the *Npr1* promoter as shown by sequential ChIP assay. There was increased acetylation of histones H3 and H4 in ATRA-treated cells and enhanced recruitment to Ets-1 and Sp1 binding sites within the *Npr1* promoter as summarized in Figure 3 [82]. RA receptors have been shown to interact with multitude of co-regulators, chromatin modifiers, and transcription machinery and regulate gene transcription [10, 31, 44, 73, 163].

4.3. Glucocorticoid Receptor

Interaction of glucocorticoids with the cardiovascular and renal system has been shown to regulate cardiovascular function but the precise mechanism is unclear [51, 96, 167]. Glucocorticoids are known to activate transcription by interacting with glucocorticoid RE(s) and associated chromatin proteins [30, 117, 130]. Previous evidences showed that glucocorticoids could upregulate ANP mRNA expression in the atrial myocytes and increase its levels in the circulation and it has been suggested that glucocorticoids stimulate expression of the ANP gene by transcriptional regulation [34, 55, 56]. Earlier it has been shown that stimulation of rat mesenteric artery cells with dexamethasone (Dex, a highly specific synthetic glucocorticoid agonist), increased the levels of *Npr1* mRNA in a time-dependent manner and reached a plateau after 48 h of Dex administration [116]. The induction of *Npr1* mRNA was confirmed by cortisol treatment and repressed when both Dex and the glucocorticoid receptor antagonist RU 38486 were present in the cell culture. Increase in *Npr1* mRNA levels by Dex and cortisol stimulation was followed by an increase in cGMP levels over a 48 hours period. Recently, it has been suggested that glucocorticoid mediate inhibition of dehydration-induce water intake via NPRA activation in the hypothalamus of rat [97]. The study showed that 24 h Dex-pretreatment remarkably increased hypothalamic NPRA expression in the water-deprived rats as determined by Western blot. Moreover, immunofluorescence data showed dramatic increase in NPRA positive neurons in the hypothalamus by Dex treatment. Increase in NPRA activity by Dex was confirmed by increase in hypothalamic cGMP content. A recent study has shown that Dex significantly increased expression of NPRA in IMCD cells in a dose- and time-dependent manner via glucocorticoid receptor and increased cGMP generation in these cells [96]. This study suggests that glucocorticoids could improve renal responsiveness to ANP by upregulating NPRA expression in the renal IMCD and induce potent diuretic action in rats with decompensated heart failure.

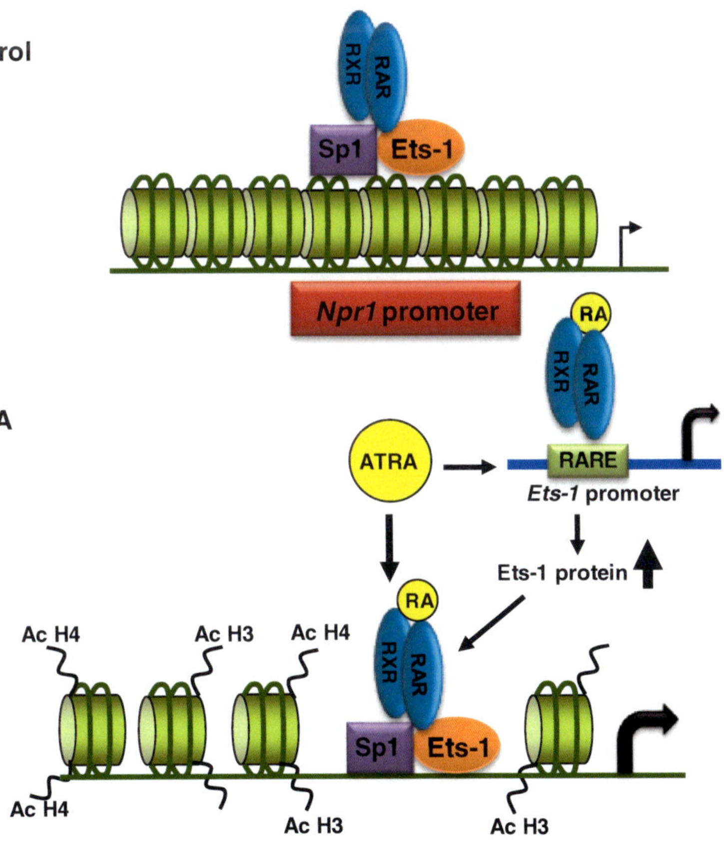

Figure 3. Schematic presentation of interactions among transcription factors during ATRA-mediated *Npr1* gene transcription. The proposed diagram shows that ATRA increases *Npr1* gene transcription via Ets-1, Sp1, and histone acetylation. ATRA induces Ets-1 gene transcription and protein expression via its receptor RAR and enhances Ets-1 and Sp1 binding to the *Npr1* promoter. In the absence of functional RARE on the *Npr1* promoter, RAR associates with the Ets-1.Sp1 complex at the *Npr1* promoter and mediates ATRA effects. Furthermore, ATRA induces histone H3 and H4 acetylation near Ets-1 and Sp1 binding sites, promotes localized unwinding of DNA, and allows transcription factors to bind in the region.

5. Peptide Hormones and Kinase-Response Regions

5.1. ANP/cGMP Response Element

Natriuretic peptides such as ANP, BNP, CNP, and urodilatin have been shown to down-regulate ligand-dependent NPRA activity in cultured cells through a post transcriptional mechanism [49, 60, 127, 131, 132]. On the other hand, sustained decrease in NPRA activity was associated with a reduction in steady state levels of *Npr1* mRNA and transcriptional activity of *Npr1* gene in 100 nM ANP-treated cells for 48 h [18, 19]. ANP effected a dose- and time-dependent reduction in a transiently transfected *Npr1* promoter-luciferase reporter (-1575LUC). Analysis of 5' deletion mutants of the *Npr1* promoter demonstrated that the ANP-dependent sequence lies between -1575 and -1290 relative to the TSS [19]. The reduction in *Npr1* mRNA levels and promoter activity was also mimicked by 8-bromo-cGMP, suggesting that second messenger cGMP through its RE was responsible for downregulation of *Npr1*gene expression [18, 19]. To further localize and characterize the putative cis acting cGMP-RE, several deletion constructs of *Npr1* promoter fragment -1520-bp were generated by exonuclease III-directed deletions and transiently transfected into cells to measure their promoter activity [66]. Preincubation of cells with 100 nM of ANP or 100 μM 8-bromo3-cGMP transfected with *Npr1* promoter-deletion construct identified the cGMP-RE to be localized between positions -1396 and -1307. Gel shift assays and footprinting further narrowed the cGMP regulatory region to position -1372 to -1354 from the TSS of *Npr1* promoter and indicated its interaction with TF(s). Cross-competition experiments with mutated oligonucleotides led to identify consensus sequence (-1372 AaAtRKaNTTCaAcAKTY -1354] for the novel cGMP-RE, which is conserved in human (75% identity) and mouse (95% identity) *Npr1* promoters [66]. Recently, a new protein GREBP (for cGMP-RE-binding protein) has been shown to bind cGMP-RE by yeast one-hybrid technique with a human kidney cDNA library [105]. EMSA and ChIP assay showed a specific binding for the cGMP-RE under physiological conditions. By acting on cGMP-RE, GREBP inhibited the expression of a luciferase-coupled *Npr1* promoter construct. Treatment with ANP for 3 h increased GREBP expression by 60% and inhibited *Npr1* expression by 30%. Silencing of GREBP with siRNA increased *Npr1* promoter activity and mRNA levels. GREBP is a nuclear protein mainly expressed in the heart that acts as a transcriptional repressor of the *Npr1* gene [105].

5.2. Angiotensin II Response Region

Angiotensin II (Ang II) is considered the most potent component of the renin-angiotensin system (RAS) and exerts its pleotropic actions such as vasoconstriction, blood pressure regulation, and cell proliferation in many tissues, including the kidney, adrenal glands, brain, and vasculature. The vasoconstrictor hormone Ang II and the vasodilator hormone ANP interact and mutually antagonize the biological actions of each other at various levels. Previously, it has been shown that Ang II inhibits GC activity of NPRA by activation of protein kinase C [62] and/or by stimulation of phosphotyrosine phosphatase activity via AT_2 receptors [15]. Moreover, Ang II has been shown to exert its effect on *Npr1* gene at the transcription/promoter level [58]. Treatment with Ang II in MMCs transiently transfected with *Npr1* gene promoter construct (-1982 to +55) significantly inhibited the promoter activity in a dose-dependent (0.1–100 nM) and time-dependent [8–48 h) manner with a maximal inhibition at 10 nM Ang II concentration after 24 h. The mRNA level of *Npr1* was significantly downregulated by Ang II treatment as determined by reverse-transcription PCR (RT-PCR) assay. Promoter deletion analysis showed an Ang II responsive region -1346 to -916 upstream of the TSS in the *Npr1* promoter and contains a putative cAMP-RE, which can be recognized by CREB. CREB has been identified as a positive and negative regulator of various gene promoters [70, 119], but its involvement in the Ang II response of *Npr1* promoter is yet to be investigated. Ang II-dependent transcriptional repression of *Npr1* promoter and downregulation of GC activity has been shown to be mediated by PKC pathways [5]. The *Npr1* promoter activity containing the region -1182 to -918 from TSS was reduced by 60% with Ang II treatment in MMCs. It was also demonstrated that Ang II in the presence of PKC agonist phorbol ester reduced *Npr1* mRNA levels and intracellular accumulation of cGMP levels by 70%, whereas PKC antagonist staurosporine completely reversed the effect of Ang II and phorbol ester. All these studies suggest the mechanism of Ang II effect on *Npr1* expression, at least in part, to be the inhibition of gene transcription.

5.3. Serum and Glucocorticoid-Inducible Kinase (Sgk)-Mediated Regulation

It has been suggested that chronic hyperosmolality enhances ANP-dependent cGMP production via stimulation of transcription and protein synthesis in cultured rat IMCD cells [84]. A considerable number of studies have shown that GC activity of NPRA is either activated [12, 75] or repressed [67, 146] by an increase in extracellular tonicity. In these studies the mechanisms involved were not elucidated and it has been suggested that activation predominates with longer exposure (~ 24 h) and inhibition with short-term exposure (minutes) to osmotic stimulus. More recent investigations have studied the mechanisms involved in activation of NPRA expression by osmotic stimulus. It was shown in the cultured rat IMCD cells that NaCl (75 mM) and sucrose (150 mM), but not urea (150 mM), increased NPRA activity, gene expression, and *Npr1* promoter activity from as early as 4 h reaching a maximum at 24 h [24]. There was also increase in phosphorylated MAP kinases by the osmotic stimulus and inhibition of p38 MAPK with SB203580 blocked the osmotic induction of NPRA activity and expression, as well as *Npr1* promoter activity. Cotransfection of p38 beta MAPK together with the *Npr1* promoter construct showed an amplification of the osmotic stimulation of the promoter indicating that extracellular osmolality stimulates receptor activity and gene expression through a specific p38 beta-dependent mechanism. Increased extracellular tonicity has also been shown to mediate a time-dependent reduction in endothelial nitric-oxide synthase (eNOS) and inducible nitric-oxide synthase (iNOS) protein levels in IMCD cells [22]. Reduction in eNOS expression together with the concomitant decline in intracellular cyclic GMP levels appears to account for a significant portion of the p38 MAPK-dependent osmotic stimulation of *Npr1* gene expression. These investigators have further examined the involvement of Sgk in the osmotic stimulation of *Npr1* promoter activity as Sgk participates in the regulation of sodium handling in distal tubular segments [26]. Role of Sgk1 in inducing *Npr1* promoter was confirmed by cotransfection of -1590 *Npr1* luciferase reporter construct with Sgk1 expression plasmid which increased reporter activity by 3-fold.

On the other hand, kinase-dead Sgk1mutant plasmid decreased the promoter activity and also partially blocked the NaCl-dependent *Npr1* promoter induction. Moreover, Sgk1 and its kinase-dead mutant had no effect on eNOS promoter activity implying that Sgk1 operates over an eNOS-independent, p38 MAPK-dependent pathway in mediating osmotic induction of the *Npr1* gene promoter [26]. It has also been shown that water restriction of rats or mice led to increased urine osmolality, increased Sgk1 and *Npr1* expression, and dehydration natriuresis [25]. In cultured rat renal medullary cells, siRNA-mediated knockdown of Sgk1 blocked the osmotic induction of *Npr1* gene expression. Furthermore, it was observed that *Npr1* gene knockout mice were resistant to dehydration natriuresis, which suggests that Sgk1-dependent activation of the NPRA pathway may contribute to this response and play an important role in promoting the physiological response of the kidney to elevations in extracellular tonicity.

6. Epigenetic Regulation

The expression of cell type specific genes is tightly regulated by a hierarchical mechanism composed of genetic and epigenetic factors. Epigenetic changes influence gene expression by changes in chromatin structure via modification of DNA (methylation of CpG islands or CG islands) and/or histones (e.g., methylation, acetylation, and phosphorylation) [33]. Epigenetic mechanisms including changes in histone acetylation/deacetylation and DNA methylation have been shown to be involved in altered gene expression under various pathological conditions including cancer [137], cardiac hypertrophy [64], renal fibrosis [106, 115], and inflammation [164].

6.1. CpG Island and DNA Methylation

CpG islands are stretches of DNA that have a higher CpG (CG dinucleotides) density in the vicinity of TSS of the genes and are evolutionary conserved to promote gene expression by regulating the chromatin structure and TF binding [13, 71, 110]. Approximately 70% of human genes are linked to promoter CpG islands and most CpG island promoters remain unmethylated even in the cell types that do not express the gene [138, 161]. The sequence analysis of mouse *Npr1* promoter has shown to have 34 CpG sites between -356 to +55 region [81]. To examine the effect of methylation on these CpG sites, pGL3-Basic and *Npr1* promoter -356/+55 was methylated *in vitro* by *Hha*I, *Hpa*II, and *Sss*I methylase. Partial methylation by *Hha*I (7 sites) reduced the promoter activity by 70%, whereas partial methylase *Hpa*II (5 sites) having 2 of its sites present within Ets-1 binding site, reduced the luciferase activity by almost 80%, suggesting the inhibition of Ets-1 binding to the promoter. Methylation of all 34 sites by *Sss*I methylase dramatically reduced the promoter activity by almost 95%. The gel shift analysis showed that methylation of CpG binding sites present in the Ets-1 recognition site significantly reduced its binding to *Npr1* promoter. Methylation of binding sites of various transcription factors, such as Sp1, Sp3, Ap-2, and CCAAT-binding factor have a direct influence on the expression of downstream genes and is a mechanism for methylation-induced gene repression [3, 4, 42, 171].

It was reported that *Npr1* gene was highly methylated in hepatitis C virus (HCV) -related hepatocellular carcinoma (HCC) as determined by methylated DNA IP-chip analysis, bisulphite treatment, and quantitative methylation analysis [43]. Moreover, 75% of *Npr1* promoter was found to be methylated in five liver tumor cell lines and showed very low expression levels of *Npr1* mRNA which was reversed by treatment with 5-aza-2' deoxycytidine (DNA demethylation agent) and trichostatin A (histone deactylase inhibitor) suggesting that gene was silenced by promoter methylation in HCC. Furthermore, a recent study has shown *Npr1* promoter methylation to be a later event associated with lymph node metastasis [21]. Real-time PCR-based DNA methylation assay showed 42.5% methylation frequency for *Npr1* promoter in 120 gastric cancer samples and DNA methylation correlated with low expression of mRNA as detected by real-time RT-PCR. In this study *Npr1* gene was included as one of the marker in high CpG island methylator phenotype associated with lymph node metastasis and prognosis in gastric cancer. These recent studies indicate *Npr1* as methylation sensitive gene which shows hypermethylation in cancer.

6.2. Histone Modifications

Histone acetylation can modulate gene transcription at two levels: global histone acetylation and promoter-specific histone acetylation. Global histone acetylation correlates with general transcriptional activity, promoter-specific acetylation is a critical mechanism for control of specific gene activity [158]. Epigenetic regulation of *Npr1* gene has not been studied in detail. Previous study has shown an increase in global histone H3 and H4 acetylation levels in the extracted histones from ATRA-treated MMCs [82]. Furthermore, ChIP assay showed an increased binding of acetylated histones H3 and H4 at the *Npr1* promoter region (-120 to + 73) from the TSS containing Ets-1 and Sp1 binding sites in presence of ATRA. Evidence suggests that ATRA treatment induces epigenetic modification at the target loci and regulates gene transcription by DNA demethylation and histone modifications, which include acetylation, methylation, and phosphorylation at the promoter level [50, 73, 99]. As a consequence of histone acetylation and decompacted chromatin structure, transcription factors and basal transcriptional machinery are recruited to the region and activate gene transcription [14, 61, 90]. DNA methylation and histone acetylation are major epigenetic modifications that are most intensively studied in the context of gene transcription. Since both DNA methylation and histone acetylation are strongly implicated in regulating gene expression, it is plausible to predict that interplay may exist between these two epigenetic mechanisms.

7. Polymorphisms and Regulatory Variants

Polymorphism in genes is increasingly associated to the genetic susceptibility to complex diseases such as essential hypertension and cardiac hypertrophy. Common quantitative variants in the regulatory region of *Npr1* gene have been found which modulate the expression of the gene. A novel microsatellite marker of dinucleotide (TC/GA) repeat was identified approximately 300 bp upstream of the start codon [114]. An association study in patients with essential hypertension (EH) found 10-12 repeats of TC/GA in the three alleles in unrelated Japanese subjects but no association was found with EH. However, in a different study with hypertensive patients showed a significant association of this microsatellite marker with left ventricular mass index and left ventricular septal thickness suggesting its contribution to ventricular remodeling in human EH [134]. An association study between (CT) dinucleotide repeat polymorphism in *Npr1* promoter and EH in a group of Japanese subjects identified 5 allele types with 6, 9, 10, 11, and 12 CT repeats around position –293, upstream of ATG codon in the human *Npr1* gene of normotensive and hypertensive subjects [157]. The frequency of the (CT)n=6 allele was significantly higher among hypertensive than normotensive subjects. An analysis of linkage between G/A polymorphism at position -77, downstream of the (CT)n polymorphism found that the (CT)n=6 allele was tightly linked to an A at position -77, while all other (CT)n alleles were linked to G. Promoter-reporter analyses revealed that the promoter containing (CT)n=6 was less transcriptionally activity than that containing (CT)n=10. Moreover, site-directed mutation showed that the (CT)n and G/A polymorphisms act synergistically to affect GC-A promoter activity suggesting that (CT)n polymorphism in the 5'-flanking region of the GC-A gene as a potent and novel susceptibility marker for hypertension [157]. On the other hand, no association was found between (CT)n polymorphism of *Npr1* gene and hypertension in a Hellenic population [155] or with presence and extent of coronary artery disease in Greek patients with a history of myocardial infarction [72]. In another study an insertion/deletion mutation in the 5'-flanking region of the *Npr1* promoter was identified [113]. The deletion allele lacks 8 nucleotides at position -60 of the *Npr1* promoter and alters binding sites for the activator protein-2 (AP-2) and Zeste transcriptional factors. Moreover, a reduced receptor activity and transcriptional activity of the deletion allele was found to be 30% lower than that of the wild-type allele which may increase susceptibility to developing EH or left ventricular hypertrophy. In another study a single-stranded conformational polymorphism caused by a repetitive dinucleotide segment of 10 TA in BN-Lx (normotensive genotype) and of 40 TA in SHR (hypertensive genotype) was localized at position -943 relative to the TSS of the *Npr1* gene, downstream of a putative cGMP-regulatory region [154]. Transient transfections of *Npr1* promoter construct -1520 to -920 showed that the construct from BN-Lx stimulated the luciferase activity, whereas that from SHR slightly inhibited it. Significant correlations of the TA repeat with basal guanylyl cyclase activity and *Npr1* mRNA levels were found in recombinant inbred strains (RIS) derived from BN-Lx and SHR crosses suggesting that the longer TA repeat unit in the promoter of *Npr1* gene of SHR, in tandem with a putative cGMP responsive element, regulates the transcription of the *Npr1* gene with consequences on diastolic blood pressure [154].

Sequencing of entire human *Npr1* gene identified 10 polymorphic sites in the 5' region, intron, and 3' UTR by using DNA from 34 unrelated human individuals [77]. The proximal promoter region showed length polymorphism a $(CT)_n$ dinucleotide repeat at nucleotide -293 designated as CT6, CT10, and CT11 and are present with overall frequency of 6, 44, and 50% respectively. A base substitution G/A was identified at nucleotide -77 of the 5' UTR. The length polymorphism at the 3' UTR at nucleotide 14,319 constituted three common haplotypes 3-plus, 4-plus, and 4-minus with frequencies of 39, 57, and 4%, respectively. Transient transfection analysis of reporter plasmid with *Npr1* promoter sequences and 3' UTR showed that these polymorphisms have functional effects and could significantly affect the genetic risks for EH and cardiac hypertrophy [77]. A novel length (3C/4C) polymorphism at position 15129 in the 3' UTR of *Npr1* mRNA, 30 bp downstream of stop codon has been identified and is associated with hypertensive family history (129]. Moreover, the functional *Npr1* 3C variant, when homozygous was also associated with higher systolic blood pressure and prolonged ventricular relaxation. A novel missense mutation has been identified methionine (ATG) to isoleucine (ATC) at nucleotide 1023 (M341I), in exon 3 of human *Npr1* gene which is found to be associated with EH and risk for myocardial infarction [111, 112]. Recent studies have shown that the non-coding regions of the genome including 5' and 3' UTRs and introns are responsible for regulation, facilitating complex sequential gene expression through the combinatorial effect of numerous mechanisms, and interactions working together to fine-tune gene expression [9].

Conclusion and Future Directions

Regulation of the *Npr1* gene appears to be tightly controlled by transcriptional mechanisms that respond to external and internal stimuli like hormonal stimulation, osmoregulation, as well as transcription factors that are important for basal promoter activity and have cardio and renal-protective effects. Our current knowledge of *Npr1* gene regulation clearly demonstrates that the regulation is multimodal and can be highly context-dependent. The emerging evidence on 5' and 3' UTR target sites as potential sites of polymorphisms that influence variation in *Npr1* expression add other layers of complexity to how *Npr1* gene is regulated with altered NPRA activity. The non-coding regions of the genome, including the 5' and 3' UTRs, introns, and intergenic regions are vital for the precise regulation of gene expression and have evidently expanded during the evolution of complex organisms. Additional research into the mechanisms controlling the expression of the *Npr1* gene can be pictured at several levels. Direct molecular and biochemical studies may uncover further regulatory elements in the 5' flanking regulatory region of the *Npr1*gene. As the functional interpretations of the genome become richer, the increased density of contact sites for DNA-interacting proteins and chromatin modifications is likely to point towards further important regulatory sites. A better understanding of transcriptional and epigenetic control mechanisms regulating *Npr1* gene expression will improve therapeutic strategies to control *Npr1* gene expression and function in pathophysiological conditions. Moreover, an ongoing search for variants in *Npr1* regulatory sequences may also identify polymorphisms that modulate NPRA levels. As more genomes are sequenced the number of annotated variants will increase and also the number of interrogated polymorphisms in population studies and functional regulatory variants may be recognized. While the study of transcriptional control of *Npr1* gene is a fascinating biology, it is also important to understand *Npr1* gene regulation which stems from the association of cardiovascular disease risk with altered expression and activity of NPRA.

Acknowledgment

The authors wish to thank Kamala Pandey for assistance during the preparation of this manuscript.

REFERENCES

[1] Anand-Srivastava MB, and Trachte GJ. Atrial natriuretic factor receptors and signal transduction mechanisms. *Pharmacol Rev* 45: 455-497, 1993.

[2] Anderson JL, May HT, Horne BD, Bair TL, Hall NL, Carlquist JF, Lappe DL, Muhlestein JB, and Intermountain Heart Collaborative Study G. Relation of vitamin D deficiency to cardiovascular risk factors, disease status, and incident events in a general healthcare population. *Am J Cardiol* 106: 963-968, 2010.

[3] Aoyama T, Okamoto T, Kohno Y, Fukiage K, Otsuka S, Furu M, Ito K, Jin Y, Nagayama S, Nakayama T, Nakamura T, and Toguchida J. Cell-specific epigenetic regulation of ChM-I gene expression: crosstalk between DNA methylation and histone acetylation. *Biochem Biophys Res Commun* 365: 124-130, 2008.

[4] Aoyama T, Okamoto T, Nagayama S, Nishijo K, Ishibe T, Yasura K, Nakayama T, Nakamura T, and Toguchida J. Methylation in the core-promoter region of the chondromodulin-I gene determines the cell-specific expression by regulating the binding of transcriptional activator Sp3. *J Biol Chem* 279: 28789-28797, 2004.

[5] Arise KK, and Pandey KN. Inhibition and down-regulation of gene transcription and guanylyl cyclase activity of NPRA by angiotensin II involving protein kinase C. *Biochem Biophys Res Commun* 349: 131-135, 2006.

[6] Arman M, Calvo J, Trojanowska ME, Cockerill PN, Santana M, Lopez-Cabrera M, Vives J, and Lozano F. Transcriptional regulation of human CD5: important role of Ets transcription factors in CD5 expression in T cells. *J Immunol* 172: 7519-7529, 2004.

[7] Azakie A, Fineman JR, and He Y. Myocardial transcription factors are modulated during pathologic cardiac hypertrophy *in vivo*. *J Thorac Cardiovasc Surg* 132: 1262-1271, 2006.

[8] Azakie A, Fineman JR, and He Y. Sp3 inhibits Sp1-mediated activation of the cardiac troponin T promoter and is downregulated during pathological cardiac hypertrophy *in vivo*. *Am J Physiol Heart Circ Physiol* 291: H600-611, 2006.

[9] Barrett LW, Fletcher S, and Wilton SD. Regulation of eukaryotic gene expression by the untranslated gene regions and other non-coding elements. *Cell Mol Life Sci* 69: 3613-3634, 2012.

[10] Bastien J, and Rochette-Egly C. Nuclear retinoid receptors and the transcription of retinoid-target genes. *Gene* 328: 1-16, 2004.

[11] Benfante R, Antonini RA, Vaccari M, Flora A, Chen F, Clementi F, and Fornasari D. The expression of the human neuronal alpha3 Na+,K+-ATPase subunit gene is regulated by the activity of the Sp1 and NF-Y transcription factors. *Biochem J* 386: 63-72, 2005.

[12] Berl T, Mansour J, and Veis JH. Regulation of atrial natriuretic peptide-stimulated cGMP production in the inner medulla. *Kidney Int* 41: 37-42, 1992.

[13] Bird A. DNA methylation patterns and epigenetic memory. *Genes Dev* 16: 6-21, 2002.

[14] Bjorklund S, Almouzni G, Davidson I, Nightingale KP, and Weiss K. Global transcription regulators of eukaryotes. *Cell* 96: 759-767, 1999.

[15] Bottari SP, King IN, Reichlin S, Dahlstroem I, Lydon N, and de Gasparo M. The angiotensin AT2 receptor stimulates protein tyrosine phosphatase activity and mediates

inhibition of particulate guanylate cyclase. *Biochem Biophys Res Commun* 183: 206-211, 1992.

[16] Brady M, Koban MU, Dellow KA, Yacoub M, Boheler KR, and Fuller SJ. Sp1 and Sp3 transcription factors are required for trans-activation of the human SERCA2 promoter in cardiomyocytes. *Cardiovasc Res* 60: 347-354, 2003.

[17] Briggs MR, Kadonaga JT, Bell SP, and Tjian R. Purification and biochemical characterization of the promoter-specific transcription factor, Sp1. *Science* 234: 47-52, 1986.

[18] Cao L, Chen SC, Cheng T, Humphreys MH, and Gardner DG. Ligand-dependent regulation of NPR-A gene expression in inner medullary collecting duct cells. *Am J Physiol* 275: F119-125, 1998.

[19] Cao L, Wu J, and Gardner DG. Atrial natriuretic peptide suppresses the transcription of its guanylyl cyclase-linked receptor. *J Biol Chem* 270: 24891-24897, 1995.

[20] Cederberg A, Hulander M, Carlsson P, and Enerback S. The kidney-expressed winged helix transcription factor FREAC-4 is regulated by Ets-1. A possible role in kidney development. *J Biol Chem* 274: 165-169, 1999.

[21] Chen HY, Zhu BH, Zhang CH, Yang DJ, Peng JJ, Chen JH, Liu FK, and He YL. High CpG island methylator phenotype is associated with lymph node metastasis and prognosis in gastric cancer. *Cancer Sci* 103: 73-79, 2012.

[22] Chen S, Cao L, Intengan HD, Humphreys M, and Gardner DG. Osmoregulation of endothelial nitric-oxide synthase gene expression in inner medullary collecting duct cells. Role in activation of the type A natriuretic peptide receptor. *J Biol Chem* 277: 32498-32504, 2002.

[23] Chen S, Costa CH, Nakamura K, Ribeiro RC, and Gardner DG. Vitamin D-dependent suppression of human atrial natriuretic peptide gene promoter activity requires heterodimer assembly. *J Biol Chem* 274: 11260-11266, 1999.

[24] Chen S, and Gardner DG. Osmoregulation of natriuretic peptide receptor signaling in inner medullary collecting duct. A requirement for p38 MAPK. *J Biol Chem* 277: 6037-6043, 2002.

[25] Chen S, Grigsby CL, Law CS, Ni X, Nekrep N, Olsen K, Humphreys MH, and Gardner DG. Tonicity-dependent induction of Sgk1 expression has a potential role in dehydration-induced natriuresis in rodents. *J Clin Invest* 119: 1647-1658, 2009.

[26] Chen S, McCormick JA, Prabaker K, Wang J, Pearce D, and Gardner DG. Sgk1 mediates osmotic induction of NPR-A gene in rat inner medullary collecting duct cells. *Hypertension* 43: 866-871, 2004.

[27] Chen S, Nakamura K, and Gardner DG. 1,25-dihydroxyvitamin D inhibits human ANP gene promoter activity. *Regul Pept* 128: 197-202, 2005.

[28] Chen S, Ni XP, Humphreys MH, and Gardner DG. 1,25 dihydroxyvitamin d amplifies type a natriuretic peptide receptor expression and activity in target cells. *J Am Soc Nephrol* 16: 329-339, 2005.

[29] Chen S, Olsen K, Grigsby C, and Gardner DG. Vitamin D activates type A natriuretic peptide receptor gene transcription in inner medullary collecting duct cells. *Kidney Int* 72: 300-306, 2007.

[30] Chen Y, Watson AM, Williamson CD, Rahimi M, Liang C, Colberg-Poley AM, and Rose MC. Glucocorticoid receptor and histone deacetylase-2 mediate dexamethasone-

induced repression of MUC5AC gene expression. *Am J Respir Cell Mol Biol* 47: 637-644, 2012.

[31] Cheng YH, Yin P, Xue Q, Yilmaz B, Dawson MI, and Bulun SE. Retinoic acid (RA) regulates 17beta-hydroxysteroid dehydrogenase type 2 expression in endometrium: interaction of RA receptors with specificity protein (SP) 1/SP3 for estradiol metabolism. *J Clin Endocrinol Metab* 93: 1915-1923, 2008.

[32] Chinkers M, Garbers DL, Chang MS, Lowe DG, Chin HM, Goeddel DV, and Schulz S. A membrane form of guanylate cyclase is an atrial natriuretic peptide receptor. *Nature* 338: 78-83, 1989.

[33] D'Alessio AC, Weaver IC, and Szyf M. Acetylation-induced transcription is required for active DNA demethylation in methylation-silenced genes. *Mol Cell Biol* 27: 7462-7474, 2007.

[34] Dananberg J, and Grekin RJ. Corticoid regulation of atrial natriuretic factor secretion and gene expression. *Am J Physiol* 263: H1377-1381, 1992.

[35] Danik JS, and Manson JE. Vitamin d and cardiovascular disease. *Curr Treat Options Cardiovasc Med* 14: 414-424, 2012.

[36] Das S, Periyasamy R, and Pandey KN. Activation of IKK/NF-kappaB provokes renal inflammatory responses in guanylyl cyclase/natriuretic peptide receptor-A gene-knockout mice. *Physiol Genomics* 44: 430-442, 2012.

[37] Davidson SM, Townsend PA, Carroll C, Yurek-George A, Balasubramanyam K, Kundu TK, Stephanou A, Packham G, Ganesan A, and Latchman DS. The transcriptional coactivator p300 plays a critical role in the hypertrophic and protective pathways induced by phenylephrine in cardiac cells but is specific to the hypertrophic effect of urocortin. *Chembiochem* 6: 162-170, 2005.

[38] de Bold AJ. Atrial natriuretic factor: a hormone produced by the heart. *Science* 230: 767-770, 1985.

[39] de Bold AJ, Ma KK, Zhang Y, de Bold ML, Bensimon M, and Khoshbaten A. The physiological and pathophysiological modulation of the endocrine function of the heart. *Can J Physiol Pharmacol* 79: 705-714, 2001.

[40] de The H, Vivanco-Ruiz MM, Tiollais P, Stunnenberg H, and Dejean A. Identification of a retinoic acid responsive element in the retinoic acid receptor beta gene. *Nature* 343: 177-180, 1990.

[41] Deng AY, Dene H, and Rapp JP. Congenic strains for the blood pressure quantitative trait locus on rat chromosome 2. *Hypertension* 30: 199-202, 1997.

[42] Deng G, Song GA, Pong E, Sleisenger M, and Kim YS. Promoter methylation inhibits APC gene expression by causing changes in chromatin conformation and interfering with the binding of transcription factor CCAAT-binding factor. *Cancer Res* 64: 2692-2698, 2004.

[43] Deng YB, Nagae G, Midorikawa Y, Yagi K, Tsutsumi S, Yamamoto S, Hasegawa K, Kokudo N, Aburatani H, and Kaneda A. Identification of genes preferentially methylated in hepatitis C virus-related hepatocellular carcinoma. *Cancer Sci* 101: 1501-1510, 2010.

[44] Dilworth FJ, and Chambon P. Nuclear receptors coordinate the activities of chromatin remodeling complexes and coactivators to facilitate initiation of transcription. *Oncogene* 20: 3047-3054, 2001.

[45] Dittmer J. The biology of the Ets1 proto-oncogene. *Mol Cancer* 2: 29, 2003.

[46] Dolfini D, Gatta R, and Mantovani R. NF-Y and the transcriptional activation of CCAAT promoters. *Crit Rev Biochem Mol Biol* 47: 29-49, 2012.

[47] Donaldson LW, Petersen JM, Graves BJ, and McIntosh LP. Solution structure of the ETS domain from murine Ets-1: a winged helix-turn-helix DNA binding motif. *EMBO J* 15: 125-134, 1996.

[48] Dong S, Kanno T, Yamaki A, Kojima T, Shiraiwa M, Kawada A, Mechin MC, Chavanas S, Serre G, Simon M, and Takahara H. NF-Y and Sp1/Sp3 are involved in the transcriptional regulation of the peptidylarginine deiminase type III gene (PADI3) in human keratinocytes. *Biochem J* 397: 449-459, 2006.

[49] Duda T, and Sharma RK. Regulation of guanylate cyclase activity by atrial natriuretic factor and protein kinase C. *Mol Cell Biochem* 93: 179-184, 1990.

[50] Ekici M, Hohl M, Schuit F, Martinez-Serrano A, and Thiel G. Transcription of genes encoding synaptic vesicle proteins in human neural stem cells: chromatin accessibility, histone methylation pattern, and the essential role of rest. *J Biol Chem* 283: 9257-9268, 2008.

[51] Forster A, Emmler T, Schwalm S, Ebadi M, Heringdorf DM, Nieuwenhuis B, Kleuser B, Huwiler A, and Pfeilschifter J. Glucocorticoids protect renal mesangial cells from apoptosis by increasing cellular sphingosine-1-phosphate. *Kidney Int* 77: 870-879, 2010.

[52] Fuller F, Porter JG, Arfsten AE, Miller J, Schilling JW, Scarborough RM, Lewicki JA, and Schenk DB. Atrial natriuretic peptide clearance receptor. Complete sequence and functional expression of cDNA clones. *J Biol Chem* 263: 9395-9401, 1988.

[53] Gao XH, Dwivedi PP, Omdahl JL, Morris HA, and May BK. Calcitonin stimulates expression of the rat 25-hydroxyvitamin D3-24-hydroxylase (CYP24) promoter in HEK-293 cells expressing calcitonin receptor: identification of signaling pathways. *J Mol Endocrinol* 32: 87-98, 2004.

[54] Garbers DL. Guanylyl cyclase receptors and their endocrine, paracrine, and autocrine ligands. *Cell* 71: 1-4, 1992.

[55] Gardner DG, Gertz BJ, Deschepper CF, and Kim DY. Gene for the rat atrial natriuretic peptide is regulated by glucocorticoids in vitro. *J Clin Invest* 82: 1275-1281, 1988.

[56] Gardner DG, Hane S, Trachewsky D, Schenk D, and Baxter JD. Atrial natriuretic peptide mRNA is regulated by glucocorticoids *in vivo*. *Biochem Biophys Res Commun* 139: 1047-1054, 1986.

[57] Garg R, Oliver PM, Maeda N, and Pandey KN. Genomic structure, organization, and promoter region analysis of murine guanylyl cyclase/atrial natriuretic peptide receptor-A gene. *Gene* 291: 123-133, 2002.

[58] Garg R, and Pandey KN. Angiotensin II-mediated negative regulation of Npr1 promoter activity and gene transcription. *Hypertension* 41: 730-736, 2003.

[59] Ghosh AK, and Varga J. The transcriptional coactivator and acetyltransferase p300 in fibroblast biology and fibrosis. *J Cell Physiol* 213: 663-671, 2007.

[60] Goetz KL. Renal natriuretic peptide (urodilatin?) and atriopeptin: evolving concepts. *Am J Physiol* 261: F921-932, 1991.

[61] Gregory PD, Wagner K, and Horz W. Histone acetylation and chromatin remodeling. *Exp Cell Res* 265: 195-202, 2001.

[62] Haneda M, Kikkawa R, Maeda S, Togawa M, Koya D, Horide N, Kajiwara N, and Shigeta Y. Dual mechanism of angiotensin II inhibits ANP-induced mesangial cGMP accumulation. *Kidney Int* 40: 188-194, 1991.

[63] Haussler MR, Jurutka PW, Mizwicki M, and Norman AW. Vitamin D receptor (VDR)-mediated actions of 1alpha,25(OH)(2)vitamin D(3): genomic and non-genomic mechanisms. *Best Pract Res Clin Endocrinol Metab* 25: 543-559, 2011.

[64] Heintzman ND, Stuart RK, Hon G, Fu Y, Ching CW, Hawkins RD, Barrera LO, Van Calcar S, Qu C, Ching KA, Wang W, Weng Z, Green RD, Crawford GE, and Ren B. Distinct and predictive chromatin signatures of transcriptional promoters and enhancers in the human genome. *Nat Genet* 39: 311-318, 2007.

[65] Hultgardh-Nilsson A, Cercek B, Wang JW, Naito S, Lovdahl C, Sharifi B, Forrester JS, and Fagin JA. Regulated expression of the ets-1 transcription factor in vascular smooth muscle cells *in vivo* and in vitro. *Circ Res* 78: 589-595, 1996.

[66] Hum D, Besnard S, Sanchez R, Devost D, Gossard F, Hamet P, and Tremblay J. Characterization of a cGMP-response element in the guanylyl cyclase/natriuretic peptide receptor A gene promoter. *Hypertension* 43: 1270-1278, 2004.

[67] Iimura O, Kusano E, Ishida F, Oono S, Ando Y, and Asano Y. Hyperosmolality rapidly reduces atrial-natriuretic-peptide-dependent cyclic guanosine monophosphate production in cultured rat inner medullary collecting duct cells. *Pflugers Arch* 430: 81-87, 1995.

[68] Jaubert J, Jaubert F, Martin N, Washburn LL, Lee BK, Eicher EM, and Guenet JL. Three new allelic mouse mutations that cause skeletal overgrowth involve the natriuretic peptide receptor C gene (Npr3). *Proc Natl Acad Sci U S A* 96: 10278-10283, 1999.

[69] Jayaraman G, Srinivas R, Duggan C, Ferreira E, Swaminathan S, Somasundaram K, Williams J, Hauser C, Kurkinen M, Dhar R, Weitzman S, Buttice G, and Thimmapaya B. p300/cAMP-responsive element-binding protein interactions with ets-1 and ets-2 in the transcriptional activation of the human stromelysin promoter. *J Biol Chem* 274: 17342-17352, 1999.

[70] Ji R, Sanchez CM, Chou CL, Chen XB, Woodward DF, and Regan JW. Prostanoid EP(1) receptors mediate up-regulation of the orphan nuclear receptor Nurr1 by cAMP-independent activation of protein kinase A, CREB and NF-kappaB. *Br J Pharmacol* 166: 1033-1046, 2012.

[71] Jones PA. Functions of DNA methylation: islands, start sites, gene bodies and beyond. *Nat Rev Genet* 13: 484-492, 2012.

[72] Karayannis G, Tsezou A, Giannatou E, Papanikolaou V, Giamouzis G, and Triposkiadis F. Polymorphisms of renin-angiotensin system and natriuretic peptide receptor A genes in patients of Greek origin with a history of myocardial infarction. *Angiology* 61: 737-743, 2010.

[73] Kashyap V, and Gudas LJ. Epigenetic regulatory mechanisms distinguish retinoic acid-mediated transcriptional responses in stem cells and fibroblasts. *J Biol Chem* 285: 14534-14548.

[74] Kastner P, Mark M, Ghyselinck N, Krezel W, Dupe V, Grondona JM, and Chambon P. Genetic evidence that the retinoid signal is transduced by heterodimeric RXR/RAR functional units during mouse development. *Development* 124: 313-326, 1997.

[75] Katafuchi T, Mizuno T, Hagiwara H, Itakura M, Ito T, and Hirose S. Modulation by NaCl of atrial natriuretic peptide receptor levels and cyclic GMP responsiveness to atrial natriuretic peptide of cultured vascular endothelial cells. *J Biol Chem* 267: 7624-7629, 1992.

[76] Khurana ML, and Pandey KN. Receptor-mediated stimulatory effect of atrial natriuretic factor, brain natriuretic peptide, and C-type natriuretic peptide on testosterone production in purified mouse Leydig cells: activation of cholesterol side-chain cleavage enzyme. *Endocrinology* 133: 2141-2149, 1993.

[77] Knowles JW, Erickson LM, Guy VK, Sigel CS, Wilder JC, and Maeda N. Common variations in noncoding regions of the human natriuretic peptide receptor A gene have quantitative effects. *Hum Genet* 112: 62-70, 2003.

[78] Koller KJ, and Goeddel DV. Molecular biology of the natriuretic peptides and their receptors. *Circulation* 86: 1081-1088, 1992.

[79] Kuhn M. Structure, regulation, and function of mammalian membrane guanylyl cyclase receptors, with a focus on guanylyl cyclase-A. *Circ Res* 93: 700-709, 2003.

[80] Kumar P, Arise KK, and Pandey KN. Transcriptional regulation of guanylyl cyclase/natriuretic peptide receptor-A gene. *Peptides* 27: 1762-1769, 2006.

[81] Kumar P, Bolden G, Arise KK, Krazit ST, and Pandey KN. Regulation of natriuretic peptide receptor-A gene expression and stimulation of its guanylate cyclase activity by transcription factor Ets-1. *Biosci Rep* 29: 57-70, 2009.

[82] Kumar P, Garg R, Bolden G, and Pandey KN. Interactive roles of Ets-1, Sp1, and acetylated histones in the retinoic acid-dependent activation of guanylyl cyclase/atrial natriuretic peptide receptor-A gene transcription. *J Biol Chem* 285: 37521-37530, 2010.

[83] Kumar P, and Pandey KN. Cooperative activation of Npr1 gene transcription and expression by interaction of Ets-1 and p300. *Hypertension* 54: 172-178, 2009.

[84] Kusano E, Iimura O, Ishida F, Akimoto T, Amemiya M, Ando Y, and Asano Y. Chronic hyperosmolality enhances ANP-dependent cGMP production via stimulation of transcription and protein synthesis in cultured rat IMCD cells. *Tohoku J Exp Med* 197: 209-220, 2002.

[85] Lee YS, and Jeong WI. Retinoic acids and hepatic stellate cells in liver disease. *J Gastroenterol Hepatol* 27 Suppl 2: 75-79, 2012.

[86] Lehrke I, Schaier M, Schade K, Morath C, Waldherr R, Ritz E, and Wagner J. Retinoid receptor-specific agonists alleviate experimental glomerulonephritis. *Am J Physiol* 282: F741-751, 2002.

[87] Lerner AJ, Gustaw-Rothenberg K, Smyth S, and Casadesus G. Retinoids for treatment of Alzheimer's disease. *BioFactors* 38: 84-89, 2012.

[88] Letovsky J, and Dynan WS. Measurement of the binding of transcription factor Sp1 to a single GC box recognition sequence. *Nucleic Acids Res* 17: 2639-2653, 1989.

[89] Levin ER, Gardner DG, and Samson WK. Natriuretic peptides. *N Engl J Med* 339: 321-328, 1998.

[90] Li B, Carey M, and Workman JL. The role of chromatin during transcription. *Cell* 128: 707-719, 2007.

[91] Li Y, Saito Y, Kuwahara K, Rong X, Kishimoto I, Harada M, Adachi Y, Nakanishi M, Kinoshita H, Horiuchi M, Murray M, and Nakao K. Guanylyl cyclase-A inhibits angiotensin II type 2 receptor-mediated pro-hypertrophic signaling in the heart. *Endocrinology* 150: 3759-3765, 2009.

[92] Liang F, Schaufele F, and Gardner DG. Functional interaction of NF-Y and Sp1 is required for type a natriuretic peptide receptor gene transcription. *J Biol Chem* 276: 1516-1522, 2001.

[93] Liang F, Schaufele F, and Gardner DG. Sp1 dependence of natriuretic peptide receptor A gene transcription in rat aortic smooth muscle cells. *Endocrinology* 140: 1695-1701, 1999.

[94] Lie-Venema H, Gittenberger-de Groot AC, van Empel LJ, Boot MJ, Kerkdijk H, de Kant E, and DeRuiter MC. Ets-1 and Ets-2 transcription factors are essential for normal coronary and myocardial development in chicken embryos. *Circ Res* 92: 749-756, 2003.

[95] Lindahl GE, Chambers RC, Papakrivopoulou J, Dawson SJ, Jacobsen MC, Bishop JE, and Laurent GJ. Activation of fibroblast procollagen alpha 1(I) transcription by mechanical strain is transforming growth factor-beta-dependent and involves increased binding of CCAAT-binding factor (CBF/NF-Y) at the proximal promoter. *J Biol Chem* 277: 6153-6161, 2002.

[96] Liu C, Chen Y, Kang Y, Ni Z, Xiu H, Guan J, and Liu K. Glucocorticoids improve renal responsiveness to atrial natriuretic peptide by up-regulating natriuretic peptide receptor-A expression in the renal inner medullary collecting duct in decompensated heart failure. *J Pharmacol Exp Ther* 339: 203-209, 2011.

[97] Liu C, Guan J, Kang Y, Xiu H, Chen Y, Deng B, and Liu K. Inhibition of dehydration-induced water intake by glucocorticoids is associated with activation of hypothalamic natriuretic peptide receptor-A in rat. *PloS one* 5: e15607, 2010.

[98] Lopez MJ, Wong SK, Kishimoto I, Dubois S, Mach V, Friesen J, Garbers DL, and Beuve A. Salt-resistant hypertension in mice lacking the guanylyl cyclase-A receptor for atrial natriuretic peptide. *Nature* 378: 65-68, 1995.

[99] Love WK, Berletch JB, Andrews LG, and Tollefsbol TO. Epigenetic regulation of telomerase in retinoid-induced differentiation of human leukemia cells. *Int J Oncol* 32: 625-631, 2008.

[100] Lowe DG, Chang MS, Hellmiss R, Chen E, Singh S, Garbers DL, and Goeddel DV. Human atrial natriuretic peptide receptor defines a new paradigm for second messenger signal transduction. *EMBO J* 8: 1377-1384, 1989.

[101] Lowe DG, Klisak I, Sparkes RS, Mohandas T, and Goeddel DV. Chromosomal distribution of three members of the human natriuretic peptide receptor/guanylyl cyclase gene family. *Genomics* 8: 304-312, 1990.

[102] Lucas KA, Pitari GM, Kazerounian S, Ruiz-Stewart I, Park J, Schulz S, Chepenik KP, and Waldman SA. Guanylyl cyclases and signaling by cyclic GMP. *Pharmacol Rev* 52: 375-414, 2000.

[103] Lundqvist J, Wikvall K, and Norlin M. Vitamin D-mediated regulation of CYP21A2 transcription - a novel mechanism for vitamin D action. *Biochim Biophys Acta* 1820: 1553-1559, 2012.

[104] Mantovani R. The molecular biology of the CCAAT-binding factor NF-Y. *Gene* 239: 15-27, 1999.

[105] Martel G, Hamet P, and Tremblay J. GREBP, a cGMP-response element-binding protein repressing the transcription of natriuretic peptide receptor 1 (NPR1/GCA). *J Biol Chem* 285: 20926-20939, 2010.

[106] Marumo T, Hishikawa K, Yoshikawa M, Hirahashi J, Kawachi S, and Fujita T. Histone deacetylase modulates the proinflammatory and -fibrotic changes in tubulointerstitial injury. *Am J Physiol* 298: F133-141, 2010.

[107] McGrath MF, de Bold ML, and de Bold AJ. The endocrine function of the heart. *Trends Endocrinol Metab* 16: 469-477, 2005.

[108] McKnight S, and Tjian R. Transcriptional selectivity of viral genes in mammalian cells. *Cell* 46: 795-805, 1986.

[109] Mizui M, Isaka Y, Takabatake Y, Sato Y, Kawachi H, Shimizu F, Takahara S, Ito T, and Imai E. Transcription factor Ets-1 is essential for mesangial matrix remodeling. *Kidney Int* 70: 298-305, 2006.

[110] Moore LD, Le T, and Fan G. DNA Methylation and Its Basic Function. *Neuropsychopharmacology* 2012.

[111] Nakayama T, Soma M, Mizutani Y, Xinjuan X, Honye J, Kaneko Y, Rahmutula D, Aoi N, Kosuge K, Saito S, Ozawa Y, Kanmatsuse K, and Kokubun S. A novel missense mutation of exon 3 in the type A human natriuretic peptide receptor gene: possible association with essential hypertension. *Hypertens Res* 25: 395-401, 2002.

[112] Nakayama T, Soma M, Saito S, Honye J, Sato M, Aoi N, Kosuge K, Haketa A, Kanmatsuse K, and Kokubun S. Missense mutation of exon 3 in the type A human natriuretic peptide receptor gene is associated with myocardial infarction. *Med Sci Monit* 9: CR505-510, 2003.

[113] Nakayama T, Soma M, Takahashi Y, Rehemudula D, Kanmatsuse K, and Furuya K. Functional deletion mutation of the 5'-flanking region of type A human natriuretic peptide receptor gene and its association with essential hypertension and left ventricular hypertrophy in the Japanese. *Circ Res* 86: 841-845, 2000.

[114] Nakayama T, Soma M, Takahashi Y, Rehemudula D, Sato M, Uwabo J, Izumi Y, and Kanmatsuse K. Nucleotide sequence of the 5'-flanking region of the type A human natriuretic peptide receptor gene and association analysis using a novel microsatellite in essential hypertension. *Am J Hypertens* 12: 1144-1148, 1999.

[115] Noh H, Oh EY, Seo JY, Yu MR, Kim YO, Ha H, and Lee HB. Histone deacetylase-2 is a key regulator of diabetes- and transforming growth factor-beta1-induced renal injury. *Am J Physiol* 297: F729-739, 2009.

[116] Nuglozeh E, Mbikay M, Stewart DJ, and Legault L. Rat natriuretic peptide receptor genes are regulated by glucocorticoids in vitro. *Life Sci* 61: 2143-2155, 1997.

[117] Oakley RH, Revollo J, and Cidlowski JA. Glucocorticoids regulate arrestin gene expression and redirect the signaling profile of G protein-coupled receptors. *Proc Natl Acad Sci U S A* 109: 17591-17596, 2012.

[118] Oliver PM, Fox JE, Kim R, Rockman HA, Kim HS, Reddick RL, Pandey KN, Milgram SL, Smithies O, and Maeda N. Hypertension, cardiac hypertrophy, and sudden death in mice lacking natriuretic peptide receptor A. *Proc Natl Acad Sci U S A* 94: 14730-14735, 1997.

[119] Orlando U, Cooke M, Cornejo Maciel F, Papadopoulos V, Podesta EJ, and Maloberti P. Characterization of the mouse promoter region of the acyl-CoA synthetase 4 gene: Role of Sp1 and CREB. *Mol Cell Endocrinol* 2013.

[120] Palm-Leis A, Singh US, Herbelin BS, Olsovsky GD, Baker KM, and Pan J. Mitogen-activated protein kinases and mitogen-activated protein kinase phosphatases mediate

the inhibitory effects of all-trans retinoic acid on the hypertrophic growth of cardiomyocytes. *J Biol Chem* 279: 54905-54917, 2004.

[121] Pandey KN. Biology of natriuretic peptides and their receptors. *Peptides* 26: 901-932, 2005.

[122] Pandey KN. Emerging roles of natriuretic peptides and their receptors in pathophysiology of hypertension and cardiovascular regulation *J Am Soc Hypertens* 2: 210-226, 2008.

[123] Pandey KN. The functional genomics of guanylyl cyclase/natriuretic peptide receptor-A: perspectives and paradigms. *FEBS J* 278: 1792-1807, 2011.

[124] Pandey KN. Guanylyl cyclase / atrial natriuretic peptide receptor-A: role in the pathophysiology of cardiovascular regulation. *Can J Physiol Pharmacol* 89: 557-573, 2011.

[125] Pandey KN. *Vascular action. Natriuretic peptide receptor.* Totawa, NJ: Humana Press., 1996.

[126] Pandey KN, Adamson MC, Gu YC, and Kozak CA. Genetic mapping of the gene encoding guanylate cyclase-A/atrial natriuretic factor receptor (Npra) to mouse chromosome 3. *Mamm Genome* 5: 520-522, 1994.

[127] Pandey KN, Nguyen HT, Sharma GD, Shi SJ, and Kriegel AM. Ligand-regulated internalization, trafficking, and down-regulation of guanylyl cyclase/atrial natriuretic peptide receptor-A in human embryonic kidney 293 cells. *J Biol Chem* 277: 4618-4627, 2002.

[128] Pandey KN, and Singh S. Molecular cloning and expression of murine guanylate cyclase/atrial natriuretic factor receptor cDNA. *J Biol Chem* 265: 12342-12348, 1990.

[129] Pitzalis MV, Sarzani R, Dessi-Fulgheri P, Iacoviello M, Forleo C, Lucarelli K, Pietrucci F, Salvi F, Sorrentino S, Romito R, Guida P, Rappelli A, and Rizzon P. Allelic variants of natriuretic peptide receptor genes are associated with family history of hypertension and cardiovascular phenotype. *J Hypertens* 21: 1491-1496, 2003.

[130] Polman JA, Welten JE, Bosch DS, de Jonge RT, Balog J, van der Maarel SM, de Kloet ER, and Datson NA. A genome-wide signature of glucocorticoid receptor binding in neuronal PC12 cells. *BMC Neurosci* 13: 118, 2012.

[131] Potter LR, and Garbers DL. Dephosphorylation of the guanylyl cyclase-A receptor causes desensitization. *J Biol Chem* 267: 14531-14534, 1992.

[132] Potter LR, and Garbers DL. Protein kinase C-dependent desensitization of the atrial natriuretic peptide receptor is mediated by dephosphorylation. *J Biol Chem* 269: 14636-14642, 1994.

[133] Rikitake Y, and Moran E. DNA-binding properties of the E1A-associated 300-kilodalton protein. *Mol Cell Biol* 12: 2826-2836, 1992.

[134] Rubattu S, Bigatti G, Evangelista A, Lanzani C, Stanzione R, Zagato L, Manunta P, Marchitti S, Venturelli V, Bianchi G, Volpe M, and Stella P. Association of atrial natriuretic peptide and type a natriuretic peptide receptor gene polymorphisms with left ventricular mass in human essential hypertension. *J Am Coll Cardiol* 48: 499-505, 2006.

[135] Rudge TL, and Johnson LF. Synergistic activation of the TATA-less mouse thymidylate synthase promoter by the Ets transcription factor GABP and Sp1. *Exp Cell Res* 274: 45-55, 2002.

[136] Salehi-Tabar R, Nguyen-Yamamoto L, Tavera-Mendoza LE, Quail T, Dimitrov V, An BS, Glass L, Goltzman D, and White JH. Vitamin D receptor as a master regulator of the c-MYC/MXD1 network. *Proc Natl Acad Sci U S A* 109: 18827-18832, 2012.

[137] Santos-Rosa H, Schneider R, Bannister AJ, Sherriff J, Bernstein BE, Emre NC, Schreiber SL, Mellor J, and Kouzarides T. Active genes are tri-methylated at K4 of histone H3. *Nature* 419: 407-411, 2002.

[138] Saxonov S, Berg P, and Brutlag DL. A genome-wide analysis of CpG dinucleotides in the human genome distinguishes two distinct classes of promoters. *Proc Natl Acad Sci U S A* 103: 1412-1417, 2006.

[139] Schaier M, Liebler S, Schade K, Shimizu F, Kawachi H, Grone HJ, Chandraratna R, Ritz E, and Wagner J. Retinoic acid receptor alpha and retinoid X receptor specific agonists reduce renal injury in established chronic glomerulonephritis of the rat. *J Mol Med* 82: 116-125, 2004.

[140] Schulz S. C-type natriuretic peptide and guanylyl cyclase B receptor. *Peptides* 26: 1024-1034, 2005.

[141] Schulz S, Singh S, Bellet RA, Singh G, Tubb DJ, Chin H, and Garbers DL. The primary structure of a plasma membrane guanylate cyclase demonstrates diversity within this new receptor family. *Cell* 58: 1155-1162, 1989.

[142] Sekiguchi T, Miyamoto K, Mizutani T, Yamada K, Yazawa T, Yoshino M, Minegishi T, Takei Y, Kangawa K, Minamino N, Saito Y, and Kojima M. Molecular cloning of natriuretic peptide receptor A from bullfrog (Rana catesbeiana) brain and its functional expression. *Gene* 273: 251-257, 2001.

[143] Seth A, Giunta S, Franceschil C, Kola I, and Venanzoni MC. Regulation of the human stress response gene GADD153 expression: role of ETS1 and FLI-1 gene products. *Cell Death Differ* 6: 902-907, 1999.

[144] Seth A, and Watson DK. ETS transcription factors and their emerging roles in human cancer. *Eur J Cancer* 41: 2462-2478, 2005.

[145] Shi SJ, Vellaichamy E, Chin SY, Smithies O, Navar LG, and Pandey KN. Natriuretic peptide receptor A mediates renal sodium excretory responses to blood volume expansion. *Am J Physiol* 285: F694-702, 2003.

[146] Shinohara M, Nonoguchi H, Ujiie K, Terada Y, Owada A, Tomita K, and Marumo F. Effects of hyperosmolality on ANP-stimulated cGMP generation in rat inner medullary collecting duct. *Kidney Int* 45: 1362-1368, 1994.

[147] Simonsson M, Heldin CH, Ericsson J, and Gronroos E. The balance between acetylation and deacetylation controls Smad7 stability. *J Biol Chem* 280: 21797-21803, 2005.

[148] Smale ST, Schmidt MC, Berk AJ, and Baltimore D. Transcriptional activation by Sp1 as directed through TATA or initiator: specific requirement for mammalian transcription factor IID. *Proc Natl Acad Sci U S A* 87: 4509-4513, 1990.

[149] Sterner DE, and Berger SL. Acetylation of histones and transcription-related factors. *Microbiol Mol Biol Rev* 64: 435-459, 2000.

[150] Suga S, Nakao K, Itoh H, Komatsu Y, Ogawa Y, Hama N, and Imura H. Endothelial production of C-type natriuretic peptide and its marked augmentation by transforming growth factor-beta. Possible existence of "vascular natriuretic peptide system". *J Clin Invest* 90: 1145-1149, 1992.

[151] Takahashi Y, Nakayama T, Soma M, Izumi Y, and Kanmatsuse K. Organization of the human natriuretic peptide receptor A gene. *Biochem Biophys Res Commun* 246: 736-739, 1998.

[152] Tamura N, Chrisman TD, and Garbers DL. The regulation and physiological roles of the guanylyl cyclase receptors. *Endocr J* 48: 611-634, 2001.

[153] Tremblay J, Desjardins R, Hum D, Gutkowska J, and Hamet P. Biochemistry and physiology of the natriuretic peptide receptor guanylyl cyclases. *Mol Cell Biochem* 230: 31-47, 2002.

[154] Tremblay J, Hum DH, Sanchez R, Dumas P, Pravenec M, Krenova D, Kren V, Kunes J, Pausova Z, Gossard F, and Hamet P. TA repeat variation, Npr1 expression, and blood pressure: impact of the Ace locus. *Hypertension* 41: 16-24, 2003.

[155] Tsezou A, Karayannis G, Giannatou E, Papanikolaou V, and Triposkiadis F. Association of renin-angiotensin system and natriuretic peptide receptor A gene polymorphisms with hypertension in a Hellenic population. *J Renin Angiotensin Aldosterone Syst* 9: 202-207, 2008.

[156] Umesono K, Murakami KK, Thompson CC, and Evans RM. Direct repeats as selective response elements for the thyroid hormone, retinoic acid, and vitamin D3 receptors. *Cell* 65: 1255-1266, 1991.

[157] Usami S, Kishimoto I, Saito Y, Harada M, Kuwahara K, Nakagawa Y, Nakanishi M, Yasuno S, Kangawa K, and Nakao K. Association of CT dinucleotide repeat polymorphism in the 5'-flanking region of the guanylyl cyclase (GC)-A gene with essential hypertension in the Japanese. *Hypertens Res* 31: 89-96, 2008.

[158] Vaissiere T, Sawan C, and Herceg Z. Epigenetic interplay between histone modifications and DNA methylation in gene silencing. *Mutation research* 659: 40-48, 2008.

[159] Vellaichamy E, Khurana ML, Fink J, and Pandey KN. Involvement of the NF-kappa B/matrix metalloproteinase pathway in cardiac fibrosis of mice lacking guanylyl cyclase/natriuretic peptide receptor A. *J Biol Chem* 280: 19230-19242, 2005.

[160] Vo N, and Goodman RH. CREB-binding protein and p300 in transcriptional regulation. *J Biol Chem* 276: 13505-13508, 2001.

[161] Weber M, Hellmann I, Stadler MB, Ramos L, Paabo S, Rebhan M, and Schubeler D. Distribution, silencing potential and evolutionary impact of promoter DNA methylation in the human genome. *Nat Genet* 39: 457-466, 2007.

[162] Wei JQ, Shehadeh LA, Mitrani JM, Pessanha M, Slepak TI, Webster KA, and Bishopric NH. Quantitative control of adaptive cardiac hypertrophy by acetyltransferase p300. *Circulation* 118: 934-946, 2008.

[163] Wu JB, Chen K, Ou XM, and Shih JC. Retinoic acid activates monoamine oxidase B promoter in human neuronal cells. *J Biol Chem* 284: 16723-16735, 2009.

[164] Wu M, Zhang Y, Wu NH, and Shen YF. Histone marks and chromatin remodelers on the regulation of neurogenin1 gene in RA induced neuronal differentiation of P19 cells. *J Cell Biochem* 107: 264-271, 2009.

[165] Wu N, Siow YL, and O K. Ischemia/reperfusion reduces transcription factor Sp1-mediated cystathionine beta-synthase expression in the kidney. *J Biol Chem* 285: 18225-18233, 2010.

[166] Wu YF, Matsuo N, Sumiyoshi H, and Yoshioka H. The Sp1 and CBF/NF-Y transcription factors cooperatively regulate the mouse pro-alpha3(V) collagen gene (Col5a3) in osteoblastic cells. *Acta Med Okayama* 64: 95-108, 2010.

[167] Xu B, Strom J, and Chen QM. Dexamethasone induces transcriptional activation of Bcl-xL gene and inhibits cardiac injury by myocardial ischemia. *Eur J Pharmacol* 668: 194-200, 2011.

[168] Yamagami S, Suzuki K, and Suzuki N. Expression and exon/intron organization of two medaka fish homologs of the mammalian guanylyl cyclase A. *J Biochem* 130: 39-50, 2001.

[169] Yamaguchi M, Rutledge LJ, and Garbers DL. The primary structure of the rat guanylyl cyclase A/atrial natriuretic peptide receptor gene. *J Biol Chem* 265: 20414-20420, 1990.

[170] Yu Z, Li M, Zhang D, Xu W, and Kone BC. Sp1 trans-activates the murine H(+)-K(+)-ATPase alpha(2)-subunit gene. *Am J Physiol* 297: F63-70, 2009.

[171] Zhu WG, Srinivasan K, Dai Z, Duan W, Druhan LJ, Ding H, Yee L, Villalona-Calero MA, Plass C, and Otterson GA. Methylation of adjacent CpG sites affects Sp1/Sp3 binding and activity in the p21(Cip1) promoter. *Mol Cell Biol* 23: 4056-4065, 2003.

[172] Zhuang Y, Wendt C, and Gick G. Regulation of Na,K-ATPase beta 1 subunit gene transcription by low external potassium in cardiac myocytes. Role of Sp1 AND Sp3. *J Biol Chem* 275: 24173-24184, 2000.

[173] Zile MH. Vitamin A-not for your eyes only: requirement for heart formation begins early in embryogenesis. *Nutrients* 2: 532-550, 2010.

[174] Zittermann A. Vitamin D and disease prevention with special reference to cardiovascular disease. *Prog Biophys Mol Biol* 92: 39-48, 2006.

INDEX

A

B

C

D

E

F

G

H

J

K

L

M

N

O

P

Q

R

S

T

U

V

W

X

Y

Z